Maternal Substance Abuse and the Developing Nervous System

Maternal Substance Abuse and the Developing Nervous System

Edited by

Ian S. Zagon

Department of Neuroscience and Anatomy
The Milton S. Hershey Medical Center
The Pennsylvania State University
Hershey, Pennsylvania

Theodore A. Slotkin

Department of Pharmacology
Duke University Medical Center
Durham, North Carolina

Academic Press, Inc.
Harcourt Brace Jovanovich, Publishers

San Diego   New York   Boston   London   Sydney   Tokyo   Toronto

Copyright © 1992 by ACADEMIC PRESS, INC.
All Rights Reserved.
No part of this publication may be reproduced or transmitted in any form or by any
means, electronic or mechanical, including photocopy, recording, or any information
storage and retrieval system, without permission in writing from the publisher.

Academic Press, Inc.
1250 Sixth Avenue, San Diego, California 92101-4311

United Kingdom Edition published by
Academic Press Limited
24–28 Oval Road, London NW1 7DX

Library of Congress Cataloging-in-Publication Data

Maternal substance abuse and the developing nervous system / edited by
   Ian S. Zagon, Theodore A. Slotkin.
      p.   cm.
   Includes bibliographical references and index.
   ISBN 0-12-775225-0 (hardcover)
      1. Drug abuse in pregnancy.   2. Alcoholism in pregnancy.
   3. Fetus--Effect of drugs on. 4. Nervous system--Abnormalities-
   -Etiology.   5. Nervous system--Diseases--Etiology.   6. Fetal alcohol
   syndrome.   7. Developmental neurology.   8. Neurotoxicology.
   I. Zagon, Ian S.   II. Slotkin, Theodore A.
      [DNLM: 1. Alcohol, Ethyl--adverse effects.   2. Drugs--adverse
   effects.   3. Nervous System--Drug effects.   4. Nervous System-
   -growth & development.   5. Nervous System Diseases--chemically
   induces.   6. Nervous System Diseases--in infancy & childhood
   7. Prenatal Exposure Delayed Effects.   8. Substance Abuse--in
   pregnancy.   9. Substance Dependence--in pregnancy.   WS 340 M425]
   RG580.S75M38   1992
   618.3'268--dc20
   DNLM/DLC
   for Library of Congress   92-7145
                                                                                         CIP

PRINTED IN THE UNITED STATES OF AMERICA
92 93 94 95 96 97   EB   9 8 7 6 5 4 3 2 1

For Eileen and Linda

— Contents —

−4−

Fetal Alcohol Effects: Rat Model of Alcohol Exposure during the Brain Growth Spurt

Charles R. Goodlett and James R. West

−5−

Clinical Implications of Smoking: Determining
Long-Term Teratogenicity

Peter A. Fried

−6−

Prenatal Exposure to Nicotine: What Can We Learn
from Animal Models?

Theodore A. Slotkin

−7−
Prenatal Cocaine and Marijuana Exposure: Research and Clinical Implications
Barry Zuckerman and Deborah A. Frank

−8−
Cocaine and the Developing Nervous System: Laboratory Findings
Linda Patia Spear and Charles J. Heyser

– 11 –

Maternal Exposure to Opioids and the Developing Nervous System: Laboratory Findings

Ian S. Zagon and Patricia J. McLaughlin

– 12 –

Benzodiazepines and the Developing Nervous System: Laboratory Findings and Clinical Implications

Carol K. Kellogg

— Preface —

Substance abuse by pregnant women is accompanied by passage of these agents to the fetus through the placenta, and to the infant by way of the breast milk. The repercussions of such exposure on biological development may be manifested as either dramatic or subtle changes. Major alterations, such as gross malformations—more properly the subject matter of classical teratology—are readily understood and appropriate public warnings can be issued. When alterations are more subtle and not easily categorized, yet of extraordinary importance to the well-being of the individual, consensus is less likely and the formulation of public policy becomes more difficult. Sometimes adverse influences in the embryonic and/or fetal periods may be short-term but have vital significance, such as increasing the chance of birth trauma or Sudden Infant Death Syndrome. In other cases, exposure to abused substances during gestation could leave an imprint that extends to, or first appears in, adulthood. The nervous system is particularly at risk for problems because its period of morphogenesis and functional organization extends through both prenatal and postnatal phases. Nervous system development is a complex product of gene expression, specification of cell replication and differentiation, migration, synaptogenesis, choice of neurotransmitter and receptor populations, and patterning of synaptic performance. Although this plasticity is the reason that learning and adaptability are present throughout development, the same sequences of temporal and spatial organization contribute to a lengthy window of vulnerability to external influences such as drug exposure.

Many legal and illegal substances in our society are of special importance in our understanding the etiology and pathogenesis of congenital malformations, a leading cause of death for infants under the age of one year. Of course, this is without reference to the number of infants, children, and adults who suffer functional problems not classified as "congenital malformations," which therefore go unregistered. Although the magnitude of maternal substance abuse is difficult to determine, estimates range as high as affecting 10% of all children born in the United States. The impact on health care and society can be illustrated for maternal consumption of just one class of compounds, potent opioids, such as heroin and methadone. Approximately 6000 to 9000 opioid-exposed infants are born each year in the United States, exceeding the number of individuals born with Down Syndrome or that develop childhood cancer. In the past four decades, over 300,000 individuals, or more than 1 in 1000 Americans, have been exposed to opioids prenatally. In the wake of the more recent epidemic of cocaine use in the United States, 10,000 to 100,000 of all live births annually are thought to be associated with maternal consumption of this drug. The impact of maternal substance abuse on human development is typically exacerbated by habitual or casual multiple-drug abuse, by paternal drug consumption, and by the co-abuse of legal substances, such as tobacco or alcohol. If, indeed, maternal substance consumption leads to modifications of the developing nervous system, then economic and educational priorities will require considerable adjustment for individuals who, tragically, begin life with physical, behavioral, learning, and social disabilities.

The mechanisms of drug action on the fetal and infant nervous system are therefore of considerable interest. A major aim of this book is to place basic science and clinical observations into perspective in order to formulate future scientific inquiries and public policies. A number of themes are maintained throughout the book. First, recent evidence indicates that many abused substances act directly or indirectly to mimic or influence the actions of natural neurotrophic factors, including neurotransmitter-like substances and their receptors. Second, it appears that there is transient genetic expression of peptides, classical neurotransmitters, and receptors during development, and that disturbance of this timetable by drugs can have a profound effect on neurobiological organization and function. Third, many of the studies reported here confirm the importance of animal models and tissue culture paradigms in our understanding of the etiology and consequences of nervous system damage resulting from maternal substance abuse. Such experimental approaches may aid in the design of means to ameliorate adverse actions of these drugs.

In designing this book, we have asked major figures in each field to review the literature, apprise the audience of their own latest findings, and provide a perspective on maternal substance abuse and the developing nervous system. Wherever possible, we have tried to pair basic and clinical presentations so as to indicate how observations in human populations parallel results from basic research. In some instances, chapters on the same substance may present different viewpoints or focus on related elements that may ordinarily be absent from single presentations. The book is intended to summarize not only where we have been, but also to map directions for future research efforts, with the ultimate goal of preventing or treating the effects of maternal substance abuse early in development. Given the broad scope of the problem, it is our expectation that the information provided by these experts will be of interest not only to clinical and basic researchers in the field of maternal substance abuse, but also to individuals in psychology, social work, cellular and molecular biology, embryology, neuroscience, pharmacology, and in clinical professions such as pediatrics, neonatology, and obstetrics. Public interest mandates that maternal substance abuse and its ramifications for science and society be considered by those in all of these fields.

Ian S. Zagon
Theodore A. Slotkin

— 1 —

Introduction

Duane Alexander

*National Institute of Child Health
and Human Development
Bethesda, Maryland 20892*

Picture for a moment, if you will, the profile of the woman with the maximum chance of having an adverse pregnancy outcome. First, she is likely to be in her middle or late teens and probably has not finished high school. Her pregnancy is probably unplanned and unwanted, and the father may be unknown. She is likely to be unemployed, in the lowest socioeconomic group, and may be homeless. She is probably from a minority group, is poorly nourished, and is likely to have received little or no prenatal care. She is likely to have one or more sexually transmitted diseases. In addition to this scenario, she may smoke tobacco and abuse alcohol. And in recent years, added to this profile has been drug abuse—opioids, marijuana, and most recently, cocaine.

All these factors individually portend problems for the infant, but among all of them, the one that correlates highest with immediate adverse outcome is cocaine abuse, with a fourfold higher risk of low birth weight and its attendant problems. Longer-term adverse effects have been more difficult to determine and associate with drug abuse or any one of these factors, because the most common profile seen in the substance-abusing pregnant woman combines many of the above, making correlation of adverse outcome with one factor alone virtually impossible. Further, this profile is increasing in frequency and, in the inner city, is threatening to overwhelm maternity and infant health care systems; the attendant social crisis has already begun to reverse the downward trend in infant mortality.

Indicating the extent to which the problem of drug abuse is being added to already problematic pregnancies, surveys in major cities reveal that in some hospitals 40 to 50% of women delivering test positive for drugs, usually cocaine. But the problem is not confined there—surveys in general obstetric practices have indicated that approximately 10% of pregnant women are using some illicit drug, again, usually cocaine.

1

Based on these surveys, an estimated 375,000 infants of the nearly 4,000,000 births each year in the United States have been exposed to illicit drugs prenatally.

The concerns about opioids and cocaine come in addition to relatively recent discoveries about the adverse effects of tobacco and alcohol on the fetus during pregnancy. Tobacco clearly reduces birth weight in a dose-related manner, to such an extent that of all the things we know that will lower the incidence of low birth weight, stopping smoking in pregnancy would have the greatest impact. Less is known about the long-term effect of nicotine on brain development, but the immediate adverse effect on pregnancy outcome is sufficient to have induced legislative requirements for warning labels on cigarette packs.

By contrast, alcohol use during pregnancy seems to have little effect on immediate pregnancy outcome, but has devastating consequences for the physical and mental development of the fetus. The physical anomalies and mental retardation composing the fetal alcohol syndrome and the milder fetal alcohol effect have been known since the 1970s and, as its extent has become clearer, it has become evident that fetal alcohol syndrome is currently the leading known preventable cause of mental retardation, with an incidence of more than 1% in some population groups. Still being determined is the extent of alcohol-induced adverse central nervous system effects short of mental retardation and the timing and dose of alcohol required to produce these effects. Nonetheless, as with tobacco, legislatively mandated warnings about adverse effects of alcohol use during pregnancy now appear on alcoholic beverage containers.

The cocaine epidemic has brought a double threat in pregnancy. Like tobacco, but even stronger, cocaine clearly has an immediate adverse effect by reducing birth weight and triggering premature labor. And like alcohol, but weaker, it seems to have adverse long-term effects on brain function, and may produce some physical anomalies, although evidence of anomalies is inconsistent and so far unsubstantiated. These effects are still being determined and evaluated, in part because enough time has not elapsed to assess longer-term subtle effects, and in part because the subtle effects occur amid a constellation of adverse social circumstances that make it difficult to distinguish the possible effects of the environment in which the child is raised from the effects of prenatal cocaine exposure. Conducting these studies is further complicated by the fact that drug use, unlike alcohol or tobacco use, is illegal, and obtaining information on which of many drugs is used at what time and in what amount during pregnancy is extremely difficult. The barriers to conducting definitive studies of drug use during pregnancy, and possible ways to resolve the attendant problems, have been well described in a recent report to Congress from the National Institute of Child Health and Human Development (NICHD, 1991).

The studies presented in the book describe our current state of knowledge of the effects of maternal substance abuse during pregnancy on the developing nervous system of the fetus. From both animal and human studies, they document what is known and what remains to be learned. Clearly we already know enough to justify even stronger efforts to curtail the use of tobacco, alcohol, and illicit drugs during pregnancy. It is equally clear that we have no effective means of inducing the necessary behavioral changes. It is important that we continue to expand our knowledge in this area, and that studies of the effects of maternal substance abuse on the developing central nervous system continue to receive support from federal agencies, such as the National Institute on Drug Abuse and the National Institute of Child Health and Human Development, as well as from nonfederal sources.

REFERENCE

National Institute of Child Health and Human Development. (1991). "Effects of Drug Exposure *In Utero.*" Report to Congress. Bethesda, MD.

— 2 —

Fetal Alcohol Syndrome and Fetal Alcohol Effects: A Clinical Perspective of Later Developmental Consequences

ই

Ann Pytkowicz Streissguth

Fetal Alcohol and Drug Unit
Department of Psychiatry and Behavioral Sciences
University of Washington
Seattle, Washington

INTRODUCTION

Alcohol is now well recognized as a teratogenic drug: prenatal exposure can cause death to the embryo and fetus, growth deficiency, malformations, and central nervous system aberrations that can last a lifetime. Whereas definitive documentation of the etiology and mechanisms comes from the experimental animal literature (see Goodlett and West, Chapter 4), the clinical literature is important in understanding the significant impact of this most widely used teratogen on our children. The epidemiologic literature is relevant for establishing public policy and guiding programs of prevention for this most preventable cause of mental retardation and developmental disability, and understanding the clinical phenomena of fetal alcohol syndrome (FAS) and fetal alcohol effects (FAE) is essential for developing appropriate treatment and intervention programs for affected children so that they can lead as productive and fulfilling lives as possible. There has been a shocking paucity of such research in the United States in the past 18 years since FAS was identified.

Fetal alcohol syndrome is generally recognized as the leading known cause of mental retardation (Abel and Sokol, 1987), surpassing Down's syndrome and spina bifida. (Most mental retardation cannot be attributed to a specific etiology.) Precise figures on FAS are difficult to obtain, however, and it seems likely that most attempts at estimating prevalence [including those most recently proposed by Abel and Sokol (1991)] are underestimates owing to difficulties with ascertainment and identification, confusion over diagnostic criteria, and problems in making interpretations based on literature surveys, including studies of variable validity. FAS is a clinical diagnosis (see below). The term FAS does not include *all* individuals affected by alcohol *in utero*, but rather it represents one specific and identifiable end of the continuum of disabilities caused by maternal alcohol use during pregnancy.

The clinical features of FAS were independently identified in France (Lemoine et al., 1968) and the United States (Jones et al., 1973; Jones and Smith, 1973). Most of the patients described have been infants or young children, but increasingly maladaptive behaviors among adolescents with FAS (Streissguth et al., 1991) make this an important topic for further study. A systematic examination of all school-age children with only mild mental retardation (IQ 55 to 70), born in Sweden during a 2-year period, indicated that 8% were afflicted with alcohol-related disabilities (Hagberg et al., 1981a,b). A recent report involving ophthalmology examinations of this same cohort raised the proportion with suspected FAS to 10% [a larger proportion than was identified by all known genetic disorders (Hagberg et al., 1981a,b; Stromland, 1990)]. Enough is currently known to indicate that FAS is a major health problem. The fact that precise figures are not available should not dissuade us from recognizing the urgency of the need for research on the characteristics and special needs of this underserved population of disabled persons.

FETAL ALCOHOL SYNDROME AND FETAL ALCOHOL EFFECTS: DIAGNOSTIC ISSUES AND TERMINOLOGY

Fetal alcohol syndrome is diagnosed when patients have a positive history of maternal alcohol abuse during pregnancy and (1) growth deficiency of prenatal origin (height and/or weight); (2) a pattern of specific minor anomalies that includes a characteristic facies (generally defined by short palpebral fissures, midface hypoplasia, smooth and/or long philtrum, and thin upper lip); and (3) central nervous system manifestations (including microcephaly or history of delayed development, hyperactivity, attention deficits, learning disabilities, intellectual deficits, or seizures) (Clarren and Smith, 1978; Smith, 1982). Patients exposed to alcohol *in utero* with some partial FAS phenotype, and/or central nervous system dysfunction, but without sufficient features for a firm diagnosis of FAS or strong consideration of any alternative diagnosis, are identified as *possible FAE* (Clarren and Smith, 1978).

Fetal alcohol syndrome is a specific medical diagnosis usually given by a dysmorphologist, a geneticist, or a pediatrician with special training in birth defects or dysmorphology. It is not appropriately diagnosed by checklists or without a full clinical examination by a specially trained person. Unfortunately, most physicians have not received special training in syndrome identification, and as dysmorphology is a rather new field, many persons with FAS go unrecognized. This is a particular problem with regard to persons who have not been identified before puberty, as the facial features are less distinctive in adolescence and adulthood. The optimal age for making the diagnosis is between 8 months and 8 years, although the most severely affected children can be identified at birth by a skilled diagnostician. Diagnosis in the adolescent and adult is facilitated by photographs from infancy and childhood.

Fetal alcohol effects is a term used in two different ways. In the clinical sense, possible or probable FAE refers to individual children given a clinical examination who were known to be born to an alcohol-abusing mother and who have some, but not all, of the characteristics necessary for a diagnosis of the full FAS. Partial syndrome expression is not uncommon in syndromology. The words *possible* or *probable* precede the term FAE as an expression of uncertainty that the observed characteristics (in the absence of the full syndrome) are all attributable to alcohol. From the standpoint of understanding the patient's needs, the distinction may be irrelevant. Clearly for research purposes and for making prognostications for the individual patient, the criteria used for making the diagnosis are extremely important.

Fetal alcohol effects is also a generic term used for all the "effects" or outcomes known from epidemiologic studies to be caused by alcohol (low birth weight is an example). Whereas it is clear from group studies that low birth weight is an effect associated with maternal alcohol abuse during pregnancy, it is not possible to say with any degree of certainty that low birth weight in an individual child is indeed caused by alcohol. There are, of course, many causes of low birth weight, of which prenatal alcohol exposure is only one.

Considerable confusion exists in the medical literature regarding the use of the terms FAS and FAE. It is unclear what the relationship of these clinical findings is to studies that attempt to designate children as FAS or FAE based on checklist criteria, in the absence of a full clinical examination by a qualified diagnostician. In our own experience, anomalies tallies by highly trained technicians have not correlated well with the clinical diagnosis of FAS as determined by an experienced diagnostician (unpublished data from the Seattle Longitudinal Study on Alcohol and Pregnancy (1974–1987); see also Clarren et al., 1987). We would not necessarily expect the findings from long-term studies of clinically diagnosed patients to be congruent with the findings from studies attempting to identify newborns based on anomalies tallies, particularly in view of the well-documented differences in minor anomalies among different racial groups, which are seldom discussed in the studies identifying patients with anomalies tallies.

In our opinion, the European clinicians diagnosing FAS identify children quite similar to those identified by dysmorphologists in the United States (e.g., Lemoine *et al.*, 1968, and Dehaene et al., 1977a,b, in France; Majewski and Majewski, 1988, Spohr and Steinhausen, 1987, and Löser and Ilse, 1991, in Germany; Olegard and colleagues, 1984, in Sweden; and Gairi, 1990, in Spain). However, direct extrapolation to U.S. statistics is difficult because European clinicians often use a classification system, involving light, moderate, and severe categories of FAS which do not translate easily to the nomenclature used in the United States.

For the purposes of this chapter, the terms FAS and FAE will refer to children whose diagnoses have been established by clinical examination by persons qualified to make this diagnosis. Most of these patients were referred for clinical examinations, unless otherwise specified. This chapter will focus on *later development*, meaning school-age children, adolescents, and adults. Studies on the *early* developmental effects of alcohol (infancy or preschool ages) have been reviewed earlier (Streissguth, 1986). This chapter will also not review the literature on the broader realm of nonbehavioral fetal alcohol effects deriving from epidemiologic studies, as these have been recently reviewed elsewhere (Little and Wendt, 1991; Streissguth et al., in press).

CLINICAL STUDIES OF FETAL ALCOHOL SYNDROME, FETAL ALCOHOL EFFECTS, AND CHILDREN OF ALCOHOLIC MOTHERS

Hundreds of clinical reports on individual children with FAS have appeared in the medical literature. However, only a few clinicians, mostly from Europe, have reported clinical findings on groups of school-age children with FAS/FAE or children of alcoholic mothers.

Majewski and Majewski (1988) reported on 175 children of varying ages (including about 50 aged 6 to 15 years) with alcohol embryopathy (AE) ranging from mild to severe. Patients continued to be growth deficient for age (height, weight, and head circumference), but were somewhat less deficient in terms of standard deviations at older ages of childhood. In a subset of 18 who were not severely mentally retarded and who received intellectual assessment, severity of symptoms corresponded to severity of mental handicaps. This confirmed earlier work by Streissguth et al. (1978a). In their full sample, 83% were described as mentally retarded, 81% were microcephalic, 72% were hyperactive, and 54% had muscular hypotonia. Majewski also noted that among 24 siblings, the younger ones were more likely to be impaired than the older ones, confirming earlier work by Dehaene et al. (1981) and Olegard et al. (1984). Majewski concluded that the two precipitating factors for AE are heavy exposure and chronic maternal alcoholism. The most severely affected children were born to mothers with the most severe physical manifestations of alcoholism. In unpublished electroencephalogram (EEG) studies conducted with Spohr, one fourth of 61 patients with various degrees of AE had anomalous EEG patterns, including pathologic dysrhythmias or hypersynchronic potentials (Majewski, 1984).

Fetal alcohol syndrome studies from Berlin have been reported by Spohr and Steinhausen. A recent paper from this group (Spohr and Steinhausen, 1987) describes a 3- to 4-year reevaluation of a sample of 54 patients of varying ages, with a mean age of around 8 years. They concluded that the dysmorphic signs became less apparent with increasing age, that their neurologic performance improved, that the EEG patterns were less pathologic, and that they showed an improvement in psychiatric status and cognitive functions. However, these data are difficult to interpret, as many of the earlier problems they cite (eating problems, for example) would be expected to show some amelioration with increased age. Furthermore, they used a psychiatric rating scale developed for preschool children, and only small subsets of the overall were available for each test. The IQ data are also difficult to interpret, and seem to be associated with improvement in only four children. The most compelling finding from the Berlin study was the severity of the educational difficulties of the full group of 54 patients on follow-up (see Table 1). Only 17% of the 35 children older than 7 years were able to attend normal schools. The most handicapping conditions they had were hyperactivity and distractibility. Of the entire sample of 54, 70% were described as

TABLE 1   Educational Status of 35
Patients with FAS/FAE, 7 Years
or Older in Germany[a]

Normal school	17%
Educationally subnormal schools	51%
Training centers	20%
Too subnormal for training	11%
Hyperactive[b]	70%

[a]From data reported by Spohr and Steinhausen (1987).
[b]The hyperactivity data is from a larger group of 54
patients.

hyperactive on follow-up. Spohr and Steinhausen also reported that the most severely
affected patients had the poorest educational status, and that changes in IQ and
psychopathology were not associated with age, sex, morphology, environment, or
treatment.

Aronson (1984) presented a detailed account of 99 children who constituted the
living offspring of 30 alcoholic mothers in Goteborg, Sweden (see Table 2). Evaluated
at an average age of 14 years, one half of the children had borderline or retarded
mental development (a much higher proportion of deficit than the 15% who had low
birth weight or the 10% with malformations). Furthermore, 49% had neuropsychological symptoms (defined as the presence of at least three of the following: hyperactivity, impulsiveness, distractibility, temper tantrums, short memory span, difficulties
with concentration, perseveration, and perceptual difficulties). The children had significantly lower mental abilities than their mothers, and rearing environment was not
related to mental abilities in the children (Olegard et al., 1984). Rearing environment
was, on the other hand, related to psychosocial problems in the offspring (Aronson et
al., 1984).

These Swedish studies are particularly interesting in terms of an early Russian

TABLE 2   Outcome in 99 Living
Children of 30 Alcoholic Mothers
Examined at Mean Age
14 Years in Sweden[a]

Low birth weight	15%
Malformations	10%
Borderline mental retardation	50%
Neuropsychological symptoms[b]	49%

[a]From data reported by Aronson et al. (1984).
[b]At least three of the following in an individual child:
hyperactivity, distractibility, short memory span, impulsivity, temper tantrums, perseveration, perceptual
difficulties.

study by Shruygin in 1974, which compared children who were born before and after their mothers became alcoholic. Those born before the mothers' alcoholism displayed disorders that were primarily vegetative, emotional, and behavioral with symptom onset at 9 to 10 years and symptom remission with improved social circumstances. Many of the children born after full-fledged maternal alcoholism had profound impairments of the CNS (central nervous system) that were manifest early in infancy. Of these latter, 14 of 23 were mentally retarded.

Seattle clinicians have published a series of clinical reports on children with FAS/FAE beginning with Jones, Smith, Ulleland, and Streissguth (1973), in which 11 patients were described, all preschool children or infants (Jones et al., 1973; Jones and Smith, 1973). In 1978, Streissguth and associates (1978a) expanded this group to 20, including some early school-age children, and noted a direct relationship between the severity of dysmorphology and lowered IQ. They also noted the very broad range of IQ scores associated with FAS (from severely retarded to normal), even in this small sample.

In a second paper in 1978, Streissguth and colleagues (1978b) reported on IQ data obtained 1 to 3 years after the initial evaluation and included the first two adults reported with FAS, showing IQ scores across their life span. Other clinical reports of older patients with FAS (e.g., Iosub et al., 1981) have subsequently appeared. Whereas the mean IQ for the group remained the same across time, large individual differences were observed: two children had IQ scores that increased by at least two standard deviations (30 IQ points), while two others had IQ scores that decreased by the same amount. Age appeared to be a relevant variable, with the greatest increase in IQ scores occurring for children moving between the preschool and school-age years. A more recent test–retest study of IQ in patients with FAS also shows fairly stable scores in adolescence, with a small number of patients showing either increased or decreased IQ over time (Streissguth et al., 1991).

In 1985, Streissguth, Clarren, and Jones reported a 10-year follow-up study of the first 11 children diagnosed with FAS in 1973. Of the 8 reexamined, 4 were clearly mentally retarded and in need of sheltered environments; the other 4 had IQ scores in the borderline retarded range, and were attending regular classes with remedial help. The latter 4 were in the least stable living situations. Figure 1 presents data on these 8 children, plotted in terms of standard deviations from the mean for height, weight, head circumference, and IQ. Although there was considerable individual variation, most children remained quite stable over time for height, head circumference, and IQ.

FIGURE 1   Growth and IQ curves by age in 11 children originally diagnosed with FAS in 1973. Records taken at five ages are plotted when available. (A) height; (B) weight; (C) head circumference; (D) IQ. IQ scores are derived from individual age-appropriate tests of general intelligence and mental development, including the Wechsler Intelligence Scale for Children–Revised, the Wechsler Preschool and Primary Scale of Intelligence, the Stanford–Binet Intelligence Scale, form L-M, and the Bayley Scales of Mental Development. The Stanford–Binet scores originally reported in 1973 were revised in 1985 according to norms published since 1973. The first IQ point for patient 11 is circled because it was estimated from the Vineland Social Maturity Scale and clinical observation. Reprinted by permission from *Lancet*, Streissguth, Clarren, and Jones (1985).

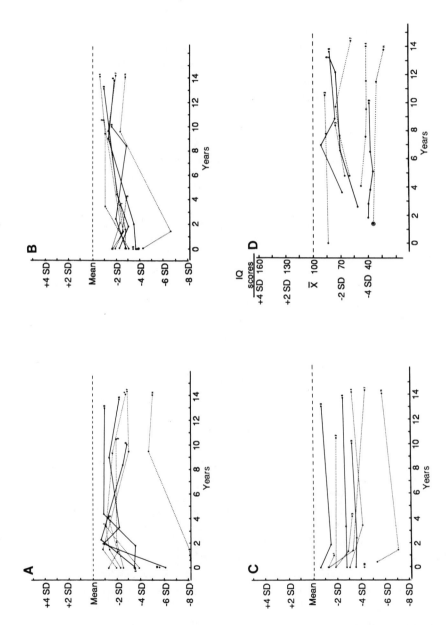

Weight showed a consistent trend toward greater normalcy with increasing age, particularly in pubescent girls. A general coarsening of facial features associated with puberty (particularly increased growth of the nose and chin) made the facial features less distinctive than in the infant and the young child. As many dysmorphology syndromes become less discernible with increasing age, this was not interpreted as an amelioration of the syndrome, but rather as a normal process in syndrome identification. In this small clinical study there was the suggestion that psychosocial problems were already increasing with the onset of adolescence, particularly in those children with near-normal intelligence.

The first major report on adolescents and adults with FAS was published by Streissguth et al. (1991a) on 61 patients from two northwestern cities and from four southwestern Indian reservations, where systematic FAS screening systems had been previously carried out. The patients were systematically examined on a large battery of psychological tests; 47% lived on reservations; 74% were American Indians; 70% had an FAS diagnosis; the rest were classified as FAE. (The small number of patients with FAE represented a constraint of the study design and should not be interpreted as a true proportion.) These patients ranged in age from 12 to 40 years, with the mean age around 17 years.

One important finding was that the physical features of FAS are less distinctive after puberty (see Fig. 2). The faces of the patients were not so characteristic as they had been in childhood. Growth deficiency for weight was not so remarkable as in infancy and childhood, although the majority remained short and had microcephaly. Whereas the weight–height proportion (weight for height age) of the younger patients with FAS ranged from 1% to 15% (Streissguth et al., 1985), in this older sample the weight–height proportion ranged from 3% (very thin) to 90% (very heavy). These findings help explain why the initial identification of persons with FAS is difficult in adolescence and adulthood. It also emphasizes the importance of early identification.

Intellectual development was extremely varied, with some patients being very mentally retarded and others having normal intelligence (see Fig. 3). The average intellectual level for the patients with FAS was in the mildly retarded range. Almost half of them, however, had an IQ of 70 or greater, so would not be technically classified as mentally retarded. This has important implications for obtaining community services, as many persons with FAS are not automatically eligible for programs designed for the mentally retarded. Although the average academic functioning of these patients was at the second to fourth grade level, some did read and spell at a fifth grade level or beyond. In general, arithmetic skills were the most deficient academic area, probably representing difficulty with abstract thought.

Unlike previous studies of younger children with FAS, which have dealt primarily with IQ and achievement scores, this study carried out systematic evaluations of the patients' level of adaptive functioning. Despite an average chronologic age of 17 years, these patients had only a 7-year level of adaptive functioning on the Vineland Adaptive Behavior Scale. Of the three domains making up this adaptive behavior score, they performed best on daily living skills (at an average 9-year level) and most poorly on socialization skills (at approximately the 6-year level). Although one or two pa-

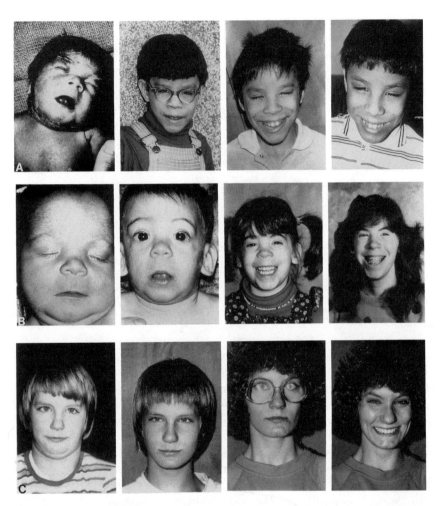

FIGURE 2   Patients with FAS photographed across the life span. (A) Severely retarded American Indian adolescent, diagnosed FAS at birth, and photographed as a neonate, and at ages 5, 10, and 14 years. He has been growth deficient and microcephalic throughout life. With increasing age, there is a considerable relative growth of the nose resulting in a high, wide nasal bridge. Note persistence of smooth philtrum. This patient was originally described in Jones and Smith (1973). (B) Adolescent girl diagnosed FAS at birth, with later intellectual functioning in the borderline range. Photographed at birth, 9 months, 5, and 14 years. While gradual maturation of facial features is taking place, note persisting small palpebral fissures, relatively long, smooth philtrum, and narrow upper vermilion. (C) Adult white woman, diagnosed FAS at 4 years of age; IQ 85–90; photographed here at 9, 13, and 19 years. Facial manifestations of FAS have evolved, in this patient, into a fairly normal facial phenotype by adult life, illustrating the value of a photographic record in the assessment of adults with FAS. At 19, her head circumference was below the first centile, height was below the fifth centile, and weight was around the tenth centile. Photographs in Fig. 2 reprinted by permission from J.A.M.A.: Streissguth et al. (1991a).

Ann Pytkowicz Streissguth

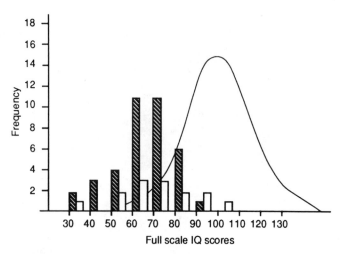

FIGURE 3 IQ scores in patients with FAS/FAE. Frequency distribution of IQ scores from the Wechsler Adult Intelligence Scale–Revised and the Wechsler Intelligence Scale for Children–Revised, whichever was age appropriate. Mean chronological age was 18 years. The bell-shaped curve represents the normal distribution. Solid bars indicate fetal alcohol syndrome (n = 38; mean IQ, 66); open bars, fetal alcohol effects (n = 14; mean IQ, 73). Reprinted by permission from *J.A.M.A.*, Streissguth et al. (1991a).

tients had age-appropriate daily living skills, none was age-appropriate in terms of socialization or communications skills. These patients with FAS/FAE, who were not technically retarded, had failed to accomplish specific types of adaptive behaviors characteristic of adolescents, such as failure to consider consequences of action, lack of appropriate initiative, unresponsiveness to subtle social cues, and lack of reciprocal friendships.

These findings underscore the critical importance of keeping adolescents with FAS/FAE in the school setting, as they certainly do not have the adaptive living skills to survive well outside of a structured environment, even when they are not technically retarded. These findings also point to the necessity of schools' taking a broad functional approach to education and the importance of job skills training and work experience. Of those patients on whom information was available, only 6% were in vocational programs, 2% were working, and none was entirely independent.

Family environments of these patients with FAS/FAE had been remarkably unstable. On average, they had lived in five different principal homes in their lifetimes. Only 9% were with both biologic parents; 3%, with their biologic mothers. For those for whom accurate data could be obtained, 69% of their biologic mothers were known to be dead. This statistic demonstrates the severe impact of alcoholism in women. (They died not only of cirrhosis, but also of many other types of alcohol-related accidents and violence.) This information leads to the conclusion that an early diagnosis of FAS in a child is important from the standpoint of both mother and child. Mothers giving birth to children with FAS are clearly at risk for alcohol-related disability and premature death. Diagnosis of FAS in the child can not only help the child to receive proper services early in life, but can also help the mother come to grips with her own alcoholism (see Giunta and Streissguth, 1988).

This study led to the conclusion that FAS is not just a childhood disorder. There is a predictable, long-term progression of the disorder into adulthood, in which maladaptive behaviors present the greatest challenge to management. As we point out in the article, however, the outcomes that we have documented represent the interactive influences of biology and environment. Most of these patients were born before mothers were generally aware that drinking during pregnancy was harmful. Most of these patients were undiagnosed as infants and young children, or if they were, this diagnostic information was not carried along with them through life. Thus, most were raised by caretakers who were unaware of their diagnosis and taught by teachers who had no knowledge that they had a life-long disability.

It is our hope that with more widespread diagnosis of FAS, and with clearer understanding of the long-term consequences of FAS, more reasonable and appropriate environmental interventions can be developed, at home, in the school, and in the broader community. Out of the realization that a child is disabled with FAS can come the help and resources for each child to develop to his or her own best potential, in an environment that is ultimately the most enhancing.

The wide variation in intellectual levels in this group of patients confirms what we have known since the beginning: namely, that the diagnosis of FAS does not carry with it any particular guarantees or inevitabilities about IQ or about academic achievement levels. A diagnosis of FAS does not mean that a person cannot graduate from high school or even attend college. It does mean that some degree of brain damage has been sustained and that the manifestations of this will be apparent in the person's adaptive behaviors. This chapter further suggests that the more serious manifestations of FAS may well be experienced at that time in life when expectations for independent functioning are the greatest. It is our hope that, out of this greater knowledge about the long-term consequences of FAS, can grow better programming, more widespread help and support for parents and teachers, and more realistic and helpful expectations for the patients themselves. Unrealistic expectations can lead to frustration, despair, and hopelessness. Knowledge about a disability should garner support for the disabled person and hope for a happier, more fulfilling future.

EPIDEMIOLOGIC OR CASE CONTROL STUDIES

Children of Alcoholic Mothers

In 1974, Jones, Smith, Streissguth, and Myrianthopoulos published the first matched comparison study of 23 children of alcoholic mothers compared to 46 matched controls. Mothers had been identified as alcoholic during pregnancy from a large national prospective study, and control mothers were matched on socioeconomic status (SES), age, education, race, parity, marital status, and geographic region of delivery. At 7 years of age, children of alcoholic mothers had significantly lower IQ scores (IQ 81 versus 95) and significantly poorer scores on tests of reading, spelling, and arithmetic (Streissguth, 1976). Furthermore, 44% of the children of the alcoholic mothers had borderline to retarded mental deficiency (IQ 79 or below), compared to 9% of the controls. Comparable decrements in height, weight, and head circumference were also noted.

In 1982, Steinhausen, Nestler, and Spohr published a paper from Berlin in which a large sample ($n = 32$ to 68, depending on the tests given) of patients with FAS were compared to a group of 28 healthy children, matched on age, sex, socioeconomic status, and foster placement. The most frequent problems were mental retardation, microcephaly, and growth deficiency as well as a variety of measures of psychopathology, including hyperactivity, eating and sleeping problems, and speech problems. This was the first study to point out the amount of psychopathology characterizing young children with FAS.

In 1985, Aronson and colleagues reported a study of 21 children born to alcoholic mothers in relation to a comparison group matched on birth weight, age, sex, and residence. On follow-up at an average age of 6 years, children of alcoholic mothers had significantly lower IQ than those of controls (IQ 95 versus 112), and significantly more visual, perceptual, and behavioral problems (see Table 3). The behavioral problems included hyperactivity, distractibility, and short attention spans. Working with the same groups of patients, Kyllerman and colleagues (1985) found that the children of alcoholic mothers failed to show a catch-up growth and had significantly poorer fine and gross motor development (see Table 3).

In a very important study of an entire Canadian Indian village with a high alcoholism level, Robinson, Conry, and Conry (1987) examined all children under the age of 18 years and interviewed all mothers regarding their drinking histories. One

TABLE 3   Outcomes in Children of Alcoholic Mothers vs. Controls in Sweden[a]

	Alcoholic mothers	Controls
Weight (SD)	−1.2	−0.1
Height (SD)	−1.1	0.0
Head circumference (SD)	−0.8	0.3
Gross motor score (mean)	87	103
Fine motor score (mean)	91	103
IQ (WISC or Griffiths) (mean)[b]	95	112
Marked visual perceptual problems (>1 year delay)	47%	0%
Behavioral problems (hyperactivity, distractibility, short attention span)	62%	0%

[a]From data reported by Aronson et al. (1985) and Kyllerman et al. (1985). Controls matched on birthweight, age, sex, and residence; $n = 21$ per group, mean age 6 years.
[b]100 = normal.

in every 8 children had FAS; two thirds of these children were mentally retarded. When these children were compared with controls matched on age and sex, the children with FAS/FAE had significantly lower IQ scores (IQ 68 versus 88) with almost no overlap between the two groups.

Population-Based Studies of School-Age Children Whose Mothers Used Alcohol during Pregnancy

In this section, we review studies that address the long-term consequences of prenatal exposure to lower levels of alcohol. The studies use groups of women whose self-reports of alcohol use during pregnancy constitute the primary outcome data. Several longitudinal prospective studies on prenatal alcohol use have been initiated, but data on school-age children are relatively recent.

The Seattle Longitudinal Prospective Study on Alcohol and Pregnancy used a cohort of approximately 500 children examined at various ages, whose mothers were interviewed during pregnancy regarding alcohol use patterns and the use of other drugs, as well as nutrition and health practices during pregnancy. Statistical analyses involved multiple regression analyses, which permit simultaneous adjustment for a variety of potentially confounding variables such as parental education, age, race, as well as smoking and use of other drugs. In more recent reports, partial least squares analyses have been used to detect the underlying patterns of relationship in these extremely multivariate data. The study design, research rationale, procedures, and early findings have been described previously (Streissguth et al., 1981, 1984, 1986).

As this study was conceptualized as human behavioral teratology research from its inception, several design features made it particularly strong: (1) The mothers were generally at low risk for adverse pregnancy outcome. They were primarily white, married, and middle class, and all in prenatal care by the fifth month of pregnancy. (2) The study began in 1974, at a time when most women (80 to 81%) drank during or before pregnancy. As there was no general knowledge that alcohol was bad for the baby, there appeared to be little denial of drinking. [In fact, test/retest studies for comparability of reporting over a 1-week interval revealed the same high test/retest reliability for alcohol as for caffeine (Streissguth et al., 1976)]. There was also little use of cocaine, heroin, and methadone in this population in 1974 and 1975. (3) The follow-up sample was selected from a larger screening sample of 1529 to maximize the number of heavier drinkers and smokers to improve the statistical leverage. (4) The follow-up sample was stratified for smoking across alcohol groups to improve the ability to separate the effects of alcohol from those of smoking. (5) The outcome variables were carefully selected based on hypotheses derived from clinical experience with patients with FAS. (6) Follow-up of the cohort has been excellent over time, with approximately 85% of the original sample seen at each follow-up examination and no systematic loss of heavily exposed subjects.

The study has been described as a study of "social drinkers" because it was population based and aimed at women who were at low risk for adverse pregnancy outcome. The study screened all available women at two representative hospitals—

none as specifically chosen for alcohol treatment or alcohol problems. The alcohol use of this population was representative of that of ordinary, low-risk women during pregnancy in 1974; we called it "social" drinking. This terminology is consistent with current usage employed by the National Institute on Alcohol Abuse and Alcoholism (1990). Fewer than 1% of the women identified themselves as having problems with drinking using a standard set of questions used in the 1970s for identifying problem drinkers. Eighty percent of them drank during pregnancy; 5% drank daily. The drinkers reported a median of about one drink per day before they knew they were pregnant and about one third drink per day at midpregnancy. Of the drinkers, 19% had a binge pattern of use, i.e., drank five or more drinks on occasion at least once during pregnancy, and 29% before pregnancy recognition. Only 30% of the group smoked, and many heavier drinkers did not smoke or take other drugs. [Statistics are from the mothers of the 482 children examined at 7 years of age with valid IQ and Wide-Range Achievement Test (WRAT) tests.] All medications used during pregnancy were also evaluated, and did not confound the alcohol results.

The primary neurobehavioral outcomes of relevance to this chapter derive from a large neurobehavioral and attentional battery administered when the children were 7 years old. These papers describe the tasks, statistical analysis, and full findings: Streissguth et al., 1986; Sampson et al., 1989; Streissguth et al., 1989a,b; Streissguth et al., 1990; Streissguth et al. (in press).

The neurobehavioral outcomes from this longitudinal study have been reported in a number of publications, and a monograph on the longitudinal evaluation is under review. After adjusting for a variety of potentially confounding variables (such as other ingestants, smoking, nutrition, demographic, and family environment variables), prenatal alcohol exposure was related to a variety of cognitive/neurobehavioral outcomes in 7-year-old children examined under highly standardized laboratory conditions and evaluated as well by their classroom teachers. Attentional and memory deficits and slow reaction time, as measured on laboratory tasks, were all related to prenatal alcohol exposure, as were a cluster of behaviors (distractibility, poor organization and cooperation, and rigid approach to problem solving) displayed in both the laboratory situation and the classroom, as measured by both psychometrists and the children's individual teachers. Deficits in arithmetic functioning, as well as word recognition, were also associated with prenatal alcohol exposure, but spelling was not, in this group of 500 children who had, for the most part, just finished the second grade in school. Recent work shows continuing alcohol-related deficits in academic performance and classroom behaviors continuing through 11 years, according to teacher reports and standardized achievement tests (Carmichael Olson et al., 1990). Physical growth, as assessed by height and weight, was no longer associated with prenatal alcohol exposure in this cohort of mild to moderately exposed 7-year-old children, and facial dysmorphology was associated with only the highest exposure levels (Clarren et al., 1987). Thus, as anticipated, neurobehavioral tests revealed the most sensitive and long-lasting effects of prenatal alcohol exposure in this population-based study of children of low-risk mothers who had all been receiving prenatal care by midpregnancy.

Two recent papers from the Atlanta Alcohol and Pregnancy studies have provided partial replication from a different population of drinking mothers, primarily lower class black mothers. Follow-up of a small cohort ($n = 68$) of prenatally identified children between 5 and 8.5 years (mean 5 years, 10 months) revealed deficits in short-term memory and encoding, overall mental processing, preacademic skills, and teacher rating of "externalizing" behaviors (overactivity, aggressiveness, inattention, and destructiveness) (Coles et al., 1991; Brown et al., 1991). The women in this pair of Atlanta studies averaged about 11 drinks per week—outcomes were generally worse for those who drank throughout pregnancy, but the children of those who stopped drinking during pregnancy were not without effects. This confirms findings reported earlier by Aronson and colleagues for heavier drinking mothers.

CONCLUSIONS

1. Patients with FAS continue to manifest developmental disabilities as they mature. Although patients with this diagnosis span a wide range of intellectual abilities, most identified to date continue to fall within the borderline to mentally retarded range of abilities in adolescence and adulthood. Those who are more severely affected with respect to physical growth and malformations tend to be more intellectually impaired.

2. Physical manifestations of FAS change with physical maturity, but remain recognizable. Shortness of stature and microcephaly remain important characteristics in the older patient, but the mature facies is less characteristic of the syndrome than that in the infant and young child, owing partly to increased growth of the nose and chin and other changes in the midfacial region.

3. Attentional deficits characterize 75 to 80% of the patients with FAS, contributing to difficulty with classroom learning during the school years and to major problems with employment during adolescence and adulthood. Significant problems with academic skills (particularly arithmetic disability) and with adaptive behaviors in the realm of daily living skills, socialization, and communication are prevalent after puberty.

4. A large proportion of patients with FAS/FAE have, in the past, become the responsibility of the community to raise and shelter, because of the high rate of maternal death, terminations of maternal rights, and abandonment of these children. There is no indication from present data that such patients will, in general, be able to be self-sufficient, although the number followed into adulthood remains small. Few children with FAS continue in the regular classroom as they get older; however, it is not known how many may simply drop out of school undiagnosed.

5. There is no systematic information available on the effects of specific remedial programs or intervention efforts with patients with FAS/FAE during the school-age years, adolescence, or adulthood. These problems should be addressed by educators, vocational and occupational therapists, and mental health specialists, from the standpoint of both individual needs and the allocation of community resources.

6. Children of alcoholic mothers, particularly those drinking during pregnancy,

are at risk for poorer outcome in terms of physical growth, mental development, academic achievement, fine and gross motor function, visual perceptual problems, and behavior problems. This group urgently needs further study, even in the absence of the physical manifestations of FAS.

7. The postnatal environment of children with FAS does not appear to be associated with changes in IQ scores over time. However, the findings are discrepant regarding the influence of the rearing environment or psychological problems or psychopathology. Relevant factors may include age of the children at the time of the study and severity of the disability in the children studied [i.e., whether the children have FAS or are children (without FAS) of alcoholic mothers]. Systematic research on remedial interventions is urgently needed, particularly in terms of preventing secondary psychopathology.

8. Recent evidence from the Seattle longitudinal prospective study on the effects of maternal social drinking during pregnancy on 7-year-old offspring indicates continued effects on attention, reaction time, intelligence, memory, learning problems (particularly arithmetic), and other neuropsychologic deficits, even after adjusting statistically for other significant determinants. A behavioral pattern including distractibility, poor organization and cooperation, and a rigid approach to problem solving was also associated with prenatal alcohol exposure. More recent work indicates continued academic and classroom behavioral problems at age 11 years, associated with prenatal alcohol exposure in this population-based study.

9. A rough dose–response association is observed, with heavier levels of exposure associated with greater offspring impairment (observable in the individual clinical case), whereas the effects of more moderate levels of exposure are observed in group studies with careful statistical adjustment of potentially confounding variables. Binge patterns of exposure are emerging as increasingly predictive of neuropsychologic outcomes in school-age offspring.

10. There is no indication from either the clinical or the epidemiologic studies that the effects of prenatal alcohol exposure diminish with the age of the child. Only the *types* of manifestations change with increasing maturity. Growth deficiency and facial dysmorphology become less marked with age. On the other hand, cognitive disabilities become more marked with the increased demands of school and society for academic performance and abstract problem solving. Problems with adaptive behavior become particularly noteworthy in adolescence as demands for independent functioning are increased.

11. Thus we see an evolving array of disabilities emerging at different ages in children who are prenatally alcohol exposed. The severity of the outcomes—and to some extent the types of outcomes—seem to relate to the severity of and pattern of exposure.

12. Very few data are available on how the postnatal environment might exacerbate or ameliorate the effect of prenatal alcohol exposure.

RECOMMENDATIONS

1. Studies of children exposed to prenatal alcohol at all levels should be extended into the later developmental ages, as it is clear that prenatal alcohol effects do not

disappear with increasing maturity of the offspring; they only change their manifestations. Some psychosocial outcomes might be modifiable by the environmental milieu in which the child lives and learns.

2. Clinical studies of children with FAS/FAE have been sorely underfunded in the United States. Natural history studies are urgently needed to examine such patients as they mature, to determine their characteristics and special needs at each developmental age. There are more studies under way on babies with acquired immunodeficiency syndrome (AIDS) than there are on patients with FAS at all ages combined.

3. Research on remedial efforts with children with FAS/FAE is desperately needed. If indeed this is the most prevalent cause of mental retardation, then it behooves us to develop remedial and intervention procedures to enable these innocent victims to live the most productive lives possible. (Development of intervention efforts for children with Down syndrome and childhood autism have been primary research foci with those children for 30 years.)

4. One of the greatest barriers to effective treatment of such children now is confusion in the field about diagnostic issues. Although the concept of FAS is well recognized, there is no commonly used term or concept for describing children without the full syndrome. As these children are more likely to pass unrecognized in society, research is urgently needed on children now loosely defined as FAE. As behavioral effects appear to be more vulnerable to prenatal alcohol than do physical effects, and as the physical effects are known to occur at heavier exposure levels, it seems obvious that more children will experience brain damage from prenatal alcohol exposure than will be diagnosed with FAS. More research is needed on the *behavioral phenotype* of FAE to develop guidelines for better clinical identification. In our own clinical experience, failure to recognize and appropriately intervene with such children can further jeopardize their developmental outcome.

5. More research is needed on children of alcoholics, specifically those with alcoholic mothers. Several studies have suggested that children of alcoholics are a particularly high-risk group, not only for alcoholism, but also for psychosocial and behavioral problems. More research is needed on the prevalence of child abuse, sexual abuse, and neglect in such homes, in order to understand the multiple risks to healthy child development and to develop appropriate interventions.

6. Finally, there has been almost no systematic study of the mothers who produce children with FAS/FAE. If we are to prevent this important cause of mental retardation, we need in-depth clinical studies of these mothers (investigating both psychosocial as well as biochemical variables).

It should be no surprise that the important clinical work on children with FAS/FAE is being carried out in countries other than the United States, or under the auspices of the Indian Health Service in this country. Long-term tracking of such children, adolescents, and adults is difficult within the context of our existing health care delivery system. Furthermore, these are among the most difficult patients to follow, not only because of their alcoholic families but because of their own transiency as they grow older. It is highly unlikely that such studies will ever be carried out in an appropriately systematic and scientific manner without special funding.

We should not let such problems interfere with our resolve to study alcohol's effects on children and to develop remedial studies to help children who have been damaged by prenatal alcohol exposure. Alcohol continues to be a bigger threat to larger numbers of children than any illicit drug known to man, including cocaine. In the 10 years since the Surgeon General recommended not drinking during pregnancy, there have been at least 70,000 children born in the United States with the full FAS. It is time to face the magnitude of the problem, continue existing prevention strategies [such as bottle labeling of alcoholic beverages (Wagner, 1991) and the development of special prevention programs (Little et al., 1984)], and to study the characteristics and needs of affected children. We know that FAS and all prenatal alcohol effects are preventable disabilities. We now need to learn how best to help the children born with these disabilities.

ACKNOWLEDGMENTS

The preparation of this manuscript and the work described was partially supported by a grant from the National Institute on Alcohol Abuse and Alcoholism (AA01455-01-016) and IHS Contracts 240-83-0035, 243-88-0166, 243-89-0019, and 282-91-0013. The technical support of Cara Ernst, Kristan Geissel, Monica Tobin, and Brian Lim is greatly appreciated.

REFERENCES

Abel, E. L., and Sokol, R. J. (1987). Incidence of fetal alcohol syndrome and economic impact of FAS-related anomalies. *Drug Alcohol Depend.* **19**, 51–70.

Abel, E. L., and Sokol, R. J. (1991). A revised conservative estimate of the incidence of FAS and its economic impact. *Alcohol. Clin. Exp. Res.* **15**(3), 514–524.

Aronson, M., Sandin, B., Sabel, K. G., Kyllerman, M., and Olegard, R. (1984). Children of alcoholic mothers: Outcome in relation to the social environment in which the children were brought up. *Rep. Depart. Appl. Psychol.* **9**(2), 3–7.

Aronson, M., Kyllerman, M., Sabel, K. G., Sandin, B., and Olegard, R. (1985). Children of alcoholic mothers: Developmental, perceptual and behavioral characteristics as compared to matched controls. *Acta Paediatr. Scand.* **74**, 27–35.

Brown, R. T., Coles, C. D., Smith, I. E., Platzman, K. A., Silverstein, J., Erickson, S., and Falek, A. (1991). Effects of prenatal alcohol exposure at school age. II. Attention and behavior. *Neurotoxicol. Teratol.* **13**, 369–376.

Carmichael Olson, H., Streissguth, A. P., and Barr, H. (1990). Effects of moderate prenatal alcohol exposure on teacher-rated classroom behavior and academic performance in children aged 11 years. *Teratology* **41**(5), 626.

Clarren, S. K., and Smith, D. W. (1978). The fetal alcohol syndrome. *N. Engl. J. Med.* **298**, 1063–1067.

Clarren, S. K., Sampson, P. D., Larsen, J., Donnell, D., Barr, H. M., Bookstein, F. L., Martin, D. C., and Streissguth, A. P. (1987). Facial effects of fetal alcohol exposure: Assessment by photographs and morphometric analysis. *Am. J. Med. Genet.* **26**, 651–666.

Coles, C. D., Brown, R. T., Platzman, K. A., Erickson, S., and Falek, A. (1991). Effects of prenatal alcohol exposure at school age. I. Physical and cognitive development. *Neurotoxicol. Teratol.* **13**, 357–367.

Dehaene, P., Samaille-Villette, C., Crepin, G., Walbaum, R., Deroubaix, P., and Blanc-Garin, A. P. (1977a). Le syndrome d'alcoolisme fœtal dans le nord de al France. *Reveu L'alcoolisme* **23**(3), 145–158.

Dehaene, P., Tritran, M., Samaille-Villette, C., Samaille, P., Crepin, G., Delahousse, G., Walbaum, R., and Fasquelle, P. (1977b). Fréquence du syndrome d'alchoolisme fœtal. *La Nouvelle Presse Médicale* **6,** 1763.

Dehaene, P., Crepin, G., Delahousse, G., Querleu, D., Walbaum, R., Titran, M., and Samaille-Villette, C. (1981). Aspects épidémiologiques du syndrome d'alcoolism fœtal: 45 observations en 3 ans. *La Nouvelle Presse Médicale* **6,** 2639–2643.

Dehaene, P., Samaille-Villette, C., Boulanger-Fasquelle, P., Subtil, D., Delahousse, G., and Crepin, G. (1991). Diagnostic et prévalence du syndrome d'alcoolisme fœtal en maternité. *La Presse Médicale* **20,** 1002.

Driscoll, C. D., Streissguth, A. P., and Riley, E. P. (1990). Prenatal alcohol exposure: Comparability of effects in humans and animal models. *Neurotoxicol. Teratol.* **12,** 231–237.

Gairi, J. M. (1990). Perinatal mortality, congenital defects and alcohol consumption in the population of Barcelona (Spain). *Proceedings of the International Workshop on Alcohol and Fetal Development* October 4–5, 1990, Valencia, Spain.

Giunta, C. T., and Streissguth, A. P. (1988, September). Patients with fetal alcohol syndrome and their caretakers. *Soc. Casework: J. Contemp. Soc. Work* **69**(7), 453–459.

Hagberg, B., Hagberg, G., Lewerth, A., and Lindberg, U. (1981a). Mild mental retardation in Swedish school children. I. Prevalence. *Acta Paediatr. Scand.* **70,** 441–444.

Hagberg, B., Hagberg, G., Lewerth, A., and Lindberg, U. (1981b). Mild mental retardation in Swedish school children. II. Etiology and pathogenetic aspects. *Acta Paediatr. Scand.* **70,** 445–452.

Iosub, S., Fuchs, M., Bingol, N., Stone, R. K., and Gromisch, D. S. (1981). Long-term follow-up of three siblings with fetal alcohol syndrome. *Alcohol. Clin. Exp. Res.* **5,** 523–527.

Jones, K. L., and Smith, D. W. (1973). Recognition of the fetal alcohol syndrome in early infancy. *Lancet 2,* 999–1001.

Jones, K. L., Smith, D. W., Ulleland, C. N., and Streissguth, A. P. (1973). Pattern of malformation in offspring of chronic alcoholic mothers. *Lancet* **1,** 1267–1271.

Jones, K. L., Smith, D. W., Streissguth, A. P., and Myrianthopoulos, N. C. (1974). Outcome in offspring of chronic alcoholic women. *Lancet* **1,** 1076–1078.

Kyllerman, M., Aronson, M., Sabel, K. G., Karlberg, E., Sandin, B., and Olegard, R. (1985). Children of alcoholic mothers: Growth and motor performance compared to matched controls. *Acta Paediatr. Scand.* **74,** 20–26.

Lemoine, P., Harrousseau, H., Borteyru, J. P., and Menuet, J. C. (1968). Les enfants de parents alcooliques: Anomalies obsérvees: a propos de 127 cas. *Ouest Med.* **8,** 476–482.

Little, R. E., and Wendt, J. K. (1991). The effects of maternal drinking in the reproductive period: An epidemiologic review. *J. Subst. Abuse* **3,** 187–204.

Little, R. E., Young, A., Streissguth, A. P., and Uhl, C. N. (1984). Preventing fetal alcohol effects: Effectiveness of a demonstration project. *In* CIBA Foundation Symposium 105: "Mechanisms of Alcohol Damage *in Utero*" (R. Porter, M. O'Connor, and J. Whelan eds.), pp. 254–274. Pitman, London.

Löser, V. H., and Ilse, R. (1991). Korperliche und geistige langzeittentwicklung bei kindern mit alkoholembryopathie. *Sonderdruck aus Sozialpadiatrie in Praxis und Klinik* **13**(1), 8–14.

Majewski, F. (1984). Two teratogens of major importance: Alcohol and anticonvulsants. *In* "The Developing Brain and Its Disorders" (M. Arima, Y. Suzuki and H. Yabuuchi eds.), pp. 223–246. University of Tokyo Press, Tokyo, Japan.

Majewski, F., and Majewski, B. (1988). Alcohol embryopathy: Symptoms, auxological data, frequency among the offspring and pathogenesis. *Exerpta Medica. International Conference Series* **805,** 837–844.

National Institute on Alcohol Abuse and Alcoholism (1990, January). Seventh Special Report to the U.S. Congress on alcohol and health. *U.S. Department of Health and Human Services.*

Olegard, R., Aronson, M., Kyllerman, M., Sabel, K. G., and Sandin, B. (1984). Children of alcoholic mothers, Part I. Pre- and perinatal conditions and size at birth in relation to later mental capacity and neuropsychological symptoms. *Reports from the Department of Applied Psychology* **9.**

Robinson, G. C., Conry, J. L., and Conry, R. F. (1987). Clinical profile and prevalence of Fetal Alcohol Syndrome in an isolated community in British Columbia. *Can. Med. Assoc. J.* **137,** 203–207.

Sampson, P. D., Streissguth, A. P., Barr, H. M., and Bookstein, F. L. (1989). Neurobehavioral effects of prenatal alcohol. Part II: Partial least squares analysis. *Neurotoxicol. and Teratol.* **11**(5), 477–491.

Shruygin, G. I. (1974). Ob osobennostyakh psikhicheskogo razvitiya detei ot materei, stradayushchikh khronicheskim alkogolizmom [Characteristics of the mental development of children of alcoholic mothers]. *Pediatria. Mosk* **11**, 71–73.

Smith, D. W. (1982). "Recognizable Patterns of Human Malformation: Genetic, Embryologic and Clinical Aspects" 3rd Ed. W.B. Saunders, Philadelphia, Pennsylvania.

Spohr, H. L., and Steinhausen, H. C. (1987). Follow-up studies of children with fetal alcohol syndrome. *Neuropediatrics* **18**, 13–17.

Steinhausen, H. C., Nestler, V., and Spohr, H. L. (1982). Development and psychopathology of children with the fetal alcohol syndrome. *Dev. Behav. Pediatr.* **3**(2), 49–54.

Streissguth, A. P. (1976). Psychologic handicaps in children with fetal alcohol syndrome. *Work Prog. Alcohol.* **273**, 140–145.

Streissguth, A. P. (1986). The behavioral teratology of alcohol: Performance, behavioral, and intellectual deficits in prenatally exposed children. *In* "Alcohol and Brain Development" (J. R. West ed.), pp. 3–44. Oxford University Press, New York.

Streissguth, A. P., Martin, D. C., and Buffington, V. E. (1976). Test-retest reliability of three scales derived from a quantity–/–frequency–/–variability assessment of self-reported alcohol consumption. *In* "Work in Progress in Alcoholism" (F. A. Seixas and S. Eggleston eds.), Vol. 273, pp. 458–466. Annals of the New York Academy of Sciences, New York.

Streissguth, A. P., Herman, C. S., and Smith, D. W. (1978a). Intelligence, behavior, and dysmorphogenesis in the fetal alcohol syndrome: A report on 20 patients. *J. Pediatr.* **92**(3), 363–367.

Streissguth, A. P., Herman, C. S., and Smith, D. W. (1978b). Stability of intelligence in the fetal alcohol syndrome: A preliminary report. *Alcohol. Clin. Exp. Res.* **2**(2), 165–170.

Streissguth, A. P., Martin, D. C., Martin, J. C., and Barr, H. M. (1981). The Seattle longitudinal prospective study on alcohol and pregnancy. *Neurobehav. Toxicol. Teratol.* **3**, 223–233.

Streissguth, A. P., Barr, H. M., and Martin, D. C. (1984). Alcohol exposure *in utero* and functional deficits in children during the first four years of life. *In* CIBA Foundation Symposium 105: "Mechanisms of Alcohol Damage *in Utero*" (R. Porter, M. O'Connor and J. Whelan eds.), pp. 176–196. Pitman, London.

Streissguth, A. P., Clarren, S. K., and Jones, K. L. (1985). Natural history of the fetal alcohol syndrome: A ten-year follow-up of eleven patients. *Lancet* **2**, 85–92.

Streissguth, A. P., Barr, H. M., Sampson, P. D., Parrish-Johnson, J. C., Kirchner, G. L., and Martin, D. C. (1986). Attention, distraction and reaction time at age 7 years and prenatal alcohol exposure. *Neurobehav. Toxicol. Teratol.* **8**(6), 717–725.

Streissguth, A. P., Barr, H. M., Sampson, P. D., Bookstein, F. L., and Darby, B. L. (1989a). Neurobehavioral effects of prenatal alcohol. Part I: Literature review and research strategy. *Neurobehav. Toxicol. Teratol.* **11**(5), 461–476.

Streissguth, A. P., Bookstein, F. L., Sampson, P. D., and Barr, H. M. (1989b). Neurobehavioral effects of prenatal alcohol. Part III: PLS analyses of neuropsychologic tests. *Neurotoxicol. Teratol.* **11**(5), 493–507.

Streissguth, A. P., Barr, H. M., and Sampson, P. D. (1990). Moderate prenatal alcohol exposure: Effects on child IQ and learning problems at age 7½ years. *Alcohol. Clin. Exp. Res.* **14**(5), 662–669.

Streissguth, A. P., Aase, I. M., Clarren, S. K., Randels, S. P., LaDue, R. A., and Smith, D. F. (1991a). Fetal alcohol syndrome in adolescents and adults. *J.A.M.A.* **265**, 1961–1967.

Streissguth, A. P., Randels, S. P., and Smith, D. F. (1991b). A test–retest study of intelligence in patients with fetal alcohol syndrome: Implications for care. *J. Am. Acad. Child Adolesc. Psychiatry* **30**(4), 584–587.

Streissguth, A. P., Barr, H. M., and Sampson, P. D. Effects of alcohol use during pregnancy on child development: Report on a longitudinal, prospective study of human behavioral teratology. *In* "Longitudinal Studies of Children Born at Psychological Risk: Cross-National Perspectives" (C. W. Greenbaum and J. G. Auerbach eds.), Ablex, Norwood, New Jersey. In press.

Streissguth, A. P., Sampson, P. D., Barr, H. M., Bookstein, F. L., and Carmichael Olson, H. Effects of prenatal alcohol vs. tobacco on children: Contributions from the Seattle Longitudinal Prospective Study and implications for public policy. *In* "Prenatal Exposure to Pollutants" (working title). (H. L. Needleman and D. Bellinger, eds.). Johns Hopkins University Press. In press.

Stromland, K. (1990). Contribution of ocular examination to the diagnosis of foetal alcohol syndrome in mentally retarded children. *J. Ment. Defic. Res.* **34,** 429–435.

Wagner, E. N. (1991). The alcoholic beverages labeling act of 1988: A preemptive shield against fetal alcohol syndrome claims? *J. Leg. Med.* **12,** 167–200.

— 3 —

Effects of Prenatal Alcohol Exposure

❧

Nancy L. Day
Western Psychiatric Institute and Clinic
University of Pittsburgh, School of Medicine
Pittsburgh, Pennsylvania

INTRODUCTION

This chapter reviews current knowledge about the effects of prenatal exposure to alcohol on the offspring. Research has explored this question from three general directions: (1) clinical evaluations and studies of the natural history of people with fetal alcohol syndrome; (2) studies of offspring of alcoholic women; and (3) assessment of the effects of alcohol use during pregnancy on the exposed offspring.

In general, the findings from these three areas of research converge to define a continuum of effects. At the most extreme end of the spectrum, people who have fetal alcohol syndrome exhibit a pattern of growth deficits, mental retardation, behavioral difficulties, and morphologic abnormalities. At the other end, offspring who are exposed to moderate doses of alcohol during pregnancy may exhibit only a few of the characteristics of the full-blown syndrome.

At both ends of the spectrum, however, the findings are often quite inconsistent between subjects and from study to study. The differences between subjects may result from different levels or patterns of exposure, exposure at different times of pregnancy, or differing vulnerability to the effects of alcohol. Methodologic issues in research design, such as case identification, measurement, and selection of covariates lead to differences in outcomes between studies.

DEFINITION OF FETAL ALCOHOL SYNDROME

Fetal alcohol syndrome (FAS) is defined by a cluster of infant characteristics. Cardinal features of the syndrome include (1) growth deficiency in both the prenatal and postnatal periods; (2) central nervous system anomalies such as microcephaly, mental retardation, and irritability; and (3) craniofacial anomalies including short palpebral fissures, frontonasal alterations, thin upper vermilion border, flat midface, and hypoplastic maxilla and/or mandible (Sokol and Clarren, 1989). Separately, each of these features is defined as an alcohol-related birth defect or a fetal alcohol effect (FAE).

METHODOLOGICAL ISSUES IN THE ASSESSMENT OF THE EFFECTS OF PRENATAL SUBSTANCE USE

A number of methodological issues are important in the evaluation of research on prenatal substance use. These include the method of identifying cases, selection of the appropriate comparison group, timing of assessment during pregnancy, the method of assessment, and measurement of other risk factors for poor pregnancy outcome.

Method of Selecting Cases

To maximize sample size and efficiency of ascertainment, researchers have used sampling sites that were convenient or that had a large number of cases. A number of studies, for example, have selected women from clinics or hospitals where they are being treated for alcoholism. However, women who are in treatment for alcohol abuse and dependence differ from those who are not in treatment in a number of ways. They have higher rates of other substance use, more physical and psychiatric comorbidity, and lower social status. These covariates are, in themselves, risk factors for problems during pregnancy or for problems in the development of the fetus. Thus, a treatment population is a biased sample of alcoholics, and the same factors that lead to treatment will also increase the probability of a less optimal outcome.

Selection of an Appropriate Comparison Group

When women who drink heavily or are alcoholic are compared to a group of nondrinking women, the probability is very high that these groups differ in a number of other important features of their lifestyles. Therefore, there is always a risk that differences are related to correlates of alcohol use during pregnancy rather than to the alcohol use by itself.

Timing of Assessment during Pregnancy

An additional methodologic factor that must be considered is the time of first interview and the intervals and timing of subsequent contacts. If women are interviewed at nonstandardized times during pregnancy, the recall period, and thus the recall error, will vary. For example, if women are interviewed at the first prenatal visit, the time of interview may vary from early in the first trimester to late in the third trimester. This is particularly a problem as the women most likely to arrive late for prenatal care are the women who are most likely to use drugs and alcohol.

In the Maternal Health Practices and Child Development (MHPCD) study, one group of women reported on their first trimester alcohol use in the fourth month of pregnancy and again at the seventh month of pregnancy (Robles and Day, 1990). A second group was asked about first trimester use at the fourth month and at delivery. The correlation between first trimester use reported at the fourth and seventh months was 0.61. The correlation between first trimester use reported at month four and at delivery was 0.53. Therefore, reporting on a complex behavior such as a drinking pattern was not particularly reliable. The direction of error was difficult to estimate, since among the women who reported differing rates at the two periods, the reported amount was more likely to increase than to decrease. This finding has been reported by other investigators as well (Ernhart et al., 1988). There are two interpretations from these results: one, that this reflects a problem of memory, and the other, that women may feel safer reporting on a labeled behavior from a greater distance in time.

Assessment of Substance Use during Pregnancy

The pattern of alcohol use can be described using three variables: quantity, frequency, and duration of use. Quantity, the amount per occasion, can be described further by three separate elements: usual, maximum, and minimum quantity. Although many studies use only usual quantity, in the MHPCD data, if we use only the usual quantity and frequency, we miss 27% of the total amount of alcohol that the women consumed. Maximum quantity per occasion adds an additional 18%, and minimum quantity adds 9% to the overall average daily volume for the cohort.

An additional problem in ascertaining alcohol use during pregnancy is that women may not think back to the beginning of pregnancy. In the MHPCD study, we developed a technique to measure alcohol use during each month of the first trimester (Day and Robles, 1989). At the beginning of the interview, women are asked to indicate on a calendar (1) when they got pregnant, (2) when they realized they were pregnant, and (3) when the pregnancy was diagnosed. After the questions on alcohol consumption, the interviewer returns to the calendar and asks, for the periods between conception and recognition and between recognition and confirmation, whether the subject was using alcohol more like her prepregnancy pattern or more like what she reported for her first trimester. Fifty percent of the women who drank before pregnancy reported that from conception to recognition of pregnancy, their alcohol use was similar to their prepregnancy pattern rather than what they had reported as their first trimester pattern. Twenty-three percent reported that from recognition to diagnosis, they were still using alcohol at their prepregnancy rate rather than at their reported first trimester rate.

The dates of conception, recognition, and diagnosis are used to calculate a month-by-month rate of alcohol use and a weighted estimate of the average daily use for the first trimester. In the MHPCD study, 11% of the women reported using alcohol at the rate of one or more drinks per day during the first trimester, whereas, when we calculated the rate based on the responses to the above questions, 20% of the women were actually drinking at this rate. Also, 65% of the women reported no use during the trimester, when only 46% were real abstainers. Thus, when women report on their first trimester use, they do not consider their intake of alcohol before the recognition of pregnancy, even though they have been instructed to think back to the very beginning of the pregnancy. This may mean that studies that have reported effects for drinking before pregnancy actually reflect the effects of alcohol exposure in early pregnancy before pregnancy recognition.

The measurement of alcohol use during pregnancy is made even more difficult because most women decrease their use during pregnancy. This decrease occurs during the first trimester, usually after the recognition of pregnancy. Because of this pattern, accurate reporting of use during the first trimester is very difficult, since the women are being asked to recall their use during a period when the use is changing.

Ascertainment and Measurement of Covariates of Substance Use

Race, education, income, marital status, and the use of tobacco, marijuana, and other illicit substances are all correlates of alcohol use during pregnancy. These variables are also independent risk factors for problems during pregnancy and should be measured with the same amount of accuracy and detail as alcohol use.

EPIDEMIOLOGY

Patterns of Alcohol Use among Women

Sixty-four percent of all women in a national survey reported that they drank currently, 25% of women drank at least once a week, and 5% drank at least 60 drinks a month (Hilton, 1988). Few of these women reported that they drank large quantities; only 20% ever drank as many as five drinks on an occasion. Younger women and unmarried women drank more than older and married women in this study, and women of higher social class were more likely to be drinkers than women of lower social class. White women were more likely to be drinkers and were heavier drinkers than black women (Herd, 1988).

During pregnancy, the majority of women decrease their drinking. This decrease usually begins at pregnancy recognition, which is most often during the first trimester, and continues through the third trimester. In the MHPCD study, 44% of the women reported drinking one or more drinks per day before pregnancy. However, after the first month of pregnancy, only 37% drank at this rate. By the second and third months of pregnancy, the proportions were 21 and 14%, respectively. By the end of the third trimester, only 5% of the women reported drinking an average of a drink a day (Day et al., 1989). However, by 8 months postpartum, the alcohol use of the women had returned to the rate reported before pregnancy (Day et al., 1990).

The covariates of drinking also change during pregnancy. In the MHPCD study, there were no significant differences, in terms of sociodemographic factors, between women who averaged a drink per day and the women who abstained during the first trimester. By the third trimester, the women who continued drinking, and particularly those who continued to drink heavily, were significantly more likely to be black, unmarried, and to use tobacco, marijuana, and other illicit drugs (Day et al., 1989). Each of these factors is, by itself, a risk factor for problems with the course and outcome of pregnancy, so the women who drink the most are at the highest risk of negative outcomes for a number of reasons.

Prevalence of Alcoholism among Women

The Epidemiologic Catchment Area (ECA) studies provide data on the prevalence of diagnoses of alcohol abuse and dependence, as defined by the Diagnostic and Statistical Manual (DSM-III) of the American Psychiatric Association (1980). The ECA investigators interviewed subjects from the general population, using a structured questionnaire, to make psychiatric diagnoses. Among women aged 18 to 29, 3% received a diagnosis of either abuse or dependence, and for women aged 30 to 44, the proportion who had ever had alcoholism (abuse and dependence combined) was the same. A total of 5.4% of women who were drinkers were diagnosed as alcoholic, and 35% of all women who were heavy or problem drinkers had a lifetime diagnosis of alcoholism (Helzer et al., 1991). Only a portion of the women had received treatment for alcoholism, and these women had the longest disease duration.

Epidemiology of Fetal Alcohol Syndrome

The risk of FAS in the general population has been estimated by Abel and Sokol (1991) to be approximately 0.33 cases per 1000. This is considerably revised from their earlier estimate of 1.9 cases per 1000, and represents data from a prospective assessment. Offset against this, however, is a report from Little et al. (1990). They studied the records of 40 infants born to 38 alcohol abusers. In no case was a diagnosis of fetal alcohol syndrome made, even though alcohol abuse was documented in the medical record. Six of the infants met criteria for fetal alcohol syndrome, and 50% of the 34 liveborn infants were judged to have poor postnatal growth and development. Therefore, FAS may be considerably underdiagnosed.

The rate of FAS also varies depending on the racial and socioeconomic characteristics of the sample. Abel and Sokol (1991) estimate the rate of FAS among whites to be 0.29 cases per 1000 and among blacks to be 0.48 per 1000 (Abel and Sokol, 1991). This disparity in rates has been noted earlier by the same authors (Sokol et al., 1986), who found that four factors predicted FAS: high proportion of drinking days, a positive score on the Michigan Alcohol Screening Test, high parity, and black race. Bingol et al. (1987) reported similar findings in a comparison of upper-middle and lower class alcoholic mothers. In the upper-middle class mothers, only 1 of 36 of the mothers had an offspring with FAS, compared to a rate of 40.5% among the 48 lower class mothers. All of the upper-middle class mothers were white; 70% of the lower class mothers were black, and 30% were Hispanic. The amount of absolute alcohol intake was equivalent in the two groups, but 8% of the upper-middle class group and 18% of the lower class group had cirrhosis. Therefore, there are factors in addition to alcohol intake that affect the occurrence of FAS in offspring of alcoholic mothers. Whether these factors that increase vulnerability are biological or psychosocial is not known.

STUDIES OF PATIENTS WITH FETAL ALCOHOL SYNDROME

Follow-up assessments of patients with FAS have been largely limited to case series of children diagnosed at birth as having FAS. Sample sizes are small, and the populations are biased, because the subjects were generally selected from clinics or institutions. These follow-up studies have found that subjects with FAS are small for their age (Streissguth et al., 1985) and mentally retarded, although the range is from severely retarded to near normal (Landesman-Dwyer, 1982; Streissguth et al., 1978). People with FAS may have a multitude of behavioral problems including stereotyped behaviors, irritability, hyperactivity, tremulousness, and hyperdistractibility (Streissguth et al., 1978; Olegard et al., 1979; Majewski, 1978a,b; Shaywitz et al., 1980; Steinhauser et al., 1982), although these problems are not present in all cases. Motor and speech development are sometimes delayed, and people with FAS are more likely to have speech and hearing impairments (Steinhauser et al., 1982; Church and Gerkin, 1988).

However, although these problems are found at higher rates among cases of FAS, the variability in cognitive development, growth and morphology, and behavioral measures demonstrates that there is a broad spectrum of severity even within the diagnostic rubric of FAS.

A recent study assessed the long-term development for people with FAS. Streissguth et al. (1991) examined 61 subjects who had been identified in childhood as having FAS or FAE. Subjects were selected from clinical case-loads and from four Native American reservations. The subjects, as they reached adolescence and adulthood, still manifested growth deficits, but the deficits were not as pronounced as they had been at younger ages; head circumference remained the parameter most affected. Dysmorphologic features also became less prominent, though they were still present. IQ scores ranged from 20 to 105; the mean for the group was 68. Reflecting this latter finding, only 6% of the subjects were in regular classes in school; the remainder were unable to achieve this level in school or to maintain regular outside employment. In addition, the subjects had poor concentration and attention, dependency, stubbornness, social withdrawal, and a number of other problems including lying, cheating, and stealing.

It should be kept in mind, however, that these subjects were selected as a sample of convenience and do not represent the entire spectrum of outcome for people with FAS. Where they belong, along the spectrum of possible long-term outcomes for people with FAS, is not known.

STUDIES OF OFFSPRING OF ALCOHOLIC WOMEN

The proportion of offspring of alcoholics who have FAS is quite small. The risk that an alcoholic woman will have a child with FAS is estimated to be approximately 6% (Abel and Sokol, 1987), although for offspring born after an FAS sibling, the risk is very high (70%) (Abel and Sokol, 1987). The high probability that all subsequent offspring will be affected after an initial case means that any cofactor must be a correlate or a consequence of chronic alcoholism. Even in the absence of FAS, the babies of alcoholic women may have decreased birth weight, an increased rate of congenital anomalies (Sokol et al., 1980; Little et al., 1989; Hollstedt et al., 1983), and an increased incidence of impaired intellectual functioning (Aronson et al., 1985).

EFFECTS OF DRINKING DURING PREGNANCY ON OFFSPRING AT BIRTH

In studies of the effect of drinking during pregnancy, each of the specific characteristics of FAS has been used independently as an outcome variable. Growth and morphology, neurobehavioral effects, and behavioral and cognitive development will be reviewed separately.

Growth and Morphology

The Maternal Health Practices and Child Development Study began in 1982 (Day et al., 1989). Women were recruited in their fourth prenatal month and followed subsequently at the seventh prenatal month, delivery, and at 8 and 18 months, and 3 and 6 years postpartum. At each point, women are interviewed, using a standard protocol, about alcohol and drug use, and sociodemographic and life-style variables. Children are examined for growth, morphology, and medical problems and are assessed with age-appropriate measures of cognitive and motor development. Among the initial cohort of 650 women selected for the study, there were 595 liveborn singleton infants. At the 3-year follow-up, 85% were still in the study.

In the MHPCD study, an increased risk of having a low-birth-weight baby was associated with alcohol consumption in the first and/or second month of pregnancy. Maternal drinking during the early first trimester was also associated with an increased risk of giving birth to an infant who was below the tenth percentile for length or head circumference (Day et al., 1989).

Smith et al. (1986) reported that birth weight, length, and head circumference were reduced in offspring of women who drank continuously throughout pregnancy when compared to offspring of nondrinkers. Further, birth weight was influenced by both dose and duration of the alcohol exposure. In other studies, a relationship between heavy alcohol use and decreased birth weight has also been reported, although in some cases the relationship is to drinking before pregnancy (Hanson et al., 1978; Little, 1977), in others to drinking during early pregnancy (Kaminski et al., 1978; Kuzma and Sokol, 1982; Mau and Netter, 1974), and in still others, to drinking during later pregnancy (Little, 1977; Rosett et al., 1983). As we discussed earlier in the methodology section, it seems likely that the correlations with drinking before pregnancy may have, instead, been the effects of drinking early in the first trimester. In addition, since few women start drinking heavily during pregnancy, it is difficult to separate the effects of drinking late in pregnancy from drinking throughout pregnancy.

Not all studies have reported an association between alcohol use and decreased birth weight. In a group of healthy full-term infants, Coles et al. (1985) found no growth differences between those exposed and those not exposed to alcohol. Ernhart et al. (1985) also found no neonatal growth effects of prenatal alcohol exposure.

In the MHPCD study, newborns were not more likely to be premature (<37 weeks), nor was gestational age decreased (Day et al., 1989). This has been reported by Kaminski et al. (1978) as well. Two other studies did find significant but small effects on gestational age (Tennes and Blackard, 1980; Hingson et al., 1982).

A positive relationship between minor morphologic malformations and alcohol use during pregnancy was found in the MHPCD study (Day et al., 1989). This finding has been replicated in most other studies of alcohol use during pregnancy (Hanson et al., 1978; Rosett et al., 1983; Ernhart et al., 1985; Russell, 1991). However, other reports have failed to find a higher rate of morphologic abnormalities related to prenatal alcohol exposure (Kaminski, 1978; Coles et al., 1985; Tennes and Blackard, 1980; Hingson et al., 1982).

In summary, results of studies of drinking practices and birth outcome are somewhat inconsistent. The relationship between alcohol exposure and malformations is strong; an increased rate of malformations was noted among children of heavy drinkers in most, although not all, studies. The preponderance of evidence also demonstrates that prenatal alcohol use is related to growth abnormalities at birth, although there are again inconsistencies in the times during pregnancy that predict growth deficits. Most studies did not find a relationship between gestational age and alcohol use during pregnancy.

Neurobehavioral Effects

Studies of neurobehavioral effects of prenatal alcohol exposure have been widely divergent in their findings. One study found that alcohol-exposed infants were less able to habituate to aversive stimuli (Streissguth et al., 1983) as measured by the Neonatal Behavioral Assessment Scale (NBAS) (Brazelton, 1984). In another study, alcohol-exposed infants exhibited changes in their reflexive behavior, state control, and motor behavior on the NBAS (Coles et al., 1985). Two additional studies reported that increased irritability on the NBAS was related to prenatal alcohol exposure (Fried and Makin, 1987), and that alcohol use was related to depression of the infant's range of state (Jacobson et al., 1984). In an investigation of the effect of dose and duration of alcohol use during pregnancy (Smith et al., 1986), differences in orientation responses on the NBAS were found to be the result of the duration of drinking, whereas differences in autonomic regulation were attributable to both duration and dose.

In contrast, Ernhart et al. (1985) found no relationship between NBAS performance and prenatal alcohol use. In the MHPCD study, there was also no association between neonatal behavior and alcohol use during any trimester of pregnancy (Richardson et al., 1989).

Investigators have monitored sleep cycling and arousal as a measure of neurophysiological development, integrity, and maturation. Rosett et al. (1979) reported that the infants of mothers who drank heavily throughout pregnancy had a greater proportion of quiet sleep, were more restless, and had more major body movements. Electroencephalogram (EEG) power spectra analyses of infants of alcoholic mothers have shown hypersynchrony of the EEG as well as an increase in the integrated power in all sleep states, with the greatest increase seen in active sleep (Chernick et al., 1983; Ioffe et al., 1984). In addition, Ioffe and Chernick (1988) reported that infant EEG maturation was affected by maternal binge drinking.

In the MHPCD study, relationships between alcohol exposure during pregnancy and abnormalities of neurophysiological status were found even after the effect of confounding variables such as other drugs, tobacco, infant gender, and demographic factors were controlled. Trimester-specific disturbances in sleep cycling, motility, and arousals were noted (Scher et al., 1988), and there was an effect of maternal alcohol use on the sleep-state cycling of the infants (Stoffer et al., 1988). It was suggested that these abnormalities on the sleep-EEG recording might reflect differences in maturation of the central nervous system.

In conclusion, there is little research on the neurobehavioral effects of prenatal exposure to potential teratogens, and the studies that are available are often contradictory. Again, some of this can be explained by methodologic problems. Some of the studies have used retrospective reports of substance use over the entire pregnancy or have had only a single assessment during pregnancy, whereas others have failed to control adequately for confounding variables in the statistical analyses.

LONG-TERM EFFECTS OF PRENATAL ALCOHOL EXPOSURE

Two separate issues must be considered in the assessment of long-term effects of prenatal exposure to alcohol. One is an assessment of the persistence of the growth deficits and morphological abnormalities over time. A second concern is the possibility that effects that were not present in the immediate neonatal period may become apparent as the child grows. For example, cognitive deficits, behavioral problems, and learning disorders all must await the appropriate developmental time points to occur.

Growth and Morphology

In our ongoing longitudinal work, we have shown that there is a significant relationship between exposure to alcohol during pregnancy and the growth of the offspring at both 8 and 18 months of age. These long-term growth deficits were related to alcohol exposure during the latter part of pregnancy, the second and third trimesters, and were not related to alcohol exposure during the first trimester or before pregnancy (Day et al., 1990; Day et al., 1991b). These deficits in growth continued to be evident at 3 years of age, when the exposed offspring in the MHPCD study were smaller in weight, length, and head circumference (Day et al., 1991a). These differences were still present even after controlling statistically for the effect of the current environment, use of other drugs during pregnancy, and a number of other significant covariates.

Therefore, unlike children who are exposed to tobacco during pregnancy, children who were exposed to alcohol did not exhibit catch-up growth. A further analysis of these data, using longitudinal data-analysis techniques, demonstrated that the lag in growth occurred subsequent to birth and before the eighth month of life. After that, the exposed offspring remained significantly smaller than nonexposed offspring, but their rate of growth did not differ (Geva et al., submitted).

Similar results have been reported by Middaugh and Boggan (1991), using the C57B1/6 strain of mice. In parallel with our findings, these investigators did not find a significant effect of moderate prenatal alcohol exposure at birth on growth. However, subsequent to birth, the exposed offspring were significantly smaller, and this decrease in rate of growth occurred at one specific point in development, the time of weaning.

In a further test of the relationship between prenatal alcohol exposure and growth, we assessed the differential effects of patterns of drinking. Research on laboratory animals has indicated that binge drinking may have a more significant impact on development than the same exposure spread out more evenly over time, although the effect was more often on head circumference and cognitive development. We tested the predictive validity of two measures of drinking: average daily volume, an averaged measure of total consumption, and frequent heavy drinking, a measure of how often a subject consumes five or more drinks per occasion. Both measures significantly predicted growth retardation. However, further analyses demonstrated that average daily volume was a stronger predictor of effects, and when average daily volume was controlled, the frequency of heavy drinking, or a pattern of binging did not contribute any additional explanation. Therefore, for growth at least, a pattern of continued toxic exposure seems to be a more important predictor of deficits than a pattern of frequent heavy or binge drinking.

Only a few other studies have followed exposed children over time. In one, growth retardation was reported at 8-month follow-up, but was not present at subsequent evaluations (Barr et al., 1984). This study did not have measures of alcohol use during the later part of pregnancy; however, the time that we have demonstrated is most likely to be vulnerable to growth deficits. O'Connor et al. (1986) found no effects of prenatal alcohol exposure on height, weight, or head circumference at 1 year of age, and Fried and O'Connell (1987) found no growth effects on offspring at 12 and 24 months of age.

In contrast, Russell (1991) found significant effects of alcohol use before pregnancy on the height and head circumference of the offspring at 6 years of age. Coles et al. (1991) reported on 68 children who had been followed from birth through 5 years 10 months and reported that children who were exposed continuously to alcohol throughout pregnancy had smaller head circumferences, although they were not significantly different in weight or height. It should be noted that the studies that found growth effects (Day et al., 1991a,b; Russell, 1991; Coles et al., 1991) were studies from lower socioeconomic status samples, whereas those that did not report effects were from middle class samples.

There was a significant correlation between alcohol exposure and the rate of minor physical anomalies in the MHPCD study (Day et al., 1990; Day et al., 1991b). Another study (O'Connor et al., 1986) found that the rate of minor physical anomalies was correlated with maximum quantity of consumption before and during pregnancy. Graham et al. (1988) also found a dose–response relationship between dysmorphology at 4 years of age and prenatal alcohol use before pregnancy recognition. Coles et al. (1991) found a higher rate of alcohol-related birth defects at 6 years of age among the offspring of both women who continued to drink heavily throughout pregnancy and those who drank heavily during early pregnancy and then quit.

Behavioral and Cognitive Effects

In the early postnatal period, Fried et al. (1987) found a significant relationship between prenatal alcohol use and neonatal reflexes as measured by the Prechtl examination at 9 days. This effect was gone by 30 days of age, however. Coles et al. (1978), in contrast, found that the infants of mothers who drank throughout pregnancy showed less improvement in reflexes and autonomic regulation over the first month of life when compared to the infants of women who stopped drinking or who never drank during pregnancy.

Cognitive Development

At 6 months of age, Coles et al. (1987) found that infants of mothers who continued to drink during pregnancy had significantly lower scores on the Bayley mental and motor scales than infants of mothers who stopped or abstained during pregnancy. Streissguth et al. (1980) found that at 8 months, alcohol use before pregnancy was correlated with decreased scores on the mental and motor scores on the Bayley scales. Gusella and Fried (1984) reported that infants of women who drank during pregnancy did less well at 13 months on the mental index and on the verbal comprehension and spoken language cluster scores of the Bayley scales. In another study (O'Connor et al., 1986), drinking before pregnancy was found to be associated with an average deficit of 24 points on the Bayley mental scale at 12 months among children of heavy drinkers, when their scores were compared to the IQ scores of their mothers. The children of light drinkers and abstainers did not differ in IQ scores from their mothers. In the MHPCD project, prenatal alcohol use did not significantly predict either the mental or the motor scores on the Bayley at 8 or 18 months (Richardson and Day, 1991).

Streissguth et al. (1989) reported that prenatal maternal alcohol use was negatively associated with IQ in the offspring at 4 years of age. Before pregnancy, consumption of 1.5 ounces of absolute alcohol per day predicted an average decrease of five points in the IQ score of the child. In the same study, when the children were 7.5 years of age, there was a decrement of seven IQ points with exposure prenatally to more than 1 ounce of absolute alcohol per day during pregnancy (Streissguth et al., 1990). Achievement scores, in this study, were significantly related to binge drinking before pregnancy.

In another study, the children who were exposed to alcohol continuously throughout pregnancy did less well on cognitive tests, exhibiting deficits in sequential processing and overall mental processing at age 5 years 10 months (Coles et al., 1991). Children who were exposed in early pregnancy, but not in later pregnancy, also did less well on some of the measures of academic skills, specifically in math and reading.

Thus, although the results of cognitive testing conflict, there is a trend toward decreases in the IQ score, although it must be noted that this statistically significant effect may not be clinically significant. These findings highlight the potential of damage to the central nervous system. This conclusion, however, can only be made after controlling for environmental factors that also affect cognitive development.

Behavior

At 4 years of age, the children who were prenatally exposed to moderate drinking were less attentive and more active during observations in their home (Landesman-Dwyer et al., 1981), and were less attentive and had longer reaction times on a vigilance task in a laboratory (Streissguth et al., 1984). In the same study, attention and reaction time as measured by a Continuous Performance Test (CPT) vigilance task, at age 7.5 years continued to be negatively related to prenatal alcohol exposure (Streissguth et al., 1986).

In another study (Brown et al., 1991), the investigators assessed the prevalence of symptoms of attention deficit hyperactivity disorder (ADHD). They found that the children who were exposed throughout pregnancy showed deficits in their ability to sustain attention and, on teacher's reports, had attentional and behavioral problems. However, none of the symptomatology reached clinical or diagnostic level for ADHD. The children who were exposed to alcohol prenatally differed in symptomatology from children with diagnosed ADHD, leading the investigators to conclude that although the alcohol-exposed children had behavior and attention problems, they differed from those found in children with a diagnosis of ADHD.

In Seattle, Barr et al. (1990) evaluated the gross and fine motor performance of the 4-year-olds. Prepregnancy alcohol exposure was related to fine motor errors, increased time to correct the errors, and poorer gross motor balance. In the MHPCD study, at 3 years of age, alcohol significantly predicted the failure of children to be willing to begin or to complete a gross motor test.

Therefore, the findings regarding the long-term effects of prenatal exposure to alcohol on the cognitive development of the child are inconsistent. This may be due to the difficulty of separating the teratogenic effects of alcohol from the environmental effects that may accompany the disorganized environments, both psychological and structural, that often accompany alcohol and drug use.

SUMMARY AND CONCLUSIONS

As noted at the beginning of the chapter, the continuum of effects for prenatal alcohol exposure ranges from the fully developed picture of fetal alcohol syndrome to the much milder expression of fetal alcohol effects. The fact that these symptoms fall along a continuum, and that this continuum correlates with the actual amount of alcohol to which the fetus was exposed prenatally, is a strong argument that the effects are related to alcohol exposure and not solely to the poor environments that many of these children experience. Further, we know that children who are exposed to alcohol during pregnancy, but not to alcoholism, have fewer effects than children of alcoholic women, and that people with FAS are usually the offspring of late-stage and chronic alcoholic women. Therefore, the continuum of effects also follows the spectrum from drinking to alcoholism.

Studies have shown that children exposed to alcohol and/or children of alcoholics are smaller at birth and that they have a higher rate of morphological abnormalities. The findings of neurological differences are less consistent and therefore less convincing, at this point.

In the long term, some studies have shown that the children who are exposed to alcohol prenatally have growth deficits, whereas others have not found this to be true. The studies that have found growth deficits longitudinally have generally been studies of low-income populations, populations with more risk factors. Thus, it may be that prenatal alcohol exposure increases vulnerability to growth deficits, but these growth deficits are expressed only in populations that have multiple risk factors. Alcohol, therefore, may contribute to a cumulative burden, and children who are exposed to alcohol prenatally may have fewer resources for offsetting the growth deficits compared to children who have fewer risk factors and more optimal environments. There is some evidence for this hypothesis in the finding that women who are black and/or lower class are more likely to have children who are more affected by alcohol prenatally, compared to women who are white and/or middle class. Thus, the impact of prenatal alcohol exposure seems to be potentiated by other factors.

People with FAS, by definition, have cognitive deficits, and these deficits have recently been shown even for adults who have FAS. The effects on cognitive development of prenatal exposure to lower levels of alcohol again demonstrate a continuum of effects, and, though not completely consistent, show that alcohol exposure is related to a small decrease in cognitive abilities. Similarly, people with FAS have behavioral problems that are present through adulthood. Behavioral problems have also been reported, though not consistently, among offspring who were exposed to lower amounts of alcohol prenatally.

Therefore, although there are a number of gaps in our knowledge of the effects of prenatal alcohol exposure on the offspring, we can say that alcohol affects both growth and development along a continuum with the least exposed offspring exhibiting the fewest effects, and the most heavily exposed offspring being most seriously affected. The expression of these effects, however, is moderated by environmental factors.

REFERENCES

Abel, E., and Sokol, R. (1987). Incidence of fetal alcohol syndrome and economic impact of FAS-related anomalies. *Drug Alcohol Depend.* **19**, 51–70.

Abel, E., and Sokol, R. (1991). A revised conservative estimate of the incidence of FAS and its economic impact. *Alcohol. Clin. Exp. Res.* **15**, 514–524.

American Psychiatric Association. (1980). "Diagnostic and Statistical Manual of Mental Disorders," DSM-III, 3rd Ed. American Psychiatric Association, Washington, D.C.

Aronson, M., Kyllerman, M., Sabel, K. G., Sandin, B., and Olegard, R. (1985). Children of alcoholic mothers: Developmental, perceptual, and behavioral characteristics as compared to matched controls. *Acta Psychiatr. Scand.* **74**, 27–35.

Barr, H. M., Streissguth, A. P., Martin, D. C., and Herman, C. S. (1984). Infant size at 8 months of age: Relationships to maternal use of alcohol, nicotine, and caffeine during pregnancy. *Pediatrics* **74**, 336–341.

Barr, H. M., Streissguth, A. P., Darby, B. L., and Sampson, P. D. (1990). Prenatal exposure to alcohol, caffeine, tobacco, and aspirin: Effects on fine and gross motor performance in 4-year-old children. *Dev. Psych.* **26**, 339–348.

Bingol, N., Schuster, C., Fuchs, M., Iosub, S., Turner, G., Stone, R., and Gromisch, D. (1987). The influence of socioeconomic factors on the occurrence of fetal alcohol syndrome. *Adv. J. Alcohol Subst. Abuse.* **6**, 105–118.

Brazelton, T. B. (1984). "Neonatal Behavioral Assessment Scale" 2nd Ed. Lippincott, Philadelphia, Pennsylvania.

Brown, R., Coles, C., Smith, I., Platzman, K., Silverstein, J., Erickson, S., and Falek, A. (1991). Effects of prenatal alcohol exposure at school age. II. Attention and behavior. *Neurotoxicol. Teratol.* **13**, 369–376.

Chernick, V., Childiaeva, R., and Ioffe, S. (1983). Effects of maternal alcohol intake and smoking on neonatal electroencephalogram and anthropometric measurements. *Am. J. Obstet. Gynecol.* **146**, 41–47.

Church, M. W., and Gerkin, K. P. (1988). Hearing disorders in children with fetal alcohol syndrome: Findings from case reports. *Pediatrics* **82**, 147–154.

Coles, C. D., Smith, I., Fernhoff, P. M., and Falek, A. (1985). Neonatal neurobehavioral characteristics as correlates of maternal alcohol use during gestation. *Alcohol. Clin. Exp. Res.* **9**, 454–460.

Coles, C. D., Smith, I. E., Lancaster, J. S., and Falek, A. (1987). Persistence over the first month of neurobehavioral differences in infants exposed to alcohol prenatally. *Inf. Behav. Dev.* **10**, 23–37.

Coles, C., Brown, R., Smith, I., Platzman, K., Erickson, S., and Falek, A. (1991). Effects of prenatal alcohol exposure at school age. I. Physical and cognitive development. *Neurotoxicol. Teratol.* **13**, 357–367.

Day, N. L., Jasperse, D., Richardson, G., Robles, N., Sambamoorthi, U., Taylor, P., Scher, M., Stoffer, D., and Cornelius, M. (1989). Prenatal exposure to alcohol: Effect on infant growth and morphologic characteristics. *Pediatrics* **84**, 536–541.

Day, N. L., Robles, N. (1989). Methodological issues in the measurement of substance use. In Prenatal Abuse of Licit and Illicit Drugs, *Ann. N.Y. Acad. Sci.* **562**, 8–13.

Day, N. L., Richardson, G. A., Robles, N., Sambamoorthi, U., Taylor, P., Scher, M., Stoffer, D., Jasperse, D., and Cornelius, M. (1990). Effect of prenatal alcohol exposure on growth and morphology of offspring at 8 months of age. *Pediatrics* **85**, 748–752.

Day, N. L., Robles, N., Richardson, G. A., Robles, N., Sambamoorthi, U., Taylor, P., Scher, M., Stoffer, D., and Cornelius, M. (1991a). The effects of prenatal alcohol use on the growth of children at three years of age. *Alcohol. Clin. Exp. Res.* **15**, 67–71.

Day, N. L., Goldschmidt, L., Robles, N., Richardson, G., and Cornelius, M. (1991b). Prenatal alcohol exposure and offspring growth at eighteen months of age: The predictive validity of two measures of drinking. *Alcohol. Clin. Exp. Res.* **15**, 914–918.

Ernhart, C. B., Wolf, A. W., Linn, P. L., Sokol, R., Kennard, M., and Filipovich, H. (1985). Alcohol-related birth defects: Syndromal anomalies intrauterine growth retardation and neonatal behavioral assessment. *Alcohol. Clin. Exp. Res.* **9**, 447–453.

Ernhart, C. B., Morrow-Tlucak, M., Sokol, R. J., and Martier, S. (1988). Underreporting of alcohol use in pregnancy. *Alcohol. Clin. Exp. Res.* **12**, 506–511.

Fried, P. A., and Makin, J. E. (1987). Neonatal behavioural correlates of prenatal exposure to marihuana cigarettes and alcohol in a low-risk population. *Neurotoxicol. Teratol.* **9**, 1–7.

Fried, P. A., and O'Connell, C. M. (1987). A comparison of effects of prenatal exposure to tobacco, alcohol, cannabis, and caffeine on birth size and subsequent growth. *Neurotoxicol. Teratol.* **9**, 79–85.

Fried, P. A., Watkinson, B., Dillon, R. F., and Dulberg, C. S. (1987). Neonatal neurological status in a low-risk population after prenatal exposure to cigarettes, marijuana and alcohol. *J. Dev. Behav. Pediatr.* **8**, 318–326.

Geva, D., Day, N., Goldschmidt, L., and Stoffer, D. (1991). A longitudinal analysis of the effect of prenatal alcohol exposure on growth. Submitted for publication.

Graham, J. M., Hanson, J. W., Darby, B. L., Barr, H., and Streissguth, A. (1988). Independent dysmorphology evaluations at birth and 4 years of age for children exposed to varying amounts of alcohol *in utero*. *Pediatrics* **81**, 772–778.

Gusella, J., and Fried, P. (1984). Effects of maternal social drinking and smoking on offspring at 13 months. *Neurobehav. Toxicol. Teratol.* **6**, 13–17.

Hanson, J., Streissguth, A., and Smith, D. (1978). The effects of moderate alcohol consumption during pregnancy on fetal growth and morphogenesis. *J. Pediatr.* **92**, 457–460.

Helzer, J. E., Burnam, A., and McEvoy, L. T. (1991). Alcohol abuse and dependence. *In* "Psychiatric Disorders in America: The Epidemiologic Catchment Area Study" (L. N. Robins and D. A. Regier, eds.), pp. 81–115. The Free Press, New York.

Herd, D. (1988). Drinking by black and white women: Results from a national survey. *Soc. Prob.* **35**, 493–505.

Hilton, M. E. (1988). The demographic distribution of drinking patterns in 1984. *Drug Alcohol Depend.* **22**, 37–47.

Hingson, R., Alpert, J., Day, N., Dooling, E., Kayne, H., Morelock, S., Oppenheimer, E., and Zuckerman, B. (1982). Effects of maternal drinking and marijuana use on fetal growth and development. *Pediatrics* **70**, 539–546.

Hollstedt, C., Dahlgren, L., and Rydberg, U. (1983). Alcoholic women in fertile age treated at an alcohol clinic. *Acta Psychiatr. Scand.* **67**, 195–204.

Ioffe, S., and Chernick, V. (1988). Development of the EEG between 30 and 40 weeks gestation in normal and alcohol-exposed infants. *Dev. Med. Child Neurol.* **30**, 797–807.

Ioffe, S., Childiaeva, R., and Chernick, V. (1984). Prolonged effects of maternal alcohol ingestion on the neonatal electroencephalogram. *Pediatrics* **74**, 330–335.

Jacobson, S. W., Fein, G. G., Jacobson, J. L., Schwartz, P., and Dowler, J. (1984). Neonatal correlates of prenatal exposure to smoking caffeine and alcohol. *Inf. Behav. Dev.* **7**, 253–265.

Kaminski, M., Rumeau, C., and Schwartz, D. (1978). Alcohol consumption in pregnant women and the outcome of pregnancy. *Alcohol. Clin. Exp. Res.* **2**, 155–163.

Kuzma, J., and Sokol, R. (1982). Maternal drinking behavior and decreased intrauterine growth. *Alcohol. Clin. Exp. Res.* **6**, 396–402.

Landesman-Dwyer, S. (1982). The relationship of children's behavior to maternal alcohol consumption. *In* "Fetal Alcohol Syndrome: Human Studies" (E. G. Abel, ed.), Vol. 2, pp. 127–148. CRC Press, Boca Raton, Florida.

Landesman-Dwyer, S., Ragozin, A., and Little R. (1981). Behavioral correlates of prenatal alcohol exposure: A four-year follow-up study. *Neurobehav. Toxicol. Teratol.* **3**, 7–193.

Little, B. B., Snell, L. M., Rosenfeld, C. R., Gilstrap, L. C., and Gant, N. F. (1990). Failure to recognize fetal alcohol syndrome in newborn infants. *Am. J. Dis. Child.* **144,** 1142–1146.

Little, R. (1977). Moderate alcohol use during pregnancy and decreased infant birth weight. *Am. J. Public Health* **67,** 1154–1156.

Little, R., Streissguth, A., Barr, H., and Herman, C. (1980). Decreased birth weight in infants of alcoholic women who abstained during pregnancy. *J. Pediatr.* **96,** 974–977.

Majewski, F. (1978a). Studies on alcohol embryopathy. *Fortschr. Med.* **96,** 2207–2213.

Majewski, F. (1978b). The damaging effect of alcohol on offspring. *Der Nervenarzt* **49,** 410–416.

Mau, G., and Netter, P. (1974). Kaffee- und alkoholkunsum–Risikofaktoren in der schwangerschaft? *Geburtshilfe Fraunheilkd.* **34,** 1018–1022.

Middaugh, L., and Boggan, W. (1991). Postnatal growth deficits in prenatal ethanol-exposed mice: Characteristics and critical periods. *Alcohol: Clin. Exp. Res.* **15,** 919–926.

O'Connor, M. J., Brill, N. J., and Sigman, M. (1986). Alcohol use in primiparous women older than 30 years of age: Relation to infant development. *Pediatrics* **78,** 444–450.

Olegard, R., Sabel, K., Aronson, M., Sandin, B., Johansson, P., Carlsson, C., Kyllerman, M., Iverson, K., and Hrben, A. (1979). Effects on the child of alcohol abuse during pregnancy. *Acta Pediatr. Scand.* (Suppl.) **275,** 112–121.

Richardson, G. A., and Day, N. L. (1991). "Prenatal Exposure to Alcohol, Marijuana, and Tobacco: Effect on Infant Mental and Motor Development." Paper presented at the Society for Research in Child Development, Seattle, Washington.

Richardson, G. A., Day, N. L., and Taylor, P. (1989). The effect of prenatal alcohol, marijuana, and tobacco exposure on neonatal behavior. *Inf. Behav. Dev.* **12,** 199–209.

Robles, N., and Day, N. (1990). Recall of alcohol consumption during pregnancy. *J. Stud. Alcohol* **51,** 403–407.

Rosett, H., Snyder, P., Sander, L., Lee, A., Cook, P., Weiner, L., and Gould, J. (1979). Effect of maternal drinking on neonate state regulation. *Dev. Med. Child Neurol.* **21,** 464–473.

Rosett, H., Weiner, L., Lee, A., Zuckerman, B., Dooling, E., and Oppenheimer, E. (1983). Patterns of alcohol consumption and fetal development. *Obstet. Gynecol.* **61,** 539–546.

Russell, M. (1991). Clinical implications of recent research on the fetal alcohol syndrome. *Bull. N.Y. Acad. Med.* **67,** 207–222.

Scher, M. S., Richardson, G. A., Coble, P. A., Day, N., and Stoffer, D. (1988). The effects of prenatal alcohol and marijuana exposure: Disturbances in neonatal sleep cycling and arousal. *Pediatr. Res.* **24,** 101–105.

Shaywitz, S., Cohen, D., and Shaywitz, B. (1980). Behavior and learning difficulties in children of normal intelligence born to alcoholic mothers. *J. Pediatr.* **96,** 978–982.

Smith, I. E., Coles, C. D., Lancaster, J., Fernhoff, P., and Falek, A. (1986). The effect of volume and duration of prenatal ethanol exposure on neonatal physical and behavioral development. *Neurobehav. Toxicol. Teratol.* **8,** 375–381.

Sokol, R. J., and Clarren, S. K. (1989). Guidelines for use of terminology describing the impact of prenatal alcohol on the offspring. *Alcoholism* **13,** 597–598.

Sokol, R., Miller, S., and Reed, G. (1980). Alcohol abuse during pregnancy: An epidemiological study. *Alcohol. Clin. Exp. Res.* **4,** 135–145.

Sokol, R. J., Ager, J., and Martier, S. (1986). Significant determinants of susceptibility to alcohol teratogenicity. *Ann. N.Y. Acad. Sci.* **477,** 87–102.

Steinhauser, H., Nester, V., and Spohr, H. (1982). Development and psychopathology of children with the fetal alcohol syndrome. *J. Dev. Behav. Pediatr.* **3,** 49–54.

Stoffer, D. S., Scher, M. S., Richardson, G. A., Day, N. L., and Coble, P. (1988). A Walsh-Fourier analysis of the effects of moderate maternal alcohol consumption on neonatal sleep-state cycling. *J. Am. Stat. Assoc.* **83,** 954–963.

Streissguth, A., Herman, C., and Smith, D. (1978). Stability of intelligence in the fetal alcohol syndrome: A preliminary report. *Alcohol: Clin. Exp. Res.* **2,** 165–170.

Streissguth, A. P., Barr, H. M., Martin, D. C., and Herman, C. S. (1980). Effects of maternal alcohol, nicotine, and caffeine use during pregnancy on infant mental and motor development at eight months. *Alcohol: Clin. Exp. Res.* **4,** 152–164.

Streissguth, A., Barr, H., and Martin, D. (1983). Maternal alcohol use and neonatal habituation assessed with the Brazelton Scale. *Child. Dev.* **545,** 1109–1118.

Streissguth, A. P., Martin, D. C., Barr, H. M., Sandman, B. M., Kirchner, G. L., and Darby, B. L. (1984). Intrauterine alcohol and nicotine exposure: Attention and reaction time in 4-year-old children. *Dev. Psychol.* **20,** 533–541.

Streissguth, A. P., Clarren, S. K., and Jones, K. L. (1985). Natural history of the fetal alcohol syndrome: A ten-year follow-up of eleven patients. *Lancet* **2,** 85–92.

Streissguth, A. P., Barr, H. M., Sampson, P. D., Parrish-Johnson, J., Kirchner, G., and Martin, D. (1986). Attention, distraction, and reaction time at seven years and prenatal alcohol exposure. *Neurobehav. Toxicol. Teratol.* **8,** 717–725.

Streissguth, A. P., Barr, H. M., Sampson, P. D., Darby, B. L., and Martin, D. C. (1989). IQ at age 4 in relation to maternal alcohol use and smoking during pregnancy. *Dev. Psychol.* **25,** 3–11.

Streissguth, A. P., Barr, H. M., and Sampson, P. D. (1990). Moderate prenatal alcohol exposure: Effects on child IQ and learning problems at age 7½ years. *Alcohol: Clin. Exp. Res.* **14,** 662–669.

Streissguth, A. P., Aase, J., Clarren, S., Randels, S., LaDue, R., and Smith, D. (1991). Fetal alcohol syndrome in adolescents and adults. *J.A.M.A.* **265,** 1961–1967.

Tennes, K., and Blackard, C. (1980). Maternal alcohol consumption, birth weight and minor physical anomalies. *Am. J. Obstet. Gynecol.* **138,** 774–780.

— 4 —

Fetal Alcohol Effects: Rat Model of Alcohol Exposure during the Brain Growth Spurt

Charles R. Goodlett and James R. West

Alcohol and Brain Research Laboratory
Department of Anatomy
University of Iowa
College of Medicine
Iowa City, Iowa

TERATOGENIC EFFECTS OF ALCOHOL IN HUMANS

Because ethanol is a legalized psychoactive drug, it has significant social, recreational, and economic importance, but it also carries a strong potential for abuse and dependence. It is now undeniably established that alcohol abuse by pregnant women can produce teratogenic effects on the developing fetus (Jones and Smith, 1973; Jones et al., 1973; Clarren and Smith, 1978; Streissguth et al., 1980; Abel, 1984; Rosett and Weiner, 1984; Clarren, 1986; Streissguth, 1986). The similarity of characteristics among the most seriously affected cases led to the recognition of the fetal alcohol syndrome (FAS), which, as first formally described in 1973, includes prenatal and postnatal growth deficiency, a particular pattern of facial dysmorphology, and central nervous system (CNS) dysfunction (Jones and Smith, 1973; Jones et al., 1973). Even in the absence of the full fetal alcohol syndrome, one or more of the effects can be manifest in exposed offspring and have been termed fetal alcohol effects or FAE (Rosett et al., 1981). Certain fetal alcohol effects, particularly growth deficiency and behavioral effects reflecting alterations in the CNS, have been described in offspring of mothers who reported only moderate drinking (Hanson et al., 1978; Little et al., 1989; Conry, 1990; Streissguth et al., 1990).

For the large majority of chemicals or products known to be teratogenic, exposure of the developing fetus can be prevented by removing them from the environment or prohibiting access to them. For ethanol, these options are not feasible alternatives. It is also unlikely that medical advice, social pressure, or legal regulation of alcohol consumption during pregnancy would be effective in eliminating fetal alcohol effects, owing to the difficulty in modifying the drinking habits of alcohol abusers. There has been recent publicity of the dangers of excessive drinking during pregnancy in medical advisories (Surgeon General, 1981), in the media, and in literary accounts (Dorris, 1989); legislation requiring the application of warning labels on alcoholic beverage containers began in November of 1989. Even with the increased public awareness of the dangers of alcohol abuse during pregnancy, FAS has been and will continue to be a major known cause of developmental disabilities, producing significant medical and economic problems for society (Abel and Sokol, 1991).

Not only do significant numbers of women continue to abuse alcohol during pregnancy, but many more also consume low to moderate amounts, presumably because of the common belief that such use constitutes no danger to their fetuses. There appears to exist widespread public sentiment that alcohol consumption during pregnancy is dangerous to the fetus only if the drinking is abusive and excessive, as found perhaps only in the most extreme chronic alcoholic populations (e.g., Kinsley, 1991). This downplaying of the potential risks of social drinking indicates that an attitude of acceptance of "moderate" alcohol consumption during pregnancy still exists.

Despite this prevailing public opinion, the question of how much alcohol exposure constitutes a danger to the developing CNS is far from resolved. In contrast to the established dangers of heavy alcohol use during pregnancy, there is presently no scientific consensus on the potential dangers of moderate drinking during pregnancy. This specific question illustrates some of the difficult problems facing investigators of human fetal alcohol effects. Although human studies were responsible for the identification of FAS, many important questions concerning the damaging effects of alcohol on the fetus cannot be adequately addressed in human populations. The different and uncontrolled patterns of alcohol consumption, the unreliability of self-report of alcohol use, and the numerous, uncontrolled, and interactive variables (environmental, nutritional, and genetic) present in human studies, combined with clear ethical prohibitions of giving controlled doses of ethanol to pregnant women, make animal studies necessary to understand the biological effects of alcohol on development.

CURRENT ROLE OF ANIMAL MODELS IN STUDIES OF FETAL ALCOHOL EFFECTS

With the evolution of relatively sophisticated animal model studies, in which nutritional factors are controlled and alcohol intake and blood alcohol concentrations are measured, it has been unambiguously documented that alcohol *per se* damages the developing organism (Chernoff, 1977; Randall et al., 1977; Randall and Taylor, 1979; Webster et al., 1980; Webster et al., 1983; Blakley and Scott, 1984), especially the developing nervous system (reviewed in Abel, 1984; West, 1986; Jones, 1988; West et al., 1989; West et al., 1990; Miller, 1992). One or more of the three components of FAS has been produced in several different animal models: (1) facial dysmorphology (Sulik et al., 1981); (2) prenatal and postnatal growth deficiencies (Abel and Greizerstein, 1982; Clarren et al., 1988), sometimes including organ malformations

(Randall and Taylor, 1979; Webster et al., 1980; Sulik et al., 1981; Blakley, 1988; Webster, 1989); and (3) alterations in brain growth, structure, and function (e.g., Diaz and Samson, 1980; Samson and Diaz, 1982; Abel, 1984; Sulik et al., 1984; Miller, 1986; West, 1986; Miller, 1987, 1988, 1989, 1990; Clarren, et al., 1990; Driscoll et al., 1990; Riley, 1990; West et al., 1990; West and Goodlett, 1990). Thus, the question of whether alcohol acts as a neuroteratogen has been resolved. In the 1990s, the issues facing investigators concerned with alcohol and brain development are questions about susceptibility, risk factors, intervention strategies, and mechanisms of damage, all of which can be addressed in animal models.

Questions Concerning Thresholds and Intervention Strategies

For convenience, current questions of greatest relevance may be lumped into three categories. The first group of questions includes what amount of alcohol is sufficient to damage brain or behavioral development, whether intervention (including strict abstinence) imposed at some point during the course of pregnancy can alleviate some of the deleterious effects of alcohol exposure, and whether any sorts of therapies (pharmacological, nutritional, educational) can alleviate the consequences of prenatal alcohol exposure. These questions all bear social and political impact, and the answers to them can help guide public policy and medical advice.

Questions Concerning Risk Factors

A second group of questions facing fetal alcohol researchers is most relevant to clinical management of populations at risk for fetal alcohol effects. Most of these questions are centered around better identification of factors influencing the severity of alcohol-related damage to developing organisms. These include better specification of maternal and fetal genetic influences modulating the severity of fetal alcohol effects (Riley and Lochry, 1982). Maternal influences, some of which are under genetic control, include factors leading to differences in blood alcohol levels (e.g., Chernoff, 1980), such as differences in alcohol acceptance, alcohol metabolism, or liver function. Other maternal influences on the severity of fetal alcohol effects are independent of the blood alcohol levels, but also can have a strong genetic component (Gilliam et al., 1988; Gilliam et al., 1989; Gilliam and Irtenkauf, 1990). These influences may include interactions of alcohol with maternal physiology, placental function, or intrauterine environment (reviewed in Schenker et al., 1990).

In addition, the genotype of the offspring can influence the severity of CNS damage (Goodlett et al., 1989a). For example, recent results from our laboratory using the neonatal rat exposure model (fully described on pages 49–50) have demonstrated significant strain differences in susceptibility to alcohol-induced brain growth restriction induced by early postnatal alcohol exposure in rats from six different inbred strains (Goodlett et al., 1990b). These differences were not accounted for by strain differences in blood alcohol profiles or clearance rates of the pups (Goodlett et al., 1991b). Thus, the strain differences in alcohol-induced brain growth restriction appear to be related to differences in vulnerability of the CNS itself. Such genetic influences illustrate risk factors that may modulate the expression of fetal alcohol effects. Identifying the mechanisms underlying such influences may provide critical information concerning the risk for FAS and FAE.

Questions Concerning Mechanisms

The third group of questions facing researchers involves identification of the basic mechanisms of alcohol-induced damage. A variety of different potential mechanisms of action have been suggested to account for different fetal alcohol effects (see O'Connor, 1984; Michaelis and Michaelis, 1986; Michaelis, 1990; Randall et al., 1990; Schenker et al., 1990, for reviews). The search for mechanisms of the teratogenic effects of ethanol has accelerated, as indicated by the increasing number of reports addressing these questions. For example, reports have proposed specific biochemical mechanisms (inhibition of retinoic acid production) to account for major teratogenic defects (Duester, 1991; Pullarkat, 1991); differential alteration of cell-cycle kinetics of the two neocortical proliferative zones to account for the aberrant cortical architecture following gestational exposure (Miller and Nowakowski, 1991); and interference with cyclic adenosine monophosphate (AMP)-dependent kinase catalytic activity to account for growth suppression (Pennington, 1990). Nevertheless, the specific mechanism of any effect of alcohol on the developing CNS has yet to be fully elucidated.

Because alcohol interferes with biological processes at many different levels (molecular, cellular, and physiological), there are likely to be many different mechanisms mediating the various effects of alcohol. It is unlikely that the multifaceted effects of alcohol on development can be uniformly accounted for under a single mechanism of action. Specific identification of the mechanism of a given fetal alcohol effect would be of great value. Such an identification of how alcohol produces its effects (e.g., impaired growth, cell loss, altered brain architecture, or changes in biochemical or functional properties) would then allow rational use of that knowledge to guide strategies for counseling, intervention, and treatment.

Animal Models and Progress in Fetal Alcohol Research

The task of making significant progress on each of these three sets of questions is the current challenge facing fetal alcohol researchers. The recent history of the field shows that progress toward understanding these types of questions can best be accomplished with judicious use of animal models. Through the use of animal model studies over the past few years, great strides have been made toward understanding some of the most important variables and conditions that influence the expression of fetal alcohol effects. Data reflecting the types of effects resulting from alcohol exposure during development have now accumulated from many different investigators, often using different experimental models or those that incorporate different animal species, and evaluating a wide range of dependent variables (reviewed in Abel, 1984; West, 1986; Randall et al., 1990; Schenker et al., 1990; Miller, 1992).

This chapter will not attempt to review exhaustively the multitudinous effects of alcohol on the developing CNS. Neither will it attempt to integrate all of the findings into a unifying account of fetal alcohol effects, which would be, at best, premature owing to the many differences in procedures and dependent variables and the lack of understanding of basic mechanisms of the damaging effects of alcohol. Instead, the chapter summarizes some of the recent findings using a rat model of alcohol exposure during the brain growth spurt, focusing on alcohol-induced alterations of morphological development of the brain. Two factors will be emphasized that have been recently demonstrated to exert critical influences on the types or extent of alterations in brain morphology or function: (1) the dependence of effects on the blood alcohol concentration, and (2) the presence of temporal windows of vulnerability to alcohol during development. Most of the supporting data are derived from studies using the neonatal rat in a model of alcohol exposure during the period of the brain growth spurt. However, data concerning prenatal exposure in rats or in other species will also be included where necessary or appropriate.

AN ANIMAL MODEL OF BINGE DRINKING DURING THE THIRD TRIMESTER

The formation of the CNS begins early in development, and its maturation extends over a longer period than that of other organ systems. Moreover, the brain undergoes various stages of development in a sequence common to all mammals, i.e., formation of neural tube, neuronal proliferation and migration, followed later by the brain growth spurt involving neurite outgrowth, synaptogenesis, glial and microneuronal proliferation, and myelination. Thus, alcohol exposure at different stages of brain development will impinge on different developmental processes, so the effects on the brain and the mechanisms mediating those effects will be a function of the stage of brain development.

In humans, the brain growth spurt begins around the twenty-fourth week of gestation and extends through the third trimester and into the first 2 years of birth (Dobbing and Sands, 1973). Clinical studies have shown that when heavy alcohol consumption continues through the third trimester of pregnancy, the severity of fetal alcohol effects is increased (Rosett et al., 1980; Rosett et al., 1983; Coles et al., 1985; Smith et al., 1986). Most women who abuse alcohol during the third trimester do so in patterns of short episodes of heavy use, i.e., binge drinking (Stephens, 1985). In rats, the period of rapid brain growth, comparable to the brain growth spurt of the third trimester in humans, occurs neonatally during the first 2 weeks after birth (Dobbing and Sands, 1979). Consequently, to model binge drinking during the third trimester in humans, rats must be exposed during the early postnatal period.

To administer alcohol to neonatal rats in a controlled manner without interfering with adequate nutritional intake, several laboratories are now using artificial rearing procedures in which alcohol (as well as a nutritionally adequate diet) is infused via a surgically implanted gastrostomy tube (Samson and Diaz, 1982; West et al., 1984).

This procedure allows precise control of the concentration and timing of the alcohol exposure, provides for adequate body growth during the period of alcohol exposure, and allows delivery of the alcohol during the period of rapid brain growth comparable to the human third trimester.

In using the artificial rearing method, we have taken advantage of the ease of controlling the time of administration and concentration of alcohol to produce a model of binge exposure during the brain growth spurt. The essence of this model is to concentrate a given amount of alcohol into just a few (usually 4 or fewer) of the 12 daily feedings, rather than giving alcohol in each of the feedings throughout each day. This concentrated exposure during just part of the day results in a cyclic daily pattern of blood alcohol concentrations that reach a peak 1.5–2 hr after the last feeding, then decline to zero over the next 12–16 hr. This cyclic pattern of alcohol exposure resulting from short but relatively concentrated alcohol intake thus models several aspects of alcohol exposure in binge drinking during the third trimester, and will be referred to as "bingelike" exposure.

ALTERATIONS IN BRAIN MORPHOLOGY

Our studies of bingelike alcohol exposure in neonatal rats [usually between postnatal day (PD) 4 and PD9] have focused on the short-term (PD10) and long-term (adult) alterations in brain growth and morphology, and depletion of neuron numbers in several different brain regions. Over the last 5 years, we have conducted numerous studies in which either the daily dose, concentration, pattern, or developmental timing of alcohol exposure was manipulated during this neonatal period. The significance and outcomes of these manipulations related to blood alcohol concentration and temporal windows of vulnerability will be discussed in the subsequent sections (pages 58–69). However, across all these studies there is a general consistency of the types of effects on brain, and these will first be reviewed and placed in the context of other known effects on brain development demonstrated in other animal models.

Microencephaly

One of the classic features of FAS is microcephaly, usually defined as a head size two standard deviations or more below the norm. The analog of microcephaly is microencephaly, or small brain for body size. The bingelike neonatal exposure in rats consistently produces microencephaly. Dose-dependent reductions in brain weight (and brain-weight to body-weight ratios) are present at PD10 (Bonthius et al., 1988; Pierce et al., 1989; Bonthius and West, 1990) and in adults (Bonthius and West, 1991; Goodlett et al., in press). There is some degree of "catch-up" in brain weights in adulthood compared to PD10; nevertheless, significant and permanent reductions in brain weight are produced without significant reductions in body weight. This microencephaly in the neonatal model contrasts with prenatal exposure in rats, in which deficits in brain weight generally are associated with deficits in body weight, and long-term brain-weight deficits may not be observed with prenatal exposure.

Gross Morphological Abnormalities

With the neonatal bingelike exposure, especially paradigms that result in relatively high peak blood alcohol concentrations (BACs), the reduction in brain weight can be accompanied by gross alterations in the morphology of various brain regions. In particular, in rats given exposure on PD4 to PD9 (resulting in mean peak BACs of 318 mg/dl), then evaluated on PD90, enlargement of the lateral ventricles and alterations in size and shape of the cerebellum were noted (Bonthius and West, 1991).

The cerebellum appears to be very susceptible to permanent morphological changes following neonatal bingelike exposure (Bonthius and West, 1991; Goodlett et al., in press). Figure 1 shows midsagittal sections through the vermis of the cerebellum, illustrating typical gross morphological changes in the cerebellum that can occur when peak BACs exceed 250 mg/dl; in the cases illustrated, the measured peak BACs on PD6 were 390 mg/dl (Fig. 1C) and 530 mg/dl (Fig. 1B). Note the reduction in overall sizes of the sections from the alcohol-exposed rats, which was particularly severe for the rat exposed to the highest peak BAC. Note also the loss of appropriate lobular shapes, especially obvious for the high-exposure rat. We have found that such alterations in cerebellar morphology can occur even when the binge exposure is limited to just 2 days of exposure (peak BACs between 350 and 400 mg/dl), on PD4 and PD5 (Hamre and West, 1990). Furthermore, when the treatment on PD4 to PD9 produces peak BACs greater than 400 mg/dl, the granule cell layer can assume abnormal shapes (e.g., Fig. 1B). When BACs are extremely high (greater than 500 mg/dl), occasional cases are found with large portions of the cerebellum missing. Figure 2 shows the most severe instance of gross morphological defects in the adult cerebellum; the rat had a peak BAC on PD6 of 530 mg/dl.

The gross morphological changes in the cerebellum in this animal model of binge exposure suggest that alterations in the cerebellum would occur in FAS cases in which binge drinking occurred during the third trimester. Such changes in the vermis could be observable with careful magnetic resonance imaging, a technique that has been used to demonstrate cerebellar morphological changes in Williams syndrome and Down's syndrome (Jernigan and Beluggi, 1990), and in autism (Courchesne et al., 1988).

Neuronal Depletion

Reduction in the numbers of neurons in the brain is thought to be a common outcome following developmental alcohol exposure, and one that often accompanies microencephaly. Studies from other laboratories using prenatal alcohol exposure have demonstrated reductions in cell numbers in the cortex (Miller and Potempa, 1990) and in CA1 in the hippocampus (Barnes and Walker, 1981; Wigal and Amsel, 1990). With postnatal exposure, other investigators have documented cell loss in the cerebellum (Bauer-Moffett and Altman, 1977; Cragg and Phillips, 1985).

The studies in our laboratory using bingelike exposure in the neonatal rat model have documented reductions in the number of cells in several different neuronal populations (reviewed in West et al., 1990). Neuronal populations known to be

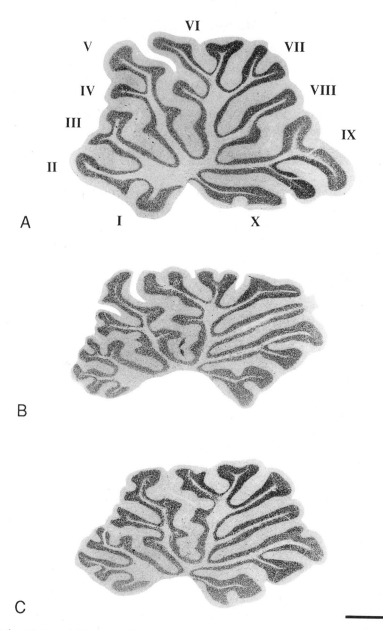

FIGURE 1 Midsagittal, Nissl-stained sections through the vermis of the cerebellum of adult rats (littermates) from (A) the gastrostomy control group, (B) from the group given neonatal bingelike exposure of 7.5 g/kg/day of alcohol (peak BAC, 530 mg/dl), and (C) another littermate from the group given neonatal bingelike exposure to 6.6 g/kg/day (peak BAC, 390 mg/dl). Roman numerals in (A) indicate the different lobules of the vermis. Note the alterations in shape and size of the lobules as a function of alcohol exposure. Magnification bar = 1000 μm.

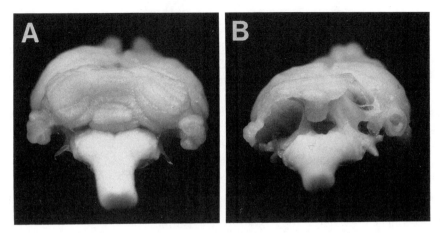

FIGURE 2   A posterior view of the cerebellum of adult rats from (A) the gastrostomy control group, and (B) its littermate from the group exposed neonatally to 7.5 g/kg/day of alcohol, a treatment that produces high peak BACs during the brain growth spurt. The peak BAC on postnatal day 6 of this alcohol-exposed rat was 531 mg/dl. Note the gross defects in the cerebellum of the alcohol-treated rat. This is the most severe cerebellar defect seen in our laboratory; many other rats have been exposed to BACs this high or even higher without exhibiting such severe cerebellar hydrocephalus.

depleted in this animal model include cerebellar Purkinje and granule cells (Pierce et al., 1989; Bonthius and West, 1990; Goodlett et al., 1990a; Hamre et al., 1991), pyramidal cells in field CA1 of the hippocampal formation (Pierce et al., 1989; Bonthius and West, 1990; Bonthius and West, 1991), and mitral and granule cells of the olfactory bulb (Bonthius and West, in press). In addition, studies of a nonhuman primate model of gestational bingelike exposure found significant cell loss also in the cerebellum (Bonthius et al., 1991).

Several general conclusions can be derived from the studies using bingelike exposure in neonatal rats. First, *neuronal populations differ in their susceptibility to alcohol-induced depletion.* Some neuronal populations (the hippocampal CA3 pyramidal cells and dentate gyrus granule cells) are relatively resistant to cell depletion, whereas others (cerebellar Purkinje cells; olfactory mitral cells) are quite vulnerable. Even within the same structure, different neuronal populations may have markedly different vulnerabilities to depletion. For example, hippocampal CA1 pyramidal neurons are significantly depleted when the neonatal alcohol treatment produces BACs greater than 280 mg/dl; in the same rats, CA3 pyramidal neuron numbers are not significantly affected. Interestingly, similar patterns of alcohol-induced cell loss in the hippocampus have been reported following prenatal alcohol exposure (Barnes and Walker, 1981; Wigal and Amsel, 1990).

Second, *certain populations of large projection neurons that are postmitotic at the time of the neonatal alcohol exposure* (i.e., cerebellar Purkinje cells, hippocampal CA1 pyramidal cells; olfactory mitral cells) *may be depleted at least as severely as (and in some cases more severely than) the proliferating microneuronal populations of the same brain regions* (i.e., cerebellar granule cells, dentate gyrus granule cells, olfactory granule cells). The hippocampal formation provides the best illustration of greater vulnerability of postmitotic projection neurons, in that alcohol can deplete CA1 pyramidal cell numbers without significantly altering dentate gyrus granule cell numbers.

Third, *within a given brain structure, the extent of loss of a given neuronal population has been shown to vary as a function of location within the structure.* For example, the extent of alcohol-induced Purkinje cell loss in the vermis of the cerebellum is less in some lobules (e.g., VI and VII) than in the others (Pierce et al., 1989; Bonthius and West, 1990; Goodlett et al., 1990a). Interestingly, the lobules that are less affected are those that mature relatively late in the neonatal exposure period. The upper panel of Fig. 3 (based on data from Bonthius and West, 1990) shows that regions of the vermis that mature relatively early (lobules I and X) suffer more severe Purkinje cell depletion (determined on PD10) than do regions that begin to mature late in the neonatal period (including distal VIA, VIB, and VII) (Bonthius and West, 1990).

The neuronal depletion in the cerebellum, hippocampus, and olfactory bulb we initially documented on PD10 has also been observed at PD21 (Hamre and West, 1990; Hamre et al., 1991), and in adults (Bonthius and West, 1991; Bonthius and West, in press; Goodlett et al., in press). For each brain region, the same pattern of selective cell loss observed at PD10 was observed in older animals. For example, the regional differences in Purkinje cell loss across the cerebellar vermis followed the same pattern in the adults (based on data from Bonthius and West, 1991) as in the PD10 animals (compare Fig. 3B with Fig. 3A). Likewise, as in the PD10 study, depletion of CA1 pyramidal cells in the hippocampal formation was observed in adults exposed neonatally to BACs greater than 250 mg/dl, but the number of CA3 pyramids and dentate gyrus granule cells was not significantly affected (Bonthius and West, 1991; Goodlett et al., in press).

Aberrant Connections

Although neuronal depletion represents one of the most extreme and perhaps most serious effects of developmental exposure to alcohol, morphological alterations in neuronal connections have also been demonstrated. In the midtemporal hippocampal formation, alcohol exposure throughout gestation results in aberrant mossy fiber projections into the infrapyramidal zone of CA3a, which normally does not receive mossy fiber afferents. Furthermore, neonatal alcohol exposure on PD1 through PD10, as compared to gestational exposure, produces even more extensive aberrant mossy fibers (West and Hamre, 1985). These aberrant mossy fiber projections are present in adulthood, and have been documented using Timm stain (West and Hodges-Savola, 1983) and with horseradish peroxidase tract-tracing (West and Pierce, 1984).

Acute Cortical Astrogliosis

Recently, neonatal bingelike alcohol exposure on PD4 to 9 has been shown to produce a large but transient increase in the size and numbers of astrocytes labeled

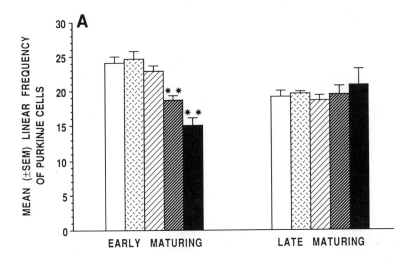

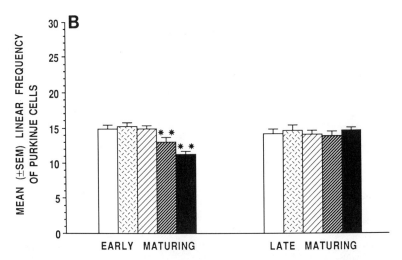

FIGURE 3   Graphs of mean (± SEM) linear frequency of Purkinje cells in the vermis of the cerebellum in groups exposed neonatally to alcohol and in control groups, evaluated on postnatal day 10 (A) or postnatal day 90 (B). Legend for (A) and (B) for the groups from left to right (treatments administered on PD 4–9)—SC, suckle control; GC, gastrostomy control; EtOH-12, group administered 6.6 g/kg of alcohol per day distributed over all 12 feedings of each day, using low concentrations each feeding (mean peak BAC ≈ 50 mg/dl); EtOH-4, group administered 4.5 g/kg of alcohol per day in four consecutive feedings each day (mean peak BAC ≈ 200 mg/dl); EtOH-2, group administered 4.5 g/kg of alcohol per day in two consecutive feedings each day (mean peak BAC ≈ 350 mg/dl); **, significantly different from all other treatment groups ($p < .01$).

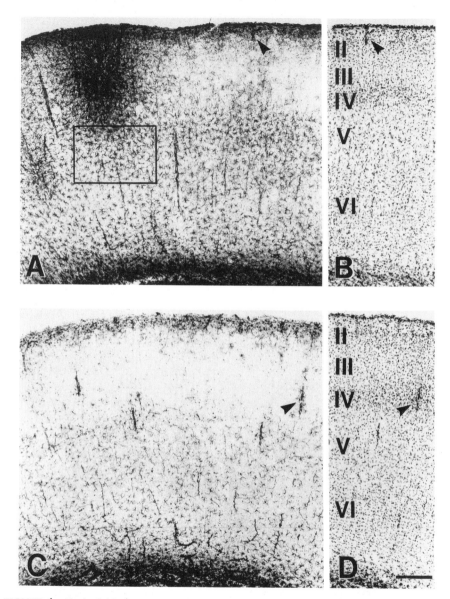

FIGURE 4 Bright field photomicrographs of sections through parietal cortex of 10-day-old rats from groups given either (A, B) bingelike alcohol exposure on postnatal days 4–9 (BACs ≈ 300 mg/dl), or (C, D) a gastrostomy control littermate. The sections were processed for GFAP peroxidase-antiperoxidase immunocytochemistry using a polyclonal antibody to GFAP (Dakopatts, Inc.), and counterstained with cresyl violet. For the alcohol-exposed pup, Panels A and B are photomicrographs from the same section: Panel A shows the GFAP immunoreactivity (cresyl violet stain filtered out with a Kodak Wratten Filter #47B); Panel B shows the cresyl violet staining from a portion of the same section (immunoreactivity photographically suppressed with a Kodak Wratten Filter #29). Likewise, for the gastrostomy control pup, Panel C and D are of the same section; Panel C shows the GFAP immunoreactivity, and Panel D shows the cresyl violet stain. Landmarks (blood vessels) denoting the same locations on the two sets of panels from the alcohol-exposed pup (A and B) and the gastrostomy control pup are indicated with arrowheads. Note the cortical astrogliosis induced by alcohol, including the massive gliosis around a disrupted blood vessel and the generalized gliosis in layer V. Magnification bar = 200 μm.

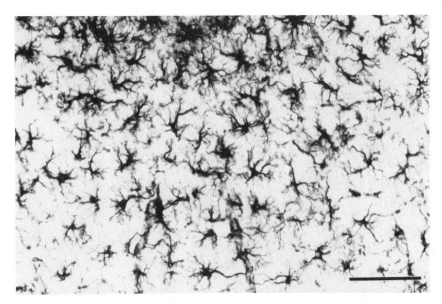

FIGURE 5   Higher magnification of boxed area of Fig. 4, showing reactive astrocytes labeled with GFAP immunocytochemistry. Note the thick, densely labeled fibrillary processes that are directed toward the area of vascular disruption, and the heavy immunoreactivity in the somas of the astrocytes. The cresyl violet counterstaining was photographically suppressed with a Kodak Wratten 47B filter. Magnification bar = 100 μm.

immunocytochemically with an antibody to glial fibrillary acidic protein (GFAP) in the cerebral cortex (West and Goodlett, 1990; Goodlett et al., submitted). Radioimmunoassays of different brain regions indicated that groups exposed to mean peak BACs of 300 mg/dl had more than 300% more GFAP in the cortex than did controls. Sections processed for GFAP immunoreactivity revealed that GFAP-positive astrocytes in alcohol-exposed pups were larger and more numerous. The alcohol-related astrogliosis was apparent in two distinct anatomical locations (see Figs. 4, 5, and 6): (1) in all alcohol-exposed pups, in layer V throughout most regions of the cortex; and (2) in pups exposed to high BACs, loci of intensely reactive glia were frequently seen surrounding blood vessels.

The astrogliosis involved an increase in density of GFAP-labeled astrocytes in the deep layers of the cortex (see Figs. 4 and 6). Perhaps more significantly, the labeled astrocytes were hypertrophied, containing thick, heavily labeled fibrillary processes, and somas that were also densely immunoreactive. The reactive astroglia were particularly hypertrophied in the loci surrounding the vascular disruptions in the cortex of pups exposed to peak BACs reaching 300 mg/dl (see Figs. 4 and 5). In contrast, the increased density and morphological hypertrophy of immunolabeled astrocytes in the deep cortical layers occurred even in the group exposed to low alcohol concentration (Fig. 6). None of the alcohol-induced changes in GFAP-labeled astroglia were observable immunocytochemically after PD12.

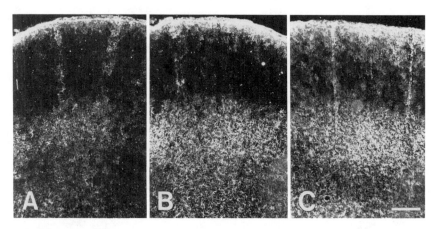

FIGURE 6   Dark-field photomicrographs of GFAP immunoreactivity in the somatosensory cortex on postnatal day 10 of a gastrostomy control rat (A), a rat given exposure producing peak BACs near 50 mg/dl (B), and a rat given bingelike exposure producing peak BACs near 300 mg/dl (C). GFAP immunoreactivity appears bright against the black background in these dark-field photomicrographs. Note that both alcohol treatments resulted in increased size and density of GFAP-labeled astrocytes in layer V. Note also that the astrogliosis in the deep cortical layers was similar in the two alcohol treatments. Magnification bar = 200 μm.

The increased size, change in morphology, and increased number of GFAP-positive astrocytes are all characteristic of reactive astrogliosis, which is usually associated with neuropathological processes (Nathaniel and Nathaniel, 1981; Eng and De-Armond, 1982; Norenberg et al., 1988). However, the mechanism of the transient alcohol-induced astrogliosis is unknown. At least three alternatives are possible: (1) breakdown of the blood–brain barrier, allowing stimulation of astrocytes by circulating cytokines or other blood-borne factors; (2) reactive gliosis in response to alcohol-induced neuronal damage or degeneration; or (3) direct stimulation of GFAP production by alcohol. Determining whether these or perhaps other possible mechanisms are responsible will require further studies.

DEPENDENCE OF EFFECTS ON BLOOD ALCOHOL CONCENTRATION

After nearly two decades of animal research concerning fetal alcohol effects, the relationship between the extent of CNS effects observed and the degree of alcohol exposure (as confirmed by determining profiles of blood alcohol concentrations) is still not well defined. Several factors complicate the specification of effects as functions of BACs in fetal alcohol animal studies. Many studies use prenatal rodent models in which alcohol is administered to pregnant dams in liquid diets or in drinking water.

The most severe limitation of these types of studies is that the dam controls the rate of administration by its pattern of intake of the liquid diet. Consequently, the experimenter has limited control over the dose and timing of alcohol exposure. The peak BACs attained with liquid diet procedures can vary widely owing to individual differences in alcohol acceptance and metabolism. In addition, many studies fail to make even cursory determinations of BAC profiles, and those that have usually do so in dams that are used solely for BAC determinations. Furthermore, the daily pattern of intake generally is not monitored, with investigators relying simply on recording the total daily intake to report daily consumption of alcohol in g/kg. However, the actual BAC attained is a function of the pattern of intake, so the daily intake gives no information about the actual BAC reaching the fetuses. Finally, even though some studies have manipulated the concentration of alcohol in liquid diets (e.g., comparing the effects of diets containing 17 versus 35% ethanol-derived calories), practical requirements, such as having adequate numbers of litters in each group, pair-feeding, and surrogate fostering, restrict the number of different dose groups that can be included in a given study.

The sum effect of these limitations is that there has been no systematic evaluation of the dependence of CNS dysfunction on the BACs attained with prenatal rodent models using liquid diet administration. Studies in which alcohol is administered to pregnant dams by intragastric gavage overcome some of these difficulties, in that dose and timing are controlled by the experimenter. Dose–effect studies of this sort have been prominent in teratology studies in mice (e.g., Webster et al., 1980; Webster, 1989). Studies of CNS dysfunction using gavage require carefully matched groups to control for stress effects and nutritional status; usually only one or two doses are feasible. Furthermore, BAC determinations on an individual or litter basis are usually lacking, thereby limiting the analysis to reporting representative means and variances of the BACs produced by a given treatment. Variation in alcohol-induced effects due to differences in BAC profiles among litters goes unrecognized.

Artificial rearing of neonatal rats is used primarily as a means to administer alcohol to rats during the brain growth spurt. However, this procedure also provides the great advantage of very precise control over the concentration, timing, and total daily dose of alcohol delivered to individual pups. Thus, the experimenter can manipulate the concentration of alcohol in the infused diet, the number of feedings that contain alcohol, the developmental timing of alcohol-containing feedings, and the daily pattern of the alcohol exposure. Furthermore, estimates of the peak blood alcohol concentrations attained with a given alcohol exposure paradigm can be obtained from *individual pups*.

These aspects of the neonatal artificial rearing procedure provide the significant advantage of using correlational and regression analyses of alcohol effects based on estimated peak BACs. Over the past few years, we have exploited the experimental control provided by the artificial rearing method to specify to what extent the alcohol-related effects (e.g., brain weight restriction; cell depletion in various brain regions; glial reactions), depend on the actual alcohol exposure (determined from estimates of peak BACs of each individual).

Brain Weight Restriction Is a Linear Function of Peak Blood Alcohol Concentrations

In nearly every study performed in our laboratory over the past 5 years, whole brain weight and regional brain weights (forebrain, cerebellum, brain stem) have been obtained following perfusion and brain extraction. Retrospectively, summarizing many studies that involved literally thousands of pups, it is clear that alcohol exposure during this period consistently restricts brain weight (but not body weight), and that the severity of restriction increases with BAC. Typically, the threshold peak BAC that produces significant brain weight restriction in any given study using bingelike exposure on postnatal days 4 to 9 lies between 140 and 180 mg/dl. Interestingly, this is approximately the same BAC found to produce neuroteratological effects in a primate model of gestational binge exposure to alcohol (Clarren et al., 1988; Clarren et al., 1990).

The experimental control inherent in the neonatal rat model using artificial rearing has allowed us to evaluate brain-weight restriction as a mathematical function of peak BACs from individual pups. Different groups of pups were given daily doses of alcohol ranging from 2.5 to 7.5 g/kg/day delivered in four consecutive feedings each day from PD4 to PD9, which produced peak BACs that ranged continuously across groups from 30 to 500 mg/dl. On PD10, whole brain weights of the alcohol-exposed pups were an inverse linear function of BAC, and BAC accounted for 84% of the variance in brain weight of alcohol-exposed pups (Bonthius and West, 1988).

A more recent study evaluated rats killed at 260 days old, following exposure to one of three different daily doses (2.5, 4.5, and 6.6 g/kg/day) administered on PD4 through PD9, which produced a BAC range from 30 to 580 mg/dl across groups. As shown in Fig. 7, the neonatal treatment also restricted total brain weight in these adult rats as an inverse linear function of estimated peak BAC ($R^2 = 0.59$). Note that the high dose produced much more variable peak BACs than did the two lower doses; consequently, the value of BAC in accounting for variance in brain-weight restriction (and also neuronal loss) was greatest for higher doses. These data show that with precise control over the neonatal alcohol exposure and estimates of the peak BACs, the effects of alcohol on brain weight can be predicted relatively accurately for Sprague–Dawley rats.

Departures from Linearity in the Cerebellar Weight-Restriction Curve

One exception to the general rule of a linear decline in brain weight as a function of increasing BAC was found for the cerebellum at extremely high doses (BACs). In the dose–response study by Bonthius and West (1988), a departure from linearity of the dose-dependent restriction of cerebellar weight was observed at the highest dose (7.5 g/kg/day). The restriction of cerebellar weight was more severe than predicted by the effects produced by the lower doses, owing to a precipitous decrease in cerebellar weight when BACs exceeded 400 mg/dl. This effect was seen only in the cerebellum, not in the forebrain or brain stem. We have also observed that gross defects in the cerebellum can occasionally occur when BACs exceed 500 mg/dl (see Fig. 2).

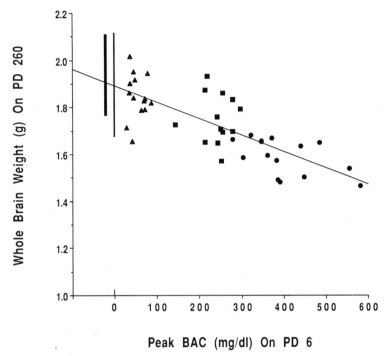

Peak BAC (mg/dl) On PD 6

FIGURE 7   Scatterplot of whole brain weight (on postnatal day 260) as a function of peak BAC (on postnatal day 6) in rats exposed to one of three doses (2.5 (▲); 4.5 (■); or 6.6 (●) g/kg/day) on postnatal days 4–9. The range of values is also indicated for the suckle controls (━━━) and gastrostomy controls (──). There was a highly significant linear decline in whole brain weight as a function of BACs [$F(1,40)$ = 56.9, $p < .0001$], and the regression line accounted for 59% of the variance of the alcohol-exposed pups ($R^2 = 0.59$).

Blood Alcohol Concentration Profiles Are the Critical Determinant of Central Nervous System Effects

One of the most important generalizations derived from our recent studies is the BAC profile—not the total daily dose—is the critical determinant of brain-weight restriction and neuronal depletion. A given daily dose condensed into only a few feedings each day (using higher concentrations of alcohol each feeding), produces peak BACs that are higher than when the same daily dose, or even a higher one, is distributed over more feedings throughout each day (less concentrated in each feeding) (reviewed in West et al., 1989). This increased damage to the CNS produced by bingelike exposure has been documented using several different dependent measures, including brain-weight restriction (Bonthius et al., 1988; Bonthius and West, 1990, 1991), loss of cerebellar Purkinje cells or granule cells (Bonthius and West, 1990, 1991), loss of hippocampal CA1 pyramidal cells (Bonthius and West, 1990, 1991), loss of olfactory bulb mitral and granule cells (Bonthius and West, in press), and behavioral outcomes (Goodlett et al., 1987; Kelly et al., 1987b).

The severity of effects of a given daily dose of alcohol depends on the peak BACs attained, which is a function of the daily pattern of exposure. Generally, as the peak BAC increases, cell loss increases, brain weight restriction becomes more severe, and behavioral deficiencies increase. These studies demonstrate that one way to increase the peak BAC and thereby to increase CNS damage is to administer a given dose in higher and higher concentrations over shorter and shorter intervals. These results clearly indicated that one of the factors that puts the CNS of the fetus at risk for damage is a pattern of binge drinking (marked by episodes of concentrated alcohol consumption) that produced BACs sufficient to induce pronounced intoxication.

Depletion of Different Neuronal Populations Can Follow Different Blood Alcohol Concentration Curves

Within-subjects analyses of cell counts in different brain regions (hippocampus, cerebellum, and olfactory bulb) in rats exposed neonatally to alcohol have indicated that neuronal populations have different thresholds of peak blood alcohol concentrations necessary to produce significant depletion (West et al., 1990). Of structures evaluated so far, the cerebellum and the olfactory bulb appear to have relatively low thresholds for significant cell loss. In the vermis, significant (and permanent) Purkinje cell and granule cell loss has been documented in groups exposed to bingelike treatments on PD4 to PD9 (producing mean peak BACs at or greater than 180 mg/dl), but not following treatments producing mean peak BACs around 50 mg/dl. In an important exception, the flocculi of the cerebellum were found to have significant cell loss even with exposures producing BACs near 50 mg/dl, and the loss was as severe as that produced with high BACs (Napper and West, 1991). In the olfactory bulb, significant loss of mitral and granule cells was also documented at 180 mg/dl. In contrast, in the hippocampal formation, significant loss of pyramidal cells in field CA1 occurred only in the group that was exposed to mean peak BACs greater than 300 mg/dl. Furthermore, no significant reductions were observed for field CA3 pyramidal cells or for granule cells of the dentate gyrus in any group.

Blood Alcohol Concentration Dependence of Behavioral Effects

In addition to the morphological studies, we and others have evaluated behavioral effects in several different tasks, including sensorimotor development, activity, motor competence, passive avoidance, spatial navigation in a water maze, and spatial learning and memory in a radial maze (see West et al., 1989; Goodlett et al., in press for reviews). Some of these tasks were specifically chosen because performance on them is known to depend on the functional integrity of the cerebellum (motor competence) or the hippocampal formation (activity, passive avoidance, and spatial learning).

Neonatal alcohol exposure delays sensorimotor development (Kelly et al., 1987a), and impairs suckling behavior on PD15 (Barron et al., 1989); the severity of the effects is a function of the BAC attained. Alcohol-related hyperactivity occurs both in developing pups (Kelly et al., 1987a) and in adult rats (Kelly et al., 1987b), but only in groups exposed to binge exposure producing relatively high peak BACs. Deficits in motor function have been revealed with gait analysis, showing that locomotor patterns of neonatally exposed rats were abnormal because of an increased stance width and a decreased stride length (Meyer et al., 1990a). It should also be noted that rats exposed to alcohol prenatally exhibit a similar disturbance in gait (Hannigan and Riley, 1988).

Several reports from different laboratories indicate that at least one test of motor competence that incorporates substantial practice effects (i.e., traversing parallel bars) consistently reveals deficient performance of rats exposed to alcohol during the brain growth spurt (Meyer et al., 1990b; Thomas et al., 1991). The deficits were BAC dependent, in that they were not significant in groups with mean peak BACs less than 100 mg/dl; the deficits were more severe in groups exposed to BACs greater than 300 mg/dl than in groups exposed to BACs between 150 and 250 mg/dl. Another test of motor competence—balancing on a rotarod—gave inconsistent results, in that rats exposed to relatively high BACs neonatally showed no impairments as young adults (Meyer et al., 1990b; Thomas et al., 1991); others tested at 13 months of age were significantly impaired (Goodlett et al., 1991b). Traversing parallel bars appeared to be more difficult to learn than balancing on the rotarod, and rotarod competence acquisition of the older (and bigger) adult rats was slower than that of younger (and smaller) adult rats. Thus, the differences in outcomes appear to be related to differences in task difficulty, suggesting that even with relatively severe cerebellar cell loss, deficits may be apparent only with relatively challenging tasks or tasks that require substantial motor learning.

Alterations in passive avoidance and spatial learning have also been described, and were especially apparent in juvenile rats following neonatal alcohol exposure. The passive avoidance deficits (Barron and Riley, 1990) were observed in 23-day-old rats, were dose dependent (found in groups given binge exposure of 6 g/kg/day but not 4 g/kg/day), and were gender dependent (found in females but not males). Severe deficits in acquisition of spatial navigation in the Morris maze swimming task were present in rats of both sexes when tested between PD19 and PD40 (Goodlett et al., 1987; Goodlett et al., in press). These spatial navigation deficits depended on the BAC achieved during the treatment period. Rats exposed to BACs less than 150 mg/dl were not significantly affected, whereas those exposed to relatively high BACs (>300 mg/dl) were severely impaired. When tested initially as adults, the spatial navigation deficits were substantially less severe, and were significant only for females (Kelly et al., 1988; Goodlett et al., in press). In contrast, acquisition of spatial learning in the radial maze as adults was significantly impaired following exposure to high peak BACs (>300 mg/dl) during PD4 to PD9 , both in males and females (Goodlett et al., in press).

Thus, effects on performance both of motor tasks and of cognitive tasks can occur following neonatal alcohol exposure. Consistent impairments, on the tasks used thus far, emerge when peak BACs during the neonatal exposure exceed 250 mg/dl. However, other tasks or other procedures may reveal deficits at lower exposure levels.

CENTRAL NERVOUS SYSTEM EFFECTS INDUCED BY LOW BLOOD ALCOHOL CONCENTRATIONS

One aspect of the discussion on page 62 is that, with the important exception of the cerebellar flocculi, significant reductions in cell numbers following neonatal exposure were not observed in groups exposed to low peak BACs (<100 mg/dl). However, neuronal depletion probably represents one of the most extreme effects of alcohol exposure during development, so cell loss may not be an appropriate indicator of susceptibility of the CNS to low to moderate levels of exposure. In that regard, two other morphological measures have revealed deleterious effects of low-level alcohol exposure during the neonatal brain growth spurt—permanent reductions in zinc density in the mossy fibers, and acute astrogliosis in the cortex.

Mossy fiber zinc density was measured in stratum lucidum of hippocampal field CA3 on PD45 using quantitative zinc:TS-Q histofluorescence (Savage et al., 1991c). Three groups of rats were exposed to alcohol on PD4 to PD9, in patterns producing mean peak BACs of 57, 229, or 430 mg/dl, respectively. Compared to suckle controls and artificially reared gastrostomy controls, all three alcohol-treated groups had significantly lower mossy fiber zinc densities, with reductions of 28 to 36% in the dorsal hippocampus and 20 to 25% in the ventral hippocampus. Importantly, the extent of mossy fiber zinc reduction did not differ among the three alcohol-exposed groups, despite the massive group differences in peak BACs. Thus, exposure to low levels of alcohol produced deficits in zinc content that were just as severe as those following heavy exposure. Interestingly, a comparable depletion of zinc density also occurred with *gestational* exposure to low BACs (Savage et al., 1989), suggesting that the developmental timing of the exposure is not a critical determinant of this effect.

The astroglial reactions in the deep layers of the cortex described earlier (pages 54–58) also are discernible on PD10 following postnatal alcohol treatments that produce BACs averaging about 50 mg/dl (Fig. 6). In layer V of the cortex of the alcohol-exposed pups, the density of astrocytes was greater, and the astrocytes were hypertrophied, having dense immunoreactivity in the soma and thick, heavily labeled fibrillary processes. The astrogliosis in the deep cortical layers in the low-dose group was qualitatively similar to that of the high-dose group having a mean BAC of 300 mg/dl (compare Fig. 6B and 6C). However, BAC-dependent differences were apparent in other aspects of the transient alcohol-induced gliosis. As mentioned earlier, the total GFAP content in the cortex, measured using radioimmunoassays, increased linearly with increasing peak BAC (ranging from 30 to 400 mg/dl). In addition, the reactive glia surrounding what appeared to be areas of vascular disruption (see Fig. 4 and pages 54–57) were seen *only* in cases exposed to high BACs (<250 mg/dl).

With regard to the controversial issue of what degree of alcohol consumption constitutes a risk to the developing CNS, these studies suggest that identifiable, presumably deleterious effects on CNS structure can result from even moderate levels of exposure. Other studies, using a prenatal exposure paradigm via a liquid diet producing low maternal BACs (peak BACs of 30 to 40 mg/dl), indicate that neurochemical and neurophysiological alterations are very sensitive markers of long-lasting neurological effects of gestational alcohol exposure.

Savage and his collaborators demonstrated that such prenatal alcohol exposure resulted in significant reductions in total glutamate binding in 45-day-old rats (Farr et al., 1988). In a follow-up study of binding to the N-methyl-D-aspartate (NMDA) glutamate receptor subtype, NMDA-sensitive glutamate binding density was also reduced in CA1 and the dentate gyrus of the dorsal hippocampal formation, in a pattern similar to that of the alcohol-induced change in total glutamate binding (Savage et al., 1991a). The reduction in NMDA-sensitive binding is apparently attributable to a long-lasting shift in the NMDA receptor population from an agonist-preferring state to an antagonist preferring state (Savage et al., 1991b). These findings converge with electrophysiological studies in hippocampal slices of adult rats given identical prenatal alcohol exposure. Morrisett et al. (1989) demonstrated reductions in responsiveness of CA1 pyramidal neurons to NMDA, and Swartzwelder et al. (1988) demonstrated decreased long-term potentiation [neuronal plasticity that is mediated in part by NMDA receptors, (Collinridge, 1985)].

Taken together, both the prenatal-exposure studies and the neonatal-exposure studies indicate that BACs in the range typically produced by low to moderate drinking can result in significant and long-lasting changes in the brain. The functional significance of these changes is presently unknown. However, it is likely that these demonstrated changes are associated with altered function or even with neuropathological processes. For example, NMDA-mediated neurotransmission and long-term potentiation in the hippocampus have been critically linked to processes of learning and memory, and disruption of these processes can interfere with cognition (Morris et al., 1986; Olds et al., 1990). Additionally, astroglial reactions often are a consequence of an underlying neuropathological process (Nathaniel and Nathaniel, 1981; Eng and DeArmond, 1982; Norenberg et al., 1988), suggesting the transient cortical reactions may mark a significant alcohol-induced neuropathological effect. Given these considerations, and in light of the limited information presently available on the mechanisms of these known effects, the only prudent conclusion is that alcohol can affect the developing brain even at low exposure levels. Abstinence during pregnancy is the only way to avoid such effects.

TEMPORAL WINDOWS OF VULNERABILITY

Because of the protracted length of mammalian CNS development and the stereotyped sequence of different developmental events, the specific effects of alcohol on the CNS and the mechanisms producing those effects are likely to depend on the developmental timing of the alcohol exposure. If the period of vulnerability to a given effect is relatively limited, it may be possible to define experimentally the "temporal window of vulnerability" for that effect by careful manipulation of the time of exposure. One such example of a temporal window of vulnerability is the induction of FAS-like facial anomalies with alcohol exposure during embryogenesis in mice (Sulik et al., 1981). Structural malformations resembling FAS-like midfacial dysmorphology were produced at the greatest frequency by intraperitoneal injections of alcohol during a window of time centered around gestational day 7. Injections only a few hours earlier or a few hours later (7 days ± 4 hr) were not nearly so effective in producing these FAS-like features.

If defined temporal windows of vulnerability can be identified for a given effect in the CNS, it would provide an essential step toward understanding the specific mechanisms of alcohol-induced damage. Knowledge of mechanisms underlying CNS dysfunction is predicated on specifying the critical developmental events associated with such vulnerability. This in turn requires knowing when a given teratogenic effect of alcohol can be produced, and whether the slopes of the temporal boundaries of the vulnerable period are steep or broad.

For the CNS, relatively little is known about potential windows of vulnerability to fetal alcohol effects, but periods of heightened susceptibility as well as temporal constraints in producing those effects have long been postulated (see West, 1987, for review). Studies in our laboratory and in others (e.g., Cragg and Phillips, 1985) have shown that even in the context of exposure during the brain growth spurt, relatively sharp windows of vulnerability to some effects, most notably cell depletion, can be established.

Purkinje Cell Depletion

A study manipulating the developmental timing of neonatal alcohol exposure indicated that cerebellar weight was significantly restricted by bingelike alcohol exposure even when given on only a single day, but only when that day was PD4 or, to a lesser extent, PD5 (Goodlett et al., 1989b). This suggested that alcohol exposure limited to one of those days could produce depletion of cerebellar neurons. This was confirmed in a subsequent study, in which significant reductions in Purkinje cells occurred following alcohol exposure on PD4 in which BACs reached 360 mg/dl (Goodlett et al., 1990a).

The alcohol-induced depletion of Purkinje cells during the brain growth spurt is now well established. Whether the brain growth spurt constitutes a period of heightened vulnerability requires additional comparisons with prenatal exposure. We compared the effects of neonatal alcohol exposure (PD4–PD9) with gestational alcohol exposure (G13–G18), in which the temporal patterning of the alcohol administration and the resulting BAC profiles in the pups and the dams, respectively, were matched as closely as possible (Marcussen et al., submitted). Pups from both treatments (and from the appropriate matched controls) were perfused on PD10, and Purkinje cells in the vermis were counted. The group exposed to alcohol neonatally suffered significant Purkinje cell loss; the group exposed prenatally were not different from controls. Thus, the postnatal brain growth spurt constitutes a period of enhanced vulnerability to Purkinje cell loss. It should be noted that during this neonatal period of vulnerability in rats, Purkinje cells are postmitotic and are undergoing differentiation. During the relatively invulnerable late gestational period, the Purkinje cells are proliferating and migrating. Thus, for this population of neurons, the differentiation phase renders them most vulnerable to alcohol-induced depletion.

Once it was established that the early postnatal period was the period of greatest

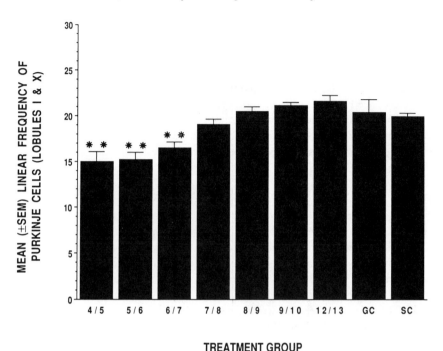

TREATMENT GROUP

FIGURE 8   Mean (± SEM) linear frequency of Purkinje cells in the lobules IX and X of the vermis of the cerebellum on postnatal day 21 of control groups (SC and GC) and groups given bingelike alcohol exposure restricted to 2 days during the brain growth spurt (on the postnatal days indicated along the abscissa). These lobules correspond to the early-maturing region of Fig. 3. Note that alcohol exposure resulted in significant Purkinje cell deficits only if treatment began before postnatal day 7. **Significantly different from the control groups and from all groups exposed to alcohol beginning on or after postnatal day 7.

vulnerability of the Purkinje cells, an extensive study of temporal windows of vulnerability over PD4 to PD13 was performed (Hamre and West, 1990; Hamre et al., 1991). Pups were administered 6.6 g/kg/day of alcohol, limited to two consecutive days beginning on one of seven postnatal ages, either PD4, PD5, PD6, PD7, PD8, PD9, or PD12. The bingelike alcohol treatments produced peak BACs between 360 and 440 mg/dl; however, the groups exposed on different days did not differ significantly in terms of BAC. Purkinje and granule cells were counted in the cerebellar vermis on PD21. Significant alcohol-induced cerebellar cell loss (relative to artificially reared or normally reared controls) was found to occur only if the treatment began before PD7 (Fig. 8). Furthermore, Purkinje cell depletion was greatest with the earliest alcohol treatment used (PD4–PD5). A dramatic and significant difference in Purkinje cell number was apparent between the groups in which treatment began before PD7 (i.e., PD4/5, PD5/6, and PD6/7) and those in which it began on PD7 or thereafter.

Considering these findings, the vulnerability of the cerebellum to alcohol-induced cell loss clearly depends on the developmental timing of exposure. Given the ineffectiveness of the late gestational exposure in producing Purkinje cell loss, and the ineffectiveness of exposure beginning on PD7, these studies appear to define a temporal window of vulnerability. Precise definition of the exact boundaries of this temporal window will require alcohol treatments that include G19–G21 and/or PD0–PD3. Furthermore, the precise temporal boundaries may depend on the peak BACs that are produced by the treatments on specific days. Nevertheless, the data support the contention that vulnerability to alcohol-related cerebellar cell loss changes over development, and the period early in the brain growth spurt that includes onset of Purkinje cell differentiation, appears the time of greatest vulnerability.

N-Methyl-D-Aspartate-Sensitive Glutamate Receptor Binding

As discussed earlier, both prenatal and neonatal alcohol exposure can reduce the number of pyramidal neurons in field CA1 of the hippocampus if the BACs are relatively high. However, as indicated on page 65, Savage and collaborators demonstrated that neurochemical and neurophysiological alterations can occur in the hippocampal formation when peak BACs are much lower (<50 mg/dl), including decreased density of N-methyl-D-aspartate-sensitive glutamate receptor binding (Savage et al., 1991a). This effect has been shown to depend on the developmental timing of alcohol exposure (Savage et al., 1991d).

Alcohol treatments that produced peak BACs under 60 mg/dl were administered during different periods of development, either during gestational periods (via liquid diets): G1–G21, G1–G11, G11–G21, OR G16–G21; or during the neonatal brain growth spurt (via artificial rearing), from PD4–PD9. Hippocampal NMDA-sensitive [^{3}H]glutamate binding site density was determined when the rats were 45 days old using in vitro radiohistochemical techniques. Significant decreases in NMDA-sensitive glutamate binding site density occurred only when the alcohol exposure included the last week of gestation. Neither the exposure during the first half of gestation nor the exposure during the neonatal brain growth spurt altered binding density. Exposure restricted to the last third of gestation (G15–G21) reduced binding density as severely as exposure throughout gestation (G1–G21) or during the last half of gestation (G11–G21). Thus, in the rat, the window of vulnerability to the decreased hippocampal NMDA agonist binding induced by low levels of alcohol exposure appears to occur during the time of neurogenesis of the hippocampal pyramidal neurons.

Aberrant Mossy Fibers

Disruptions of the hippocampal connections in the form of aberrant mossy fiber projections can also be induced by developmental alcohol exposure (see page 54). This effect also has a period of vulnerability, albeit one of degree. Gestational alcohol exposure (via liquid diets) during G1 to G10 or during G11 to G21 failed to produce aberrant mossy fibers. Exposure during PD1 to PD10 produced severe cases of aberrant projections into the midtemporal intra- and infrapyramidal region of CA3a and CA3b

(West and Hamre, 1985). Alcohol exposure throughout gestation (G1–G21) was sufficient to produce aberrant mossy fiber projections in CA3a, but to a much lesser extent than that produced by exposure during PD1 to PD10 (see West and Goodlett, 1990). Importantly, postnatal alcohol exposure limited to PD4 to PD10 does not produce aberrant mossy fiber projections (Kelly and West, unpublished observations, 1987), suggesting that the most vulnerable time for this effect in the rat occurs in the first few days of postnatal life.

Enhanced Neuroendocrine Response to Stress

Stress-induced activation of the hypothalamic–pituitary–adrenal (HPA) axis has been shown to be enhanced in juvenile and adult rats previously exposed to ethanol throughout gestation (Weinberg et al., 1986; Nelson et al., 1986). Recent studies by Rivier and her colleagues (Lee et al., 1990) demonstrated that alcohol exposure restricted to the second gestational week, but not exposure restricted to the first or the third gestational week, enhanced stress-induced HPA activation in 21-day-old rats. Compared to controls, offspring exposed during the second gestational week (but not the first or the third week) had significantly higher levels of plasma adrenocorticotrophic hormone (ACTH) after a session of inescapable footshock. Furthermore, exposure during the second gestational week also resulted in increased biosynthesis of corticotropin-releasing factor (CRF) on postnatal day 21, compared to the other groups. Northern blot analysis indicated that CRF mRNA levels were significantly increased in the hypothalamus, and in situ hybridization revealed a significant increase in CRF mRNA in the parvocellular division of the paraventricular nucleus. These data indicate that the second gestational week in rats constitutes a temporal window of vulnerability to altered development and regulation of the HPA axis, in which a primary effect appears to be a long-term compensatory increase in CRF expression in the brain.

SUMMARY

The bingelike exposure during the neonatal brain growth spurt in rats has provided an animal model system affording control over two variables that are critical for precise understanding of the teratogenic effects of the alcohol: the profile of BACs produced and the developmental timing of exposure. In terms of alcohol exposure during the brain growth spurt, use of this model has now established several fundamental characteristics of CNS effects, summarized as follows:

- certain effects in the CNS can occur with relatively low BACS;
- the number and severity of effects increase with increasing BAC;
- when the alcohol exposure is precisely controlled, many effects appear to be a linear function of peak BAC attained;
- effects differ in terms of the peak BAC at which they are consistently observable, suggesting they have different thresholds;

- some effects, even severe ones such as neuron loss, can occur with exposure restricted to a relatively short period, even 1 day; and
- temporal windows of vulnerability exist for some of the morphological effects described, and are likely to occur for others.

The studies reviewed here, as well as others using different animal models, provide the basic information necessary to begin more specific in vivo studies directed toward mechanisms of teratogenic effects of alcohol. For example, knowledge that a single day of exposure with peak BACs exceeding 300 mg/dl administered on selected postnatal days can produce Purkinje cell depletion allows one to begin to evaluate the molecular and cellular events associated with this temporal window of vulnerability. Also, alcohol-induced changes in gene expression or cellular reactions to the alcohol insult can now be evaluated to determine the time course (onset, peak, decline) of such responses. These can then begin to provide critical insight into factors controlling neuronal survival or death (in postmitotic populations), alterations in cell cycles (in proliferating populations), or changes in the synaptic organization, neurochemical constitution, or functional status of a given region following alcohol exposure. Knowledge concerning the time course and mechanisms of these (or other) effects is necessary before any rational strategy of intervention can be formulated.

ACKNOWLEDGMENTS

Excellent technical assistance in many of the studies reported herein was provided by Jolonda C. Mahoney. We also thank Paul C. Reimann for his expert assistance in producing the photomicrographs. James P. O'Callaghan of the Environmental Protection Agency performed the GFAP radioimmunoassays. We thank Jonathan Leo, Britt L. Marcussen, and Jennifer D. Thomas for providing comments on a previous version of this manuscript. This work was supported by Grants AA05523 and AA07313 from the National Institute of Alcohol Abuse and Alcoholism.

REFERENCES

Abel, E. L. (1984). "Fetal Alcohol Syndrome and Fetal Alcohol Effects." Plenum Press, New York.
Abel, E. L., and Greizerstein, H. B. (1982). Growth and development in animals prenatally exposed to alcohol. In "Fetal Alcohol Syndrome, Vol. III, Animal Studies" (E. L. Abel, ed.), pp. 39–57. CRC Press, Boca Raton, Florida.
Abel, E. L., and Sokol, R. J. (1991). A revised conservative estimate of the incidence of FAS and its economic impact. Alcohol. Clin. Exp. Res. 15, 514–524.
Barnes, D. E., and Walker, D. W. (1981). Prenatal ethanol exposure permanently reduces the number of pyramidal neurons in rat hippocampus. Dev. Brain Res. 1, 333–340.
Barron, S., and Riley, E. P. (1990). Passive avoidance performance following neonatal alcohol exposure. Neurotoxicol. Teratol. 12, 135–138.
Barron, S., Kelly, S. J., and Riley, E. P. (1989). The effects of neonatal alcohol exposure on suckling performance in rat pups. Alcohol. Clin. Exp. Res. 13, 319.
Bauer-Moffett, C., and Altman, J. (1977). The effect of ethanol chronically administered to preweaning rats on cerebellar development: A morphological study. Brain Res. 119, 249–268.
Blakley, P. M. (1988). Experimental teratology of ethanol. In "Issues and Reviews in Teratology" (H. Kalter, ed.), Vol. 4, pp. 237–282. Plenum Press, New York.

Blakley, P. M., and Scott, W. J., Jr. (1984). Determination of the proximate teratogen in the mouse fetal alcohol syndrome. 1. Teratogenicity of ethanol and acetaldehyde. *Toxicol. Appl. Pharmacol.* **72,** 355–363.

Bonthius, D. J., and West, J. R. (1988). Blood alcohol concentration and microencephaly: A dose–response study in the neonatal rat. *Teratology* **37,** 223–231.

Bonthius, D. J., and West, J. R. (1990). Alcohol-induced neuronal loss in developing rats: Increased brain damage with binge exposure. *Alcohol. Clin. Exp. Res.* **14,** 107–118.

Bonthius, D. J., and West, J. R. (1991). Permanent neuronal deficits in rats exposed to alcohol during the brain growth spurt. *Teratology* **44,** 147–163.

Bonthius, D. J., and West, J. R. (1991). Acute and long-term neuronal deficits in the rat olfactory bulb following alcohol exposure during the brain growth spurt. *Neurotoxicol. Teratol.* **13,** 611–619.

Bonthius, D. J., Goodlett, C. R., and West, J. R. (1988). Blood alcohol concentration and severity of microencephaly in neonatal rats depend on the pattern of alcohol administration. *Alcohol* **5,** 209–214.

Bonthius, D. J., Bonthius, N. E., Napper, R. M. A., Astley, S. J., Clarren, S. K., and West, J. R. (1991). Purkinje cell deficits in the pig-tailed macaque following gestational alcohol exposure. *Alcohol. Clin. Exp. Res.* **15,** 340.

Chernoff, G. F. (1977). The fetal alcohol syndrome in mice: An animal model. *Teratology* **15,** 223–230.

Chernoff, G. F. (1980). The fetal alcohol syndrome in mice: Maternal variables. *Teratology* **22,** 71–75.

Clarren, S. K. (1986). Neuropathology in fetal alcohol syndrome. *In* "Alcohol and Brain Development" (J. R. West, ed.), pp. 158–166. Oxford University Press, New York.

Clarren, S. K., and Smith, D. W. (1978). The fetal alcohol syndrome: A review of the world literature. *N. Engl. J. Med.* **298,** 1063–1067.

Clarren, S. K., Astley, S. J., and Bowden, D. M. (1988). Physical anomalies and developmental delays in nonhuman primate infants exposed to weekly doses of ethanol during gestation. *Teratology* **37,** 561–570.

Clarren, S. K., Astley, S. J., Bowden, D. M., Lai, H., Rudeen, P. K., Shoemaker, W. J., and Bunt-Milam, A. (1990). Neuroanatomic and neurochemical abnormalities in nonhuman primate infants exposed to weekly doses of ethanol during gestation. *Alcohol. Clin. Exp. Res.* **14,** 674–683.

Coles, C. D., Smith, I., Fernhoff, P. M., and Falek, A. (1985). Neonatal neurobehavioral characteristics as correlates of maternal alcohol use during gestation. *Alcohol. Clin. Exp. Res.* **9,** 454–460.

Collinridge, G. L. (1985). Long-term potentiation in the hippocampus: Mechanisms of initiation and modulation by neurotransmitters. *Trends Pharmacol. Sci.* **6,** 407–411.

Conry, J. (1990). Neuropsychological deficits in fetal alcohol syndrome and fetal alcohol effects. *Alcohol. Clin. Exp. Res.* **14,** 650–655.

Courchesne, E., Yeung-Courchesne, R., Press, G. A., Hesselink, J. R., Jernigan, T. L. (1988). Hypoplasia of cerebellar vermal lobules VI and VII in autism. *N. Engl. J. Med.* **318,** 1349–1354.

Cragg, B., and Phillips, S. (1985). Natural loss of Purkinje cells during development and increased loss with alcohol. *Brain Res.* **325,** 151–160.

Diaz, J., and Samson, H. H. (1980). Impaired brain growth in neonatal rats exposed to ethanol. *Science* **208,** 751–753.

Dobbing, J., and Sands, J. (1973). The quantitative growth and development of the human brain. *Arch. Dis. Child.* **48,** 757–767.

Dobbing, J., and Sands, J. (1979). Comparative aspects of the brain growth spurt. *Early Hum. Dev.* **3,** 79–83.

Dorris, M. (1989). "The Broken Cord." Harper and Row, New York.

Driscoll, C. D., Streissguth, A. P., and Riley, E. P. (1990). Prenatal alcohol exposure: Comparability of effects in humans and animal models. *Neurotoxicol. Teratol.* **12,** 231–237.

Duester, G. (1991). A hypothetical mechanism for fetal alcohol syndrome involving ethanol inhibition of retinoic acid synthesis at the alcohol dehydrogenase step. *Alcohol. Clin. Exp. Res.* **15,** 568–572.

Eng, L. F., and DeArmond, S. J. (1982). Immunocytochemical studies of astrocytes in normal development and disease. *In* "Advances in Cellular Neurobiology" (S. Fedoroff and L. Hertz, eds.), Vol. 3, pp. 145–171. Academic Press, New York.

Farr, K. L., Montano, C. Y., Paxton, L. L., and Savage, D. D. (1988). Prenatal ethanol exposure decreases hippocampal [^{3}H]glutamate binding in 45-day-old rats. *Alcohol* **5,** 125–133.

Gilliam, D. M., and Irtenkauf, K. T. (1990). Maternal genetic effects on ethanol teratogenesis and dominance of relative embryonic resistance to malformations. *Alcohol. Clin. Exp. Res.* **14**, 539–545.

Gilliam, D. M., Kotch, L. E., Dudek, B. C., and Riley, E. P. (1988). Ethanol teratogenesis in mice selected for differences in alcohol sensitivity. *Alcohol* **5**, 513–519.

Gilliam, D. M., Kotch, L. E., Dudek, B. C., and Riley, E. P. (1989). Ethanol teratogenesis in selectively bred long-sleep and sort-sleep mice: A comparison to inbred C57BL/6J mice. *Alcohol. Clin. Exp. Res.* **13**, 667–672.

Goodlett, C. R., Kelly, S. J., and West, J. R. (1987). Early postnatal alcohol exposure that produces high blood alcohol levels impairs development of spatial navigation learning. *Psychobiology,* **15**, 64–74.

Goodlett, C. R., Gilliam, D. M., Nichols, J. M., and West, J. R. (1989a). Genetic influences on brain growth restriction induced by developmental exposure to alcohol. *Neurotoxicology* **10**, 321–334.

Goodlett, C. R., Mahoney, J. C., and West, J. R. (1989b). Brain growth deficits following a single day of alcohol exposure in the neonatal rat. *Alcohol* **6**, 121–126.

Goodlett, C. R., Marcussen, B. L., and West, J. R. (1990a). A single day of alcohol exposure during the brain growth spurt induces microencephaly and cerebellar Purkinje cell loss. *Alcohol* **7**, 104–114.

Goodlett, C. R., Nichols, J. M., and West, J. R. (1990b). Strain differences in susceptibility to brain growth restriction induced by early postnatal alcohol exposure. *Alcohol. Clin. Exp. Res.* **14**, 293.

Goodlett, C. R., Nichols, J. M., and West, J. R. (1991a). Strain differences in susceptibility to alcohol-induced brain growth restriction are not accounted for by differences in alcohol elimination rates. *Alcohol. Clin. Exp. Res.* **15**, 343.

Goodlett, C. R., Thomas, J. D., and West, J. R. (1991b). Long-term deficits in cerebellar growth and rotarod performance of rats following "bingelike" alcohol exposure during the neonatal brain growth spurt. *Neurotoxicol. Teratol.* **13**, 69–74.

Goodlett, C. R., Bonthius, D. J., Wasserman, E. A., and West, J. R. An animal model of central nervous system dysfunction associated with fetal alcohol exposure: Behavioral and neuroanatomical correlates. *In* "Learning and Memory: Behavioral and Biological Processes" (I. Gormezano and E. A. Wasserman, eds.), pp. 183–2098. Lawrence Erlbaum, Englewood, New Jersey. In press.

Goodlett, C. R., Leo, J. T., O'Callaghan, J. P., Mahoney, J. C., and West, J. R. Astrocytic gliosis induced by alcohol exposure during the brain growth spurt. Submitted for publication.

Hamre, K. M., and West, J. R. (1990). Purkinje cell survival following ethanol exposure during the brain growth spurt depends on the timing of exposure and location in the cerebellar vermis. *Soc. Neurosci. Abstr.* **16**, 32.

Hamre, K. M., Goodlett, C. R., and West, J. R. (1991). Spatial and temporal correlation of Purkinje cell and granule cell loss after postnatal ethanol administration in the rat. *Alcohol. Clin. Exp. Res.* **15**, 341.

Hannigan, J. H., and Riley, E. P. (1988). Prenatal ethanol alters gait in rats. *Alcohol* **5**, 451–454.

Hanson, J. W., Streissguth, A. P., and Smith, D. W. (1978). The effects of moderate alcohol consumption during pregnancy on fetal growth and morphogenesis. *J. Pediatr.* **92**, 457–460.

Jernigan, T. L., and Bellugi, U. (1990). Anomalous brain morphology on magnetic resonance images in Williams syndrome and Down's syndrome. *Arch. Neurol.* **47**, 529–533.

Jones, D. J. (1988). Influence of ethanol on neuronal and synaptic maturation in the central nervous system—morphological investigations. *Prog. Neurobiol.* **31**, 171–197.

Jones, K. L., and Smith, D. W. (1973). Recognition of the fetal alcohol syndrome in early infancy. *Lancet* **2**, 999–1001.

Jones, K. L., Smith, D. W., Ulleland, C. N., and Streissguth, A. P. (1973). Pattern of malformation in offspring of chronic alcoholic mothers. *Lancet* **1**, 1267–1271.

Kelly, S. J., Hulsether, S. A., and West, J. R. (1987a). Alterations in sensorimotor development: Relationship to postnatal alcohol exposure. *Neurotoxicol. Teratol.* **9**, 243–251.

Kelly, S. J., Pierce, D. R., and West, J. R. (1987b). Microencephaly and hyperactivity in adult rats can be induced by neonatal exposure to high blood-alcohol concentrations. *Exp. Neurol.* **96**, 580–593.

Kelly, S. J., Goodlett, C. R., Hulsether, S. A., and West, J. R. (1988). Impaired spatial navigation in adult female but not adult male rots exposed to alcohol during the brain growth spurt. *Behav. Brain Res.* **27**, 247–257.

Kinsley, M. (1991). Cocktails for two. *New Republic,* p. 6, June 3, 1991.

Lee, S., Imaki, T., Vale, W., and Rivier, C. (1990). Effect of prenatal exposure to ethanol on the activity of the hypothalamic-pituitary-adrenal axis of the offspring: Importance of the time of exposure to ethanol and possible modulating mechanisms. *Mol. Cell. Neurosci.* **1**, 168–177.

Little, R. W., Anderson, K. W., Ervin, C. H., Worthington-Roberts, B., Clarren, S. K. (1989). Maternal alcohol use during breast feeding and infant mental and motor development at one year. *N. Engl. J. Med.* **321**, 425–430.

Marcussen, B. L., Goodlett, C. R., Mahoney, J. C., and West, J. R. Alcohol-induced Purkinje cell loss during differentiation but not during neurogenesis. *Alcohol,* In press.

Meyer, L. S., Kotch, L. E., and Riley, E. P. (1990a). Alterations in gait following ethanol exposure during the brain growth spurt in rats. *Alcohol. Clin. Exp. Res.* **14**, 23–27.

Meyer, L. S., Kotch, L. E., and Riley, E. P. (1990b). Neonatal ethanol exposure: Functional alterations associated with cerebellar growth retardation. *Neurotoxicol. Teratol.* **12**, 15–22.

Michaelis, E. K. (1990). Fetal alcohol exposure: Cellular toxicity and molecular events involved in toxicity. *Alcohol. Clin. Exp. Res.* **14**, 819–826.

Michaelis, E. K., and Michaelis, M. L. (1986). Molecular events underlying the effects of ethanol on the developing central nervous system. *In* "Alcohol and Brain Development" (J. R. West, ed.), pp. 277–309. Oxford University Press, New York.

Miller, M. W. (1986). Fetal alcohol effects on the generation and migration of cerebral cortical neurons. *Science* **233**, 1308–1311.

Miller, M. W. (1987). Effect of prenatal exposure to alcohol on the distribution and time of origin of corticospinal neurons in the rat. *J. Comp. Neurol.* **257**, 372–382.

Miller, M. W. (1988). Effect of prenatal exposure to ethanol on the development of cerebral cortex: I. Neuronal generation. *Alcohol. Clin. Exp. Res.* **12**, 440–449.

Miller, M. W. (1989). Effect of prenatal exposure to ethanol on the development of cerebral cortex: II. Cell proliferation in the ventricular and subventricular zones of the rat. *J. Comp. Neurol.* **287**, 326–338.

Miller, M. W. (1990). Prenatal exposure to ethanol delays the rate and schedule of migration of neurons to rat somatosensory cortex. *Alcohol. Clin. Exp. Res.* **14**, 319.

Miller, M. W. (ed.) (1992). "Development of the CNS: Effects of Alcohol and Opiates. Wiley-Liss, New York.

Miller, M. W., and Nowakowski, R. S. (1991). Effect of prenatal exposure to ethanol on the cell cycle kinetics and growth fraction in the proliferative zones of fetal rat cerebral cortex. *Alcohol. Clin. Exp. Res.* **15**, 229–232.

Miller, M. W., and Potempa, G. (1990). Numbers of neurons and glia in mature rat somatosensory cortex: Effects of prenatal exposure to ethanol. *J. Comp. Neurol.* **293**, 92–102.

Morris, R. G. M., Anderson, E., Lynch, G. S., and Baudry, M. (1986). Selective impairment of learning and blockade of long-term potentiation by an N-methyl-D-aspartate receptor antagonist, AP5. *Nature* **319**, 774–776.

Morrisett, R. A., Martin, D., Wilson, W. A., Savage, D. D., and Swartzwelder, H. S. (1989). Prenatal exposure to ethanol decreases the sensitivity of the adult rat hippocampus to N-methyl-D-aspartate. *Alcohol* **6**, 415–420.

Napper, R. M. A., and West, J. R. (1991). Decreased total number of Purkinje cells in rat cerebellar flocculus following postnatal ethanol exposure. *Alcohol. Clin. Exp. Res.* **15**, 340.

Nathaniel, E. J. H., and Nathaniel, D. R. (1981). The reactive astrocyte. *In* "Advances in Cellular Neurobiology" (S. Fedoroff and L. Hertz, eds.), Vol. 2, pp. 249–301. Academic Press, New York.

Nelson, L. R., Taylor, A. N., Lewis, J. W., Poland, R. E., Redei, E., and Branch, B. J. (1986). Pituitary-adrenal responses to morphine and footshock stress are enhanced following prenatal alcohol exposure. *Alcohol. Clin. Exp. Res.* **10**, 397–402.

Norenberg, M. D., Hertz, L., and Schousboe, A. (eds.) (1988). "The Biochemical Pathology of Astrocytes." Alan R. Liss, New York.

O'Connor, M. (ed.) (1984). "Mechanisms of Alcohol Damage *in Utero.*" Ciba Foundation Symposium 105. Pitman Books, London.

Olds, J., Golski, S., McPhie, D., Olton, D., Mishkin, M., and Alkon, D. (1990). Discrimination learning alters the distribution of protein kinase C in the hippocampus of rats. *J. Neurosci.* **10**, 3707–3713.

Pennington, S. N. (1990). Molecular changes associated with ethanol-induced growth suppression in the chick embryo. *Alcohol. Clin. Exp. Res.* **14**, 832–837.

Pierce, D. R., Goodlett, C. R., and West, J. R. (1989). Differential neuronal loss following early postnatal alcohol exposure. *Teratology* **40**, 113–126.

Pierce, D. R., Ragan, J. N., Serbus, D. C., and Light, K. E. (1991). Intragastric intubation of alcohol during postnatal development of rats results in selective cell loss in the cerebellum. *Alcohol. Clin. Exp. Res.* **15**, 175.

Pullarkat, R. K. (1991). Hypothesis: Prenatal ethanol-induced birth defects and retinoic acid. *Alcohol. Clin. Exp. Res.* **15**, 565–567.

Randall, C. L., and Taylor, W. J. (1979). Prenatal ethanol exposure in mice: Teratogenic effects. *Teratology* **19**, 305–312.

Randall, C. L., Taylor, W. J., and Walker, D. W. (1977). Ethanol-induced malformations in mice. *Alcohol. Clin. Exp. Res.* **1**, 219–224.

Randall, C. L., Ekblad, J., and Anton, R. F. (1990). Perspectives on the pathophysiology of fetal alcohol syndrome. *Alcohol. Clin. Exp. Res.* **14**, 807–812.

Riley, E. P. (1990). Long-term behavioral effects on prenatal alcohol exposure in rats. *Alcohol. Clin. Exp. Res.* **14**, 670–673.

Riley, E. P., and Lochry, E. A. (1982). Genetic influences in the etiology of fetal alcohol syndrome. *In* "Fetal Alcohol Syndrome, Vol. III. Animal Studies" (E. L. Abel, ed.), pp. 113–130. CRC Press, Boca Raton, Florida.

Rosett, H. L., and Weiner, L. (1984). "Alcohol and the Fetus." Oxford University Press, New York.

Rosett, H. L., Weiner, L., Zuckerman, B., McKinlay, S., and Edelin, K. C. (1980). Reduction of alcohol consumption during pregnancy with benefits to the newborn. *Alcohol. Clin. Exp. Res.* **4**, 178–184.

Rosett, H. L., Weiner, L., and Edelin, K. C. (1981). Strategies for prevention of fetal alcohol effects. *Obstet. Gynecol.* **57**, 1–7.

Rosett, H., L., Weiner, L., Lee, A., Zuckerman, B., Dooling, E., and Oppenheimer, E. (1983). Patterns of alcohol consumption and fetal development. *Obstet. Gynecol.* **61**, 539–546.

Samson, H. H., and Diaz, J. (1982). Effects of neonatal ethanol exposure on brain development in rodents. *In* "Fetal Alcohol Syndrome, Vol. III. Animal Studies" (E. L. Abel, ed.), pp. 131–150. CRC Press, Boca Raton, Florida.

Savage, D. D., Montano, C. Y., Paxton, L. L., and Kasarskis, E. J. (1989). Prenatal ethanol exposure decreases hippocampal mossy fiber zinc in 45-day-old rats. *Alcohol. Clin. Exp. Res.* **13**, 588–593.

Savage, D. D., Montano, C. Y., Otero, M. A., and Paxton, L. L. (1991a). Prenatal ethanol exposure decreases hippocampal NMDA-sensitive [^{3}H]glutamate binding-site density in 45-day-old rats. *Alcohol* **8**, 193–201.

Savage, D. D., Ortiz, K. A., Sanchez, C. F., and Paxton, L. L. (1991b). Prenatal ethanol exposure increases hippocampal NMDA receptor antagonist binding in 45-day-old rats. *Alcohol. Clin. Exp. Res.* **15**, 339.

Savage, D. D., Queen, S. A., Paxton, L. L., Goodlett, C. R., Mahoney, J. C., and West, J. R. (1991c). Postnatal ethanol exposure decreases hippocampal mossy fiber zinc density in 45-day-old rats. *Alcohol. Clin. Exp. Res.* **15**, 339.

Savage, D. D., Queen, S. A., Sanchez, C. F., Paxton, L. L., Mahoney, J. C., Goodlett, C. R., and West, J. R. (1991d). Prenatal ethanol exposure during the last third of gestation in rat reduces hippocampal NMDA agonist binding site density in 45-day-old offspring. *Alcohol* **9**, 1–5.

Schenker, S., Becker, H. C., Randall, C. L., Phillips, D. K., Baskin, G. S., and Henderson, G. I. (1990). Fetal alcohol syndrome: Current status of pathogenesis. *Alcohol. Clin. Exp. Res.* **14**, 635–647.

Smith, I. E., Coles, C. D., Lancaster, J., Fernhoff, P. M., and Falek, A. (1986). The effect of volume and duration of prenatal ethanol exposure on neonatal physical and behavioral development. *Neurobehav. Toxicol. Teratol.* **8**, 375–381.

Stephens, C. J. (1985). Alcohol consumption during pregnancy among southern city women. *Drug Alcohol Depend.* **16**, 19–29.

Streissguth, A. P. (1986). The behavioral teratology of alcohol: Performance, behavioral, and intellectual deficits in prenatally exposed children. *In* "Alcohol and Brain Development" (J. R. West, ed.), pp. 3–44. Oxford University Press, New York.

Streissguth, A. P., Landesman-Dwyer, S., Martin, J. C., and Smith, D. W. (1980). Teratogenic effects of alcohol in humans and laboratory animals. *Science* **209**, 353–361.

Streissguth, A. P., Barr, H. M., and Sampson, P. D. (1990). Moderate prenatal alcohol exposure: Effects on child IQ and learning problems at age 7.5 years. *Alcohol. Clin. Exp. Res.* **14**, 662–669.

Sulik, K. K., Johnston, M. C., and Webb, M. A. (1981). Fetal alcohol syndrome: Embryogenesis in a mouse model. *Science* **214**, 936–938.

Sulik, K. K., Lauder, J. M., and Dehart, D. (1984). Brain malformation in prenatal mice following acute maternal ethanol administration. *Int. J. Dev. Neurosci.* **2**, 203–214.

Surgeon General of the United States. (1981). Surgeon General's advisory on alcohol and pregnancy. *FDA Drug Bull.* **11**, 1–2.

Swartzwelder, H. S., Farr, K. L., Wilson, W. A., and Savage, D. D. (1988). Prenatal exposure to ethanol decreases physiological plasticity in the hippocampus of the adult rat. *Alcohol* **5**, 121–124.

Thomas, J. D., Goodlett, C. R., Wasserman, E. A., and West, J. R. (1991). Motor coordination deficits associated with cerebellar damage in adult rats induced by alcohol exposure during the brain growth spurt. *Alcohol. Clin. Exp. Res.* **15**, 336.

Webster, W. S. (1989). Alcohol as a teratogen: A teratological perspective of the fetal alcohol syndrome. In "Human Metabolism of Alcohol, Vol. 1: Pharmacokinetics, Medicolegal Aspects, and General Interest" (K. E. Crow and R. D. Batt, eds.), pp. 133–155. CRC Press, Boca Raton, Florida.

Webster, W. S., Walsh, D. A., Lipson, A. H., and McEwen, S. E. (1980). Teratogenesis after acute alcohol exposure in inbred and outbred mice. *Neurobehav. Toxicol. Teratol.* **2**, 227–234.

Webster, W. S., Walsh, D. A., McEwen, S. E., and Lipson, A. H. (1983). Some teratogenic properties of ethanol and acetaldehyde in C57BL/6J mice: Implications for the study of the fetal alcohol syndrome. *Teratology* **27**, 231–243.

Weinberg, J., Nelson, L. R., and Taylor, A. N. (1986). Hormonal effects of fetal alcohol exposure. In "Alcohol and Brain Development" (J. R. West, ed), pp. 310–342. Oxford University Press, New York.

West, J. R. (ed.) (1986). "Alcohol and Brain Development." Oxford University Press, New York.

West, J. R. (1987). Fetal alcohol-induced brain damage and the problem of determining temporal vulnerability: A review. *Alcohol Drug Res.* **7**, 423–441.

West, J. R., and Goodlett, C. R. (1990). Teratogenic effects of alcohol on brain development. *Ann. Med.* **22**, 319–325.

West, J. R., and Hamre, K. M. (1985). Effects of alcohol exposure during different periods of development: Changes in hippocampal mossy fibers. *Dev. Brain Res.* **17**, 280–284.

West, J. R., and Hodges-Savola, C. A. (1983). Permanent hippocampal mossy fiber hyperdevelopment following prenatal ethanol exposure. *Neurobehav. Toxicol. Teratol.* **5**, 139–150.

West, J. R., and Pierce, D. R. (1984). The effects of *in utero* ethanol exposure on hippocampal mossy fibers: An HRP study. *Dev. Brain Res.* **15**, 275–279.

West, J. R., Hamre, K. M., and Pierce, D. R. (1984). Delay in brain growth induced by alcohol in artificially reared rat pups. *Alcohol* **1**, 83–95.

West, J. R., Goodlett, C. R., Bonthius, D. J., and Pierce, D. R. (1989). Manipulating peak blood alcohol concentrations in neonatal rats: Review of an animal model for alcohol-related developmental effects. *Neurotoxicology* **10**, 347–366.

West, J. R., Goodlett, C. R., Bonthius, D. J., Hamre, K. M., and Marcussen, B. L. (1990). Cell population depletion associated with fetal alcohol brain damage: Mechanisms of BAC-dependent cell loss. *Alcohol. Clin. Exp. Res.* **14**, 813–818.

Wigal, T., and Amsel, A. (1990). Behavioral and neuroanatomical effects of prenatal, postnatal, or combined exposure to ethanol in weanling rats. *Behav. Neurosci.* **104**, 116–126.

— 5 —

Clinical Implications of Smoking: Determining Long-Term Teratogenicity

ès

Peter A. Fried

Department of Psychology
Carleton University
Ottawa, Ontario
Canada

INTRODUCTION

In the past two decades there has been an explosion in the number of investigations inquiring into the effects on development of prenatal exposure to a variety of substances. Of the various potentially harmful agents that pregnant women may ingest or otherwise absorb, cigarette smoking ranks very high among the most widely researched topics in teratology. A review of the literature suggests that a conservative figure for the number of babies who have been examined with respect to birth weight would be approximately 140,000 newborns studied, whose mothers smoked during pregnancy. The voluminous literature presumably reflects both the relative ease in collecting information with respect to smoking habits of the mothers-to-be (in contrast to gathering facts pertaining to usage of illegal drugs or substances that are not taken in a fairly regular, habitual manner) and the significant number of women who do smoke during pregnancy.

Maternal cigarette use has consistently been established as a reproductive toxin and a teratogenic agent. Perinatal exposure has been shown consistently, in a very large number of reports, to be related to a variety of negative pregnancy outcomes, particularly with respect to the course of pregnancy itself and the growth of the fetus and infant. The literature pertaining to neurobehavioral outcomes is not nearly so extensive and is not so unequivocal in its findings. The focus of this chapter is an examination of the potential long-term effects of perinatal exposure to maternal smoking. At the outset, it must be noted that the approach to be taken is one that emphasizes that effects of fetal exposure to cigarettes are likely to emerge in complex

interactions with life-style, environmental factors that make up the child's environment—the transactional framework of Sameroff and Chandler (1975).

THE OTTAWA PRENATAL PROSPECTIVE STUDY

As much of the data that will be described in this chapter is derived from mothers-
to-be, infants, and children taking part in the Ottawa Prenatal Prospective Study
(OPPS), the protocol and the limitations of this work will be described in some detail.
Additional procedural information can be found in an earlier report (Fried et al.,
1980).

In 1978, the OPPS was initiated to examine the effects of marijuana when used
during pregnancy. As part of the protocol, extensive background information was
gathered, including data pertaining to cigarette use. Table 1 outlines the general
testing procedures that have been and are being followed in this work, portions of
which are described in this chapter.

Data have been, and continue to be, collected in a prospective fashion from
approximately 700 women residing in the Ottawa, Canada region. Pregnant women
were informed of the study by their physicians, by notices in waiting rooms of obstetri-
cians, or by notices in reception rooms of prenatal clinics in the three major hospitals
in Ottawa. The information that was imparted at this point discussed, in general
terms, how life-style habits followed during pregnancy may influence the fetus. If she

TABLE 1   Protocol of the Ottawa Prenatal Prospective Study (OPPS)

Birth Date	Postnatal Questionnaire (after 1st and 5th birthday)	Tactile Form Recognition Test (48, 60, 72 months)
Brazelton Neonatal Assessment (Day 4)	Physical Anomaly Assessment (one examination between 18 months and 4 years)	Peabody Picture Vocabulary Test (48, 60, 72 months)
Prechtl Neurological Test (Day 9 and 30)	Reynell Expressive Language & Verbal Comprehensive Scale (18, 24, 36, 48 months)	Conners' Parent Rating Scale (48, 60, 72 months)
Neonatal Perception Inventory (Day 9 and 30)	Home Inventory (24 and 48 months)	Gordon Diagnostic System (vigilance and impulsivity (60 and 72 months)
Bayley Scales of Infant Development (6, 12, 18, 24 months)	McCarthy Scales or Children's Abilities (36, 48, 60, 72 months)	Neuropsychological Battery (60 and 72 months
Visual & Auditory Evoked Potentials (Once at 3–5 Years)	Pegboard fine motor coordination test (48, 60, 72 months)	Wide Range Achievement Test (60 and 72 months)

desired, the mother-to-be could contact our research facility at the psychology department of Carleton University either by telephone or by returning a prepaid postcard that was attached to an information pamphlet. At this time, the potential subject was given further details about the particular habits we were interested in—marijuana, alcohol, and cigarette use. It was emphasized that, for purposes of comparison, we also wished to recruit women who did not use any of these substances during pregnancy. After volunteering and signing an informed consent, the mother-to-be was interviewed once by a female during each of the trimesters remaining in her pregnancy, typically in the subject's home.

This volunteer method of recruiting subjects had both strengths and weaknesses. The self-selection procedure places a limit on the degree of generalization that can be made in terms of the epidemiological information collected, the possibility of selection bias being obvious. However, as reported elsewhere (Fried et al., 1980, 1984) on several key demographic variables (e.g., parity, age, family income), the volunteer sample participating in the OPPS is quite similar to nonparticipating women living in the Ottawa area giving birth in the hospitals involved in our study.

The recruitment procedure that was utilized has the advantage of increasing the likelihood of the reliability of self-report and to increase the probability of a long-term commitment to the study. Aside from subjects moving from the Ottawa region (about one third), a retention rate of over 95% has been maintained over the past decade.

In each of the interviews conducted during the pregnancy, information was collected on such variables as socioeconomic status, mother's health (both current and before pregnancy), the health history of the father, obstetrical history of previous pregnancies, a 24-hour dietary recall (including an assessment of caffeine intake), as well as past and present drug-use patterns, with particularly detailed information being gathered regarding marijuana, alcohol, and cigarettes. For determination of these three drugs, information was gathered for the year preceding pregnancy, as well as for each trimester of the pregnancy.

The average number of cigarettes smoked per day was recorded, and a nicotine score was derived by multiplying the number of cigarettes smoked daily by the nicotine content of the brand specified. During each pregnancy interview, women were also asked whether they were regularly exposed to a smoked-filled environment. Information was also obtained on the father's and current partner's smoking habits. Marijuana use was calculated in terms of the average number of joints smoked per week in the year before pregnancy and during each trimester of pregnancy. Alcohol consumption was determined using a quantitative-frequency formula and reported as average ounces of absolute alcohol per day.

For descriptive and some statistical analyses, the cigarette data were treated categorically. Smoking was categorized into three groups: nonsmoking, light, and heavy. The latter were women who smoked the equivalent of 16 mg nicotine/day or more. The minimum amount in this category approximates one package of cigarettes of average strength.

The women who smoked cigarettes during their pregnancy tended to differ from nonsmokers on a number of variables that had the potential of influencing the development of the offspring. Decreasing trends of age, formal education, maternal and paternal occupational status, and family income were among possible confounding factors noted.

The self-report procedure utilized to assess drug habits raises the fundamental issue of validity and reliability. Despite the obvious limitations of this method of determining drug use, at the time of the collection of data, no practical alternative was available. In the OPPS, a number of procedures were employed to enhance the likelihood of accurate data collection. A nonthreatening, nonjudgmental relationship in a comfortable environment (typically the home of the mother) was established between the subject and the female interviewer with the same interviewer following the mother-to-be throughout her entire pregnancy. A second procedure designed to enhance the accuracy of the responses to the orally posed questions centered on the number of times the same drug-related questions were asked. The interview took place once during each trimester, and during each of the interviews, the queries pertaining to drug use for each 3-month period of pregnancy that had passed and for the year before pregnancy were repeated, permitting test–retest reliability measurement.

PREVALENCE AND PATTERNS OF SMOKING

Data collected during the latter half of the past decade from a number of countries have reported relatively consistent findings with respect to smoking trends. In Canada, in the most recent national survey (Health and Welfare, 1988), the overall prevalence of smoking by individuals 15 years and older was found to be declining in all age groups. The figures indicated that 30% of the adult population smoked on a daily basis in the mid-1980s compared to 41% in 1970. Males accounted for the largest share of this reduction, with a diminution from 49 to 32% between those two periods. Although the numbers for the females also follow this downward trend, the rate of change was substantially less. The smoking rate among the overall female population was 32% in 1970 and had decreased to only 29% by the mid-1980s. Particularly germane to this chapter, among women of reproductive age (between 15 and 44 years) one third reported that they smoked on a regular basis. These figures are within one or two percentages of those noted in the United States (Williamson et al., 1989) and Sweden (Cnattingius, 1989).

Another statistic having bearing on smoking habits and pregnancy is that the proportion of heavy smokers (a package a day or more) has increased in the past decade. For example, in Canada, the relative increase in the number of heavy smokers was 31% for males and 57% for females. In Sweden, the proportion of heavy smokers has almost doubled. Many smokers appear to have switched to cigarettes with lower tar and nicotine levels, but smoke more of them. This has important implications since the relationship between the consequences of maternal smoking and the immediate and long-term effects on the offspring may be related to carbon monoxide, which increases directly with the number of cigarettes smoked.

The extent of cigarette use by women during pregnancy in nonghetto, urban regions has been reported to be between 22 and 28% (Fried et al., 1984, 1985; Stewart and Dunkley,1985; Streissguth et al., 1983; Williamson et al., 1989). Among the subjects in the OPPS, the percentage of heavy smokers diminished from 12% in the year before pregnancy to 7% in the first trimester, and to 5% by the third trimester (Fried et al., 1985). In the OPPS we have noted that 50% of the light smokers (less than a package a day) and 90% of the heavy smokers continued to smoke to some degree once pregnancy was confirmed. However, given the dose-dependent risk posed by cigarette use during pregnancy, even a reduction in smoking is of benefit to the fetus. As such, the success or failure of programs designed to encourage pregnant women to alter their smoking habits ought not to be evaluated solely in terms of the number of women who give up cigarettes entirely.

Interestingly, cigarette use that had declined during pregnancy, continued to decline during the first month after birth, with the nonsmoking rate being 80% and the heavy smoking rate being 4%. One year following the birth of the baby, the nonsmoking rate was 75%, whereas the heavy smoking rate was approximately equal to that noted in the third trimester. The reasons for the continued reduction in smoking in the first month after delivery may reflect, in part, a concern among women about the effects of smoking on breast feeding. In the OPPS sample 82% of the sample were breast feeding 1 month postpartum (Mackey and Fried, 1981). Parenthetically, cigarette smoking was associated with the type of feeding method employed. Nonsmokers were significantly more likely to breast feed than were heavy smokers. A similar finding was reported in an Irish sample by Hill (1988). In the OPPS sample, among the smokers who did breast feed, there was a dose-dependent tendency to wean earlier (Goodine and Fried, 1984).

That the decline in smoking associated with pregnancy lasts for a least a year is, clinically, very encouraging. It has obvious benefits for the mother and, in terms of secondhand smoke, for the child. Additionally, knowing that there is a fair probability that the reduction during pregnancy will persist beyond the birth of the baby may provide additional motivation for the smoker to reduce her habit before becoming pregnant.

COURSE OF PREGNANCY AND BIRTH WEIGHT

Maternal cigarette smoking has been linked to a number of adverse pregnancy outcomes. In well-controlled large studies, these include spontaneous abortion (e.g., Risch et al., 1988), still births and perinatal death (e.g., Cnattingius et al., 1988; Malloy et al., 1988), and sudden infant death syndrome (e.g., Hagland and Cnattingius, 1990).

Birth weight has been of particular interest as a measure of assessing the success of pregnancy because of the well-documented association between small-for-date babies and morbidity at birth or in the neonatal period. Maternal smoking has been consistently reported, in many countries and across many social strata, to be associated with a lowered birth weight. The effect is dose dependent and is not the result of a shortened gestation period. Compared to nonsmokers, light and heavy smokers have 54 and 130% increases in prevalence of newborns weighing less than 2500 g (the operational definition of low-birth-weight infant), respectively (Meyer et al., 1976).

In the OPPS sample we noted an interaction between the timing of cigarette use during pregnancy and its relative differential effects on growth parameters at birth (Fried and O'Connell, 1987). Although both first and third trimester smoking significantly influenced birth weight (after statistically adjusting for a host of possible confounding variables), the effect of third trimester use was more marked, with a decrease of 181 g associated with a pack-a-day use versus a decrease of 134 g with a similar use in the first trimester.

The importance of maternal weight gain during pregnancy in determining birth weight has been noted (Rush, 1976). In the OPPS this variable was found to be an important contributor to the variance in growth parameters at birth. However, even after controlling for the influence of maternal weight gain, a significant proportion of the variance in birth weight was explained by the smoking habits of the mothers. As the maternal nourishment in the OPPS sample did not differ between the heavy smokers and the rest of the sample, it was hypothesized that the lowered birth weight might reflect an impairment in the fetal ability to utilize nourishment in utero (Fried and O'Connell, 1987).

NEONATAL BEHAVIORAL CORRELATES OF PRENATAL EXPOSURE TO CIGARETTES

Although not totally consistent (e.g., Richardson et al., 1989), a number of studies have observed relationships between maternal smoking during pregnancy and altered behavior in the neonate. In the OPPS, prenatal cigarette exposure was associated with decreased auditory habituation in babies between 3 and 6 days of age (Fried and Makin, 1987). Habituation of various responses to repeated stimuli is thought to be an indication of nervous system integrity. The observation of altered auditory habituation is consistent with the observations of a number of workers who have described a relationship between the auditory senses and smoking during pregnancy. Saxton (1978) noted a decreased hearing response as well as a decreased rate of habituation in 4- to 6-day-old babies born to smokers, whereas Picone et al. (1982) observed decreased auditory responsiveness but faster auditory habituation in 2- and 3-day-old offspring of maternal smokers. Jacobson et al. (1984) also reported, at marginal statistical significance, poorer auditory responsiveness in the 3- and 4-day-old babies born to smokers.

In the OPPS work (Fried and Makin, 1987), maternal use of cigarettes was associated with increased tremors, and Picone et al. (1982) noted poorer autonomic regulation in the offspring exposed prenatally to constituents of cigarette smoke. Interestingly, in both the OPPS and in the work of Jacobson et al. (1984), lower irritability seemed to be associated with maternal smoking. Jacobson attributed the finding to a delayed onset of crying in his 3- and 4-day-old subjects. This hypothesis is consistent with the observations of Fried and Makin (1987) in the OPPS work. Infants were tested between 3 and 6 days of age, and when age was included in the regression equation, the decreased irritability did not remain significant.

A further observation with regard to the irritability dimension is that, in statistical terms, the regression equations revealed that nicotine scores appeared to suppress a portion of error variance in alcohol so that the relationship of alcohol with irritability was enhanced. This emphasizes the importance of the statistical control of cigarettes when examining alcohol effects.

In the OPPS the neurological status of 9- and 30-day-old infants was examined (Fried et al., 1987) using a slightly modified version of the Prechtl neurological examination (Prechtl, 1977). The examination comprised 140 items including tremor incidence, limb power and resistance, various reflexes, posture in various positions, eye movements, and spontaneous motor movements. The data were analyzed using discriminant function analysis, controlling for potentially confounding variables. For statistical reasons (Fried et al., 1987), the sample was dichotomized into smoking versus nonsmoking. At 9 days of age, the primary Prechtl variables distinguishing between the smoking and nonsmoking group were the increased incidence of fine tremor and some heightened motoric reflexes in the infants born to maternal smokers.

At 30 days of age, the Prechtl variables that assessed aspects of muscle tonus differentiated the two groups. Maternal smoking was associated with increased tonicity and heightened motoric reflexivity. The hypertonicity correlation was maintained after controlling for the birth weight of the baby, indicating that the relationship was not mediated by the infant's weight at birth. Increased startles and tremors were also associated with prenatal cigarette exposure, and the combination of these observations, coupled with the hypertonicity, was interpreted as suggestive of central nervous system excitation (Fried et al., 1987). The increased incidence of tremors at this age is consistent with the observations previously described in newborns from both the OPPS and from the laboratory of Picone et al. (1982).

A number of reports have linked maternal smoking and hyperactivity in young children (e.g., Denson et al., 1975; Nicols and Chen, 1981) and poor orientation and impaired attention in a vigilance task (Streissguth et al., 1984; Kristjansson and Fried, 1989). These will be discussed later in this chapter, but at this juncture, it appears appropriate to indicate the consistency between these long-term associations of maternal smoking and the notion of increased nervous system excitation already mentioned.

However, one must be extremely cautious in postulating that the same fundamental processes underlie the infant's neurological functioning and the behavior of the young child. Even if one makes the reasonable assumption that development is a continuous process with common substrates underlying behavioral performances (McCall et al., 1977), the impact of the child's environment must not be negated or neglected (Sameroff and Chandler, 1975). As described earlier in this chapter, we have noted that environmental factors may well be different in the home of the smoker as compared to the home of the nonsmoker.

INFANT BEHAVIORAL CORRELATES OF PRENATAL EXPOSURE
TO CIGARETTES

The Bayley Scales of Mental and Motor Development (Bayley, 1969) have typically been used to examine the relationship between maternal smoking and effects on the offspring. The Bayley test consists of three scales. The Mental Developmental Index (MDI) assesses sensory perceptual abilities, early acquisition of object constancy, memory, problem solving, and vocalization. The Psychomotor Developmental Index (PDI) assesses gross and fine motor movements. The Infant Behavior Record (IBR) evaluates the infant's attitudes, interests, and temperament.

Streissguth and her co-workers (1980), after adjusting for alcohol, caffeine, and gestational age, found no effects of maternal nicotine in 8-month-old babies. However, Gusella and Fried (1984), in a pilot study using a sample derived from the OPPS, observed a decrease in verbal comprehension (assessed by a cluster derived from the mental portion of the Bayley scales) after adjusting for confounding variables.

We have examined 217 children from the OPPS at 12 months of age using a multiple regression analysis (Fried and Watkinson, 1988). Cigarette use contributed significantly to the prediction of MDI scores. Further, when examining the inverse relationship between maternal smoking and lower MDI values, 39% of the heavy smokers had infants with MDI scores under 85 in contrast to 6% of the remaining subjects. No relationship was observed between the PDI and maternal smoking. Interestingly, there was a significant negative association between auditory-related items on the IBR and prenatal exposure to cigarette smoke, consistent with the neonatal observations described earlier, and the continuity supports the interpretation of a direct effect of prenatal maternal smoking rather than postnatal, maternal life-style habits, including postnatal secondhand smoke exposure.

Before leaving the data derived from infants approximately 1 year of age, an interpretative issue ought to be raised—one that becomes more and more an issue as the child being assessed gets older, and one that underlies much of teratological research in which the drugs in question have relatively subtle effects. Although statistical significance may be found, the amount of variance accounted for by the drugs is relatively small. In the OPPS work, the behavioral effects uniquely associated with maternal drug use (tobacco, alcohol, or marijuana) range from 1.5 to 5% after the variance due to other potentially confounding factors is partialled out. In Streissguth's work (1980), where alcohol significantly contributed to mental and motor scores in 8-month-old offspring, the amount of variance accounted for by that drug was approximately 1%.

However, whether the unique contribution of the drugs, in fact, reflects the "true" relationship between maternal drug use and the outcome observed in offspring is problematic. We have argued (Fried and Watkinson, 1988) that it is more likely that the real association of the drug with the behavioral outcomes in question may lie between the unique contribution of the drug (in a standard regression model) and its zero-order correlation (with no potential confounds considered). In the latter approach, variance attributable to drugs may be as high as 12% whereas, as stated before, the unique contribution is often in the region of 1 or 2%. The likely contribution or influence of the drug may well fall between these two figures.

At 2 years of age, children from the OPPS were examined (Fried and Watkinson, 1988) using the Bayley test and the Reynell Developmental Language Scale (1977). This latter test was used to measure language comprehension and expression. At this age, maternal smoking continued to have a significant relationship with auditory-related behaviors, as assessed by the IBR of the Bayley, but it no longer contributed a unique significant explanation of variance of the MDI or the two Reynell language scores.

The question arises as to why maternal smoking was no longer associated with cognitive functioning at 2 years of age. It may be that as the child gets older, postnatal factors play an increasing role in determining the offspring's performance, particularly when the outcome measures are relatively global. Several reports have noted an increased predictive power on mental test scores of variables that assess the child's home environment (e.g., Elardo et al., 1975; Gottfried and Gottfried, 1984). In the OPPS examination of 24-month-old infants (Fried and Watkinson, 1988), the assessment of the home environment and parental involvement with the baby was carried out using the HOME (Home Observation for Measurement of the Environment) scale (Caldwell and Bradley, 1979). The scores on the HOME were highly positively associated with the 24-month MDI and the Reynell tests and negatively correlated with maternal smoking. This combination resulted in the loss of unique explanatory power of the mother's prenatal smoking habits. It is not possible to identify the contribution of maternal smoking to this shared variance.

Although it may be that prenatal smoking has no effect on 24-month cognitive measures, this would appear unlikely. A more plausible interpretation is that the postnatal environment may act as an intervening variable between the cognitive outcomes of the 2-year-old and maternal smoking. The information collected via the HOME instrument when the child was 24 months of age, coupled with information collected earlier (Goodine and Fried, 1984; Mackey and Fried, 1981), suggests that the woman who smokes during pregnancy may have a life style that included less maternal/child involvement, which may have an impact on the child's cognitive development.

There is some evidence that the decreased involvement is seen very early. For example, as mentioned earlier, maternal smokers are less likely to breast feed. Importantly, the decision as to the method of infant feeding is typically made before the birth of the child (Mackey and Fried, 1981), and the causal sequence involves the mother's life style influencing her behavior toward the infant, rather than the baby's behavior causing a maternal reaction. However, this interpretation does not preclude the additional likelihood that the infant's behavior (as outlined in the preceding sections) affected by the in utero exposure to cigarettes will have an influence on the mother's behavior—the transactional relationship described by Sameroff and Chandler (1975).

CHILDHOOD BEHAVIORAL CORRELATES OF PRENATAL EXPOSURE TO CIGARETTES

The relationship between lowered cognitive skills in childhood and maternal smoking has been relatively consistent (recently reviewed by Rush and Callahan, 1989), with the majority of studies reporting the association either to be statistically significant or to approach significance. Recently, two studies focusing on prenatal alcohol exposure have reported being unable to find cognitive deficits in young children associated with prenatal cigarette exposure (Streissguth et al., 1989; Greene et al., 1991). In both of these longitudinal studies, environmental variables above and beyond the cigarette use may have masked the opportunity to detect maternal smoking effects, as the women who did smoke were subject to other risk factors (e.g., poorly educated, socioeconomically disadvantaged).

In the OPPS, 3-year-old children were examined (Fried and Watkinson, 1990) using the Reynell language test plus the McCarthy Scales of Children's Abilities (McCarthy, 1972). The latter test is based on six scales: verbal, perceptual, performance, quantitative, memory, and motor. A general cognitive score is derived from the first three scales. A stepwise discriminant function analysis and multivariate analyses of covariance were carried out with such potential confounding variables as other drug use, family income and education, nutrition, maternal age, parity, birth weight, and home environment being considered.

The results revealed a negative dose–response association between cognitive scores, particularly the verbal subscale of the McCarthy test and levels of maternal smoking (Table 2). Even though, as at 24 months, the HOME scores were highly related to both the cognitive measures and maternal smoking, unique discriminating power was observed. Although the verbal component of the McCarthy test was, statistically, the most significant, on each of the six subscales and on the expressive portion of the Reynell test, the children of the heavy smokers had means that were lower than did the children born to the light smokers, who, in turn did not perform so well as those born to nonsmokers.

At 4 years of age, children from the OPPS were examined once more (Fried and Watkinson, 1990) using the same battery as was used at 36 months, with the addition of three tests. One was the Tactile Form Recognition Task (Jarvis and Barth, 1984) in which the child was required to identify flat plastic shapes that were placed in the subject's hand out of sight. Recognition of the stimuli was indicated by the child pointing to the one of a set of four stimulus shapes that was identical to the one being felt. This was carried out with both the dominant and nondominant hand. Also administered was the Pegboard test (Knights and Norwood, 1980), which measures speed and accuracy in an eye–hand coordination task. The requirement of this test is for the child to place, as rapidly as possible, keyhole-shaped pegs into rows of holes in a board. Both speed and dropped pegs for each hand were recorded. The final additional test was the Peabody Picture Vocabulary Test (Dunn and Dunn, 1981), a widely used measure of receptive vocabulary.

The results were strikingly similar to those obtained when the children were a year

TABLE 2 Outcome Variables at 3 and 4 years of age[a]

Smoking group	3-year-olds			4-year-olds		
	None	Light	Heavy	None	Light	Heavy[b]
Number of subjects	61	51	21	66	47	17
McCarthy variables						
Verbal	62.7	58.1	54.1**	65.0	64.9	57.2**
Perceptual	60.0	58.2	54.3*	64.2	62.7	59.8
Quantitative	58.9	56.8	52.7*	62.2	60.7	57.3
Cognitive index	119.5	113.7	107.2**	125.7	124.4	115.3*
Memory	61.1	57.1	55.0*	60.1	59.8	54.2*
Motor	56.6	53.8	51.5	55.1	55.2	54.6
Reynell Language						
Expressive	0.53	0.20	−0.17*	0.72	0.49	−0.09*
Comprehension	1.16	1.18	0.82	1.35	1.30	1.01*
Peabody						
Vocabulary	—	—	—	119	114.2	109.2**

[a]The statistical tests used for this table were one way between-subjects ANOVAs. Light smokers are >0 to <0.16 mg nicotine/day; heavy smokers are >0.16 mg nicotine/day. Adapted from Fried and Watkinson (1990).
[b]*, $p < 0.05$; **, $p < 0.01$.

younger. Employing the same statistical procedures as described above for the 36-month-old subjects, the discriminant function differentiated among the smoking groups in a dose-related fashion (Table 2). The primary variables responsible for the discrimination were three language-associated tests: the Peabody test, the McCarthy verbal subscale, and the Reynell expressive component, all negatively related to smoking. As with the data derived from the 3-year-old children, on each of the six McCarthy subscales and on the two Reynell scales, a dose–response relationship was noted. No significant association was found on the motor and tactile tests.

As mentioned in the section of this chapter describing neonatal findings, researchers (e.g., Denson et al., 1975; Nicols and Chen, 1981) have noted an association between maternal smoking and children diagnosed as hyperactive or attention-deficit disordered. Recently, in order to evaluate and quantify sustained attention in children, computer-controlled continuous performance tasks have been utilized. Children who have been diagnosed as hyperactive or as having an attention deficit perform more poorly on these vigilance tasks than do normal children. The impairment can manifest itself as fewer correct hits (more omission errors) and/or more false alarms (more commission errors).

Using a computer-controlled performance task, Streissguth and her co-workers (1984) required 4-year-old children to press a button whenever a particular stimulus appeared. The researchers note that, when confounding variables were statistically controlled, maternal smoking was significantly related to a poorer performance on the vigilance task. With more or less the same sample at 7 years of age and using a computerized continuous performance task, prenatal cigarette exposure was not found to be related to attentional behavior (Streissguth et al., 1986).

Recently the attentional behavior of 4- to 7-year-old children from the OPPS was examined using both visual and auditory continuous performance tasks (Kristjansson and Fried, 1989). In addition to the vigilance tasks, each child's activity level was monitored using a sensor placed under the cushion of the subject's chair. Linear and logistic regression analyses were carried out on errors of omission, commission and activity. Potentially confounding variables that were considered were similar to those described in the earlier OPPS studies plus postnatal cigarette exposure, computer experience, the child's behavior (as coded by the experimenter), and the child's age.

Maternal cigarette use during pregnancy was more strongly related to the objective measure of activity level than any other dependent variable and accounted for 8% of the variance in that dimension. As discussed earlier, the typical explanatory power of prenatal drug exposure on long-term effects in offspring is in the area of 1 to 5%, and thus the 8% figure is quite robust indeed.

After statistically controlling for potential confounds, the analyses of the performance on the vigilance tasks suggested a significant association between maternal cigarette use and errors of auditory commission and a marginally significant relationship between visual errors of commission and prenatal cigarette exposure. Omission errors were not significantly related to maternal cigarette use after control variables were included in the regression equation.

Whereas omission errors are thought to reflect lapses in attention, commission errors are thought to be related to both poor attention and impulsivity (e.g., Douglas and Peters, 1979). The subjective observations recorded by an observer blind to the maternal drug history on the children's behavior during testing are consistent with this interpretation. The children who made the most commission errors tended to be quick to respond to the stimuli and appeared to have difficulty in inhibiting themselves from responding to noncritical stimuli. Commission errors were also made by children whose attention wandered and who then responded after missing the critical stimulus.

Commission errors may be more reflective of impulsive responding and increased overall activity rather than of inattentiveness. This conclusion would be consistent with the two observations in the work by Kristjansson and Fried (1989), that a relationship existed both between maternal smoking and activity and between maternal smoking and errors of commission (particularly in the auditory modality).

INVOLUNTARY SMOKING DURING PREGNANCY

As mentioned earlier in this chapter, the proportion of women who smoke during pregnancy is reported to be between 22 and 28%. Estimates indicate that between one third and one half of nonsmoking pregnant women are involuntary or passive smokers (Karakostov, 1985; Martin and Bracken, 1986; Smith et al., 1982). Involuntary smoking occurs when nonsmokers are exposed to the sidestream smoke of smokers in enclosed environments. Sidestream smoke and the smoke inhaled by the smoker (mainstream smoke) contain a number of similar components, including nicotine and carbon monoxide, the two constituents identified as the most likely to be the causative agents in producing effects in the offspring of maternal smokers (Fried and Oxorn 1980). Further, as summarized in several reports (Department of Health and Human Services, 1986; National Research Council, 1986; Fielding and Phenow, 1988) the carbon monoxide and nicotine levels in sidestream smoke are of a higher concentration than that found in mainstream smoke.

Published reports describing the consequences on the offspring of involuntary smoking by nonsmoking mothers-to-be are quite limited, with the majority considering paternal smoking as the only source of environmental smoke. Approximately 75% of these studies have noted a relationship between this source of secondhand smoke and reduced fetal growth and higher perinatal mortality (e.g., Karakostov, 1985; Rubin et al., 1986; Koo et al., 1988). Martin and Bracken (1986), examining passive smoking exposure arising from both the home and workplace, reported that involuntary smoking doubled the risk of delivering a low-birth-weight baby.

In a study in which children were followed to the age of 14 years (Rantakallio, 1983), paternal smoking showed about as strong a negative association with mental and physical development as did maternal smoking during pregnancy, although it is not clear whether or not this was after controlling for the maternal smoking.

Using children from the OPPS between the ages of 6 to 9, we have attempted to compare the effects of maternal active and passive smoking using mental development, language development, and behavior as outcome measures (Makin et al., 1991). A three-group design was employed in this study, utilizing women who smoked during pregnancy, women who were passively exposed (either at home and/or "regularly exposed to a smoke-filled environment"), and women who were neither active or passive smokers during their pregnancy.

For almost half of the women in the passive smoking condition, the involuntary exposure occurred outside the home; i.e., paternal smoking was not the source. This is an important fact as, by failing to consider nonpaternal sources—as indeed is the case in all but Martin and Bracken's (1986) work—an inadequate index of involuntary exposure is likely to be obtained. Further, problems are created with respect to the potential comparison groups who may well be exposed during pregnancy to sources of smoke that are unknown to the researcher.

In addition to the collection of background variables described earlier in this chapter, at the time of testing of the child, the mother completed a behavioral check list (Conners, 1970) and a short questionnaire pertaining to the child's school history, current smoking habits of the family, and a yearly account from birth of the child's exposure to secondhand smoke. Additionally, the examiner completed a rating scale that considered such variables as attention to the task, amount of motor activity, manual dexterity, and degree of cooperation.

Using a stepwise discriminant analytical approach, the results of this study indicated that the consequences of passive smoking were similar, in some respects, to active maternal smoking, but smaller in magnitude. The measures of oral language and speech articulation followed this pattern, with both smoking groups performing more poorly than the nonsmoking group in all four of the language measures, with the children of the passive smokers demonstrating a smaller deficit than the children of active smokers on three of the four tests.

Children in both smoking groups were rated on all scales in the Conners Parent Questionnaire as showing more behavioral problems than the nonsmoking group. These scales included the areas of conduct problems, learning problems, psychosomatic symptoms, impulsivity, anxiety, and hyperactivity. On five of the six scales, the active group was rated, by the mothers, as having the most behavioral problems. The exception was the anxiety subscale, on which the passive group was rated slightly higher than the children in the active group.

Compared to children born to nonsmokers, twice as many children in the active smoking group were rated by their mothers as having problems at school. The perception of the mothers is consistent with several reports that have noted an association between poorer adjustment at school and the mother's active smoking during pregnancy (Denson et al., 1975; Dunn et al., 1977) and is also congruent with the increased impulsivity and activity described earlier in this chapter (Kristjansson and Fried, 1989) associated with maternal smoking.

To examine cognitive functioning the Wechsler Intelligence Scale for Children–Revised (WISC-R) (Wechsler, 1974) was used employing the three factor scores developed by Kaufman (Kaufman, 1979), i.e., verbal comprehension, perceptual organization, and freedom from distractibility. This latter factor score has been found to be sensitive to attentional and information-processing problems in children. The Wide Range Achievement Test (Jastak and Wilkinson, 1984), with its subtests for reading, spelling, and arithmetic, was also administered. The results were quite consistent with the scores on the verbal comprehension and freedom from distractibility factors of the WISC-R and the scores on the achievement tests, being lower for the active-smoking group, intermediate for the passive group, and higher for the nonsmoking group.

Overall, this work suggests that maternal passive exposure to cigarettes is associated with long-term negative effects that are similar to but milder in degree than the long-term effects of maternal active smoking. This linear trend was most evident in the specific areas of speech and language development, intellectual ability, and behavioral problems.

The results also highlight an interpretative aspect concerning the findings of children born to women who actively smoked during their pregnancy. In the work by Makin et al. (1991), 97% of children born to women who had smoked during pregnancy continued to be exposed, for various lengths of time, to secondhand smoke after birth. Thus, the marked effects found in the offspring of active smokers may be the result of the combination of in utero exposure and postnatal secondary smoke. The role of the passive exposure in these children remains to be determined and will require studies with large numbers of subjects in order to evaluate this variable.

This work on passive exposure by Makin et al. (1991) represents a first attempt at delineating the long-term cognitive and behavioral consequences of being an involuntary smoker during pregnancy. Clearly, the quantification of secondhand smoke exposure, although somewhat more comprehensive than that reported previously in the literature, still leaves much room for improvement. In spite of this, however, the results, combined with the number of pregnant women exposed to the smoke of others, emphasize the importance of and the need for further research in this area.

SUMMARY AND CONCLUSIONS

The work that has been described in this chapter—both that of the OPPS and that of other studies—is an example of the relatively new field of behavioral teratology (Riley and Vorhees, 1987). Animal models have been used extensively in this area in order to experimentally control such variables as genetic background, timing and type of drug administered, and the environmental milieu during pregnancy and during the rearing of the offspring.

In work with humans the enormous number of uncontrollable events that occur between the putative teratogenic exposure and the identification of the behavioral deficit make the task of establishing a cause-and-effect relationship very difficult. Many of the complex variables (e.g., social, environmental, demographic, psychological) that influence human behavior are difficult to quantify or even identify, and those that we do know about can often not be controlled in an experimental sense. In the absence of being able to manipulate such variables, factors known to influence the outcomes of interest must, if possible, be measured and, in the statistical sense, adjusted for as much as is feasible.

Despite the impediments to ideal experimental control and the consequent difficulty in determining cause-and-effect relationships, research in this area is essential and can be informative. However, the interpretation of the findings must be made in light of the caveats associated with the research designs employed.

In assessing the possible role of smoking during pregnancy, it is important to recognize that the decrements in performance on the various outcome measures are relatively small, particularly in relationship to other predictors of child behavior. For example, as mentioned earlier, the drug variable typically accounted for less than 5% of the variance associated with cognitive outcome, whereas other life-style habits accounted for up to 35% of the cognitive outcome variability (Fried and Watkinson, 1988).

In the work of the OPPS described in this chapter, in addition to the attention paid to the potential confounding factors such as other drug use, maternal and paternal health, nutrition, and demographic variables, the prospective longitudinal design is a vital factor in contributing to the understanding of the long-term consequences of in utero exposure to cigarettes. One of the principal reasons for carrying out this difficult design is to determine whether there is a continuity in the observations made at different ages in the neonates, infants, and young children in the different drug groups. In attempting to establish whether there are effects of prenatal exposure to cigarettes beyond the neonatal period, the repeated measurements over time add a degree of credence to the interpretation of a cause-and-effect role.

If, on related behaviors, consistencies are observed from the earliest measurements until childhood, the link between prenatal cause(s) and the recurring outcome measure can be made with more confidence than if only a single measure were taken. Obviously, however, there are many interpretative problems with this approach. As often alluded to earlier in this chapter, there is a reciprocal relationship between the offspring's behavior and the parents' interaction with the child. This transactional state of affairs (Sameroff and Chandler, 1975) may well serve to exacerbate certain in utero effects. In terms of the actual world in which the child is raised, the effects noted in the offspring are not only owing solely to fetal cigarette exposure but also to the lifestyle and parent–child interaction, which are integral parts of homes in which the mother smokes. Thus, as described earlier in this chapter, looking for the statistically unique contribution of maternal cigarette smoking during pregnancy after controlling for so-called confounding factors may well obscure the reality of the total drug effect.

The results of the OPPS suggest considerable continuity in aspects of behavior that appear associated with prenatal exposure to cigarettes. As outlined in this chapter, at less than 1 week of age, altered auditory responsiveness was noted (Fried and Makin, 1987) in the babies born to mothers who smoked during pregnancy. At 1 year of age, maternal cigarette smoking was uniquely significantly related to lowered cognitive scores and auditory measures (Fried and Watkinson, 1988). At 24 months, cigarette use retained a significant association with some auditory-related behaviors, although its explanation of variance of cognitive scores was attenuated by the strong relationship between postnatal environmental factors and prenatal maternal smoking (Fried and Watkinson, 1988). In the 36-month-old children, a negative dose–response association was observed between cognitive scores, particularly the verbal subscale of the McCarthy test, and levels of smoking (Fried and Watkinson, 1990). Statistically unique discriminating power was retained after controlling for postnatal environmental factors, even though the latter were highly related to both cognitive measures and maternal smoking. In a similar fashion, at 48 months, after controlling for the home environment, the McCarthy variables, particularly the verbal components, the Peabody, and the Reynell expressive component, were all negatively associated with maternal smoking (Fried and Watkinson, 1990).

In the study on attention and impulsiveness in 4- to 7-year-old children (Kristjansson and Fried, 1989), deficits in the auditory mode seemed particularly associated with maternal smoking during pregnancy. The examination of the long-term effects of maternal passive smoking and a comparison of its effects to active smoking during pregnancy (Makin et al., 1991) also found that speech and language were among the most vulnerable to maternal secondhand smoke exposure.

The continuity over approximately 7 years of the relationship between auditory/language outcome variables and maternal smoking suggests that these aspects of behavior are mediated, at least to some degree, by the direct consequences of in utero exposure to constituents of cigarettes. However, as emphasized throughout the chapter, this does not preclude the possible effects of postnatal life-style habits and/or exposure to secondhand smoke that may also influence directly or indirectly the child's performance on the language-related variables.

The overall poorer cognitive performance of the children born to mothers who had smoked heavily during pregnancy found at all the ages tested and the intermediate effects noted in the offspring of passive smokers are consistent with an interpretation that at least a partial etiology of the longitudinal observations is the result of maternal smoking habits during pregnancy.

Although a causal relationship between maternal smoking during pregnancy and adverse behavioral outcomes cannot be posited with certainty, the evidence, as discussed in this chapter, certainly makes such a hypothesis tenable. The observations are consistent enough to warrant advising prospective parents that, at this time, the state of knowledge implicates prenatal exposure to cigarettes as limiting the child's potential for achieving optimal cognitive development.

ACKNOWLEDGMENTS

Fried's research described in this chapter was supported by grants from the National Research Council of Canada, Social Science Research Fund of Carleton University, and the Canadian Department of Health and Welfare, and is currently being supported by a National Institute on Drug Abuse grant.

Over the years the cooperation of the families participating in the OPPS has been wonderful, and I express my sincere appreciation. I thank C. O'Connell and R. Gray for editorial assistance in the preparation of the chapter.

REFERENCES

Bayley, N. (1969). "Bayley Scales of Infant Development." Psychological Corporation, New York.

Caldwell, B. M., and Bradley, R. H. (1979). "Home Observation of Measurement of the Environment." University of Arkansas, Little Rock, Arkansas.

Cnattingius, S. (1989). Smoking habits in early pregnancy. *Addict. Behav.* **14,** 453–457.

Cnattingius, S., Haglund, B., and Meirik, O. (1988). Cigarette smoking as risk factor for late fetal and early neonatal death. *Br. Med. J.* **297,** 258–261.

Conners, C. K. (1970). Parent questionnaire in Symptom patterns in hyperkinetic, neurotic, and normal children. *Child Dev.* **41,** 667–682.

Denson, R., Nanson, J. L., and McWatters, M. A. (1975). Hyperkinesis and maternal smoking. *Can. Psychiatr. Assoc. J.* **20,** 183–187.

Department of Health and Human Services. (1986). "The Health Consequences of Involuntary Smoking: A Report of the Surgeon General." Government Printing Office, Washington, D.C. (Publication no. DHHS (CDC) 87-83948).

Douglas, V. I., and Peters, K. G. (1979). Towards a clearer definition of the attentional deficit of hyperactive children. *In* "Attention and Cognitive Development" (G. A. Hale and M. Lewis, eds.), pp. 173–247. Plenum Press, New York.

Dunn, H. G., McBurney, A. K., Ingram, S., and Hunter, C. M. (1977). Maternal cigarette smoking during pregnancy and the child's subsequent development: II Neurological and intellectual maturation to the age of 6.5 years. *Can. J. Pub. Health* **68,** 43–50.

Dunn, L. M., and Dunn L. M. (1981). "Peabody Picture Vocabulary Test—Revised." American Guidance Service, Circle Pines, Minnesota.

Elardo, R., Bradley, R., and Caldwell, B. M. (1975). The relation of infants' home environments to mental test performance from six to thirty-six months: A longitudinal analysis. *Child Dev.* **46**, 71–76.

Fielding, J. E., and Phenow, K. J. (1988). Health effects of involuntary smoking. *N. Engl. J. Med.* **319**, 1452–1460.

Fried, P. A., and Makin, J. E. (1987). Neonatal behavioral correlates of prenatal exposure to marihuana, cigarettes, and alcohol in a low-risk population. *Neurotoxicol. Teraol.* **9**, 1–7.

Fried, P. A., and O'Connell, C. M. (1987). A comparison of the effects of prenatal exposure to tobacco, alcohol, cannabis, and caffeine on birth size and subsequent growth. *Neurotoxicol. Teratol.* **9**, 79–85.

Fried, P. A., and Oxorn, H. (1980). "Smoking for Two. Cigarettes and Pregnancy." Free Press, New York.

Fried, P. A., and Watkinson, B. (1988). 12- and 24-month neurobehavioral follow-up of children prenatally exposed to marihuana, cigarettes, and alcohol. *Neurotoxicol. Teratol.* **10**, 305–313.

Fried, P. A., and Watkinson, B. (1990). 36- and 48-month neurobehavioral follow-up of children prenatally exposed to marijuana, cigarettes, and alcohol. *J. Dev. Behav. Pediatr.* **11**, 49–58.

Fried, P. A., Watkinson, B., Grant, A., and Knights, R. M. (1980). Changing patterns of soft drug use prior to and during pregnancy: A prospective study. *Drug Alcohol Depend.* **6**, 323–343.

Fried, P. A., Innes, K. E., and Barnes, M. V. (1984). Soft drug use prior to and during pregnancy: A comparison of samples over a four-year period. *Drug Alcohol Depend.* **13**, 161–176.

Fried, P. A., Barnes, M. V., and Drake, E. R. (1985). Soft drug use after pregnancy compared to use before and during pregnancy. *Am. J. Obstet. Gynecol.* **151**, 787–792.

Fried, P. A., Watkinson, B., Dulberg, C. S., and Dillon, R. F. (1987). Neonatal neurological status in a low-risk population following prenatal exposure to cigarettes, marihuana, and alcohol. *J. Dev. Behav. Pediatr.* **8**, 318–326.

Goodine, L. A., and Fried, P. A. (1984). Infant feeding practices: Pre- and postnatal factors affecting choice of method and the duration of breastfeeding. *Can. J. Public Health* **75**, 439–444.

Gottfried, A. W., and Gottfried, A. E. (1984). Home environment and cognitive development in young children in middle-socioeconomic-status families. *In* "Home Environmental and Early Cognitive Families" (A. W. Gottfried, ed.), Academic Press, Orlando, Florida.

Greene, T., Ernhart, C. B., Ager, J., Sokol, R., Martier, S., and Boyd, T. (1991). Prenatal alcohol exposure and cognitive development in the preschool years. *Neurotoxicol. Teratol.* **13**, 57–68.

Gusella, J. L., and Fried, P. A. (1984). Effects of maternal social drinking and smoking on offspring at 13 months. *Neurobehav. Toxicol. Teratol.* **6**, 13–17.

Haglund, B., and Cnattingius, S. (1990). Cigarette smoking as a risk factor for sudden infant death syndrome: A population-based study. *Am. J. Pub. Health* **80**, 29–32.

Health and Welfare (1988). "Canada's Health Promotion Survey: Technical Report" Statistics Canada, Ottawa, Ontario, Canada.

Hill, A. E. (1988). Considerations for smoking advice in pregnancy. *Ulster Med. J.* **57**, 22–27.

Jacobson, S. W., Fein, G. G., Jacobson, J. L., Schwartz, P. M., and Dowler, J. K. (1984). Neonatal correlates of prenatal exposure to smoking, caffeine, and alcohol. *Inf. Behav. Dev.* **7**, 253–265.

Jarvis, P. E., and Barth, J. T. (1984). "Halstead–Reitan Test Battery: An Interpretive Guide." Psychological Assessment Resources, Odessa, Florida.

Jastak, S., and Wilkinson, G. S. (1984). "The Wide-Range Achievement Test—Revised Administration Manual." Jastak Associates, Wilmington, Delaware.

Karakostov, P. (1985). Passive smoking among pregnant women and its effects on the weight and height of the newborn. Abstract translated *in* "Smoking Bulletin Abstr. No. 31644." Office of Smoking and Health, Rockville, Maryland.

Kaufman, A. S. (1979). "Intelligence Testing with the WISC-R." Wiley, New York.

Knights, R. M., and Norwood, J. A. (1980). "A Neuropsychological Test Battery for Children: Examiners Manual." R. M. Knights, Psychological Consultants, Ottawa, Ontario, Canada.

Koo, L. C., Ho, J. H. C., and Rylander, R. (1988). Life history correlates of environmental tobacco smoke: A study on nonsmoking Hong Kong Chinese wives. *Soc. Sci. Med.* **26**, 751–760.

Kristjansson, B., and Fried, P. A. (1989). Maternal smoking during pregnancy affects children's vigilance performance. *Drug Alcohol Depend.* **24**, 11–19.

Mackey, S., and Fried, P. A. (1981). Infant breast- and bottle-feeding practices: Some related factors and attitude. *Can. J. Pub. Health* **72,** 312–318.

Makin, J., Fried, P. A., and Watkinson, B. (1991). A comparison of active and passive smoking during pregnancy: Long-term effects. *Neurotoxicol. Teratol.* **13,** 5–12.

Malloy, M. H., Kleinman, J. C., Land, G. H., and Schramm, W. F. (1988). The association of maternal smoking with age and cause of infant death. *Am. J. Epidemiol.* **128,** 46–55.

Martin, T. R., and Bracken, M. B. (1986). Association of low birth weight with passive smoke exposure in pregnancy. *Am. J. Epidemiol.* **124,** 633–642.

McCall, R. B., Eichorn, D. H., and Hogarty, P. S. (1977). Transition in early mental development. *Monogr. Soc. Res. Child. Dev.* **42,** 1–108.

McCarthy, P. (1972). "McCarthy Scales of Children's Abilities." The Psychological Corporation, New York.

Meyer, M. B., Jonas, B. S., and Tonascia, J. A. (1976). Perinatal events associated with maternal smoking during pregnancy. *Am. J. Epidemiol.* **103,** 464–476.

National Research Council, Committee on Passive Smoking. (1986). "Environmental Tobacco Smoke: Measuring Exposure and Assessing Health Effects." National Academy Press, Washington, D.C.

Nichols, P. L., and Chen, T. C. (1981). "Minimal Brain Dysfunction: A Prospective Study." Erlbaum, Hillsdale, New Jersey.

Picone, T. A., Allen, L. H., Olsen, P. N., and Ferris, M. E. (1982). Pregnancy outcome in North American women. II. Effects of diet, cigarette smoking, stress, and weight gain on placentas, and on neonatal physical and behavioral charcteristics. *Am. J. Clin. Nutr.* **36,** 1214–1224.

Prechtl, H. F. R. (1977). "The Neurological Examination of the Full-Term Newborn Infant." 2nd Ed., Clinics in Developmental Medicine, no. 63. Lippincott, Philadelphia, Pennsylvania.

Rantakallio, P. (1983). A follow-up study up to the age of 14 of children whose mothers smoked during pregnancy. *Acta Paediatr. Scand.* **72,** 747–753.

Reynell, J. (1977). "Reynell Developmental Language Scales—Revised." NFER Publishing, Windsor, England.

Richardson, G., Day, N. L., and Taylor, P. M. (1989). The effect of prenatal alcohol, marijuana, and tobacco exposure on neonatal behavior. *Inf. Behav. Dev.* **12,** 199–209.

Riley, E. P., and Vorhees, C. V. (1987). "Handbook of Behavioral Teratology." Plenum Press, New York.

Risch, H. A., Weiss, N. S., Clarke, E. A., and Miller, A. B. (1988). Risk factors for spontaneous abortion and its recurrence. *Am. J. Epidemiol.* **128,** 420–430.

Rubin, D. H., Craig, G. F., Gavin, K., and Sumner, D. (1986). Prospective survey of use of therapeutic drugs, alcohol, and cigarettes during pregnancy. *Br. Med. J.* **292,** 81–83.

Rush, D. (1976). Cigarette smoking during pregnancy: The relationship with depressed weight gain and birthweight. An updated report. In "Birth Defects: Risks and Consequences" (S. Kelly, E. B. Hook, D. T. Janerich, and I. H. Porter, eds.), pp. 161–172. Academic Press, New York.

Rush, D., and Callahan, K. R. (1989). Exposure to passive cigarette smoking and child development. A critical review. Prenatal abuse of licit and illicit drugs. *Ann. N.Y. Acad. Sci.* **562,** 74–100.

Sameroff, A. F., and Chandler, M. J. (1975). Reproductive risk and the continuum of caretaking causality. In "Review of Child Development Research" (F. Horowitz, E. M. Hetherington, S. Scarr-Salapatek, and G. Siegel, eds.), Vol. 4, pp. 187–243. University of Chicago Press, Chicago, Illinois.

Saxton, D. (1978). The behavior of infants whose mothers smoke in pregnancy. *Early Hum Dev.* **2/4,** 363–369.

Smith, N., Austen, J., and Rolles, C. J. (1982). Tertiary smoking by the fetus. *Lancet* **1,** 8283.

Stewart, P. J., and Dunkley, G. C. (1985). Smoking and health care patterns among pregnant women. *Can. Med. Assoc. J.* **133,** 989–994.

Streissguth, A. P., Barr, H. M., Martin, D. C., and Herman, C. S. (1980). Effects of maternal alcohol, nicotine, and caffeine use during pregnancy on infant mental and motor development at eight months. *Alcohol. Clin. Exp. Res.* **4,** 152–164.

Streissguth, A. P., Darby, B. L., Barr, H. M., Smith, J. R., and Martin, D. C. (1983). Comparison of drinking and smoking patterns during pregnancy over a six-year interval. *Am. J. Obstet. Gynecol.* **145,** 716–724.

Streissguth, A. P., Martin, D. C., Barr, H. M., Sandman, B. M., Kirchner, G. L., and Darby, B. L. (1984).

Intraterine alcohol and nicotine exposure: Attention and reaction time in 4-year-old children. *Dev. Psychol.* **20,** 533–541.

Streissguth, A. P., Barr, H. M., Sampson, P. D., Parrish-Johnson, J. C., Kirchner, G. L., and Martin, D. C. (1986). Attention, distraction, and reaction time at age 7 years and prenatal alcohol exposure. *Neurobehav. Toxicol. Teratol.* **8,** 717–725.

Streissguth, A. P., Barr, H. M., Sampson, P. D., Darby, B. L., and Martin, D. C. (1989). IQ at age 4 in relation to maternal alcohol use and smoking during pregnancy. *Dev. Psychol.* **25,** 3–11.

Wechsler, D. (1974). "Manual for the Wechsler Intelligence Scale for Children–Revised." Psychological Corporation, New York.

Williamson, D. F., Serdula, M. K., Kendrick, J. S., and Binkin, N. J. (1989). Comparing the prevalence of smoking in pregnant and nonpregnant women, 1985 to 1986. *J.A.M.A.* **261,** 70–74.

— 6 —

Prenatal Exposure to Nicotine: What Can We Learn from Animal Models?

Theodore A. Slotkin

Department of Pharmacology
Duke University Medical Center
Durham, North Carolina

INTRODUCTION

It has become generally accepted that cigarette smoking during pregnancy results in intrauterine growth retardation and a resultant shortfall of birth weight (Butler and Goldstein, 1973; Naeye, 1978; Eriksson et al., 1979; Meyer and Carr, 1987). Maternal tobacco use has also been implicated as the leading contributory factor in sudden infant death syndrome (Haglund and Cnattingius, 1990). Because of the profound effects of nicotine on the nervous system, a number of investigations have been undertaken to define potentially adverse behavioral effects on offspring of smoking women, and evidence for neonatal hyperactivity (Longo, 1977) and later learning deficits (Butler and Goldstein, 1973; Longo, 1977; Eriksson et al., 1979) has been presented. Be that as it may, it has been difficult to attribute these alterations to nicotine exposure *per se* because of the presence of substantial hypoxic/ischemic insult during cigarette smoking. Indeed, hypoxia/ischemia produces fetal effects quite similar to those reported for smoking (Cole et al., 1972; Dow et al., 1975), and these variables thus probably contribute to the net deleterious effect.

In light of the large population at risk, it has been critical to examine animal models of nicotine exposure with regard to identification of the specific factors that elicit growth retardation and nervous system alterations, in particular the roles of acute hypoxia/ischemia. Until recently, nearly all investigations have utilized acute injections of nicotine in fairly high doses; in these cases, growth shortfalls and behavioral alterations have been obtained that approximate those in humans whose mothers smoked cigarettes. Intrauterine growth retardation and postnatal behavioral hyperactivity have been reported in offspring of pregnant rats given 3 mg/kg twice daily throughout the gestational and nursing periods (Sonawane, 1982; Nasrat et al., 1986;

Slotkin et al., 1986b; Slotkin et al., 1987a), and learning deficits could also be readily identified (Martin and Becker, 1971); this exposure level, after correction for metabolic differences between human and rat, is often presented as representative of heavy cigarette smoking (Martin and Becker, 1971; Sonawane, 1982; Nasrat et al., 1986), but we have been able to show, as described in this chapter, that the underlying biochemical and neuronal mechanisms most closely resemble those of nonspecific damage, such as that associated with heavy metal intoxication or hypoxia (Slotkin et al., 1986a; Slotkin et al., 1986b; Slotkin and Bartolome, 1987; Slotkin et al., 1987a; Slotkin et al., 1987d; Navarro et al., 1988; Seidler and Slotkin, 1990). Indeed, behavioral similarities between the effects of maternally injected nicotine and hypoxia alone are marked (Martin and Becker, 1970; Martin and Becker, 1971). This does not mean that these effects are unimportant: such treatments clearly result in increased neuronal death (Krous et al., 1981) as well as alterations in synapse formation (Jonsson and Hallman, 1980; Slotkin et al., 1986a; Slotkin et al., 1987a) and could contribute significantly to the perinatal effects of maternal smoking.

The evidence thus seems convincing that cigarette smoking in humans (which represents the combined effects of acute nicotine with hypoxia/ischemia) does influence nervous system development adversely; similarly animal models that duplicate these circumstances (acute nicotine injections in doses that cause hypoxia/ischemia) may produce the same types of effects. Nevertheless, because of these confounding factors in the nicotine-injection model, the pursuit of the actual mechanism(s) by which smoking alters perinatal nervous system development has remained elusive: stated more explicitly, until the last 5 years, there has been no clear-cut demonstration that nicotine itself, acting on the fetus, is responsible for any of these effects. This question becomes more relevant in view of modern alternatives to smoking, including smokeless tobacco, nicotine skin patches, or nicotine-containing chewing gum. It has thus proven critical to develop animal models in which nicotine exposure occurs in the absence of hypoxia/ischemia and other variables relevant to cigarette smoking. These potential confounds are illustrated in Fig. 1. Whereas it is of key interest to identify neurobehavioral effects attributable to nicotine acting directly on the fetal brain, nicotine also can influence fetal growth and general development, which can contribute to growth retardation, which itself might be responsible for behavioral abnormalities. In parallel with this problem, actions of smoking in the maternal–fetal unit are likely to be of importance: an acute bolus of nicotine produces ischemia, attendant placental blood-flow restriction, and fetal hypoxia. In addition, smoke itself produces hypoxia due to carbon monoxide, and the anorexic effects of nicotine contribute to maternal nutritional impairment. Again, these factors can all contribute to growth retardation and neurobehavioral deficits.

An appropriate animal model of fetal nicotine exposure would thus have to separate the contributing variables of actions in the maternal–fetal unit and growth retardation from effects attributable to nicotine acting in fetal brain, presumably by administering nicotine itself as opposed to exposing gravid animals to cigarette smoke. Initially, attempts were made to include nicotine in the drinking water of pregnant rats

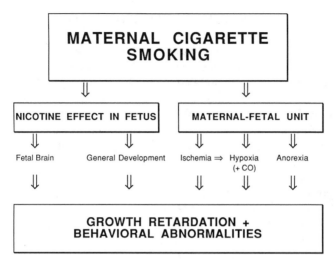

FIGURE 1   Variables contributing to neurobehavioral abnormalities associated with maternal cigarette smoking.

(Peters et al., 1979; Peters, 1984), but most investigators now agree that this manipulation is aversive and will reduce animals' fluid intake drastically rather than eliciting drinking of the nicotine-laced water (Naquira and Arqueros, 1978; Murrin et al., 1987; Slotkin et al., 1987d). This problem has been circumvented with the advent of the implantable osmotic minipump, which has enabled development of models of continuous exposure to nicotine at doses that simulate plasma levels found in human smokers, but without contributions of hypoxia/ischemia (Lichtensteiger and Schlumpf, 1985; Murrin et al., 1985; Murrin et al., 1987; Slotkin et al., 1987b; Slotkin et al., 1987d; Navarro et al., 1988; Navarro et al., 1989a; Navarro et al., 1989b; Ribary and Lichtensteiger, 1989; Navarro et al., 1990b; Slotkin et al., 1990b). Using this paradigm, we have been able to separate the direct effects of nicotine from those of nicotine plus hypoxia/ischemia on cellular development in brain regions, on synaptogenesis and synaptic activity in the central nervous system, and on development of central nervous system nicotinic receptors (Slotkin et al., 1987b; Slotkin et al., 1987d; Navarro et al., 1988; Navarro et al., 1989a; Navarro et al., 1989b; Ribary and Lichtensteiger, 1989; Navarro et al., 1990a,b; Slotkin et al., 1990b). The results summarized in this chapter thus will compare and contrast the effects of prenatal exposure to nicotine via the injection route, which include the effects of hypoxia/ischemia, as well as by the infusion route, which do not. The rat has been chosen as the species of interest because it is born with a relatively immature nervous system, thus permitting postnatal study of phases of neural maturation corresponding to late gestation through neonatal development in humans (Dobbing and Sands, 1979; Reinis and Goldman, 1980).

THE NICOTINE-INJECTION MODEL

Injection of nicotine into pregnant animals produces many of the features common to cigarette smoking. Nicotine injections, like smoking, produce ischemia and resultant tissue hypoxia (Martin and Becker, 1970; Martin and Becker, 1971; Cole et al., 1972; Dow et al., 1975; Slotkin et al., 1986b; Slotkin et al., 1987d; Navarro et al., 1988; McFarland et al., 1991), effects that are exacerbated in smoking by the presence of carbon monoxide, cyanide, and other chemical components of tobacco smoke. Nevertheless, there are limitations to the utility with which the injection route can be applied to animal models of smoking. Most important, smoking involves multiple small doses of nicotine throughout the day, an impractical situation for animal administration. Instead, most investigations have involved relatively large doses of nicotine administered once or twice a day during pregnancy, typically 3 mg/kg (Martin and Becker, 1970; Martin and Becker, 1971; Nasrat et al., 1986; Slotkin et al., 1986b; Slotkin et al., 1987a; Navarro et al., 1988). Although this delivers net daily nicotine doses (corrected for metabolic differences between species) comparable to that in heavy smokers, it does so over a much more restricted period, producing far more intense short-term ischemia and hypoxia than would be experienced in humans. Even when an injection route is selected with a slowed rate of drug entry into the circulation (such as subcutaneous administration), these doses clearly produce blanching of the skin, cyanosis, and respiratory depression (McFarland et al., 1991). As such, injection regimens may be viewed as examinations at the top of the dose–response curve, representative of a worst-case scenario of the fetal toxicity of nicotine. Nevertheless, studies using injected nicotine have provided the pioneering work in this field, and significant information has been obtained that has led to the development of animal models more suitable for defining the effects of nicotine on development; it is therefore worthwhile to examine the results obtained with the injected-nicotine model.

Approximately 20 years ago, nearly coincidental with the first reports of the effects of maternal smoking on perinatal morbidity, mortality, and neurobehavioral development in humans (Cole et al., 1972; Butler and Goldstein, 1973; Dow et al., 1975; Naeye, 1978; Eriksson et al., 1979; Meyer and Carr, 1987), Martin and Becker (1970, 1971) demonstrated that injection of nicotine into pregnant rats results in behavioral alterations in their offspring. Equally important, they pointed out that the effects are similar to those obtained with exposure to hypoxia alone, thus indicating a possible underlying mechanism relevant to both the animal model and human smoking, and opening the door to studies of the underlying biochemical and physiological variables. Because neural development in the neonatal rat corresponds to late gestational development in humans (Dobbing and Sands, 1979; Reinis and Goldman, 1980), some of these effects are elicited even with nicotine given postnatally: nicotine injected directly into newborn rats results in a pattern of noradrenergic hyperinnervation comparable to that seen in animals receiving drugs that cause lesions in specific neuronal tracts within the central nervous system (Jonsson and Hallman, 1980). Subsequently, our group found that a single episode of neonatal hypoxia is also sufficient to reproduce this effect (Slotkin et al., 1986a). The groundwork was thus laid for subsequent mechanistic studies of gestational exposure to maternally injected nicotine, its relationship to the development of nerve projections, and the role of hypoxia in the effect.

Administration of 3 mg/kg of nicotine subcutaneously twice daily to pregnant rats, beginning before the implantation of the embryo in the uterine wall and continued to the end of gestation, produces many of the features common to heavy cigarette smoking, including retarded maternal weight gain, significant fetal loss, and intrauterine growth retardation (Slotkin et al., 1986b; Slotkin et al., 1987a; Navarro et al., 1988). Although growth retardation persists into the postnatal period, weights generally normalize by weaning. Importantly, brain growth is spared relative to body weight, a finding common to a wide variety of nonselective fetal insults (Brazier, 1975; Reinis and Goldman, 1980; Seidler et al., 1982; Bell et al., 1987; Bell and Slotkin, 1988; Bell et al., 1988; Seidler et al., 1990) although brain region weights are significantly subnormal after maternal nicotine administration, the effects are much smaller than those on non-neural tissues and tend to resolve earlier.

In light of the significant growth retardation and brain sparing associated with fetal nicotine exposure, there are two issues to be addressed in determining the mechanisms underlying neurobehavioral effects: first, whether the generalized abnormalities of growth are associated with more selective alterations in nervous system development; second, whether these effects are directly attributable to nicotine penetrating to the fetal brain, or instead represent epiphenomena of growth retardation or hypoxia/ischemia caused by injected nicotine. A number of sensitive biochemical tools have been developed that permit us to probe cell differentiation and synaptogenesis within the fetal and neonatal nervous system. One of the most useful is the activity of ornithine decarboxylase (ODC), the rate-limiting enzyme in the synthesis of the polyamines, key intracellular regulators of cell replication and differentiation (Slotkin, 1979; Slotkin and Thadani, 1980; Heby, 1981; Heby and Emanuelsson, 1981; Bell and Slotkin, 1986; Slotkin and Bartolome, 1986). The ODC activity peaks in each brain region coincidentally with cell replication and differentiation, and interference with this enzyme during those phases invariably results in structural and behavioral deficits. Because it is extremely responsive to environmental and drug-induced perturbations of cell development, ODC provides a sensitive marker for alterations in the timing of replication/differentiation as well as for establishing patterns of cell damage and death. It is therefore particularly important that prenatal exposure of rats to nicotine via maternal injections (Fig. 2) results in a persistent postnatal elevation of ODC that is most prominent in the cerebellum (Slotkin et al., 1986b), a pattern known to be associated with fetal/neonatal cell damage. When conventional markers of cell development are examined, such as nucleic acids and proteins, a similar regional targeting is evident: cerebellum is affected more than other regions (Slotkin et al., 1986b). It is particularly worthwhile to examine regional DNA content: because each neural cell has only one nucleus, the amount of DNA gives an index of the total number of cells. Cerebellar DNA in nicotine-exposed animals is substantially elevated during the phase of rapid cell division, when this region is

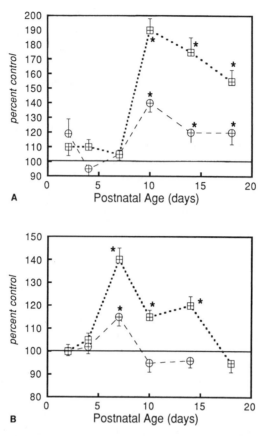

FIGURE 2 Effects of maternal nicotine injections on ODC (A) and DNA (B) in developing brain regions (Slotkin et al., 1986b). Data demonstrate a preferential increase in both variables in the cerebellum of the nicotine-exposed animals. (A) Cerebral cortex (⊕): CON < NIC, $p < 0.01$, NIC × Age, $p < 0.01$; cerebellum (⊞): CON < NIC, $p < 0.01$, NIC × Age, $p < 0.01$. (B) Cerebral cortex (⊕): NIC × Age, $p < 0.01$; cerebellum (⊞): CON < NIC, $p < 0.01$, NIC × Age, $p < 0.01$.

acquiring most of its cells. From prior work with a variety of toxic substances (Slotkin and Thadani, 1980; Bartolome et al., 1982a; Bartolome et al., 1982c; Bartolome et al., 1984; Slotkin et al., 1985b; Slotkin et al., 1985a; Slotkin et al., 1986a; Slotkin et al., 1986b; Slotkin and Bartolome, 1987; Slotkin et al., 1987a; Slotkin et al., 1987c; Navarro et al., 1988; Slotkin, 1988; Seidler and Slotkin, 1990), this pattern is known to be associated with reactive gliosis, wherein dead or absent neurons are replaced with glia; because glia are much smaller cells than neurons, more of them pack into a given volume of tissue, leading to elevations in DNA. Maternal nicotine injections thus produce central nervous system cell damage and reactive changes in the type of cell present.

Despite these dramatic effects on cell development, alterations in cell number and type do not necessarily dictate behavioral abnormalities, but rather, changes in synaptic populations and function are required. Although it is reasonable to assume that the large shifts in markers of cell maturation are likely to cause functional disturbances, it is necessary to prove this by direct experiment. Catecholaminergic systems provide a useful framework within which to examine synaptic function: noradrenergic projections are prominent throughout the brain, thus permitting evaluations across multiple regions; readily available biochemical probes give indications of the numbers of noradrenergic synaptic terminals, their ability to synthesize norepinephrine, the degree of synaptic activity, and the receptive state of the postsynaptic site juxtaposed to the terminals; and finally, studies in the mature nervous system show that these pathways are particularly sensitive to nicotine (Balfour, 1973; Yoshida et al., 1980).

Three relevant measures to be discussed in this chapter appear schematically in Fig. 3. First, the noradrenergic nerve terminal possesses a transporter for norepinephrine that is responsible for removing the transmitter from the synapse. Because the number of transport sites on each terminal is fixed, measurements of uptake capabilities give a rough measure of the number of nerve endings; this is generally studied in the *synaptosome* preparation, where pinched-off terminals are isolated and incubated with radiolabeled norepinephrine *in vitro*. Prior work has established that the development of synaptosomal uptake parallels synaptic proliferation in each brain region (Coyle and Axelrod, 1971; Kirksey and Slotkin, 1979; Slotkin et al., 1979; Singer et al., 1980). The second measure is the activity of the enzyme tyrosine hydroxylase, the rate-limiting factor in norepinephrine and dopamine biosynthesis (Nagatsu, 1973). In addition to controlling the ability of the neuron to make transmitter, the level of tyrosine hydroxylase is controlled, in part, through impulse flow, thus providing a measure of nerve activity integrated over a period of many hours or days. Again, development of tyrosine hydroxylase generally parallels both nerve terminal proliferation and the onset of tonic activity of the neuron (Breese and Traylor, 1972; Coyle and Axelrod, 1972; Slotkin et al., 1979). However, the higher degree of association of tyrosine hydroxylase with dopamine terminals than with norepinephrine terminals is a potential confound, requiring a more specific index of noradrenergic activity. This is provided by the third measure, norepinephrine turnover; turnover is assessed by inhibiting tyrosine hydroxylase with the specific competitive substrate, α-methyl-*p*-tyrosine, and measuring the subsequent fall-off in norepinephrine levels; because *de novo* transmitter synthesis has been prevented, the decline in norepinephrine will depend primarily on how fast the transmitter is being lost by release into the synapse (Brodie et al., 1966; Costa et al., 1972; Deskin and Slotkin, 1981; Seidler and Slotkin, 1981; Slotkin et al., 1981; Bartolome et al., 1982b; Navarro et al., 1988; Seidler et al., 1990; Seidler and Slotkin, 1990). This index, the "turnover" of norepinephrine, thus provides a measure of impulse activity that is selective for norepinephrine versus dopamine and is integrated over a much shorter period (minutes to hours) than that for tyrosine hydroxylase.

Examination of markers for synaptic development and activity in animals exposed prenatally to nicotine via maternal injections (Fig. 4) reveals at least two interesting

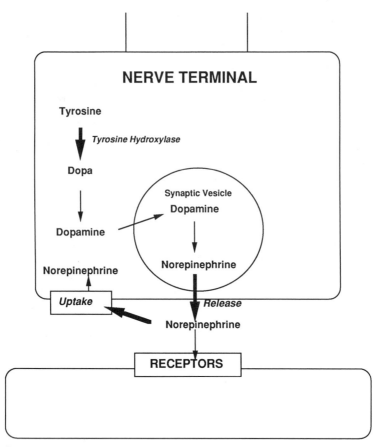

FIGURE 3   Schematic representation of a noradrenergic nerve terminal, showing three relevant variables for assessment of development. The ability of the terminal to take up norepinephrine from the synapse is a measure of the number of nerve terminals. Tyrosine hydroxylase, the enzyme that converts tyrosine to Dopa, is rate limiting in transmitter biosynthesis and is responsive to nerve impulse activity. Release of norepinephrine into the synapse is measured by inhibiting tyrosine hydroxylase with α-methyl-*p*-tyrosine and measuring the decline in transmitter levels.

features (Slotkin et al., 1987a; Navarro et al., 1988). First, there is substantial hyperproliferation of nerve terminals, evidenced by elevations in synaptosomal uptake; this is accompanied by profound neuronal hyperactivity, as assessed by transmitter turnover. The effect is selective for norepinephrine, as no changes in dopamine turnover are detected, and consequently, there is little or no change in tyrosine hydroxylase activity. Second, the effects on synaptic development extend beyond the cerebellum, the region showing the greatest perturbations of cell replication and differentiation. This indicates that actions on the programming of terminal development are distinct from more general toxic effects on cell proliferation and survival. The proof of this

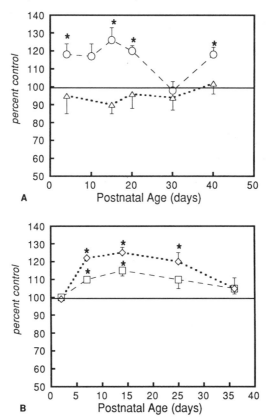

FIGURE 4 (A) Effects of maternal nicotine injections on norepinephrine uptake into synaptosomes, tyrosine hydroxylase activity, and norepinephrine levels and (B) turnover in cerebral cortex (Slotkin et al., 1987; Navarro et al., 1988). Data demonstrate the emergence of a hyperinnervation (elevated uptake) and hyperactivity (elevated turnover) characteristic of the injection model. Uptake (○): *CON < NIC, $p < 0.01$; tyrosine hydroxylase (△): CON vs. NIC, N.S. Content (□): *CON < NIC, $p < 0.05$; turnover (◇): *CON < NIC, $p < 0.05$, NIC × Age, $p < 0.05$.

interpretation is that similar findings are obtained outside the central nervous system, in peripheral noradrenergic projections (Slotkin et al., 1987a). Synaptic hyperactivity is present in renal, cardiac, and sympatho-adrenal pathways, which all share the adrenergic phenotype, but contain cell populations that proliferate and differentiate on distinctly different timetables from those of central neurons.

It is thus evident that fetal exposure to nicotine via maternal injections does produce a wide spectrum of nervous system alterations, ranging from cell damage and gliosis, to more selective changes in neuronal activity of specific projections. However, the pattern obtained in these studies speaks against the likelihood that these reflect actions of nicotine directly on the fetal nervous system: a comparison with the

TABLE 1   Comparison of Fetal Nicotine Exposure via Maternal Injections with Hypoxia or Methylmercury Exposure[a]

Measures	Nicotine	Hypoxia	Methylmercury
ODC and DNA	Elevation (cell damage + gliosis)	Elevation (cell damage + gliosis)	Elevation (cell damage + gliosis)
Synaptosomal uptake	Increase (hyperinnervation)	Increase (hyperinnervation)	Increase (hyperinnervation)
Norepinephrine turnover	Increase (hyperactivity)	Increase (hyperactivity)	Increase (hyperactivity)

[a]The findings and conclusions summarized in this table are from Bartolome et al. (1982a); Bartolome et al. (1982c); Bartolome et al. (1984); Slotkin et al. (1985b); Slotkin et al. (1985a); Slotkin et al. (1986a); Slotkin et al. (1986b); Slotkin and Bartolome (1987); Slotkin et al. (1987a); Slotkin et al. (1987c); Navarro et al. (1988); Seidler and Slotkin (1990).

effects of hypoxia or generalized toxic insult with heavy metals, such as methylmercury, indicates virtually identical spectra of effects (Table 1). In each case, there is growth retardation, postnatal elevation of ODC activity, and increased DNA indicative of cerebellar gliosis, as well as noradrenergic synaptic hyperproliferation and hyperactivity in both the central and peripheral nervous system (Bartolome et al., 1982a; Bartolome et al., 1982c; Bartolome et al., 1984; Slotkin et al., 1985b; Slotkin et al., 1985a; Slotkin et al., 1986a; Slotkin et al., 1986b; Slotkin and Bartolome, 1987; Slotkin et al., 1987a,d; Slotkin et al., 1987c; Navarro et al., 1988; Seidler and Slotkin 1990). Acute administration of nicotine invariably causes ischemia/hypoxia, and hypoxia alone can produce similar behavioral alterations in the offspring, thus pointing to significant contributions from these epiphenomena in the neurobehavioral deficits attributable to the nicotine-injection model. This does not obviate the seminal importance of these earlier findings: hypoxia and ischemia are significant components of human cigarette smoking; in addition, these studies point out the need for development of models that successfully separate the various pharmacologic contributions of nicotine acting on the maternal–fetal unit, as distinct from nicotine acting within the fetal brain. Accordingly, the nicotine-infusion model has been developed to address these issues.

THE NICOTINE-INFUSION MODEL

In contrast to the obvious and immediate effects of a bolus of injected nicotine on the maternal and fetal vasculature, the slow infusion of nicotine via subcutaneously implanted, osmotic minipumps provides a means of continuous delivery of drug without the peaks and valleys of plasma levels characteristic of the injection model; accordingly, there is no evidence of ischemia or hypoxia during infusion, despite the fact that total doses delivered can match the 6 mg/kg/day achieved with twice-daily injections (Slotkin et al., 1987b; Slotkin et al., 1987d; Navarro et al., 1988; Navarro et al., 1989a; Navarro et al., 1990a; Navarro et al., 1990b; Slotkin et al., 1990b). Plasma levels achieved with the infusion model generally match those seen in heavy cigarette smokers (Isaac and Rand, 1972; Murrin et al., 1987).

Nicotine infusions of 6 mg/kg/day, given to pregnant rats over the same period of gestation as described for the injection model, still cause some signs of gestational toxicity, including reduced maternal weight gain and significant incidence of fetal resorption (Slotkin et al., 1987b; Slotkin et al., 1987d); similarly, intrauterine growth retardation is present with the infusion model, along with impaired brain growth, brain sparing and other general features common to the nicotine-injection model and to maternal cigarette smoking in humans (Slotkin et al., 1987b; Slotkin et al., 1987d; Navarro et al., 1988; Navarro et al., 1989a; Navarro et al., 1990a; Navarro et al., 1990b; Slotkin et al., 1990b). Notably, there are some differences in general toxicity markers between the injection and infusion models: maternal weight gain, fetal/neonatal body and tissue weight gains, and maternal/perinatal mortality are less affected by the infusion regimen, despite a higher overall incidence of fetal resorption in the surviving dams and litters. It is, however, in the effects on nervous system development that the most marked route-related differences can be seen. Despite the smaller degree of maternal–fetal mortality and growth retardation with maternal nicotine infusions, there is a more pronounced and widespread elevation of ODC activity in all brain regions (Slotkin et al., 1987d; Navarro et al., 1989b); in this case, there is no apparent preference for cerebellum as was found with nicotine injections (Fig. 5). The pattern of alteration of ODC, initial postnatal elevations of activity, and a subsequent additional phase of supranormal ODC during the second postnatal week, is known to be associated with delays in cell development (Slotkin and Thadani, 1980; Bell and Slotkin, 1986; Slotkin and Bartolome, 1986). Indeed, examination of macromolecules indicates a similar, widespread interference with cell replication resulting in shortfalls in total cell numbers (DNA), again, not restricted to the cerebellum. It is particularly important that this pattern is the opposite of that seen with the nicotine-injection paradigm, in which elevations in cerebellar DNA were seen, consequent to neuronal cell damage and reactive gliosis. By implication, the infusion route is *not* producing the overt neuronal cell death seen with injected nicotine.

Even greater differences between the two nicotine-exposure models are apparent if the effects on nerve terminal proliferation and nerve activity are considered (Fig. 6). Nicotine infusions result in initial hypoactivity of noradrenergic and dopaminergic projections throughout the brain, as evidenced by reduced transmitter levels and turnover rates (Navarro et al., 1988). After recovery to near-normal activity by weaning, the nicotine-exposed animals again show deficits in young adulthood. The corresponding enzymatic measure, tyrosine hydroxylase activity, also shows subnormal values in the nicotine group, but notably, synaptosomal uptake does not. Unlike nicotine injections, in which synaptic hyperproliferation and hyperactivity are the major findings, nicotine infusions instead slow the development of nerve impulse activity without altering terminal number. Thus, for the first time, it is possible to state unequivocally that nicotine, without participation of hypoxia and ischemia,

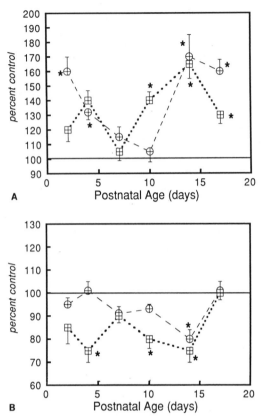

FIGURE 5   Effects of maternal nicotine infusions on ODC (A) and DNA (B) in developing brain regions (Slotkin et al., 1987d). Data demonstrate widespread elevations in ODC activity and suppression of cell acquisition. (A) Cerebral cortex (⊕): CON < NIC, $p < 0.01$, NIC × Age, $p < 0.01$; cerebellum (⊞):* CON < NIC, $p < 0.01$, NIC × Age, $p < 0.01$. (B) Cerebral cortex (⊕): CON > NIC, $p < 0.05$, NIC × Age, $p < 0.01$; cerebellum (⊞):* CON > NIC, $p < 0.01$, NIC × Age, $p < 0.01$.

interferes with cell maturation and alters synaptic performance in the developing brain.

The neurochemical abnormalities associated with prenatal exposure to nicotine via maternal infusions have been shown to correlate with perturbed behaviors and neuroendocrine parameters linked to dopaminergic and noradrenergic pathways (Lichtensteiger and Schlumpf, 1985; Ribary and Lichtensteiger, 1989). Whereas it is tempting to attribute the functional deficits simply to decreased impulse activity, there is a logical inconsistency in such a conclusion. In the mature nervous system, long-term decreases in impulse activity are compensated by up-regulation of postsynaptic receptor sites juxtaposed to the nerve terminal, with the decrease in input and increased responsiveness offsetting each other to result in normal functional status

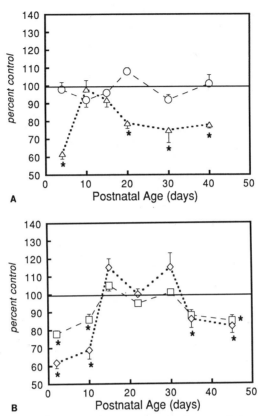

FIGURE 6 Effects of maternal nicotine infusions on norepinephrine uptake into synaptosomes, tyrosine hydroxylase activity (A) and norepinephrine levels and turnover in cerebral cortex (B) (Navarro et al., 1988). Data demonstrate no alterations in the numbers of nerve terminals (normal values for uptake), but two phases of neuronal hypoactivity (reduced levels and turnover and impaired tyrosine hydroxylase activity). Uptake (O): CON vs. NIC, N.S.; tyrosine hydroxylase (△); *CON > NIC, $p < 0.01$; content (□): *CON > NIC, $p < 0.01$, NIC × Age, $p < 0.01$; turnover (◇): *CON > NIC, $p < 0.01$, NIC × Age, $p < 0.01$.

(Charney et al., 1981; Davies and Lefkowitz, 1984; Kunos and Ishac, 1987; Weiss et al., 1988). However, direct measurements of adrenergic receptor binding sites in brain regions of animals exposed to maternal nicotine infusions reveal no significant changes in the β- or α_2-receptor subclasses, and only minor, transient increases in the α_1-receptor population (Navarro et al., 1990b). Why then are theses animals functionally subresponsive? Early in synaptic development, exposure of postsynaptic receptor sites to neurotransmitter provides information that enables the target cells to "program" their subsequent function. Accordingly, early denervation produces permanent shortfalls in receptor-mediated responses, rather than causing up-regulation of receptor numbers and responses, as would be the case with denervation in adulthood

(Deskin et al., 1981; Duncan et al., 1987; Criswell et al., 1989; Hou et al., 1989a; Hou et al., 1989b; Kostrzewa and Saleh, 1989; Kudlacz et al., 1990b; Kudlacz et al., 1990d; Kudlacz et al., 1991; Navarro et al., 1991; Wagner et al., 1991). Transmitter programming of response development appears to be widespread, shared by all adrenergic receptor subtypes (Deskin et al., 1981; Hou et al., 1989a; Hou et al., 1989b; Kudlacz et al., 1990b; Kudlacz et al., 1990d; Kudlacz et al., 1991; Navarro et al., 1991; Wagner et al., 1991), cholinergic receptors (Slotkin et al., 1987b; Hohmann et al., 1988; Navarro et al., 1989a; Navarro et al., 1989b) and dopaminergic receptors (Duncan et al., 1987; Criswell et al., 1989; Kostrzewa and Saleh, 1989). The early deficits in noradrenergic activity caused by maternal nicotine infusions would thereby reduce exposure of the sites to transmitter precisely during the critical period in which responses are being imprinted. Recent work in the peripheral nervous system has provided key information to illustrate this phenomenon. Over the first 3 weeks postnatally, cardiac β-adrenergic receptors increase in their number and linkage to heart-rate control, coincidentally with a developmental surge in noradrenergic neuronal activity (Seidler and Slotkin, 1981; Slotkin, 1986). When this surge is blunted by prenatal nicotine exposure, the number of β-receptor sites develops more slowly (Fig. 7); the functional correlate is that heart-rate responses to a β-adrenergic agonist (isoproterenol) are subnormal, requiring six times the dose to achieve a 50 beat/min increase (Navarro et al., 1990a). Although receptor numbers resolve to normal by young adulthood, the performance deficits do not: heart-rate responses, elicited either with an adrenergic drug or physiologically through central stimulation of the nerve supply to the heart, remain subnormal. Because neuronal input was reduced during the critical period of response programming, the nicotine-exposed animals will never achieve proper reactivity to neuronal stimuli. The studies with maternal nicotine infusions thus not only provide a demonstration that nicotine itself is a neurochemical/neurobehavioral teratogen, but also provide evidence for the underlying cellular mechanisms by which behavioral patterns are programmed by early neuronal input, findings that may be of considerable importance in the general understanding of behavioral teratogenesis.

To this point, none of the studies described here has addressed one of the key questions in nicotine-induced neurobehavioral anomalies, namely whether these effects occur independent of gross morphological change and growth impairment. In the clinical realm, the criterion of low birth weight is considered to be one of the most critical factors in assessing relative injury associated with fetal drug exposure; in the case of cigarettes, for example, the Surgeon General's warning to pregnant women points out the relationship of smoking to "fetal injury, premature birth, and low birth weight." Whereas in clinical studies, factors like growth impairment are difficult to separate from drug effects targeted to the nervous system, animal models are ideal for separation of these variables. Using the nicotine-infusion paradigm in pregnant rats, we have been able to evaluate the dose–effect relationships of prenatal nicotine exposure on general growth and development as compared to effects on nervous system development (Navarro et al., 1989b). By lowering the daily dose to 2 mg/kg/day, a regimen that produces plasma nicotine levels comparable to those in cigarette smokers

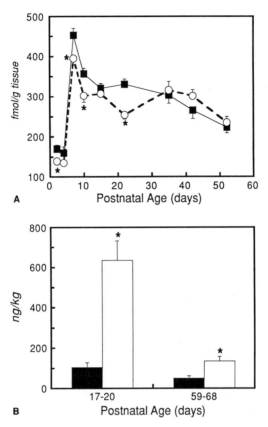

FIGURE 7 Effects of maternal nicotine infusions on development of cardiac β-adrenergic receptors (A) and on the heart rate response to β-receptor stimulation with isoproterenol (B) (Navarro et al., 1990a). Data demonstrate the slowing of receptor development and subsensitivity of responsiveness that persists past the point at which receptor sites have normalized (deficient "programming" of responsiveness). (A) CON (■); NIC (○). *CON > NIC, $p < 0.01$, NIC × Age interaction, $p < 0.01$. (B) CON (■); NIC (□). *CON < NIC, $p < 0.01$, NIC × Age interaction, $p < 0.01$.

consuming approximately one pack per day, drug effects on maternal weight gain, fetal resorption, litter size, and all standard morphological variables are eliminated from the model. The offspring also show no evidence of postnatal retardation of body or brain region growth. Nevertheless, examination of the key biochemical indices of cellular and synaptic development still indicate profound effects of prenatal nicotine exposure. Animals in the low-dose nicotine group exhibit the same biphasic elevation of brain region ODC activity as was seen at the higher dose, accompanied by shortfalls in brain cell number (assessed by DNA content) and retardation of development of noradrenergic projections (assessed by neurotransmitter levels). Especially important is that all of these effects are just as robust as those seen with the 6 mg/kg/day regimen, but in this case, in the complete absence of the growth effects. Thus, doses of nicotine that are at the bottom of the dose–response curve for eliciting growth impairment are already at the top of the dose–response curve for altering nervous system development. Comparable behavioral alterations at the lower dose of nicotine have also been reported (Lichtensteiger and Schlumpf, 1985; Ribary and Lichtensteiger, 1989).

These results indicate that, unlike those in standard teratogenesis, in which brain development is spared relative to effects on the rest of the organism, nicotine targets the nervous system. The important corollary is that, for nicotine (and by implication, for cigarette smoking), the use of a standard marker such as prematurity or birth weight is inappropriate for assessing whether there is a "safe" level of exposure. Equally vital, the question of why nicotine is an atypical teratogen opens the door to mechanistic studies in a relatively new area in the study of birth defects: the targeting of a teratogen to a highly specific population of cell receptors.

THE ROLE OF NICOTINIC CHOLINERGIC RECEPTORS

The lower dose threshold for adverse effects of fetal nicotine exposure on neuronal development, as compared to growth impairment, suggests that specific elements within the nervous system are targeted by nicotine. The most likely candidate is the nicotinic cholinergic receptor that transduces the cholinergic neuronal signal into postsynaptic cell depolarization. The presence of nicotinic receptors in the fetal nervous system has been demonstrated biochemically by in vitro radioligand binding techniques and morphologically by in situ receptor autoradiography (Hagino and Lee, 1985; Larsson et al., 1985; Lichtensteiger et al., 1987; Slotkin et al., 1987b; Cairns and Wonnacott, 1988). However, the presence of receptor-binding sites does not necessarily imply that the receptors are functionally connected to membrane depolarization, an event that must occur before prenatal nicotine exposure can influence neuronal development. Fortunately, it is relatively straightforward to determine whether nicotinic receptors have been stimulated chronically by nicotine exposure: because the receptors are desensitized by long-term stimulation, the cellular response to prolonged nicotine-induced depolarization is overproduction of the receptor sites (Schwartz et al., 1982; Clarke et al., 1985; Schwartz and Kellar, 1985). Thus, chronic nicotine administration in adult rats, where the receptors are definitively coupled to cellular responses, results in increases in the number of nicotinic receptors, readily measured with [³H]nicotine binding. Examination of fetal rat brain on gestational day 18 (Fig. 8) indicates that, with either repeated maternal nicotine injections or infusions, substantial increases in [³H]nicotine binding are occurring, indicative of long-term cellular stimulation by the drug (Slotkin et al., 1987b). Importantly, the same degree of receptor stimulation is obtained even with the 2 mg/kg/day infusion paradigm (Navarro et al., 1989b), a finding that explains why neuronal effects are already maximal at a nicotine exposure level that is not growth suppressing.

But why should cell depolarization by nicotine lead to abnormalities of nervous system development? The answer is that the act of neurotransmission is a key element

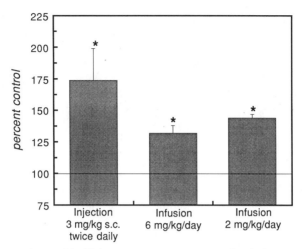

FIGURE 8   Up-regulation of fetal brain nicotinic receptors, assessed with [³H]nicotine binding, resulting from prenatal exposure to maternal nicotine injections or infusions (Slotkin et al., 1987b; Navarro et
al., 1989b). Note that the degree of increase in receptors is the same with the growth-suppressing infusion
regimen (6 mg/kg/day) and at the lower, non-growth-suppressing dose (2 mg/kg/day). *CON < NIC, $p <$
0.01.

enabling developing cells to decide between replication and differentiation. A vast
literature describes the role of neurotransmitters as trophic factors in developing cells.
The ability of neurally active substances to alter membrane polarity in cells that
contain specific receptors for these compounds has been demonstrated to control the
expression of genes that are ordinarily turned on only when the cell enters differentiation (Patterson and Chun, 1977; Black, 1978; Mytilineou and Black, 1978; Black,
1980). Recent work shows that acetylcholine, released by developing cholinergic
neuronal projections, provides such trophic signals to selective target areas within the
brain, so that formation of lesions in these tracts leads to architectural disorganization
(Hohmann et al., 1988; Navarro et al., 1989a). It is therefore possible to propose a
model that explains both the exquisite sensitivity of the developing brain to prenatal
nicotine exposure, as well as the pattern of disrupted cell and synaptic development:
fetal stimulation of nicotinic receptors by the drug results in premature onset of cell
differentiation at the expense of replication, leading to shortfalls in cell number and
aberrant patterns of synaptogenesis and synaptic activity.

   In order to test this hypothesis, it is necessary to identify when the cholinergic
signal controlling cell differentiation is ordinarily passed to target cells, to evaluate
how prenatal nicotine exposure alters this signal, and to demonstrate conclusively that
premature cholinergic stimulation will directly affect the replication/differentiation
decision. We have recently evaluated cholinergic neuronal activity in developing
brain, using a combination of two biochemical markers related to nerve terminal
development and activity. The enzyme, choline acetyltransferase, converts choline to

acetylcholine and hence plays a vital role in transmitter synthesis; however, this synthetic step is not the rate-limiting factor in acetylcholine formation, and choline acetyltransferase activity is unresponsive to nerve impulse activity (Cooper et al., 1986). Instead, the relatively constant concentration of this enzyme within the nerve ending provides a reliable index of the number of cholinergic terminals and correlates well with the development of projections to target sites (Coyle and Yamamura, 1976; Ross et al., 1977; Singh and McGeer, 1977; Bell and Lundberg, 1985; Dori and Parnavelas, 1989; Geula et al., 1990). Acetylcholine synthesis is limited by the ability of the terminals to take up the precursor, choline, via an active presynaptic membrane transport system; importantly, it is this transport process that *is* responsive to nerve impulse activity and can change acutely over a course of minutes to hours to compensate for increased demand for transmitter release (Simon et al., 1976; Klemm and Kuhar, 1979; Murrin, 1980). By combining these two factors to evaluate the ratio of choline uptake: choline acetyltransferase activity, we obtain a measure of the net impulse activity per nerve terminal. This index has already proven useful in evaluations of central cholinergic tone in developing brain and in neurodegenerative diseases such as Alzheimer's disease (Navarro et al., 1989a; Slotkin et al., 1990c).

Applying the activity ratio method to developing rat cerebral cortex (Fig. 9) indicates that cholinergic tone does not develop monotonically from low to high values, but rather displays a distinct peak centered around postnatal day 10 (Navarro et al., 1989a). Accordingly, the finding that nicotinic receptors are present in fetal brain and that prenatal exposure to nicotine stimulates these receptor to produce depolarization and receptor up-regulation indicates that nicotine does indeed transmit the cholinergic signal to target cells prematurely. Examination of the postnatal peak of nerve activity in the nicotine-exposed cohort confirms that the natural signal has been preempted, and the surge of cholinergic tone, correspondingly blunted.

If receptor stimulation thus occurs prematurely in fetuses exposed to nicotine, does it have the ability to initiate the switchover from target cell replication to differentiation? Recent studies indicate that it does (McFarland et al., 1991). Administration of nicotine to neonatal rats leads to an acute decrease in [^{3}H]thymidine incorporation, a standard biochemical index of cell replication (Fig. 10). The effect on various brain regions exhibits a hierarchy corresponding to the concentration of nicotinic receptors in each region (Slotkin et al., 1987b); thus, within 30 min of exposure to a single dose of nicotine, and persisting for several hours, DNA synthesis is reduced by as much as 50%, with the rank order of effect midbrain + brainstem $\geq$ cerebral cortex > cerebellum. The effect is restricted to replicating cells, as there are no alterations in factors reflecting general macromolecule synthesis, such as [^{3}H]leucine incorporation into proteins. Extensive studies demonstrate that the effect is, in fact, mediated by nicotinic receptors within the developing central nervous system. First, peripheral cardiovascular effects of nicotine, such as ischemia or altered cardiac performance, have been ruled out by the inability of autonomic blocking drugs to prevent the actions of nicotine on DNA synthesis. Second, the potential role of hypoxia has also been eliminated by the lack of protection afforded by oxygen-enriched breathing mixtures. Most conclusively, the direct introduction of small doses

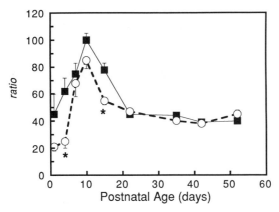

FIGURE 9 Development of cholinergic nerve activity in cerebral cortex, assessed with the ratio of choline uptake: choline acetyltransferase activity (Navarro et al., 1989a). Activity peaks at about postnatal day 10, and prenatal exposure to nicotine infusions reduces the magnitude. CON (■); NIC (○). *CON > NIC, $p < 0.01$, NIC × Age interaction, $p < 0.01$.

of nicotine (2 μg) directly into the brain causes identical reductions in [³H]thymidine incorporation into DNA, despite the fact that this dose is too low to elicit peripheral cardiovascular or respiratory alterations. Finally, maternal nicotine administration also produces the same acute arrest of cell replication in fetal brain, thus indicating that the fetal nicotinic receptors are fully capable of eliciting the same developmental switchover to differentiation (McFarland et al., 1991).

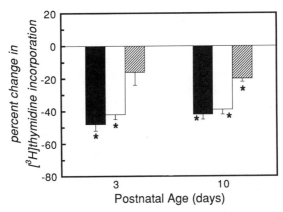

FIGURE 10 Effects of acute nicotine administration on DNA synthesis in developing brain regions, assessed with [³H]thymidine incorporation over a 30-min period after nicotine (McFarland et al., 1991). The selectivity indicates greater effects in regions enriched in nicotinic cholinergic receptors (midbrain + brain stem, cerebral cortex) as compared to a region poor in receptors (cerebellum). Midbrain plus brain stem (■); cerebral cortex (□); cerebellum (▨). *CON > NIC, $p < 0.01$, NIC × Region interaction, $p < 0.01$.

A critical test of the hypothesis that fetal nicotine exposure affects brain development through prematurely eliciting receptor stimulation to alter cell replication/differentiation patterns, has been provided through studies in which endogenous cholinergic activity is altered through manipulations that do not involve administration of nicotine. For this purpose, we can take advantage of the fact that the uptake of choline into the terminal is rate limiting in acetylcholine biosynthesis and availability of transmitter for release into the synapse (Simon et al., 1976; Klemm and Kuhar, 1979; Murrin, 1980); accordingly, large increases in dietary choline actually lead to elevations of cholinergic synaptic activity. By enriching the maternal diet with a choline-rich lecithin supplement, such that choline intake is increased by approximately fivefold, behavioral development in the offspring can be affected, with generally adverse effects but precocious maturation of some indices (Bell and Slotkin, 1985; Bell et al., 1986a; Bell et al., 1986b). These findings suggest that choline enrichment has a dual effect on nervous system development, similar to that proposed here for nicotine. Indeed, examination of the cellular mechanisms underlying the effects of dietary choline on behavior reveals a pattern identical to that seen for nicotine: reduced cortical cell numbers (measurements of DNA), reduced synaptic activity in noradrenergic projections (measurements of tyrosine hydroxylase activity), and desensitized cholinergic target cell responses, indicative of chronic overstimulation. Thus, a treatment that enhances fetal/neonatal cholinergic activity, leading to stimulation of nicotinic receptors, produces exactly the same syndrome as fetal exposure to nicotine. These findings confirm conclusively that the selectivity and high sensitivity of the developing central nervous system to fetal nicotine exposure is a direct consequence of a specific, receptor-mediated process normally intended to control the timing of cell differentiation, but elicited prematurely by the drug.

CONCLUSIONS

The use of animal models of fetal nicotine exposure allows the separation of the multiple variables that operate in producing structural and functional abnormalities in the offspring of women who smoke. First and foremost, manipulation of the route of administration allows the investigator to separate components related to nicotine's acute actions in the maternal–fetal unit, notably hypoxia and ischemia, from those attributable to nicotine penetrating the placenta and entering the fetus itself. As demonstrated with the injected-nicotine model, abnormalities of nervous system development are readily produced when nicotine is present, along with the associated changes in cardiovascular status and oxygenation that are a common component of smoking or injected nicotine. The resemblance of the effects of injected nicotine to those associated with hypoxic brain damage, including cellular damage and reactive gliosis with cerebellum as the most sensitive region, as well as the subsequent appearance of hyperinnervation and neuronal hyperactivity, all indicate that the hypoxic/ischemic changes are significant contributors to neurobehavioral alterations. However, the fact that injected nicotine also leads to fetal brain nicotinic receptor stimulation (as evidenced by receptor binding site up-regulation) indicates that the net actions must represent effects that are additive to those of nicotine itself.

The infusion model, in contrast, does not actually duplicate the actions associated with smoking, but provides the best means of determining the mechanisms participating in the adverse effects of nicotine on nervous system development. By controlling the rate of infusion so as to deliver relatively large doses of nicotine in the absence of cardiovascular changes, hypoxia, or other potentially confounding variables, it has been possible to demonstrate that nicotine by itself impairs acquisition of cells throughout the central nervous system, leading to suppressed synapse formation and chronic deficits of impulse activity. With this model, we have been able to show that alterations in the delivery of neural impulses to target tissues are directly responsible for functional shortfalls in performance that persist into adulthood, even when structural markers for cell numbers or postsynaptic receptor sites have normalized. Equally important, we have demonstrated that nicotine acts through its specific cholinergic receptors present in fetal brain tissue. By elucidating the role of acteylcholine acting as a natural neural signal that enables its target cells to terminate replication and initiate differentiation, we have established that nicotine is duplicating the effects of this ontogenetic process, but at inappropriately premature stages. Because of the involvement of highly specific neural receptors in the transduction of this developmental signal, the central nervous system is much more sensitive to disruption of development by nicotine than is the rest of the fetus, and consequently standard teratogenic markers such as premature delivery, low birth weight, or fetal/neonatal morbidity and mortality, are all inappropriate to predict the sensitivity of the nervous system; unlike that for malnutrition, hypoxia, or nonselective teratogens, the threshold for nicotine-induced impairment of nervous system development lies *below* that for alterations of general growth and development.

Animal models are obviously critical because they allow separation of the relevant variables contributing to teratogenesis in a way difficult or impossible in human substance-abuse settings. However, such exercises are constructive only if the findings can then be translated back to the human condition, to enable us to make predictions and recommendations. In the case of smoking and nicotine, recent commercial developments render such decision making important: if a pregnant woman will not or cannot cease smoking, would nicotine substitution be beneficial? Based on the animal models described here, the answer is a qualified "yes." Certainly, the associated hypoxia and ischemia produced by smoking, both as a result of nicotine received in a bolus and the carbon monoxide and cyanide found in cigarette smoke, will contribute significantly to alterations in nervous system development. For these epiphenomena of smoking, the variable of intrauterine growth retardation may provide a partial index of the degree of sensitivity of the fetus. However, for the factors related to nicotine itself, the standard indices of safety are inappropriate, because the dose–response curve for impaired nervous system development lies below that for growth suppression. Furthermore, if nicotine substitutes are chosen that deliver more nicotine than does smoking, the nicotine-related aspects of fetal nervous system damage may be as bad or worse; this is especially true when typical abuse patterns include smoking in addition to the noncigarette nicotine source. In light of our findings in rats, the only truly safe course is to discontinue nicotine exposure entirely.

The finding that nicotine acts on developing neural tissue through specific nicotinic receptors does, however, produce one particular advantage in dealing with women who are smoking at the time they become pregnant. Because nervous system differentiation to produce specific neuron types and receptors is a relatively late gestational event, the receptor-mediated changes are not likely to be prominent with exposure during early pregnancy. Unlike most structural teratogens, behavioral teratogens that operate on receptors appearing late in differentiation would have their most sensitive periods in the second and third trimesters. Thus, receptor mediation of the effects of nicotine means that there is a significant period after the initial recognition of pregnancy in which the mother can taper off and quit smoking before the elicitation of the neural changes described here. Careful explanation of these factors to the expectant mother may provide an additional incentive to quit smoking in a timely fashion.

Finally, it is worthwhile pursuing the relevance of the nicotine model by exploration of the principles behind behavioral teratogenesis in general. Virtually all psychoactive compounds act through stimulation or antagonism of specific neurotransmitter receptors, or by altering the actions or release of endogenous transmitters acting at receptor sites. In light of the fact that nicotine, through its actions on cholinergic nicotinic receptors, acts on developing cells through mimicking a natural process by which neural cells control their differentiation, then it is likely that all psychoactive compounds will affect their targets similarly. The corollary is that psychoactive drugs, unlike standard structural teratogens, will all target the nervous system more sensitively than the rest of the organism and will also display sensitive periods that occupy later developmental stages than classical teratogens that act typically during first-trimester organogenesis. Studies from a variety of laboratories now indicate that drugs acting on neurotransmitter systems or receptors for serotonin, norepinephrine, dopamine, or endogenous opiate peptides all display critical developmental periods in which they affect the "programming" of cellular responses. We have particularly been able to show that, just as with cholinergic systems, noradrenergic pathways exhibit a spike of activity that initiates the cessation of target cell replication and onset of differentiation (Slotkin et al., 1987e; Slotkin et al., 1988c; Duncan et al., 1990) and that permanent alterations in cell number and activity result from either blocking this spike or eliciting the stimulation prematurely with noradrenergic agonist drugs (Slotkin et al., 1988a; Slotkin et al., 1988b; Hou et al., 1989a; Hou et al., 1989b; Kudlacz et al., 1989; Kudlacz et al., 1990a; Kudlacz et al., 1990b; Kudlacz et al., 1990c; Kudlacz et al., 1990d; Slotkin et al., 1990a; Slotkin et al., 1990b; Navarro et al., 1991; Wagner et al., 1991). The effects of nicotine thus provide the underpinnings for more generalized models to establish how functional perturbations result from fetal drug exposures that do not cause overt structural damage. We may thus be on the threshold of understanding the principles of behavioral teratogenesis rather than simply describing, one by one, the behavioral alterations evoked by fetal exposure to each individual drug.

ACKNOWLEDGMENT

The research described in this chapter was supported by a grant from the Smokeless Tobacco Research Council.

REFERENCES

Balfour, D. J. K. (1973). Effects of nicotine on the uptake and retention of ^{14}C-noradrenaline and ^{14}C-5-hydroxytryptamine by rat brain homogenates. *Eur. J. Pharmacol.* **23**, 19–26.

Bartolome, J., Chait, E. A., Trepanier, P., Whitmore, W. L., Weigel, S., and Slotkin, T. A. (1982a). Organ specificity of neonatal methyl mercury hydroxide poisoning in the rat: Effects on ornithine decarboxylase activity in developing tissues. *Toxicol. Lett.* **13**, 267–276.

Bartolome, J., Lau, C., and Slotkin, T. A. (1982b). Neonatal hyperthyroidism causes premature development of baroreceptor-mediated cardiac sympathetic reflexes. *Dev. Neurosci.* **5**, 208–215.

Bartolome, J., Trepanier, P., Chait, E. A., Seidler, F. J., Deskin, R., and Slotkin, T. A. (1982c). Neonatal methylmercury poisoning in the rat: Effects on development of central catecholamine neurotransmitter systems. *Toxicol. Appl. Pharmacol.* **65**, 92–99.

Bartolome, J., Whitmore, W. L., Seidler, F. J., and Slotkin, T. A. (1984). Exposure to methylmercury *in utero*: Effects on biochemical development of catecholamine neurotransmitter systems. *Life Sci.* **35**, 657–670.

Bell, J. M., and Lundberg, P. K. (1985). Effects of a commercial soy lecithin preparation on development of sensorimotor behavior and brain biochemistry in the rat. *Dev. Psychobiol.* **18**, 59–66.

Bell, J. M., and Slotkin, T. A. (1985). Perinatal dietary supplementation with a commercial soy lecithin preparation: Effects on behavior and brain biochemistry in the developing rat. *Dev. Psychobiol.* **18**, 383–394.

Bell, J. M., and Slotkin, T. A. (1986). Polyamines as intermediates in developmental neurotoxic events. *Neurotoxicology* **7**, 147–160.

Bell, J. M., and Slotkin, T. A. (1988). Postnatal nutritional status influences development of cardiac adrenergic receptor binding sites. *Brain Res. Bull.* **21**, 893–896.

Bell, J. M., Whitmore, W. L., Barnes, G., Seidler, F. J., and Slotkin, T. A. (1986a). Perinatal dietary exposure to soy lecithin: Altered sensitivity to central cholinergic stimulation. *Int. J. Dev. Neurosci.* **4**, 497–501.

Bell, J. M., Whitmore, W. L., Cowery, T., and Slotkin, T. A. (1986b). Perinatal dietary supplementation with a soy lecithin preparation: Effects on development of central catecholaminergic neurotransmitter systems. *Brain Res. Bull.* **17**, 189–195.

Bell, J. M., Whitmore, W. L., Queen, K. L., Orband-Miller, L., and Slotkin, T. A. (1987). Biochemical determinants of growth sparing during neonatal nutritional deprivation or enhancement: Ornithine decarboxylase, polyamines, and macromolecules in brain regions and heart. *Pediatr. Res.* **22**, 599–604.

Bell, J. M., Whitmore, W. L., and Slotkin, T. A. (1988). Neonatal nutritional deprivation or enhancement: The cardiac–sympathetic axis and its role in cardiac growth and stress responses. *Pediatr. Res.* **23**, 423–427.

Black, I. B. (1978). Regulation of autonomic development. *Annu. Rev. Neurosci.* **1**, 183–214.

Black, I. B. (1980). Developmental regulation of neurotransmitter phenotype. *Curr. Top. Dev. Biol.* **15**, 27–40.

Brazier, M. A. B. (1975). "Growth and Development of the Brain: Nutritional, Genetic and Environmental Factors." Raven Press, New York.

Breese, G. R., and Traylor, T. D. (1972). Developmental characteristics of brain catecholamines and tyrosine hydroxylase in the rat: Effects of 6-hydroxydopamine. *Br. J. Pharmacol.* **44**, 210–222.

Brodie, B. B., Costa, E., Dlabac, A., Neff, N. H., and Smookler, H. H. (1966). Application of steady-state kinetics to the estimation of synthesis rate and turnover time of tissue catecholamines. *J. Pharmacol. Exp. Ther.* **154**, 493–498.

Butler, N. R., and Goldstein, H. (1973). Smoking in pregnancy and subsequent child development. *Br. Med. J.* **4**, 573–574.

Cairns, N. J., and Wonnacott, S. (1988). [³H](−)nicotine-binding sites in fetal human brain. *Brain Res.* **475**, 1–7.

Charney, D. S., Menkes, D. B., and Heninger, G. R. (1981). Receptor sensitivity and the mechanism of action of antidepressant treatment. *Arch. Gen. Psychiatr.* **38**, 1160–1180.

Clarke, P. B. S., Schwartz, R. D., Paul, S. M., Pert, C. B., and Pert, A. (1985). Nicotinic binding in rat brain: Autoradiographic comparison of [³H]acetylcholine, [³H]nicotine, and [¹²⁵I]-α-bungarotoxin. *J. Neurosci.* **5**, 1307–1315.

Cole, P. V., Hawkins, L. H., and Roberts, D. (1972). Smoking during pregnancy and its effect on the fetus. *J. Obstet. Gynæcol. Br. Commonw.* **79**, 782–787.

Cooper, J. R., Bloom, F. E., and Roth, R. H. (1986). "The Biochemical Basis of Neuropharmacology," 5th Ed. Oxford University Press, New York.

Costa, E., Green, A. R., Koslow, S. H., LeFevre, H. F., Revuelta, A. V., and Wang, C. (1972). Dopamine and norepinephrine in noradrenergic axons: A study *in vivo* of their precursor product relationship by mass fragmentography and radiochemistry. *Pharmacol. Rev.* **24**, 167–190.

Coyle, J. T., and Axelrod, J. (1971). Development of the uptake and storage of L-[³H]norepinephrine in the rat brain. *J. Neurochem.* **18**, 2061–2075.

Coyle, J. T., and Axelrod, J. (1972). Tyrosine hydroxylase in rat brain: Developmental characteristics. *J. Neurochem.* **19**, 1117–1123.

Coyle, J. T., and Yamamura, H. I. (1976). Neurochemical aspects of the ontogenesis of cholinergic neurons in the rat brain. *Brain Res.* **118**, 429–440.

Criswell, H., Mueller, R. A., and Breese, G. R. (1989). Priming of D_1-dopamine receptor responses: Long-lasting behavioral supersensitivity to a D_1-dopamine agonist following repeated administration to neonatal 6-OHDA-lesioned rats. *J. Neurosci.* **9**, 125–133.

Davies, A. O., and Lefkowitz, R. J. (1984). Regulation of β-adrenergic receptors by steroid hormones. *Annu. Rev. Physiol.* **46**, 119–130.

Deskin, R., and Slotkin, T. A. (1981). Central catecholaminergic lesions in the developing rat: Effects on cardiac noradrenaline levels, turnover, and release. *J. Autonom. Pharmacol.* **1**, 205–210.

Deskin, R., Seidler, F. J., Whitmore, W. L., and Slotkin, T. A. (1981). Development of α-noradrenergic and dopaminergic receptor systems depends on maturation of their presynaptic nerve terminals in the rat brain. *J. Neurochem.* **36**, 1683–1690.

Dobbing, J., and Sands, J. (1979). Comparative aspects of the brain growth spurt. *Early Hum. Dev.* **3**, 79–83.

Dori, I., and Parnavelas, J. G. (1989). The cholinergic innervation of the rat cerebral cortex shows two distinct phases in development. *Exp. Brain Res.* **76**, 417–423.

Dow, T. G. B., Rooney, P. J., and Spence, M. (1975). Does anemia increase the risks to the fetus caused by smoking in pregnancy? *Br. Med. J.* **4**, 253–254.

Duncan, C. P., Seidler, F. J., Lappi, S. E., and Slotkin, T. A. (1990). Dual control of DNA synthesis by α- and β-adrenergic mechanisms in normoxic and hypoxic neonatal rat brain. *Dev. Brain Res.* **55**, 29–33.

Duncan, G. E., Criswell, H. E., McCown, T. J., Paul, I. A., Mueller, R. A., and Breese, G. R. (1987). Behavioral and neurochemical responses to haloperidol and SCH-23390 in rats treated neonatally or as adults with 6-hydroxydopamine. *J. Pharmacol. Exp. Ther.* **243**, 1027–1034.

Eriksson, M., Larsson, G., and Zetterstrom, R. (1979). Abuse of alcohol, drugs, and tobacco during pregnancy: Consequences for the child. *Paediatrician* **8**, 228–242.

Geula, C., Tokuno, H., Hersh, L., and Mesulam, M.-M. (1990). Human striatal cholinergic neurons in development, aging, and Alzheimer's disease. *Brain Res.* **508**, 310–312.

Hagino, N., and Lee, J. W. (1985). Effect of maternal nicotine on the development of sites for [³H]nicotine binding in the fetal brain. *Int. J. Dev. Neurosci.* **3**, 567–571.

Haglund, B., and Cnattingius, S. (1990). Cigarette smoking as a risk factor for sudden infant death syndrome: A population-based study. *Am. J. Pub. Health* **80**, 29–32.

Heby, O. (1981). Role of polyamines in the control of cell proliferation and differentiation. *Differentiation* **19**, 1–20.

Heby, O., and Emanuelsson, H. (1981). Role of the polyamines in germ cell differentiation and in early embryonic development. *Med. Biol.* **59,** 417–422.

Hohmann, C. F., Brooks, A. R., and Coyle, J. T. (1988). Neonatal lesions of the basal forebrain cholinergic neurons result in abnormal cortical development. *Dev. Brain Res.* **42,** 253–264.

Hou, Q.-C., Baker, F. E., Seidler, F. J., Bartolome, M., Bartolome, J., and Slotkin, T. A. (1989a). Role of sympathetic neurons in development of β-adrenergic control of ornithine decarboxylase activity in peripheral tissues: Effects of neonatal 6-hydroxydopamine treatment. *J. Dev. Physiol.* **11,** 139–146.

Hou, Q.-C., Seidler, F. J., and Slotkin, T. A. (1989b). Development of the linkage of β-adrenergic receptors to cardiac hypertrophy and heart rate control: Neonatal sympathectomy with 6-hydroxydopamine. *J. Dev. Physiol.* **11,** 305–311.

Isaac, P. F., and Rand, M. J. (1972). Cigarette smoking and plasma levels of nicotine. *Nature* **236,** 308–310.

Jonsson, G., and Hallman, H. (1980). Effects of neonatal nicotine administration on the postnatal development of central noradrenaline neurons. *Acta Physiol. Scand.* (Suppl.) **479,** 25–26.

Kirksey, D. F., and Slotkin, T. A. (1979). Concomitant development of [^{3}H]-dopamine and [^{3}H]-5-hydroxytryptamine uptake systems in rat brain regions. *Br. J. Pharmacol.* **67,** 387–391.

Klemm, N., and Kuhar, M. J. (1979). Post-mortem changes in high affinity choline uptake. *J. Neurochem.* **32,** 1487–1494.

Kostrzewa, R. M., and Saleh, M. I. (1989). Impaired ontogeny of striatal dopamine D_1 and D_2 binding sites after postnatal treatment of rats with SCH-23390 and spiroperidol. *Dev. Brain Res.* **45,** 95–101.

Krous, H. F., Campbell, G. A., Fowler, M. W., Carton, A. C., and Farber, J. P. (1981). Maternal nicotine administration and fetal brain stem damage: A rat model with implications for sudden infant death syndrome. *Am. J. Obstet. Gynecol.* **140,** 743–746.

Kudlacz, E. M., Navarro, H. A., Eylers, J. P., Lappi, S. E., Dobbins, S. S., and Slotkin, T. A. (1989). Effects of prenatal terbutaline exposure on cellular development in lung and liver of neonatal rat: Ornithine decarboxylase activity and macromolecules. *Pediatr. Res.* **25,** 617–622.

Kudlacz, E. M., Navarro, H. A., Eylers, J. P., and Slotkin, T. A. (1990a). Adrenergic modulation of cardiac development in the rat: Effects of prenatal exposure to propranolol via continuous maternal infusion. *J. Dev. Physiol.* **13,** 243–249.

Kudlacz, E. M., Navarro, H. A., Eylers, J. P., and Slotkin, T. A. (1990b). Prenatal exposure to propranolol via continuous maternal infusion: Effects on physiological and biochemical processes mediated by beta-adrenergic receptors in fetal and neonatal rat lung. *J. Pharmacol. Exp. Ther.* **252,** 42–50.

Kudlacz, E. M., Navarro, H. A., Kavlock, R. J., and Slotkin, T. A. (1990c). Regulation of postnatal β-adrenergic receptor/adenylate cyclase development by prenatal agonist stimulation and steroids: Alterations in rat kidney and lung after exposure to terbutaline or dexamethasone. *J. Dev. Physiol.* **14,** 273–281.

Kudlacz, E. M., Navarro, H. A., and Slotkin, T. A. (1990d). Regulation of β-adrenergic receptor-mediated processes in fetal rat lung: Selective desensitization caused by chronic terbutaline exposure. *J. Dev. Physiol.* **14,** 103–108.

Kudlacz, E. M., Spencer, J. R., and Slotkin, T. A. (1991). Postnatal alterations in β-adrenergic receptor binding in rat brain regions after continuous prenatal exposure to propranolol via maternal infusion. *Res. Commun. Chem. Pathol. Pharmacol.* **71,** 153–161.

Kunos, G., and Ishac, E. J. N. (1987). Mechanism of inverse regulation of alpha$_1$- and beta-adrenergic receptors. *Biochem. Pharmacol.* **36,** 1185–1191.

Larsson, C., Nordberg, A., Falkeborn, Y., and Lundberg, R.-Å. (1985). Regional [^{3}H]acetylcholine and [^{3}H]nicotine binding in developing mouse brain. *Int. J. Dev. Neurosci.* **3,** 667–671.

Lichtensteiger, W., and Schlumpf, M. (1985). Prenatal nicotine affects fetal testosterone and sexual dimorphism of saccharin preference. *Pharmacol. Biochem. Behav.* **23,** 439–444.

Lichtensteiger, W., Schlumpf, M., and Ribary, U. (1987). Modifications pharmacologiques de l'ontogenèse neuroendocrine. *Ann. Endocrinol.* **48,** 393–399.

Longo, L. O. (1977). The biological effects of carbon monoxide on the pregnant woman, fetus, and newborn infant. *Am. J. Obstet. Gynecol.* **129,** 69–103.

Martin, J. C., and Becker, R. F. (1970). The effects of nicotine administration *in utero* upon activity in the rat. *Psychoneuron. Sci.* **19,** 59–60.

Martin, J. C., and Becker, R. F. (1971). The effects of maternal nicotine absorption or hypoxic episodes upon appetitive behavior of rat offspring. *Dev. Psychobiol.* **4**, 133–147.

McFarland, B. J., Seidler, F. J., and Slotkin, T. A. (1991). Inhibition of DNA synthesis in neonatal rat brain regions caused by acute nicotine administration. *Dev. Brain Res.* **58**, 223–229.

Meyer, D. C., and Carr, L. A. (1987). The effects of perinatal exposure to nicotine on plasma LH levels in prepubertal rats. *Neurotoxicol. Teratol.* **9**, 95–98.

Murrin, L. C., (1980). High-affinity transport of choline in neuronal tissue. *Pharmacology* **21**, 132–140.

Murrin, L. C., Ferrer, J. R., Zeng, W. (1985). Nicotine administration during pregnancy and its effect on striatal development. *Neurosci. Abs.* **11**, 69.

Murrin, L. C., Ferrer, J. R., Wanyun, Z., and Haley, N. J. (1987). Nicotine administration to rats: Methodological considerations. *Life Sci.* **40**, 1699–1708.

Mytilineou, C., and Black, I. B. (1978). Development of adrenergic nerve terminals: The effects of decentralization. *Brain Res.* **158**, 259–268.

Naeye, R. L. (1978). Effects of maternal cigarette smoking on the fetus and placenta. *Br. J. Obstet. Gynæcol.* **85**, 732–737.

Nagatsu, T. (1973). "Biochemistry of Catecholamines." University Park Press, Tokyo.

Naquira, D., and Arqueros, L. (1978). Water intake in rats affected by addition of nicotine to drinking water. *Acta Physiol. Lat. Am.* **28**, 73–76.

Nasrat, H. A., Al-Hachim, G. M., and Mahmood, F. A. (1986). Perinatal effects of nicotine. *Biol. Neonate* **49**, 8–14.

Navarro, H. A., Seidler, F. J., Whitmore, W. L., and Slotkin, T. A. (1988). Prenatal exposure to nicotine via maternal infusions: Effects on development of catecholamine systems. *J. Pharmacol. Exp. Ther.* **244**, 940–944.

Navarro, H. A., Seidler, F. J., Eylers, J. P., Baker, F. E., Dobbins, S. S., Lappi, S. E., and Slotkin, T. A. (1989a). Effects of prenatal nicotine exposure on development of central and peripheral cholinergic neurotransmitter systems. Evidence for cholinergic trophic influences in developing brain. *J. Pharmacol. Exp. Ther.* **251**, 894–900.

Navarro, H. A., Seidler, F. J., Schwartz, R. D., Baker, F. E., Dobbins, S. S., and Slotkin, T. A. (1989b). Prenatal exposure to nicotine impairs nervous system development at a dose which does not affect viability or growth. *Brain Res. Bull.* **23**, 187–192.

Navarro, H. A., Mills, E., Seidler, F. J., Baker, F. E., Lappi, S. E., Tayyeb, M. I., Spencer, J. R., and Slotkin, T. A. (1990a). Prenatal nicotine exposure impairs β-adrenergic function: Persistent chronotropic subsensitivity despite recovery from deficits in receptor binding. *Brain Res. Bull.* **25**, 233–237.

Navarro, H. A., Slotkin, T. A., Tayyeb, M. I., Lappi, S. E., and Seidler, F. J. (1990b). Effects of fetal nicotine exposure on development of adrenergic receptor binding in rat brain regions: Selective changes in α_1-receptors. *Res. Commun. Substance Abuse* **11**, 95–103.

Navarro, H. A., Kudlacz, E. M., Kavlock, R. J., and Slotkin, T. A. (1991). Prenatal terbutaline treatment: Tissue-selective dissociation of perinatal changes in β-adrenergic receptor binding from regulation of adenylate cyclase activity. *Life Sci.* **48**, 269–274.

Patterson, P. H., and Chun, L. L. Y. (1977). The induction of acetylcholine synthesis in primary cultures of dissociated rat sympathetic neurons: II. Developmental aspects. *Dev. Biol.* **60**, 473–481.

Peters, D. A. V. (1984). Prenatal nicotine exposure increases adrenergic receptor binding in the rat cerebral cortex. *Res. Commun. Chem. Pathol. Pharmacol.* **46**, 307–317.

Peters, D. A. V., Taub, H., and Tang, S. (1979). Postnatal effects of maternal nicotine exposure. *Neurobehav. Toxicol.* **1**, 221–225.

Reinis, S., and Goldman, J. M. (1980). "The Development of the Brain: Biological and Functional Perspectives." Charles C Thomas, Springfield, Illinois.

Ribary, U., and Lichtensteiger, W. (1989). Effects of acute and chronic prenatal nicotine treatment on central catecholamine systems of male and female rat fetuses and offspring. *J. Pharmacol. Exp. Ther.* **248**, 786–792.

Ross, D., Johnson, M., and Burge, R. (1977). Development of cholinergic characteristics in adrenergic neurones is age dependent. *Nature* **267**, 536–639.

Schwartz, R. D., and Kellar, K. J. (1985). *In vivo* regulation of [³H]acetylcholine recognition sites in brain by nicotinic cholinergic drugs. *J. Neurochem.* **45,** 427–433.

Schwartz, R. D., McGee, R., and Kellar, K. J. (1982). Nicotinic cholinergic receptors labeled by [³H]acetylcholine in rat brain. *Mol. Pharmacol.* **22,** 56–62.

Seidler, F. J., and Slotkin, T. A. (1981). Development of central control of norepinephrine turnover and release in the rat heart: Responses to tyramine, 2-deoxyglucose and hydralazine. *Neuroscience* **6,** 2081–2086.

Seidler, F. J., and Slotkin, T. A. (1990). Effects of acute hypoxia on neonatal rat brain: Regionally selective, long-term alterations in catecholamine levels and turnover. *Brain Res. Bull.* **24,** 157–161.

Seidler, F. J., Whitmore, W. L., and Slotkin, T. A. (1982). Delays in growth and biochemical development of rat brain caused by maternal methadone administration: Are the alterations in synaptogenesis and cellular maturation independent of reduced maternal food intake? *Dev. Neurosci.* **5,** 13–18.

Seidler, F. J., Bell, J. M., and Slotkin, T. A. (1990). Undernutrition and overnutrition in the neonatal rat: Long-term effects on noradrenergic pathways in brain regions. *Pediatr. Res.* **27,** 191–197.

Simon, J. R., Atweh, S., and Kuhar, M. J. (1976). Sodium-dependent high-affinity choline uptake: A regulatory step in the synthesis of acetylcholine. *J. Neurochem.* **26,** 909–922.

Singer, H. S., Coyle, J. T., Vernon, N., Kallman, C. H., and Price, D. L. (1980). The development of catecholaminergic innervation in chick spinal cord. *Brain Res.* **191,** 417–428.

Singh, V. K., and McGeer, E. G. (1977). Choline acetyltransferase in developing rat brain and spinal cord. *Brain Res.* **127,** 159–163.

Slotkin, T. A. (1979). Ornithine decarboxylase as a tool in developmental neurobiology. *Life Sci.* **24,** 1623–1630.

Slotkin, T. A. (1986). Endocrine control of synaptic development in the sympathetic nervous system: The cardiac–sympathetic axis. *In* "Developmental Neurobiology of the Autonomic Nervous System" (P. M. Gootman, ed.), pp. 97–133. Humana Press, Clifton, New Jersey.

Slotkin, T. A. (1988). Perinatal exposure to methadone: How do early biochemical alterations cause neurofunctional disturbances? *Prog. Brain Res.* **73,** 265–279.

Slotkin, T. A., and Bartolome, J. (1986). Role of ornithine decarboxylase and the polyamines in nervous system development: A review. *Brain Res. Bull.* **17,** 307–320.

Slotkin, T. A., and Bartolome, J. (1987). Biochemical mechanisms of developmental neurotoxicity of methylmercury. *Neurotoxicology* **8,** 65–84.

Slotkin, T. A., and Thadani, P. V. (1980). Neurochemical teratology of drugs of abuse. *In* "Advances in the Study of Birth Defects, Vol. 4: Neural and Behavioural Teratology." (T. V. N. Persaud, ed.), pp. 199–234. MTP Press, Lancaster, England.

Slotkin, T. A., Cho, H., and Whitmore, W. L. (1987a). Effects of prenatal nicotine exposure on neuronal development: Selective actions on central and peripheral catecholaminergic pathways. *Brain Res. Bull.* **18,** 601–611.

Slotkin, T. A., Whitmore, W. L., Salvaggio, M., and Seidler, F. J. (1979). Perinatal methadone addiction affects brain synaptic development of biogenic amine systems in the rat. *Life Sci.* **24,** 1223–1230.

Slotkin, T. A., Weigel, S. J., Barnes, G. A., Whitmore, W. L., and Seidler, F. J. (1981). Alterations in the development of catecholamine turnover induced by perinatal methadone: Differences in central vs. peripheral sympathetic nervous systems. *Life Sci.* **29,** 2519–2525.

Slotkin, T. A., Pachman, S., Kavlock, R. J., and Bartolome, J. (1985a). Early biochemical detection of adverse effects of a neurobehavioral teratogen: Influence of prenatal methylmercury exposure on ornithine decarboxylase in brain and other tissues of fetal and neonatal rat. *Teratology* **32,** 195–202.

Slotkin, T. A., Pachman, S., Kavlock, R. J., and Bartolome, J. (1985b). Effects of neonatal methylmercury exposure on development of nucleic acids and proteins in rat brain: Regional specificity. *Brain Res. Bull.* **14,** 397–400.

Slotkin, T. A., Cowery, T. S., Orband, L., Pachman, S., and Whitmore, W. L. (1986a). Effects of neonatal hypoxia on brain development in the rat: Immediate and long-term biochemical alterations in discrete regions. *Brain Res.* **374,** 63–74.

Slotkin, T. A., Greer, N., Faust, J., Cho, H., and Seidler, F. J. (1986b). Effects of maternal nicotine injections on brain development in the rat: Ornithine decarboxylase activity, nucleic acids, and proteins in discrete brain regions. *Brain Res. Bull.* 17, 41–50.

Slotkin, T. A., Orband-Miller, L., and Queen, K. L. (1987b). Development of [³H]nicotine-binding sites in brain regions of rats exposed to nicotine prenatally via maternal injections or infusions. *J. Pharmacol. Exp. Ther.* 242, 322–237.

Slotkin, T. A., Orband-Miller, L., and Queen, K. L. (1987c). Do catecholamines contribute to the effects of neonatal hypoxia on development of brain and heart? Influence of concurrent α-adrenergic blockade on ornithine decarboxylase activity. *Int. J. Dev. Neurosci.* 5, 135–143.

Slotkin, T. A., Orband-Miller, L., Queen, K. L., Whitmore, W. L., and Seidler, F. J. (1987d). Effects of prenatal nicotine exposure on biochemical development of rat brain regions: Maternal drug infusions via osmotic minipumps. *J. Pharmacol. Exp. Ther.* 240, 602–611.

Slotkin, T. A., Whitmore, W. L., Orband-Miller, L., Queen, K. L., and Haim, K. (1987e). Beta-adrenergic control of macromolecule synthesis in neonatal rat heart, kidney, and lung: Relationship to sympathetic neuronal development. *J. Pharmacol. Exp. Ther.* 243, 101–109.

Slotkin, T. A., Lau, C., Kavlock, R. J., Gray, J. A., Orband-Miller, L., Queen, K. L., Baker, F. E., Cameron, A. M., Antolick, L., Haim, K., Bartolome, M., and Bartolome, J. (1988a). Role of sympathetic neurons in biochemical and functional development of the kidney: Neonatal sympathectomy with 6-hydroxydopamine. *J. Pharmacol. Exp. Ther.* 246, 427–433.

Slotkin, T. A., Lau, C., Kavlock, R. J., Whitmore, W. L., Queen, K. L., Orband-Miller, L., Bartolome, M., Baker, F. E., Cameron, A. M., Antolick, L., and Bartolome, J. V. (1988b). Trophic control of lung development by sympathetic neurons: Effects of neonatal sympathectomy with 6-hydroxydopamine. *J. Dev. Physiol.* 10, 577–590.

Slotkin, T. A., Windh, R., Whitmore, W. L., and Seidler, F. J. (1988c). Adrenergic control of DNA synthesis in developing rat brain regions: Effects of intracisternal administration of isoproterenol. *Brain Res. Bull.* 21, 737–740.

Slotkin, T. A., Kudlacz, E. M., Lappi, S. E., Tayyeb, M. I., and Seidler, F. J. (1990a). Fetal terbutaline exposure causes selective postnatal increases in cerebellar α-adrenergic receptor binding. *Life Sci.* 47, 2051–2057.

Slotkin, T. A., Navarro, H. A., McCook, E. C., and Seidler, F. J. (1990b). Fetal nicotine exposure produces postnatal up-regulation of adenylate cyclase activity in peripheral tissues. *Life Sci.* 47, 1561–1567.

Slotkin, T. A., Seidler, F. J., Crain, B. J., Bell, J. M., Bissette, G., and Nemeroff, C. B. (1990c). Regulatory changes in presynaptic cholinergic function assessed in rapid autopsy material from patients with Alzheimer disease: Implications for etiology and therapy. *Proc. Natl. Acad. Sci. U.S.A.* 87, 2452–2455.

Sonawane, B. R. (1982). Effects of prenatal nicotine exposure on reproductive function of rat offspring. *Teratology* 25, 77A.

Wagner, J. P., Seidler, F. J., and Slotkin, T. A. (1991). Presynaptic input regulates development of β-adrenergic control of rat brain ornithine decarboxylase: Effects of 6-hydroxydopamine or propranolol. *Brain Res. Bull.* 26, 885–890.

Weiss, E. R., Kelleher, D. J., Woon, C. W., Soparkar, S., Osawa, S., Heasley, L. E., and Johnson, G. L. (1988). Receptor activation of G proteins. *FASEB J.* 2, 2841–2848.

Yoshida, K., Kato, Y., and Imura, H. (1980). Nicotine-induced release of noradrenaline from hypothalamic synaptosomes. *Brain Res.* 182, 361–368.

— 7 —

Prenatal Cocaine and Marijuana Exposure: Research and Clinical Implications

Barry Zuckerman and Deborah A. Frank

Division of Developmental and
Behavioral Pediatrics
Boston City Hospital
Boston University
Boston, Massachusetts

INTRODUCTION

Those who study illicit drug use during pregnancy in human populations cannot attain the scientific rigor of the animal researcher, at whose disposal are genetically homogeneous subjects, random assignment to drug exposure, known dose and known dosing interval, and the ability to control confounding variables through techniques such as pair feeding. Unfortunately, although often helpful in defining mechanisms of toxicity, animal studies cannot fully answer questions important for public policy. Drug toxicity and metabolism vary greatly among species (Hutchings, 1990). Animal tests are often poorly predictive of human teratogenicity; thalidomide, the most powerful human teratogen ever identified, was tested without ill effect in a wide range of laboratory animals, whereas aspirin and caffeine are potent teratogens in animals, but not in humans (Lewis, 1987). Moreover, no precise animal analog exists for sudden infant death syndrome (SIDS) or for higher-level cognitive and socioaffective functions, such as verbal memory or attachment, which are of theoretical concern following the utero exposure of illicit drugs. Thus, to address critical questions, researchers must accept a level of methodologic uncertainty intrinsic to the epidemiologic and clinical methods available for the study of illicit drug use as it actually occurs in free-living human samples.

Before summarizing the findings of research on in utero cocaine and marijuana exposure in humans, we shall review the methodologic difficulties that must be considered before valid conclusions may be drawn to guide public health and clinical interventions. The methodologic critique outlined here is not unique to the study of cocaine and marijuana, reflecting rather the thinking of those who have sought to evaluate the potentially subtle effects of other possible human teratogens and neurotoxins, including the work of Needleman and colleagues (1979) on childhood lead toxicity, of the Jacobsons and colleagues (1984) regarding prenatal exposure to polychlorinated biphenyls, and or Streissguth and associates on prenatal alcohol exposure (1989a,b; Sampson et al., 1989) as well as the recent review of methodologic issues of cocaine research by Zuckerman (1991).

After reviewing methodologic issues, we shall describe the transactional model of Sameroff and Chandler (1975) as a context for exploring complex interactions of biologic and social factors that must be considered in evaluating suspected long-term effects of prenatal cocaine and marijuana exposure on children. Maternal nutrition will serve as an example of an important interrelated biologic factor, and domestic violence and maternal depression, as examples of often unmeasured social issues. We shall then outline the results of studies on cocaine and marijuana individually, including possible mechanisms, effects on intrauterine growth, questions of association with birth defects, and controversies regarding short and long-term behavioral consequences of prenatal exposure. Finally we shall propose a clinical model for serving mothers and infants burdened by drug use.

METHODOLOGIC ISSUES

Identification of Exposure

In any study of possible teratogens, issues of accurate identification of gestational timing of exposure, as well as intensity and frequency of exposure, are critical. The illegal nature of cocaine and marijuana make accurate identification of exposure particularly difficult. The first difficulty is determining to which psychoactive substances the maternal–fetal unit has been exposed. There is no standardization of dosage or determination of purity for illegal drugs. Potency may vary, over time, by geographic location, and from exposure to exposure. For example, the potency of confiscated marijuana specimens, as measured by concentration of delta-9-THC (tetrahydrocannabinol), increased steadily from 1975 to 1984 (Elsohly and Elsohly, 1989). Thus, generalizations from one region of the country to another or from experience in the 1970s and 1980s to formulations of illicit drugs currently available may not be valid. Moreover, the user may be unknowingly exposed to multiple adulterants, including for example, in the case of cocaine use, amphetamines, lidocaine, ether, procaine, or talcum, which may influence outcomes of interest in unknown ways (Washton, 1989).

Because use of cocaine and marijuana are illegal and highly stigmatized during pregnancy, determination of use by maternal self-report alone will fail to identify many exposed pregnancies. Women who admit to cocaine and marijuana use are probably providing truthful information, within the limits of recall bias. In one study, approximately half of all cocaine users were identified on the basis of interview, in spite of negative urine assays (Frank et al., 1988). Interviews are currently the only method of assessing patterns of drug use, distinguishing, for example, binge from daily use. In addition, interviews are critical for exploring intrauterine exposures in the period before the pregnancy is recognized and the mother has come to medical attention (Day and Richardson, 1991).

Knowing that one's urine will be tested for drugs enhances the likelihood that respondents will report illicit drug use in pregnancy, without altering reporting of use of legal substances such as cigarettes and alcohol (Hingson et al., 1986). However, our own work suggests that even when urine is being tested, reliance on self-report alone leads to significant misclassification of pregnant cocaine and marijuana users as non-users (Frank et al., 1988; Zuckerman et al., 1989c). In a sample of 1226 mothers consecutively recruited during prenatal care, 24% of those who used cocaine and 16% of those who used marijuana denied use during a detailed confidential interview and were identified solely on the basis of urine assays (Zuckerman et al., 1989c). Interviews alone are probably even less able to identify users in clinical, rather than research settings where confidentiality and protection from prosecution and from protective service intervention cannot be assured. In one clinically based study, only 43% of cocaine-exposed pregnancies identified by peripartum urine screening were identified by clinical history taking (Neerhof et al., 1989).

The importance of biologic markers for identification of illicit drug use is now well recognized by both clinicians and researchers. Most published work and clinical practice rely on enzyme-mediated immunoassay technique (EMIT) (Syva Company, Palo Alto, California) of either maternal or infant urine. The duration of positive urine findings following use of either cocaine or marijuana varies with the amount, mode, and chronicity of use, as well as with the fluid intake and renal function of the person furnishing the specimen. Benzoylecgonine, a cocaine metabolite, may be detected in the urine of adults for 24 to 72 hr after last use by EMIT (Hamilton et al., 1977). Use of more sensitive radioimmunoassay methods extends the period of potential detection after last use to 7 days (Hamilton et al., 1977). Marijuana metabolites are excreted more slowly, and in chronic users may be detected as long as 70 days after last use (Nahas, 1976) by EMIT.

No large neonatal samples have been studied, so that duration of neonatal excretion following prenatal exposure are unknown, but assumed to be prolonged toward the outer limits of adult norms (Chasnoff et al., 1986), at least for cocaine. Fulroth and colleagues (1989), using thin-layer chromatography with EMIT confirmation, reported marked discrepancies in detection of cocaine metabolites measured in urines obtained at time of hospital admission for mothers in labor and immediately following delivery for newborns. Among 52 mother–infant pairs, cocaine metabolites were found in the urine of both mother and infant in only 26, whereas in 11 exposed dyads, maternal urines were positive, but infants', negative. Conversely, 17 infants were excreting cocaine metabolites even though their mother's urines were negative. It may be that when mothers use very close to delivery, insufficient transfer to the fetus occurs to be detected right after birth. On the other hand, one may speculate that immature metabolism may result in continued excretion by the newborn of drugs no longer detectable in maternal urine. Thus, although urine assays detect more drug-exposed pregnancies than self-report alone, reliance on either maternal or newborn urine alone for identification of illicit drug exposure may still miss exposed cases.

Because of the relatively short duration of excretion of cocaine and marijuana metabolites in urine following last use, identification of cocaine or marijuana use on the basis of urine assays obtained at delivery biases research samples toward women who use these substances late in pregnancy. Our data (Frank et al., 1988) suggest that urine assays positive at delivery identify relatively heavy and frequent users, whose outcomes may not be representative of those who use early in pregnancy and discontinue use. Most other studies have relied on peripartum assays (e.g., Burkett et al., 1990; Bingol et al., 1987) to identify exposure and thus are biased toward heavier users. Ideally, for research purposes, urine assays for illicit drugs should be obtained frequently throughout pregnancy to identify accurately timing and frequency of drug exposure. However, users of illicit drugs, particularly cocaine, often present late or not at all for prenatal care and thus can pragmatically be identified only at delivery.

The unreliability of self-report and the importance of characterizing the effects of drug exposure during the vulnerable period of early pregnancy have aroused interest in the measurement of drug metabolites in mediums other than urine. Analysis of maternal and neonatal hair for cocaine metabolites has been proposed (Graham et al., 1989) as a method for identifying previous cocaine exposure among mother–infant pairs whose urines are negative for drug metabolites at delivery. The process is technically complex. The unknown risks of contamination of hair with illicit drugs by passive environmental exposure and of alteration of drug identification by chemicals applied for cosmetic purposes, renders current methods of hair analysis of questionable usefulness for most research and clinical purposes (Bailey, 1989). More promising is the radioimmunoassay of meconium (the first neonatal stools) for cocaine, marijuana, and opiate metabolites (Ostrea et al., 1989; Ostrea et al., 1992). Although analyses of meconium will not be able to pinpoint gestational week of exposure, they may eventually be able to quantify cumulative dosage to the fetus from the period of meconium production, approximately 18 to 20 gestational weeks through delivery, on the basis of drug concentration per gram of meconium. Many questions remain, however, regarding distribution of drug metabolites in any given specimen of meconium, preparation of samples for assay, and how many stools must be sampled, particularly if the first stool cannot be assessed because it is passed in utero (Maynard et al., 1991). Further validation of meconium assays is needed before they enter widespread research and clinical use.

The issues of the effects of prenatal exposure to cocaine and marijuana are further complicated by the fact that exposure may not end at delivery, but may continue via breast feeding, passive inhalation, accidental poisoning, or intentional administration (Bateman and Heagarty, 1989; Shannon et al., 1989; Rivkin and Gilmore, 1989; Chasnoff et al., 1987c; Chaney et al., 1988; Tennes et al., 1985; Astely and Little, 1990). Prenatal and postneonatal exposure are often highly correlated. For example, a recent study could not distinguish statistically between the developmental effects of marijuana exposure via breast feeding in the first postpartum month and those of in utero exposure in the first trimester (Astely and Little, 1990). Nothing is known about potentially differential effects to the developing infant nervous system of exposure to any illicit drug by postneonatal inhalation or ingestion, compared to the effect of exposure prenatally.

In sum, optimal characterization of illicit drug use in pregnancy requires skilled and confidential interview complemented by the use of one or more biologic markers of drug exposure. Even with best available methods of interview and bioassay, there is currently an irreducible level of uncertainty in accurate identification of dose and gestational timing of cocaine and marijuana exposure. Moreover, the possible effects of concurrent postneonatal exposure have not been considered in assessing the health and behavior of prenatally exposed infants and toddlers.

Failure to identify users accurately may bias outcome data toward either over-estimation or underestimation of drug effects. If users are misclassified as nonusers, differences in outcomes of interest that may be truly attributable to drug exposure may be minimized and thus overlooked. Conversely, since heaviest users are probably most readily identified on clinical grounds, the outcomes of light or moderate users may not be assessed, leading to overgeneralization of the risk of adverse outcome to all users on the basis of findings noted only among the heaviest users.

Sample Selection Bias

As Needleman and colleagues have pointed out (1979), for research to be applicable to the community at large, samples must be representative of the population. In the field of cocaine and marijuana research, there has been an unfortunate tendency to make sweeping generalizations about the short- and long-term outcomes of drug-exposed infants on the basis of biased samples, including women referred prenatally to drug treatment programs (Chasnoff et al., 1985), women who are concurrent users of opiates (Chasnoff et al., 1985) and women whose drug use is obvious enough to be suspected by clinical staff at delivery (Madden et al., 1986; Bingol et al., 1987; Oro and Dixon, 1987). Samples identified on clinical grounds alone may reflect prevailing ethnic and socioeconmic prejudices regarding drug use during pregnancy. Chasnoff and colleagues (1990) have recently shown that clinicians are ten times more likely to identify and report illicit substance abuse among black and low-income women than among white private patients at the time of delivery, although actual rates of prenatal use of illicit drugs are not significantly different between the two groups. Thus women identified solely on clinical grounds may not be truly representative of the spectrum of women who use cocaine or marijuana during pregnancy.

Even greater biases are introduced when generalizations are made from samples of children referred for developmental problems, whose in utero cocaine exposure is then retrospectively determined. For example, on the basis of only ten cases from five different clinical genetics referral centers, with no controls, it has been suggested that in utero cocaine exposure is associated with congenital anomalies from fetal vascular disruption (Hoyme et al., 1990). Similar claims were made regarding the teratogen-icity of prenatal marijuana use based on infants referred to a genetic counseling service (Qazi et al., 1985). When large prospective samples are studied, however, (Zuckerman et al., 1989; Day and Richardson, 1991), independent effects of cocaine and marijuana on either major or minor anomalies cannot be shown. Generalizations from populations of children with clinically obvious deficits will tend to overestimate the risks of drug exposure.

The issue of sample bias is further complicated by the necessity of long-term follow-up studies to ascertain potential effects of prenatal drug exposure that may not be obvious in the newborn period. Children and families retained in long-term studies may not be typical of those who formally withdraw or are lost to research follow-up. On the one hand, parents with obviously impaired children may cooperate more readily with research, biasing results toward less optimal outcomes. On the other hand, it may be that better-functioning families, least impaired by ongoing substance abuse, may be more likely to continue to cooperate with research follow-up, thus biasing results toward children of lower environmental risk.

Even attempts at meta-analysis, involving results from multiple published samples, cannot currently overcome the problem of sample selection bias, because the samples on which most published studies are based may not be truly representative. Negative studies regarding cocaine effects are far less likely to be selected for presentation at national scientific meetings than positive studies, even when the methodology of negative studies is technically superior (Koren et al., 1989). Thus results available in the scientific literature may not accurately represent drug effects in the population at large.

In the field of prenatal cocaine and marijuana use, sample selection bias cannot be fully overcome. If drug use is ascertained prospectively from samples presenting for prenatal care (Zuckerman et al., 1989c; Fried and Watkinson, 1988), one cannot generalize to users who do not seek prenatal care, who may be at greater risk of adverse outcomes. Conversely, if users are identified at delivery or by referral to drug treatment programs, findings can be generalized only to very heavy users, usually from socially disadvantaged backgrounds. In the absence of a nationally representative sample, with ascertainment of pregnant women who have not yet presented for prenatal care, investigators can delineate only the unavoidable biases inherent in the selection of their particular sample and avoid making generalizations to other noncomparable populations. Researchers engaged in longitudinal studies must use intensive outreach and participant reinforcement to minimize loss to follow-up. Since some loss to follow-up is inevitable, one must delineate the characteristics of those lost to follow-up compared to those retained to identify possible threats to the validity of outcome data.

Confounding Variables

To date no hypothesized or demonstrated effect of in utero cocaine or marijuana exposure has been found to be specific to that drug. Outcomes such as low birth weight or impaired attention are final common pathways of any number of possible risk factors. Thus identification and control of potential confounders is critical to permit valid inference regarding drug effects.

A confounding variable is one that is related to both the independent and dependent variables. We and other investigators have shown that cocaine and marijuana use during pregnancy are highly correlated with a number of other maternal characteristics and health habits that are recognized correlates of adverse pregnancy outcome and later developmental impairment. Failure to identify and control for these variables, as for example, in the studies of cocaine exposure and SIDS (Davidson et al., 1990; Durand et al., 1990), which do not control for maternal cigarette use, may lead to misattribution of the effects of the unmeasured variable to cocaine or marijuana. Lack of control for potential confounders clearly leads to overestimation of drug effects. For example, in our own work, infants of mothers whose urines were positive for cocaine during pregnancy were approximately 400 grams smaller than infants of mothers who did not use cocaine, but only 25% of the weight decrement could be statistically attributed to cocaine exposure, rather than other factors, such as cigarette use (Zuckerman 1989c). Moreover, factors such as poor maternal nutrition, or use of other psychoactive substances such as alcohol may not only statistically confound the effects of cocaine and marijuana, but may also interact physiologically with these substances in ways as yet unmeasured in humans. Church, Dincheff, and Gessner (1988) have shown in Long–Evans rats that alcohol plus cocaine exposure poses greater risk to pregnancy outcome than either drug alone. Similar interactive effects have been shown for marijuana and alcohol.

Random assignment of pregnant women to cocaine or marijuana exposure is clearly impossible. Therefore, to derive valid estimates of cocaine and marijuana effects on infant outcome, samples must be large enough to permit multivariate statistical consideration of many potentially confounding variables, including use of other psychoactive substances, medical risk factors such as maternal undernutrition and infections, and social risk factors such as poverty and maternal depression (Frank et al., 1988). In postneonatal follow-up studies, further confounders must be considered, including the quality of the child's home environment, and other possible postnatal biologic threats to child development, including for example, active human immunodeficiency virus (HIV) infection (Brouwers et al., 1991), chronic otitis, iron deficiency (Palti et al., 1983), malnutrition (Galler, 1984), and lead poisoning (Needleman et al., 1979). Results from small samples without characterization of potential confounders and statistical assessment of their impact should not be scientifically accepted unless configured in larger, controlled studies.

Lack of Blind Assessment and Comparison Populations

Research on prenatal cocaine and marijuana exposure has been further complicated by uncritical acceptance of findings made without reference to usual standards of scientific technique. It is axiomatic in social science research that unless outcomes of interest are blindly assessed, there is risk of the investigators' expectations unconsciously altering subject behavior as well as biasing investigators' interpretation of otherwise equivocal measures (Rosenthal and Rosenow, 1969). To assure blind interpretations it is essential that both exposed and unexposed subjects be presented to an examiner who is unaware of their exposure status. Moreover, ideally the comparison sample should be comparable to the exposed sample of major factors that are likely to influence development. For example, since most studies of cocaine-exposed infants have been drawn from minority, economically deprived populations, comparison with children of more privileged backgrounds would be inappropriate, since the effects of cocaine could not be distinguished from those of pervasive economic deprivation.

Although the expectations of the investigator are unlikely to distort evaluation of easily measured outcomes such as birth weight, more subtle outcome parameters such as fetal behavior (Hume et al., 1989), neonatal abstinence scores (Dixon and Bejar, 1989), or the quality of the child's play (Roding et al., 1989) may be distorted by nonblind assessments, as well as by use of noncontemporaneous controls and noncomparable measures (Roding et al., 1989).

Selection of Appropriate Outcome Measures

The potentially adverse effects of in utero exposure to psychoactive substances, whether legal or illegal, may vary over the life span by stage of development and extend over many areas of function, including among others, prenatal and post-neonatal growth, cognition, regulation of activity, attention, and affect, and response to stress. Thresholds of vulnerability vary from individual to individual and from organ system to organ system (Hutchings, 1990). As the Jacobsons and colleagues (1984) have outlined, at any given level of exposure to a potential toxin, different individuals may exhibit different effects, some totally unaffected, some showing impairment in only one area, and other showing multiple effects. Lack of effect in one area, for example intrauterine growth, does not assure that another area, such as neurobehavior, is also unaffected. Only at the highest levels of exposure are multiple possible effects likely to cluster in an identifiable syndrome, as is probably the case with fetal alcohol syndrome, in which the typical amount of maternal alcohol consumption is 14 drinks per day (Abel and Sokol, 1987).

Of all outcomes, physical growth is probably the easiest to measure, but not necessarily of greatest functional significance. No single battery of neurobehavioral tests is accepted for use to measure the effect of any potential behavioral teratogen. Areas thought to be potentially at risk from in utero cocaine exposure, such as developmental alterations in neurotransmitters, have no standardized associated functional measures. Many of the structural lesions noted on neonatal ultrasound following in utero cocaine exposure occur in the basal ganglia and frontal lobes (Dixon and Bejar, 1989), brain areas that are clinically silent in infancy and for which no well-standardized evaluations exist even in early childhood, although a number of measures are being evaluated (Welsh and Pennington, 1988).

Thus in evaluating possible drug effects, researchers are struggling with the problem that global neurobehavioral measures that are widely recognized, well standardized, and easily obtained (e.g., the Bayley Scales of Infant Development or the Achenbach Childhood Behavioral Checklist) may not tap into important areas of potential dysfunction. On the other hand, methods possibly more sensitive to subtle impairments, such as observations of spontaneous play, do not have established validity or reliability, may be difficult to obtain in large samples, and require specialized expertise not found in many centers. As Streissguth and colleagues have pointed out (1987a) it is critical to select age-appropriate measures that are neither so difficult that many children cannot attempt them nor so easy that there is no range of test responses. Current optimal strategy dictates that both standardized methods and evolving measures selected with clear theoretical justification be used in follow-up studies.

A MULTIFACTORIAL DEVELOPMENT MODEL

Primary versus Secondary Effects

The potential effects of prenatal cocaine and marijuana exposure may be conceptualized as primary or secondary. Primary effects are those ascribed to direct drug toxicity on developing organ systems. For example, if cocaine directly alters neurotransmitters which then alter the developing central nervous system, this would be a primary effect. Secondary effects are those attributable to changes in maternal physiology with subsequent alterations in fetal perfusion and in transfer of oxygen and nutrients to the fetus. Such effects would presumably be seen in infants whose intrauterine growth is affected by hypoxia, decreased perfusion, or malnutrition for any reason, including maternal hypertension, cigarette smoking, or maternal undernutrition. As the Jacobsons and colleagues (1984) and Lester and colleagues (1991) have shown, primary and secondary effects of in utero exposures can be differentiated statistically by various forms of multivariate analyses. These identify which behavioral findings in infancy can be explained solely on the basis of shortened gestation or low birth weight and which cannot be so explained and are therefore ascribed to primary toxic effects of exposure to a potential neuroteratogen.

Transactional Model

Whether primary or secondary, the developmental sequelae of prenatal exposure to drugs are best understood through a multifactorial transactional model (Sameroff and Chandler, 1975), which predicts that outcomes will not be related in any simple linear fashion to exposure, but will reflect interactions between the biologic correlates of exposure and the quality of the environment. The primary and secondary effects of prenatal drug exposure may create biologic vulnerability for neurobehavioral dysfunction, dysfunction that may be completely or in part compensated for by brain plasticity and by competent caretaking, but heightens the child's vulnerability to the effects of poor caretaking. We now look at the interaction between maternal drug use and maternal nutrition as an example of the issue of primary and secondary effects, and at issues of caretaker responsivity, family violence, and parental depression as examples of components of the transactional model.

Role of Nutrition

The relationship between marijuana and cocaine use, maternal nutritional status, and infant outcome is complex (Zuckerman et al., 1989c). One investigator of marijuana use in pregnant women found that users had higher weight for height at conception than did nonusers (Linn et al., 1983); another reported enhanced pregnancy weight gain among users (Tennes et al., 1985). In contrast, we (Zuckerman et al., 1989c) and other investigators (Fried et al., 1984; Greenland et al., 1982) have found depressed pregnancy weight gain among marijuana users compared to that of nonusers.

Although cocaine is associated with anorexia, some researchers (Chasnoff et al., 1989a) have found no difference in pregnancy weight gain between women who do and do not use cocaine during pregnancy. However, we (Zuckerman et al., 1989c) and others (Petiti and Coleman, 1990) found deficits in both weight for height at conception and pregnancy weight gain among users compared to those of nonusers. Marijuana and cocaine used during pregnancy are each associated with depressed birth weight when maternal nutritional status is analytically controlled (Frank et al., 1990). Thus, although use of marijuana or cocaine during pregnancy is sometimes associated with impairment of maternal nutritional status, a known determinant of birth weight, it is not at all clear that this alone explains the detrimental effect of these substances. Thus use of marijuana and cocaine during pregnancy appear to exert a primary growth-retarding effect on the fetus, as well as effects secondary to impairment of maternal nutritional status. The question then arises whether cocaine or marijuana exposure prenatally will alter neurobehavioral outcomes in ways different from those identified among infants with similar intrauterine growth retardation. The symmetrical pattern of intrauterine growth retardation found in marijuana- and cocaine-exposed neonates (Frank et al., 1990) has been found in other populations to be associated with a poorer prognosis for postneonatal growth than either asymmetric intrauterine growth retardation or normal intrauterine growth (Holmes et al., 1977; Davies et al., 1979; Villar et al., 1984). Whether this pattern will have similar implications for marijuana- and cocaine-exposed newborns is unknown.

Postnatal Influences

The newborn brain has a significant capacity for adaptation. Animal studies show that even though damaged nerve cells are not replaced, new synaptic connections are made, and/or certain areas of the brain develop new functions to replace those lost from the damaged area (Anastasiow, 1990). Recovery, or plasticity, is greater in the newborn than in the adult and is facilitated by a favorable caretaking environment.

Consistent with this animal research, research on humans during the past 20 years confirms the importance of the social environment and responsive caretaking in determining the developmental outcome of biologically vulnerable newborns. For example, among premature infants, IQ scores (at 7 years of age) were lower among those who at 1 month of age were neurologically immature, as indicated by a decreased percentage of an electroencephalogram (EEG) pattern called trace alternans. However, among those infants with low trace alternans, only those who had less responsive caretaking in the first 2 years of life had the lower IQs; infants with low trace alternans and responsive caretaking developed IQs similar to those of children who were not neurologically immature (Beckwith and Pasrmalee, 1986). In another study, the combination of high perinatal stress and low family stability impaired children's developmental functioning more severely than did the effects of perinatal stress or family instability alone (Werner, 1989).

Observations of humans also suggest that perinatal factors exert their influence primarily in early infancy, whereas social/environmental factors become predominant in subsequent development (Bee et al., 1982). Outcome of infants exposed prenatally to narcotics also appears to rely at least in part on their environment, especially after the first year of life. Compared with unexposed infants, methadone-exposed infants had poor motor coordination at 4 months; however, this difference almost disappeared by 12 months, except among infants from families at high social risk (Marcus et al., 1982). At 2 years, these same infants demonstrated impaired development compared with a control group only when prenatal methadone exposure was combined with low social class (Hans, 1989). Finally, Lifschitz, Wilson, and colleagues (1985) showed that, among infants exposed to opiates in utero, the quality of the postnatal environment and not the amount of maternal opiate use appeared to be the more important determinant of outcome (Lifschitz et al., 1985). Whereas it is likely the effects of prenatal cocaine use will be similarly modulated by the child's environment, this remains to be determined.

Interactions between drug-using mothers and their infants affect the infants' developmental functioning (Bernstein et al., 1986). Dysfunctional interactions may interfere with an infant's ability to recover from a biologic vulnerability caused by prenatal drug exposure. Clinically, a common problem of infants exposed to cocaine in utero is difficulty in regulating arousal. During infancy, caregivers ideally provide stimulation when an infant is underaroused and reduce it when the infant is overexcited, so that the child develops his or her capacity to regulate arousal. Among nondrug-using mothers, those who are intrusive at 6 months and overstimulating at 3.5 years are more likely to have hyperactive children compared to mothers who were not intrusive or overstimulating (Jacobvitz and Sroufe, 1987). Such nonadaptive maternal behavior, which is clinically seen with drug-abusing mothers, will be more likely to adversely affect a biologically vulnerable, drug-exposed infant, contributing to impulsivity, distractibility, and restlessness.

Infants with poor arousal may not elicit sufficient caretaking from their mothers. If the mother has a drug problem, the effects of drugs, drug-seeking behavior, and withdrawal from drugs are likely to render her less sensitive to her infant's signals for stimulation and nutrition. This combination of poor arousal due to the direct effect of prenatal drugs and less sensitive caretaking may result in a cycle of neglect that leads to failure to thrive.

Among the problems associated with drug and alcohol use, violence is significant and deserves special environmental attention. Drug and alcohol use by a pregnant woman and the father of her infant provide a context that increases the likelihood that the woman will be the victim of violence during pregnancy (Amaro et al., 1990). Her children are likely to witness violence in their home and/or be the victims of violence. Among the many developmental and behavioral consequences of exposure to violence is the development of post-traumatic stress disorder (PTSD). Behavioral changes of PTSD noted in very young children include decreased impulse control, attraction to danger, and exhibitionism, or conversely, emotional withdrawal and increased inhibition. Neurophysiological disturbances associated with PTSD are increased sympathetic arousal and elevated ratio of urine norepinephrine to free cortisol; reduced α_2- and β_2-adrenergic binding and down-regulation of central adrenergic receptors have also been described (Pynoss, 1990). Some of these findings are similar to those ascribed to prenatal cocaine exposure. Thus it is possible that exposure to violence may contribute to the behavioral outcomes anecdotally described as characteristic of children exposed prenatally to cocaine. It is also possible that prenatal cocaine exposure and family and community violence synergistically impair children's behavior and neurophysiological functioning.

Depression frequently occurs among drug-using and addicted women (Zuckerman et al., 1989a). When observed in interactions with their infants, depressed mothers are observed to be less verbal, more irritable or hostile, inconsistent in discipline, and less affectionate than nondepressed mothers. Children raised by depressed mothers are more irritable at birth, have more behavior problems, more accidents, more learning problems, and more affective disease than children raised by nondepressed mothers (Zuckerman and Beardslee, 1987).

COCAINE

Pharmacology

Cocaine is a tropane alkaloid that is derived from the leaves of *Erythoxylan cocoa* plant from the mountain slopes of Central and South America. Classified as a stimulant drug, cocaine is similar in structure and in neurochemical and clinical actions to amphetamines. It affects multiple neurotransmitter systems in the central nervous system (CNS) including the dopaminergic (DA), and norepinephrine (NE) systems: cocaine blocks the reuptake of these neurotransmitters. The excess of these agents at the postsynaptic membrane results in an exaggerated signal.

By blocking the presynaptic reuptake of DA in the CNS, cocaine produces a neurochemical magnification of the pleasure response (Spitz and Rosecan, 1987), creating a heightened sense of power, euphoria, and sexual excitement. This action on the pleasure center is considered the basis of its abuse potential. Cocaine also stimulates DA synthesis and causes an up-regulation of postsynaptic DA receptors. These factors augment the heightened DA pleasure response that may play a role in the self-administration of cocaine. With chronic use, cocaine depletes DA in the CNS, which is thought to cause depressive symptoms and is considered an important component of cocaine withdrawal and addiction (Gawin and Kleber, 1984).

Norepinephrine modulates global alertness and vigilance in the CNS; in the peripheral autonomic nervous system it increases heart rate and contractility, blood pressure, peripheral muscle contractility, and blood glucose. In addition, it stimulates behavioral arousal due to acute stress or fear. Cocaine blocks reuptake of NE at the presynaptic terminal, increases NE synthesis, and up-regulates NE receptors on the postsynaptic nerve. Norepinephrine accumulates in the nerve synapse, and NE-mediated nerve signals propagate. The alertness and hypervigilance induced with cocaine use as well as the peripheral sympathetic arousal are related to cocaine's effect on the NE transmitter system. During pregnancy the vasoactive consequences studied in animals include decreased uterine blood flow (Evans et al., 1988) and constriction of umbilical arteries. Cocaine is highly water and lipid soluble and has a wide volume of distribution. It crosses the placenta by simple diffusion and readily crosses the blood–brain barrier. Cocaine has been found in the brain in concentrations four times higher than the peak plasma concentration (Farrar and Kearns, 1989). Liver and plasma cholinesterases metabolize cocaine into the inactive, water-soluble metabolites benzoylecgonine and ecgonine methyl ester. It is excreted via the kidney into urine and via bile into the gastrointestinal tract. Cholinesterase activity is generally lower in the fetus, in pregnant women, and in people with liver disease. Therefore these individuals would be expected to be more sensitive to small doses of cocaine because decreased metabolism prolongs exposure. For some women, however, cholinesterase levels stay the same or actually increase during pregnancy (Evans et al., 1988), which may explain some of the variability in the maternal and fetal effects of cocaine exposure.

Newborn Outcome

Cocaine's vasoactive effects have been postulated to contribute to the etiology of such complications as abruptio placentae, preterm labor and delivery, poor fetal growth, congenital abnormalities, and hemorrhagic and cystic lesions in the CNS. Whereas maternal cocaine use has been associated with prematurity in some studies (Chasnoff et al., 1989b; Chasnoff et al., 1987b; Cherukuri et al., 1988; Handler et al., 1991; Little et al., 1989; MacGregor et al., 1987; Phibbs et al., 1992), this finding is not universal (Hadeed and Siegel, 1989; Ryan et al., 1987; Zuckerman et al., 1989b). Cocaine constricts placental vessels in pregnant ewes, leading to fetal hypoxemia and presumably decreased nutrient transfer (Woods et al., 1987).

Consistent with cocaine's vasoconstrictive action, numerous studies have shown an association between cocaine use during pregnancy and a decrease in birth weight and head circumference (Chasnoff et al., 1989b; Chasnoff et al., 1987b; Chasnoff et al., 1986, Chasnoff et al., 1985b; Chasnoff et al., 1989a; Cherukuri et al., 1988; Chouteau et al., 1988; Fulroth et al., 1989; Hadeed and Siegel, 1989; Handler et al., 1991; MacGregor et al., 1987; Ostrea et al., 1992; Phibbs, 1992; Ryan et al., 1987; Zuckerman et al., 1989b). However, this effect on growth is probably compounded by maternal undernutrition and use of other drugs (Frank et al., 1988; Zuckerman et al., 1989c). Only one of the studies demonstrating poor fetal growth and small head circumference controlled for multiple potentially confounding variables (Zuckerman et al., 1989b). Infants exposed to cocaine showed a symmetric pattern of growth retardation (Frank et al., 1990) suggesting a chronic growth impairment or one starting early in gestation. These infants also showed depressed fat stores and lean body mass, a pattern commonly associated with maternal malnutrition, even after nutritional markers such as maternal weight for height and pregnancy weight gain were statistically controlled (Frank et al., 1990). Cocaine-induced vasoconstriction may impair transfer of nutrients as well as oxygen to the fetus, regardless of maternal nutritional status. Moreover, by activating the sympathetic nervous system in the fetus (Fiks et al., 1985), cocaine may increase fetal metabolism and thus cause depletion of fetal nutrient stores, leading to evidence of fetal malnutrition in excess of that predicted by maternal malnutrition.

Rare but serious congenital abnormalities such as urogenital anomalies (Chavez et al., 1989; Hoyme et al., 1990), distal limb deformities (Hoyme et al., 1990), cardiac lesions (Lipshultz, 1990), and CNS and ocular malformations (Dominguez et al., 1991) have been noted in clinical series. The Chavez and Lipshultz studies employed retrospective case controls; the other studies were not controlled. Whereas it has not been proven that cocaine causes congenital abnormalities, cocaine is a plausible etiologic agent because the defects detected can be caused by vasoconstriction. Decreased fetal blood flow can result in disruption of existing structures or altered morphogenesis of developing structures, leading to these anatomic abnormalities. Structural abnormalities may also result when placental emboli interrupt the blood flow to distal structures. The retrospective population-based case control study (Chavez et al., 1989) of renal malformations was limited by the small number (three) of urinary tract abnormalities in the cocaine-exposed group. Furthermore, the findings may have been influenced by exposure recall bias, especially since the exposure is illegal. Infrequently, occurring birth defects of the bowel, heart, and skeletal systems are unlikely to be proven in epidemiologic studies unless a large number of subjects are studied. Cigarette smoking also causes vasoconstriction and may have played a role in early case description of prenatal cocaine use and gastroschisis (Goldbaum et al., 1990). Other studies (Zuckerman et al., 1989b; Rosenstein, 1990) have not shown cocaine to be independently associated with congenital abnormalities.

A small number of uncontrolled studies describe a variety of transient abnormalities. The presence of an abnormal dilatation of iris blood vessels in cocaine-exposed infants compared to nonexposed infants has been reported (Isenberg et al., 1987). Similar changes in iris blood vessels have been reported in infants of diabetic mothers (Ricci and Molle, 1987) and may be a marker of intrauterine stress, perhaps vasomotor in origin. The vascular abnormalities regressed with time and had no long-term effect.

An ototoxic effect of prenatal cocaine exposure has been suggested (Shih et al., 1988). Auditory brainstem responses in neonates exposed to cocaine showed prolonged interpeak latencies and prolonged absolute latencies, which indicate neurologic impairment or dysfunction in the auditory system. Another study (Salamy et al., 1990) also showed delayed auditory brain stem transmission time in cocaine-exposed newborns, which reverted to normal by 3 to 6 months.

Case studies have described 16 infants with seizures in the neonatal period (Kramer et al., 1990). A focal seizure was described in one infant with an infarct on computerized tomography (CT) scan. Six patients continued to have seizures after 6 months of age. One other study reported electroencephalogram (EEG) changes at 1 month of age in a small group of cocaine-exposed infants, which returned to normal by 6 months of age (Doberczak et al., 1988). Cocaine use is associated with lowered seizure threshold in adults (Gawin and Kleber, 1984). The true incidence of seizures in cocaine-exposed neonates has not been determined; clinical experience suggests that seizures are a relatively rare complication but may occur in infants with no other risk factors.

One case report of a significant cerebral infarction associated with prenatal cocaine exposure has been published (Chasnoff, 1986). Another recent uncontrolled report describes congenital, cerebral, and ocular abnormalities in seven cocaine-exposed neonates, which may be attributed to early vascular insult (Dominguez et al., 1991). A systematic study reported that 35% of stimulant-exposed (cocaine, methamphetamine), asymptomatic infants undergoing cranial ultrasound in the first 3 days of life showed either echodensities or echolucencies indicating CNS vascular injury (Dixon and Bejar, 1989). This rate of abnormal findings was comparable to that in ill term infants and much greater than that in healthy term newborns, although this finding was not replicated in another study (Frank et al., 1992). In the Dixon study the distribution of lesions in the cocaine-exposed newborns (basal ganglia, frontal lobes, and the posterior fossa) differed from the distribution seen in ill infants. This difference may reflect vessels that were muscularized early in gestation and therefore were more vulnerable to the vasoactive effects of cocaine. A possible mechanism entails vasoconstriction leading to ischemia causing small hemorrhages that result in cyst formation. In addition, a transient increase in cerebral blood flow velocity has been described among infants prenatally exposed to cocaine, raising the risk of intracranial hemorrhage (Van de Bor, Walthers, and Simons, 1990). The clinical significance of these lesions is unknown, but their existence warrants further investigation and follow-up.

Neonatal neurobehavioral abnormalities have been reported in some but not all studies. Firm conclusions about the effect of prenatal cocaine exposure on neurobehavioral functioning cannot be drawn, owing to the inconsistency of the findings. Methodologic problems (lack of control for potential confounding variables and biased sample populations) associated with the research raise important questions regarding the validity of findings (Neuspiel and Hamel, 1991).

Assuming that termination of cocaine exposure would result in a withdrawal syndrome, many studies have assessed cocaine-exposed infants using the Neonatal Abstinence Scale (NAS) that was developed by Finnegan to describe withdrawal among opiate-exposed infants. One study (Fulroth et al., 1989) shows higher NAS scores for cocaine/opiate-exposed infants than either infants exposed to cocaine or opiate-exposed infants alone, suggesting a possible synergistic effect between cocaine and opiates. However, a larger study failed to identify any cocaine effects on the NAS scores of opiate-exposed newborns (Doberczak, 1991). Two other studies (Ryan et al.,

1987; Hadeed and Seigel, 1989) found no differences in NAS scores between cocaine-exposed and comparison groups.

The NAS scale, designed to detect effects of opiate withdrawal, is not appropriate for assessment of potential prenatal cocaine effects. Clinically, cocaine-exposed newborns are poorly responsive and sleepy and do not show systemic signs of withdrawal (e.g., gastrointestinal, sneezing) that are seen among opiate-exposed infants. When alert they are easily overstimulated; they therefore become irritable and quickly return to sleep. A number of studies have evaluated the neurobehavioral functioning of cocaine-exposed infants using the Brazelton Neonatal Behavioral Assessment Scale, which is a much more sensitive assessment of newborn behavior than the NAS scale. These studies have the same methodologic problems as studies using the NAS scale, including difficulty controlling for confounding variables, and possible misclassification of cocaine users. Results of these studies are inconsistent regarding both the presence or absence of an association between prenatal cocaine use and neurobehavioral dysfunction, and the type of dysfunction identified. Chasnoff, in two different reports (Chasnoff et al., 1985b; Chasnoff et al., 1986) that may have included some of the same infants, shows cocaine-exposed newborns have increased tremulousness and startles, decreased interactive behaviors, and increased state lability. Women from these studies were drawn from a special perinatal drug treatment program, and the effects of other drugs, especially opiates as well as other risk factors, contribute to the findings. In a third report, Chasnoff (Chasnoff et al., 1989a) showed that even among infants whose mothers stopped using drugs in the first trimester, infants showed impaired orientation, motor function, reflexes, and state regulation compared to infants whose mothers used no drugs. A study conducted by another research team shows impaired habituation on the NBAS and more stress behaviors among cocaine-exposed infants compared to controls (Eisen et al., 1991). Multivariate analysis showed that other factors (obstetric complications and maternal alcohol use) and not cocaine were associated with stress behaviors. However, prenatal cocaine continued to be associated with poor habituation when these and other factors were controlled analytically. A final study showed no differences on NBAS scores between cocaine-exposed and unexposed newborns in the first 3 days of life. Whereas a second examination between 11 and 30 days of life showed cocaine infants had significantly lower scores in motor functioning, the magnitude of effect decreased by over 50%, and the association was no longer statistically significant when confounding variables were controlled (Neuspiel et al., 1990).

Other psychoactive substances as well as other factors, especially intrauterine growth retardation, lead to neurobehavioral dysfunction that may contribute to the variability in outcomes of cocaine-exposed newborns. One study (Lester et al., 1991) using cry characteristics supports this possibility by identifying two neurobehavioral profiles among cocaine-exposed newborns. One profile characterized as "excitable" is hypothesized to be owing to the direct, primary effect of cocaine exposure. The other profile characterized as "depressed" is thought to be owing to the secondary effect of intrauterine growth retardation. Possible opposite effects of cocaine and undernutrition may help explain the variability in newborn behavior seen clinically and in these studies.

Whether alterations in neonatal behavior will be confirmed with more methodologically rigorous studies is unknown. Currently, special attention focuses on whether behavioral changes reflect neurotransmitter alterations. In the fetal brain, neurotransmitters contribute to brain development by influencing neuronal migration and differentiation as well as synaptic proliferation (Lauder, 1988). During the prenatal period, neurotransmitters also affect the development of receptor sites (Miller and Friedhof, 1988). In animals, for example, an induced decrease in dopamine prenatally results in a significant decrease in the number of dopamine receptors postnatally. Following birth these receptors do not appear to have the normal capacity to up-regulate or increase in number in response to decreases in dopamine. Since the time course of the development of receptors differs in different parts of the brain, timing of cocaine use may be critical in determining which aspects of the brain are altered, leading to different functional and behavioral difficulties. A role for neurotransmitter changes was suggested by a preliminary study showing lower levels of homovanillic acid (HVA), a principal metabolite of dopamine in the cerebral spinal fluid of cocaine-exposed compared to unexposed newborns (Needlman et al., 1992). Another pilot study shows that blood levels of the NE precursor dihydroxyphenalanine were higher in cocaine-exposed than in unexposed newborns (Mirochnick et al., 1991). Among the cocaine-exposed newborns, high NE concentrations in blood were associated with poor responsivity to auditory and visual stimuli (orientation subscale) as measured by the NBAS. However, blood NE levels do not necessarily indicate levels in the CNS. The neurotransmitter changes shown in this preliminary study might be owing entirely to chronic stress associated with cocaine-induced vasoconstriction and hypoxia in utero.

Probable fetal hypoxia and neurotransmitter abnormalities associated with in utero cocaine exposed have raised concerns that such exposure may increase the risk of SIDS, concerns heightened by reports of respiratory-pattern abnormalities in cocaine-exposed infants compared to methadone-exposed infants (Chasnoff et al., 1989c).

An early study, using a convenience sample of infants exposed prenatally to cocaine, found that 10 of 66 (15%) died of SIDS (Chasnoff et al., 1987b). However, three subsequent studies using nonbiased samples have shown only a slightly increased rate of SIDS. Combined data from the three studies yielded a risk of 8.5 of 1000 (Bauchner and Zuckerman, 1990), uncontrolled for cigarette use. Although this risk is somewhat elevated, it does not approach the level of risk found among heroin- or methadone-exposed infants (15–20 of 1000), and is only slightly higher than that reported for children living in poverty (4–5 of 1000). It is far below the risk associated with conditions that the National Institutes of Health Consensus Statement (1987) indicated may warrant apnea monitoring at home.

Postneonatal Exposure to Cocaine

Postnatal exposure to cocaine has been reported via breast feeding, by direct ingestion, and by passive inhalation. Like all highly lipid-soluble substances, cocaine crosses readily into breast milk. Chasnoff and associates (1987c) reported finding cocaine metabolites in breast milk for 36 hr after them other used 0.5 g of cocaine intranasally. The infant in this case developed hypertension and irritability and ex-creted cocaine metabolites for 60 hr after the last breast feeding. The number of infants at risk nationally for exposure to cocaine via breast feeding has not been determined. We have reported (Frank et al., 1992) that 12% of women intending to breast feed in our inner city hospital use cocaine prenatally, compared to 23% of those intending to bottle feed. However, only 2% of those intending to breast feed were excreting cocaine metabolites in their urine at delivery, so that only a small number of infants were at risk for exposure via breast milk immediately following birth. There is no information regarding the prevalence of postpartum use of cocaine by mothers, whether breast feeding or not. The long-term risk to the growth and development of the infant breast fed by a cocaine-using mother is also unknown. Prudent clinical practice suggests that women who are known active cocaine users be advised not to breast feed to avoid the dangers of acute toxicity.

Oral ingestion of cocaine associated with seizures in infants and toddlers has been described in several case reports (Ernst and Sanders, 1989; Rivkin and Gilmore, 1989), including one case in which the breast-feeding mother applied cocaine directly to her sore nipple (Chaney et al., 1988). Bateman and Heagarty (1989) have de-scribed four children hospitalized with neurologic symptoms and cocaine metabolites in their urine, following time spent in ill-ventilated rooms with adults who were smoking crack cocaine.

The number of young children at risk nationally for involuntary postnatal cocaine exposure via passive inhalation or direct ingestion, not related to breast feeding, is unknown. Shannon and colleagues (1989) found cocaine metabolites in 4.2% of toxic screens obtained for clinical purposes in a large children's hospital. Only 7 of the 52 positive screens were from children younger than 1 year; the rest were from intoxi-cated or suicidal adolescents. A study from our institution (Kharasch et al., 1991) found cocaine metabolites in 2.4% of 250 urines obtained from children younger than 5 visiting an emergency department. Only one child with a urine positive for cocaine metabolites was breast fed. Children with neurologic or cardiac symptoms possibly suggestive of cocaine exposure were excluded from the sample, so that true prevalence of passive exposure may be higher.

Whereas acute toxicity for infants and toddlers postnatally exposed to cocaine has been documented clinically, there are no data regarding long-term developmental risk following such exposure. Whether the risks of symptomatic as opposed to asymptomat-ic exposure are of differing magnitude is also unknown. Clearly environments that permit passive exposure of children to cocaine are probably not providing optimal care in many areas, so that the long-term biologic versus the social risks of postnatal passive exposure will be difficult to evaluate.

Developmental Outcome

Findings from the newborn period such as small head circumference raise con-cerns regarding longterm developmental effects. The only study reporting child out-comes of prenatal cocaine exposure involve mothers and infants who are involved in a comprehensive clinical intervention study. The results of this study show no mean differences on the Bayley Scales of Infant Development at two years of age when cocaine-exposed children are compared with social class match controls (Chasnoff et al., 1992). There is, however, a higher rate of scores more than one standard deviation below the test mean for children exposed prenatally to cocaine, although this may be explained by parents of developmentally delayed children being more readily compliant with follow-up developmental assessments (Zuckerman and Frank, 1992). As may be unavoidable among a cocaine-using population, there is a high attrition rate, higher among cocaine-using women compared to non-cocaine-using mothers. Only 27% of the 106 children recruited by birth remain in the cocaine-exposed sample at 24 months, compared with 62% of the 81 children in the nonexposed sample. In spite of this and other limitations, these findings suggest that the global developmental functioning of infants whose mothers remain in an intervention program is similar to social class matched controls.

It will be critical to replicate this study in larger samples, with intensive efforts to minimize attrition. Due to cocaine's action of altering neurotransmitter levels and the preliminary evidence of early perturbation of neurotransmitters in the peripheral circulation and in cerebral spinal fluid following prenatal cocaine-exposure (Mirochnick et al., 1991; Needlman et al., 1992), there is at least theoretical reason to be concerned about the exposed child's capacity for attention and self-regulation. Since these functions cannot be evaluated accurately in infants and young children, there is a potential to underestimate the possible effect of prenatal cocaine exposure unless children are observed to school age. Even if behavior and developmental problems are identified, the relative contributions due to caretaker dysfunction or biologic vulnerability created by prenatal cocaine exposure will need to be determined. Research to identify the independent effects of prenatal and postnatal cocaine on developmental outcome must control for adverse circumstances in the growing child's environment, especially dysfunctional parenting by caretakers who abuse drugs, multiple caretakers, violence, and adverse health effects associated with poverty such as poor nutrition, lead poisoning, etc. Studies must also employ measures sensitive to the behavior that might not be identified by traditional structured developmental tests alone.

MARIJUANA

Pharmacology

Marijuana use is most common among individuals in their late teens and early twenties. The rate of women reporting use of marijuana during pregnancy varies from 5 to 34% (Zuckerman, 1988). The principle psychoactive chemical of marijuana is 1-delta-9-tetrahydrocannabinol (THC). Approximately one half of the THC present in a marijuana cigarette is absorbed following inhalation (Renault et al., 1971). Following hydroxylation in the liver, most metabolites of THC are eliminated in the feces and urine (Abel, 1983). Because it has a strong affinity for lipids, THC is stored in fatty tissues of the body (Kruez and Axelrod, 1973). A single dose of cannabis has a tissue half-life in humans of 7 days and may take up to 30 days to be completely excreted (Nahas, 1976). For this reason, marijuana accumulates in the body during chronic use.

Like all psychoactive drugs, marijuana crosses the placenta. Placental transfer is highest early in gestation and diminishes as pregnancy progresses (Indanpaan-Heikkila et al., 1969). In addition to its potential direct effect, marijuana has the indirect effect of decreasing fetal oxygenation. Inhalation of marijuana smoke by animals has been shown to produce maternal ventilation/perfusion abnormalities, with subsequent fetal hypoxia that lasted approximately 60 min (Clapp et al., 1986). Fetal hypoxia may also result from maternal inhalation of carbon monoxide, which is present at higher levels in marijuana smoke than that in cigarette smoke (Wu et al., 1988).

Newborn Outcomes

Only seven studies with large enough sample sizes to control for confounding variables have investigated the effects of marijuana on human fetal growth, and their results are conflicting (Zuckerman et al., 1988). Only one study used both self-report and urine assay to evaluate marijuana use during pregnancy (Zuckerman et al., 1989c), and the results indicated that reliance on self-report may have contributed to the inconsistent findings of other studies. Women who had a positive urine assay for marijuana during pregnancy bore infants with lower birth weight and shorter length, when confounding variables were controlled. Had urine tests for marijuana not been performed and identification of users been based solely on self-report, 16% of marijuana users during pregnancy would not have been identified. This misclassification of marijuana users as nonusers would have obscured the significant association between prenatal marijuana use and decreased fetal growth (Zuckerman et al., 1989c). Prenatal marijuana use, like that of cigarettes, spares fetal fat deposition while depressing indicators of lean body mass, suggesting an hypoxic or other non-nutritional mechanism of effect that preferentially alters lean tissue growth (Frank et al., 1990). This finding, which is found after controlling for maternal cigarette use, is consistent with research on ewes, which shows that inhalation of marijuana produces pulmonary ventilation/perfusion abnormalities in the mother with associated prolonged hypoxia in the fetus (Abrams et al., 1985; Clapp et al., 1986). Moreover, marijuana smoking elevates blood carbon monoxide to levels higher than that attained by smoking cigarettes (Wu et al., 1988), an effect that would also impair fetal oxygenation (Longo, 1977).

Studies of neurobehavioral functions of newborns exposed prenatally to marijuana have been few. One research group showed that moderate and heavy maternal marijuana use during pregnancy was associated with infants who had increased tremors, decreased responsiveness to visual stimuli during sleep, and a higher-pitched cry (Fried and Makin, 1987). Another research group found no such correlation (Tennes et al., 1985). Heavy marijuana smoking during pregnancy altered the computer-measured acoustic characteristics of the newborn cry, consistent with patterns that, in other studies, had been related to perinatal risk factors and to later poor developmental outcome (Lester and Dreher, 1989). Another study, which assessed neonatal sleep cycling and arousal, showed that marijuana use during pregnancy was associated with a decrease in the amount of trace alternans quiet sleep (Scher et al., 1988). Day and Richardson (1991) have recently reported that prenatal marijuana exposure is correlated also with lower sleep efficiency and maintenance on sleep EEG studies at age 3 years. These findings suggests that heavy marijuana use may affect the neurophysiologic integrity of the newborn.

Postnatal Marijuana Exposure

Passive inhalation of marijuana metabolites has been described in adults (Cone and Johnson, 1986), but not, to our knowledge, in children. However, marijuana metabolites have been identified in the milk of animals (Jakubovic et al., 1974; Chao et al., 1976) and of humans (Perez-Reyes and Wall, 1982), and in the urine of breast-fed infants (Perez-Reyes and Wall, 1982). In contrast to the immediate toxicity associated with cocaine exposure via breast milk, acute effects of maternal marijuana use on the breast-fed infant have not been reported. However, two pilot studies suggest potential subtle developmental effects with heavy postpartum exposure via breast milk. Tennes and colleagues (1985) reported delayed growth and development in one of six 1-year-old infants breast fed for 3 to 12 months by daily marijuana users. In a larger multivariate study of 136 breast fed infants, marijuana exposure via breast milk in the first postpartum month was associated in a dose-related fashion with decreased motor development at 1 year (Astley and Little, 1990). However, this effect may have been confounded by exposure in the first prenatal trimester, which was closely correlated with heavy postpartum exposure. Developmental effects of marijuana exposure via breast milk beyond the first year of life have never been studied.

The national prevalence of marijuana use by postpartum women in general and by breast-feeding women in particular is unknown. At our hospital, 21% of the women intending to breast feed used marijuana during pregnancy, compared to 33% of those intending to bottle feed (Frank et al., 1992). Five percent of women intending to breast feed had marijuana metabolites in their urine at delivery, indicating that their infants may have been at risk of immediate postpartum exposure via breast milk. Because marijuana metabolites are stored in maternal fat, which may be mobilized during lactation, the duration of potential excretion via breast milk following last maternal use is of concern, but has not been clearly delineated.

Later Developmental Outcome

Only two studies have evaluated developmental and behavioral functioning among children exposed prenatally to marijuana. In one, no independent association was seen between prenatal marijuana use and developmental scores at 12 and 24 months (Freid and Watkinson, 1988). However, when these children were 4 years old, heavy prenatal marijuana use (more than six joints per week) during pregnancy was associated with poor scores on memory and verbal subscales of the McCarthy Test, compared with scores of children of mothers who did not smoke marijuana. This finding remained after controlling for confounding variables, including the home environment (Freid and Watkinson, 1990). Another study found no effects on IQ at age 4 (Streissguth et al., 1989). Since there are only two follow-up studies of prenatal marijuana exposure, and a correlation was found in one but not the other one at 4 years, and not at 1 or 2 years, interpretation is difficult. It is possible that the effect of prenatal exposure to marijuana is subtle and that its effects on complex behavior cannot be identified before 4 years of age. It seems almost equally likely that the finding may be the result of chance or some unmeasured postnatal environmental variable.

CLINICAL IMPLICATIONS

Infants born to drug- and alcohol-abusing mothers are clearly at risk for developmental and behavioral problems due to both prenatal exposure to these substances and dysfunctional parenting. However, no studies have consistently shown that an individual drug causes a specific pathognomonic developmental dysfunction.

All children at biological and social risk benefit from developmental and behavioral assessment and educational programs designed to meet their individual needs. Unfortunately, measures of developmental outcome may not provide sufficiently specific information on children's developmental and behavioral functioning. For example, by altering neurotransmitter activity in the developing nervous system, we hypothesize chronic prenatal exposure to cocaine may adversely affect autonomic function, state regulation, and response to sensory stimuli, potentially leading to impulsivity and instability of mood as children get older. Global outcome measures such as developmental quotients and behavior problem scales may not identify all areas of dysfunction or suggest specific intervention plans.

Following an assessment of all areas of a child's functioning including interaction with parents and unstructured play, a comprehensive intervention program addressing the child's needs should be developed. However, as the multifactorial developmental model discussed in the first part of this chapter implies, intervention for the drug-exposed child needs to be accompanied by intervention for the parents. Drug-abusing families need more than the counseling and support provided in most child development or early intervention programs; they need drug treatment and comprehensive medical care.

We are currently conducting a small pilot program that provides pediatric care, child development services, and drug treatment in the pediatric primary care clinic. By collocating key services in this way, we developed a less fragmented, nonstigmatizing system that is easier for families with small children to use. Developmentally appropriate stimulation is both modeled and verbally presented to parents to help them support their child's development. We support parents' self-image by identifying and acknowledging positive interactions with their children. Our experience shows parents, particularly mothers, are more likely to comply with drug treatment if they feel it is part of their efforts to care for their children. Children have the best opportunity to recover from effects of prenatal drug exposure when child development services are combined with drug treatment in a family context.

ACKNOWLEDGMENTS

This work was supported by grants from the Harris Foundation, Bureau of Health Care Delivery and Assistance, Maternal and Child Health Branch (MCJ 009094), and National Institute for Drug Abuse (RO1-DA 06532-01). We thank Jeanne McCarthy for her help in preparing the manuscript.

REFERENCES

Abel, E. (1983). Marihuana, Tobacco, Alcohol and Reproduction. CRC Press, Boca Raton, Florida.

Abel, E. L., and Sokol, R. J. (1987). Incidence of fetal alcohol syndrome and economic impact of FAS-related anomalies. *Drug Alcohol Depend.* **19**, 51–70.

Abel, E. L., and Sokol, R. J. (1990). Is occasional light drinking during pregnancy harmful? In "Controversy in the Addiction Field" (R. C. Engs, ed.), pp. 158. Kendall-Hunt Publishing, Dubuque, Iowa.

Abrams, R. M., Cook, C. E., Davis, K. H., Niederreither, K., Jaeger, M. J., and Szeto, H. H. (1985). Plasma Δ-9-tetrahydrocannabinol in pregnant sheep and fetus after inhalation of smoke from a marijuana cigarette. *Alcohol Drug Res.* **7**, 85–92.

Amaro, H., Fried, L. E., Cabral, H., and Zuckerman, B. (1990). Violence during pregnancy and substance use. *Am. J. Public Health* **80,** 575–579.

Anastasiow, N. J. (1990). Implications of the neurological model for early intervention. In "Handbook of Early Childhood Intervention" (S. J. Mersels and J. P. Shonkoff, eds.), pp. 196–216. Cambridge University Press, New York.

Aronson, M., Kyllerman, M., Sabel, K. G., et al. (1985). Children of alcoholic mothers: Developmental, perceptual, and behavioral characteristics as compared to matched controls. *Acta Paediatr. Scand.* **74,** 27–35.

Astley, S. J., and Little, R. E. (1990). Maternal marijuana use during lactation and infant development at one year. *Neurotoxicol. Teratol.* **12**, 161–168.

Bailey, D. N. (1989). Drug screening in an unconventional matrix: Hair analysis (Editorial). *J.A.M.A.* **262,** 3331.

Bateman, D. A., and Heagarty, M. C. (1989). Passive freebase cocaine ("crack") inhalation by infants and toddlers. *Am. J. Dis. Child.* **143**, 25.

Bauchner, H., and Zuckerman, B. (1990). Cocaine, sudden infant death syndrome, and home monitoring. *J. Pediatr.* **117,** 904–906.

Beckwith, K., and Parmalee, A. (1986). EEG patterns in preterm infants, home environment, and later I.Q. *Child Dev.* **57,** 777–789.

Bee, H., Barnard, K., Ayres, S., et al. (1982). Prediction of IQ and language skill from perinatal status, child performance, family characteristics, and mother–infant interaction. *Child Dev.* **6,** 1134–1156.

Bernstein, V. J., Jeremy, R. J., and Marcus, J. (1986). Mother–infant interaction in multiproblem families: Finding those at risk. *J. Am. Acad. Child Psychiatry* **5,** 631–640.

Bingol, N., Fuchs, M., Diaz, V., Stone, R. K., and Gromisch, D. S. (1987). Teratogenicity of cocaine in humans. *J. Pediatr.* **110,** 93–96.

Black, R., and Mayer, J. (1980). Parents with special problems: Alcoholism and opiate addiction. *Child Abuse Negl.* **4**, 45.

Bracken, M. B., and Holford, T. R. (1981). Exposure to prescribed drugs in pregnancy and association with congenital malformations. *Obstet. Gynecol.* **58**, 336–344.

Brouwers, P., Belman, A. L., and Epstein, L. G. (1991). "Central Nervous System Involvement: Manifestations and Evaluation in Pediatric AIDS." P. Pizzo and C. M. Wilfert, eds., pp. 318–335. Williams & Wilkins, Baltimore, Maryland.

Brown, E., and Zuckerman, B. The infant of the drug-abusing mother. *Pediatr. Am.* In press.

Burkett, G., Bandstra, E., Cohen, J., Steele, B., and Palow, D. (1990). Cocaine related maternal death. *Am. J. Obstet. Gynecol.* **163**, 40–41.

Burns, K., Melamed, J., Burns, W., et al. (1985). Chemical dependence and clinical depression. *J. Clin. Psychol.* **41**, 851–854.

Chao, F., Green, D. E., Forrest, I. S., Kaplan, J. N., Winship-Ball, A., and Braude, M. (1976). The passage of ^{14}C delta-9-THC into the milk of lactating squirrel monkeys. *Res. Commun. Chem. Pathol. Pharmacol.* **15**, 303–317.

Chaney, N. E., Franke, J., and Wadlington, W. B. (1988). Cocaine convulsions in a breast-feeding baby. *J. Pediatr.* **112**, 134–135.

Chasnoff, I. J. (1985a). "Effects of Maternal Narcotic vs. Nonnarcotic Addiction on Neonatal Neurobehavior and Infant Development." NIDA Research Monograph Series. **59**, 73–83. Rockville, Maryland.

Chasnoff, I. J., Burns, W. J., Schnoll, S. H., et al. (1985b). Cocaine use in pregnancy. *N. Engl. J. Med.* **313**, 666–669.

Chasnoff, I. J., Burns, K. A., Burns, W. J., et al. (1986). Prenatal drug exposure: Effects on neonatal and infant growth and development. *Neurobehav. Toxicol. Teratol.* **8**, 357–362.

Chasnoff, I. J., Bussey, M. E., Savich, R., and Stack, C. M. (1987a). Perinatal cerebral infarction and maternal cocaine use. *J. Pediatr.* **111**, 571–8.

Chasnoff, I. J., Burns, K. A., and Burns, W. J. (1987b). Cocaine use in pregnancy: Perinatal morbidity and mortality. *Neurotoxicol. and Teratol.* **9**, 291–293.

Chasnoff, I. J., Lewis, D. E., and Squires, L. (1987c). Cocaine intoxication in a breast-fed infant. *Pediatrics* **80**, 836–838.

Chasnoff, I. J., Griffith, D. R., MacGregor, S., Dirkes, K., and Burns, K. A. (1989a). Temporal patterns of cocaine use in pregnancy. *J.A.M.A.* **261**, 1741–1744.

Chasnoff, I. J., Lewis, D. E., Griffith, D. R., et al. (1989b). Cocaine and pregnancy: Clinical toxicological implications for the neonate. *Clin. Chem.* **35**, 1276–1278.

Chasnoff, I. J., Hunt, C. E., Kletter, R., and Kaplan, D. (1989c). Prenatal cocaine exposure is associated with respiratory pattern abnormalities. *Am. J. Dis. Child.* **143**, 583.

Chasnoff, I. J., Landress, H., and Barett, M. (1990). The prevalence of illicit-drug or alcohol use during pregnancy and discrepancies in mandatory reporting in Pinellas County, Florida. *N. Engl. J. Med.* **322**, 1202–1206.

Chasnoff, I. J., Griffith, D. R., Freier, C., and Murray, J. (1992). Cocaine/polydrug use in pregnancy: Two-year follow-up. *Pediatrics,* **89**, 284–289.

Chavez, G. F., Mulinare, J., and Cordero, J. F. (1989). Maternal cocaine use during early pregnancy as a risk factor for congenital urogenital anomalies. *J.A.M.A.* **262**, 795–798.

Cherukuri, R., Minkoff, H., Feldman, J., et al. (1988). A cohort study of alkaloidal cocaine ("crack") in pregnancy. *Obstet. Gynecol.* **72**, 147–151.

Chokshi, S. K., Gal, D., Whelton, J. A., and Isner, J. M. (1989). Evidence that fetal distress in newborns of cocaine users is due to vascular spasm and may be attenuated by pretreatment with diltiazem. *Circulation* (Suppl. II) **80**, 11–185. Abstr.

Chouteau, M., Namerow, P. B., and Leppert, P. (1988). The effect of cocaine abuse on birth weight and gestational age. *Obstet. Gynecol.* **72**, 351–354.

Church, M. W., Dincheff, B. A., and Gessner, P. K. (1988). Interactive affect of alcohol and cocaine on maternal and fetal toxicity in long-Evans rats. *Neurotoxicology and Teratology* **10**, 355–361.

Clapp, J., Wesley, M., Cooke, R., Pekals, R., and Holstein, C. (1986). The effects of marijuana smoke on gas exchange in ovine pregnancy. *Alcohol. Drug. Res.* **7**, 85–92.

Coles, C. D., Smith, I. E., Fernhoff, P. M., et al. (1984). Neonatal ethanol withdrawal: Characteristics in clinically normal, nondysmorphic neonates. *J. Pediatr.* **105**, 445–451.

Cone, E. J., and Johnson, R. E. (1986). Contact highs and urinary cannabinoid excretion after passive exposure to marijuana smoke. *Clin. Pharmacol. Ther.* **40**, 247–256.

Davidson, S. L., and Bautista, D., Chan, L., et al. (1990). Sudden infant death syndrome in infants of substance-abusing mothers. *J. Pediatr.* **117**, 876–881.

Davies, D. P., Platts, P., Prichard, J. M., and Wilkinson, P. W. (1979). Nutritional status of light-for-date infants at birth and its influence on early postnatal growth. *Arch. Dis. Child.* **54**, 703–706.

Day, N. L., and Richardson, G. A. (1991). Prenatal marijuana use: Epidemiology, methodologic issues, and infant outcome. *Clin. Perinatol.* **18**, 77.

Dixon, S. D. (1989). Effects of transplacental exposure to cocaine and methamphetamine on the neonate. *West. J. Med.* **150**, 436–442.

Dixon, S. D., and Bejar, R. (1989). Echoencephalographic findings in neonates associated with maternal cocaine and methamphetamine use: Incidence and clinical correlates. *J. Pediatr.* **115**, 7709.

Doberczak, T. M., Shanzer, S., Senie, R. T., and Kandall, S. R. (1988). Neonatal neurologic and electroencephalographic effects of intrauterine cocaine exposure. *J. Pediatr.* **113**, 354–358.

Doberczak, T. M., Kandall, S. R., and Wilets, I. (1991). Neonatal opiate abstinence syndrome in term and preterm infants. *J. Pediatr.* **118**, 933–7.

Dominguez, R., Vila-Coro, A. A., Slopis, J. M., and Bohan, T. P. (1991). Brain and ocular abnormalities in infants with *in utero* exposure to cocaine and other street drugs. *Am. J. Dis. Child.* **145**, 688–695.

Durand, D. J., Espinoza, A. M., and Nickerson, B. G. (1990). Association between prenatal cocaine exposure and sudden infant death syndrome. *J. Pediatr.* **117**, 909–911.

Eisen, L. N., Field, T. M., Bandstra, E. S., Roberts, J. P., Morrow, C., and Larson, S. K. (1991). Perinatal cocaine effects on neonatal stress behavior and performance on the Brazelton Scale. *Pediatrics.* In press.

Elsohly, M. A., and Elsohly, H. N. (1989). Marijuana: Analysis and detection of use through urinalysis. *In* "Cocaine, Marijuana, Designer Drugs: Chemistry, Pharmacology, and Behavior" (E. E. Redda, C. H. Walker, and G. Barnett, eds.), pp. 145. CRC Press, Boca Raton, Florida.

Ernhart, C. B., Wolf, A. W., Linn, P. L., et al. (1985). Alcohol-related birth defects: Syndromal anomalies, intrauterine growth retardation, and neonatal behavioral assessment. *Alcohol. Clin. Exp. Res.* **9**, 447–453.

Ernst, A. A., and Sanders, W. M. (1989). Unexpected cocaine intoxication presenting as seizures in children. *Ann. Emerg. Med.* **18**, 774–777.

Evans, R. T., O'Callaghan, F., and Norman, A. (1988). A longitudinal study of cholinesterase changes in pregnancy. *Clin. Chem.* **34**, 2249–2252.

Farrar, H. C., and Kearns, G. L. (1989). Cocaine: Clinical pharmacology and toxicology. *J. Pediatr.* **115**, 665–675.

Fiks, K., Johnson, H., and Rosen, T. (1985). Methadone-maintained mothers: Three-year follow-up of parental functioning. *Int. J. Addict.* **20**, 651–60.

Finnegan, L. P. (1986). Neonatal abstinence syndrome: Assessment and pharmacotherapy. *In* "Neonatal Therapy: An Update" (F. F. Frubaltelli and B. Granati, eds.), pp. 122–146. Elsevier Science Publishers, New York.

Frank, D., Zuckerman, B., Amaro, H., Aboagye, K., Bauchner, H., Cabral, H., Fried, L., Hingson, R., Kayne, H., Levenson, S. M., Parker, S., Reece, H., and Vinci, R. (1988). Cocaine use during pregnancy: Prevalence and correlates. *Pediatrics* **82**, 888–895.

Frank, D., Bauchner, H., Parker, S., Huber, A. M., Kyel-Aboagye, K., Cabral, H., and Zuckerman, B. (1990). Neonatal body proportionality and body composition following *in utero* exposure to cocaine and marijuana. *J. Pediatr.* **117**, 622–626.

Frank, D., Bauchner, H., Zuckerman, B., and Fried, L. (1992). Cocaine and marijuana use during pregnancy by women intending and not intending to breast feed. *J. Amer. Dietetic Assn.* **92**, 215–217.

Frank, D. A., McCarten, K., Cabral, H., Levenson, S. M., and Zuckerman, B. (1992). Cranial ultrasounds in term newborns: Failure to replicate excess abnormalities in cocaine exposed. *Pediatric Research* (in press).

Fried, P. A., and Makin, J. E. (1987). Neonatal behavioral correlates of prenatal exposure to marijuana, cigarettes, and alcohol in a low-risk population. *Neurobehav. Toxicol. Teratol.* **9**, 1–7.

Fried, P. A., and Watkinson, B. (1988). 12- and 24-month neurobehavioral follow-up of children prenatally exposed to marihuana, cigarettes, and alcohol. *Neurotoxicol. Teratol.* **10**, 305–313.

Fried, P. A., and Watkinson, B. (1990). 36- and 48-month neurobehavioral follow-up of children prenatally exposed to marijuana, cigarettes, and alcohol. *J. Dev. Behav. Pediatr.* **11**, 49–58.

Fried, P. A., Watkinson, B., and Willan, A. (1984). Marijuana use during pregnancy and decreased length of gestation. *Am. J. Obstet. Gynecol.* **150**, 23–27.

Fulroth, R., Phillips, B., and Durand, D. J. (1989). Perinatal outcome of infants exposed to cocaine and/or heroin in utero. *Am. J. Dis. Child.* **143**, 905–910.

Galler, J. R. (ed.) (1984). "Nutrition and Behavior." Plenum Press, New York.

Gawin, F. H., and Kleber, H. D. (1984). Cocaine abuse treatment. *Arch. Gen. Psychiatry* **41**, 903–909.

Gawin, F. H., and Kleber, H. D. (1988). Evolving conceptualizations of cocaine dependence. *Yale J. Biol. Med.* **61**, 123–136.

Goldbaum, G., Daling, J., and Milham, S. (1990). Risk factors for gastroschisis. *Teratology,* **42**, 397–403.

Graham, K., Koren, G., Klein, J., Schneiderman, J., and Greenwald, M. (1989). Determination of gestational cocaine exposure by hair analysis. *JAMA* **262**, 3328–3330.

Greenland, S., Staisch, K. J., Brown, N., and Gross, S. (1982). Effects of marijuana on human pregnancy, labor, and delivery. *Neurobehav. Toxicol. Teratol.* **4**, 447–450.

Hadeed, A. J., and Siegel, S. R. (1989). Maternal cocaine use during pregnancy: Effect on the newborn infant. *Pediatrics* **84**, 205–210.

Hamilton, H. E., Wallace, J. E., Shimek, E. L., Land, P., Harris, S. C., and Christenson, J. G. (1977). Cocaine and benzoylecgonine excretion in humans. *J. Forensic. Toxicol.* **22**, 697–702.

Handler, A., Kistin, N., Davis, F., and Ferre, C. (1991). Cocaine use during pregnancy: perinatal outcomes. *Am. J. Epidemiol.* **133**, 818–25.

Hans, S. L. (1989). Developmental consequences of prenatal exposure to methadone. Prenatal abuse of licit and illicit drugs. *Ann. N.Y. Acad. Sci.* **562**, 195–207.

Hingson, R., Zuckerman, B., Amaro, H., Frank, D. A., Kayne, H., Sorenson, J. R., Mitchell, J., Parker, S., Morelock, S., and Timperi, R. (1986). Maternal marijuana use and neonatal outcome: Uncertainty posed by self-reports. *Am. J. Public Health* **76**, 667–669.

Holmes, G. E., Miller, H. C., Hassanein, K., Lansky, S. B., and Goggin, J. E. (1977). Postnatal somatic growth in infants with atypical fetal growth patterns. *Am. J. Dis. Child.* **131**, 1978–1983.

Hoyme, H. E., Jones, K. L., Dixon, S. D., et al. (1990). Prenatal cocaine exposure and fetal vascular disruption. *Pediatrics* **85**, 743–747.

Hume, R. F., O'Donnell, K. J., Stanger, C. L., Killam, A. P., and Gingras, J. L. (1989). In utero cocaine exposure: Observations of fetal behavioral state may predict neonatal outcome. *Am. J. Obstet. Gynecol.* **161**, 685–690.

Hutchings, D. E. (1990). Issues of risk assessment: Lessons from the use and abuse of drugs during pregnancy. *Neurotoxicol. Teratol.* **12**, 183–189.

Indanpaan-Heikkila, J., Fritchie, G., Englert, L., et al. (1969). Placental transfer of tritiated I-tetrahydrocannabinol. *N. Engl. J. Med.* **281**, 330.

Isenberg, S. J., Spierer, A., and Inkelis, S. H. (1987). Ocular signs of cocaine intoxication in neonates. *Am. J. Opthalmol.* **103**, 211–214.

Jacobson, J. L., Jacobson, S. W., Schwartz, P. M., Fein, G. G., and Dowler, J. K. (1984). Prenatal exposure to an environmental toxin: A test of the multiple effects model. *Dev. Psychol.* **20**, 523–532.

Jacobvitz, D., and Sroufe, L. A. (1987). The early caregiver–child relationship and attention-deficit disorder with hyperactivity in kindergarten: A prospective study. *Child Dev.* **58**, 1488–1495.

Jakubovic, A., Tait, R., and McGeer, P. (1974). Excretion of the THC metabolites in ewes' milk. *Toxicol. Appl. Pharmacol.* **28**, 38–43.

Koren, G., Graham, K., Shear, H., and Einarson, T. (1989). Bias against the null hypothesis: The reproductive hazards of cocaine. *Lancet* **2**, 1440–1442.

Kharasch, S. J., Glotzer, D., Vinci, R., Weitzman, M., and Sargent, J. (1991). Unsuspected cocaine exposure in young children. *Am. J. Dis. Child.* **145**, 204–206.

Kramer, L. D., Locke, G. E., Ogunyemi, A., and Nelson, L. (1990). Neonatal cocaine-related seizures. *J. Child Neurol.* **5,** 60–64.

Kruez, D., and Axelrod, J. (1973). Delta-9-tetrahydrocannabinol: Localization in body fat. *Science* **179,** 391.

Lauder, J. M. (1988). Neurotransmitters as morphogens. *Prog. Brain Res.* **74,** 365.

Lester, B. M., and Dreher, M. (1989). Effects of marijuana use during pregnancy on newborn cry. *Child Dev.* **60,** 765–771.

Lester, B. M., Corwin, M. J., Sepkoski, C., et al. (1991). Neurobehavioral syndromes in cocaine exposed newborn infants. *Child Dev.* **62,** 694–705.

Lewis, P. (1987). Animal tests for teratogenicity: Their relevance to clinical practice. In "Drugs and Pregnancy" (D. F. Harkins, ed.), p. 22. Churchill Livingstone, Edinburgh.

Lifschitz, M. H., Wilson, G. S., Smith, E. O., et al. (1985). Factors affecting head growth and intellectual function in children of drug addicts. *Pediatrics* **75,** 269–274.

Linn, S., Schoenbaum, S. C., Monson, R. R., Rosner, R., Stubblefield, P. C., and Ryan, K. J. (1983). The association of marijuana use with outcome of pregnancy. *Am. J. Public Health* **73,** 1161–1164.

Lipshultz, S. E., Frassica, J. J., and Orav, E. J. (1991). Cardiovascular abnormalities in infants prenatally exposed to cocaine. *J. Pediatr.* **118,** 44–51.

Little, B. B., Snell, L. M., Klein, V. R., et al. (1989). Cocaine abuse during pregnancy: Maternal and fetal implications. *Obstet. Gynecol.* **73,** 157–160.

Longo, L. D. P. (1977). The biological effects of carbon monoxide on the pregnant woman, fetus, and newborn infants. *Am. J. Obstet. Gynecol.* **129,** 69–103.

MacGregor, S. N., Keith, L. G., Chasnoff, I. J., et al. (1987). Cocaine use during pregnancy: Adverse perinatal outcome. *Am. J. Perinatal Outcome* **157,** 686–690.

Marcus, J., Hans, S. L., and Jeremy, R. J. (1982). Differential motor and state functioning in newborns of women on methadone. *Neurobehav. Toxicol. Teratol.* **4,** 459–462.

Maynard, E. C., Amoruso, L. P., and Oh, W. (1991). Meconium for drug testing. *Am. J. Dis. Child.* **145,** 650–652.

Miller, J. C., and Friedhoff, A. J. (1988a). Prenatal neurotransmitter programming of postnatal receptor function. In (Boer, G. J., Feenstra, M. G. P., Mirmiran, M., Swaab, D. F., Van Haaren, F. eds.), *Progress in Brain Research.* Elsevier Science Publishers B.V., New York.

Miller, J. C., and Friedhoff, A. J. (1988b). Prenatal neurotransmitter programming of postnatal receptor function. *Prog. Brain Res.* **73,** 509.

Mirochnick, M., Meyer, J., Cole, J., et al. (1991). Circulating catecholamine in cocaine-exposed neonates: A pilot study. *Pediatrics.* In press.

Nahas, G. (1976). "Marijuana: Chemistry, Biochemistry, and Cellular Effects." Springer-Verlag, New York.

N.I.H. (1987). Consensus Statement: National Institutes of Health Consensus Development Conference on Infantile Apnea and Home Monitoring, Sept. 29 to Oct. 1, 1986. *Pediatrics* **79,** 292–299.

Needleman, H. L., Gunnoe, C., Leviton, A., Reed, R., Peresie, H., Maher, C., and Barrett, P. (1979). Deficits in psychologic and classroom performance of children with elevated dentine lead levels. *N. Engl. J. Med.* **300,** 689–695.

Needlman, R., Zuckerman, B., Anderson, G., Mirochnick, M., and Cohen. (1992). CSF monoamine precursors and metabolites in human neonates. *Pediatric Research* (in press).

Neerhof, M. G., MacGregor, S. N., Retzky, S. S., and Sullivan, T. P. (1989). Cocaine abuse during pregnancy: Peripartum prevalence and perinatal outcome. *Am. J. Obstet. Gynecol.* **161,** 633–638.

Neuspiel, D. R., and Hamel, S. C. (1991). Cocaine and infant behavior. *J. Dev. Behav. Pediatr.* **12,** 55–64.

Neuspiel, D. R., Hamel, S. C., Hochberg, E., Greene, J., and Campbell, D. (1990). Maternal cocaine use and infant behavior. *Neurotoxicol. and Teratol.* **13,** 229–233.

Niederreither, K., Jaeger, M. C., and Abrams, R. M. (1985). Cardiopulmonary effects of marijuana and delta-9-tetrahydrocannabinol in sheep. *Res. Commun. Substances Abuse* **6,** 87–98.

Oro, A., and Dixon, S. D. (1987). Perinatal cocaine and methamphetamine exposure: Maternal and neonatal correlates. *J. Pediatr.* **111,** 571–578.

Ostrea, E. M., Brady, M. J., Parks, P. M., Asensio, D. C., and Naluz, A. (1989). Drug screening of meconium in infants of drug-dependent mothers: An alternative to urine testing. *J. Pediatr.* **115**, 474–477.

Ostrea, E. M., Brady, M., Gause, S., Raymundo, A., and Steven M. (1992). Drug screening of newborns by meconium analysis: A large scale prospective epidemiologic study. *Pediatrics* **89**, 107–113.

Palti, H., Pevsner, B., and Adler, B. (1983). Does anemia in infancy affect achievement on developmental and intelligence tests? *Hum. Biol.* **55**, 183–194.

Perez-Reyes, M., and Wall, M. (1982). Presence of delta-9-tetrahydrocannabinol in human milk. *N. Engl. J. Med.* **307**, 819.

Petiti, D. B., and Coleman, C. (1990). Cocaine and the risk of low birth weight. *Am. J. Public Health.* **80**, 25–28.

Phibbs, C., Bateman, D., and Schwartz, R. (1991). The neonatal costs of maternal cocaine use. *JAMA* **266**, 1521–1526.

Pizzo, P. A., Eddy, J., Falloon, J., Balis, F., Murphy, R. F., Moss, H., Wolters, P., Brouwers, P., Jaronsinski, P., Rubin, M., Broder, S., Yarchoan, R., Brunetti, A., Maha, M., Nusinoff-Lehrman, S., and Poplack, D. G. (1988). Effect of continuous infusion of zidovudine (AZT) in children with symptomatic HIV infection. *N. Engl. J. Med.* **319**, 889–896.

Pynoss, K. S. (1990). Post-traumatic stress disorder in children and adolescents. In "Psychiatric Disorders in Children and Adolescents" (B. D. Garfinkle, G. A. Carlson, and E. B. Weller, eds.), pp. 48–63. Saunders, Philadelphia, Pennsylvania.

Qazi, Q. H., Mariano, E., Milman, D. H., Beller, E., and Crombleholme, W. (1985). Abnormalities in offspring associated with prenatal marihuana exposure. *Dev. Pharmacol. Ther.* **8**, 141–148.

Renault, P., Shuster, C., and Heinrich, R. (1971). Marijuana: Standardized smoke administration and dose–effect curves on heart rate in humans. *Science* **174**, 589.

Ricci, B., and Molle, F. (1987). Ocular signs of cocaine intoxication in neonates. *Am. J. Opthalmol.* **104**, 550–551.

Rivkin, M., and Gilmore, H. E. (1989). Generalized seizures in an infant due to environmentally acquired cocaine. *Pediatrics.* **84**, 1100–1102.

Rodning, C., Beckwith, L., and Howard, J. (1990). Characteristics of attachment organization and play organization in prenatally drug-exposed toddlers. *Dev. Psychopathol.* **1**, 277–289.

Rosenstein, B. J., Wheeler, J. S., and Heid, P. L. (1990). Congenital renal abnormalities in infants with *in utero* cocaine exposure. *J. Urol.* **144**, 11–112.

Rosenthal, R., and Rosenow. (1969). In "Behavioral Research" Academic Press, New York.

Rosenthal, R. (1969). Interpersonal Expectations: Effects of the experimenters hypothesis. In *Artifact in Behavioral Research* R. Rosenthal and R. L. Rosnow, ed. (pp. 181–277). Academic Press, New York.

Rosett, H. L., Werner, L., Lee, A., et al. (1983). Patterns of alcohol consumption and fetal development. *Obstet. Gynecol.* **61**, 539.

Ryan, L., Ehrlich, S., and Finnegan, L. (1987). Cocaine abuse in pregnancy: Effects on the fetus and newborn. *Neurotoxicol. Teratol.* **9**, 295–299.

Salamy, A., Eldredge, L., Anderson, J., and Bull, D. (1990). Brain stem transmission time in infants exposed to cocaine *in utero. J. Pediatr.* **17**, 627–629.

Sameroff, A., and Chandler, M. (1975). Reproductive risk and the continuum of caretaking casualty. In (Horowitz, F., Hetherington, M., Scarr-Salanatek, S., et al., eds.). *Review of Child Development Research,* University of Chicago Press, Chicago. 187–144.

Sampson, P. D., Streissguth, A. P., Barr, H. M., and Bookstein, F. L. (1989). Neurobehavioral effects of prenatal alcohol: Part II. Partial least squares analysis. *Neurotoxicol. Teratol.* **11**, 477–491.

Scher, M. S., Richardson, G. A., Coble, P. A., et al. (1988). The effects of prenatal alcohol and marijuana exposure: Disturbances in neonatal sleep cycling and arousal. *Pediatr. Res.* **24**, 101–105.

Shannon, M., Lacouture, P. G., Roa, J., and Woolf, A. (1989). Cocaine exposure among children seen at a pediatric hospital. *Pediatrics* **83**, 337–342.

Shih, L., Cone-Wesson, B., and Reddix, B. (1988). Effects of maternal cocaine abuse on the neonatal auditory system. *Int. J. Pediatr. Otorhinolaryngol.* **15**, 245–251.

Spitz, H. I., and Rosecan, J. S. (1987). "Cocaine Abuse: New Directions in Treatment and Research." Bruner/Mazel, New York.

Streissguth, A. P., Barr, H. M., Sampson, P. D., Bookstein, F. L., and Darby, B. L. (1989a). Neurobehavioral effects of prenatal alcohol: Part I. Research strategy. *Neurotoxicol. Teratol.* **11,** 461–476.

Streissguth, A. P., Bookstein, F. L., Sampson, P. D., and Barr, H. M. (1989b). Neurobehavioral effects of prenatal alcohol: Part III. PLS analyses of neuropsychologic tests. *Neurotoxicol. Teratol.* **11,** 493–507.

Streissguth, A. P., Barr, H. M., Sampson, P. D., et al. (1989c). IQ at age 4 in relation to maternal alcohol use and smoking during pregnancy. *Dev. Pharmacol.* **25,** 3.

Tennes, K., Avitable, N., Blackard, C., Boyles, C., Hassoun, B., Holmes, L., and Kreye, M. (1985a). Effects of marijuana and other illicit drugs on neonatal and maternal health. In "Current Research on the Consequences of Maternal Drug Abuse" (T. M. Pinkert, ed.), National Institute for Drug Research: Monograph No. **59,** 48–60. Rockville, Maryland.

van de Bor, M., Walther, F. J., and Sims, M. E. (1990a). Increased cerebral blood flow velocity in infants of mothers who abuse cocaine. *Pediatrics* **85,** 733–736.

van den Bor, M., Walther, F. J., and Ebrahimi, M. (1990b). Decreased cocaine input in infants of mothers who abused cocaine. *Pediatrics* **85,** 30–32.

Villar, J., Smerigilio, V., Martorell, R., Brown, C. H., and Klein, R. E. (1984). Heterogeneous growth and mental development of intrauterine growth-related infants during the first 3 years of life. *Pediatrics* **74,** 783–791.

Washton, A. M. (1989). Cocaine addiction: treatment, recovery, and relapse prevention (p. 11). W. W. Norton and Company, New York.

Welsh, M. C., and Pennington, B. F. (1988). Assessing frontal lobe functioning in children: Views from developmental psychology. *Dev. Neuropsychol.* **4,** 199–230.

Werner, E. (1989). Children of the garden island. *Sci. Am.* **106,** 111.

Woods, J. R., Jr., Plessinger, M. A., and Clark, K. E. (1987). Effect of cocaine on uterine blood flow and fetal oxygenation. *J.A.M.A.* **257,** 957–961.

Wu, T., Tashkin, D., Djahed, B., and Rose, J. E. (1988). Pulmonary hazards of smoking marijuana as compared to tobacco. *N. Engl. J. Med.* **318,** 347–351.

Zuckerman, B. (1988). Marijuana and cigarette smoking during pregnancy: Neonatal effects. In "Drugs, Alcohol, Pregnancy, and Parenting" (I. Chasnoff, ed.), pp. 73–89. Kluwer Academic Publishers, Dordrecht/Boston/London.

Zuckerman, B. S., and Beardslee, W. (1987). Maternal depression: An issue for pediatricians. *Pediatrics* **79,** 110–117.

Zuckerman, B., Amaro, H., Bauchner, H., and Cabral, H. (1989a). Depressive symptoms during pregnancy: Relationship to poor health behavior. *Am. J. Obstet. Gynecol.* **160,** 1107–1111.

Zuckerman, B. S., Amaro, H., and Cabral, H. (1989b). The validity of self-reported marijuana and cocaine use among pregnant adolescents. *J. Pediatr.* **115,** 812–815.

Zuckerman, B., Frank, D., Hingson, R., Amaro, H., Levenson, S. M., Kayne, H., Parker, S., Vinci, R., Aboagye, K., Fried, L. E., Cabral, H., Timperi, R., and Bauchner, H. (1989c). Effects of maternal marijuana and cocaine use on fetal growth. *N. Engl. J. Med.* **320,** 762.

Zuckerman, B., Maynard, E. C., and Cabral, H. (1991). A preliminary report of prenatal cocaine exposure

Zuckerman, B. (1991). "Selected Methodological Issues in Investigations of Prenatal Effects of Cocaine; Lessons from the Past." National Institute of Drug Abuse Research Monograph **114,** 45–54. Rockville, Maryland.

Zuckerman, B., and Frank, D. A. (1992). "Crack kids:" Not broken. *Pediatrics,* **89,** 337–339.

— 8 —

Cocaine and the Developing Nervous System: Laboratory Findings

ફે

Linda Patia Spear and Charles J. Heyser

Department of Psychology
and Center for Developmental Psychobiology
State University of New York, Binghamton
Binghamton, New York

INTRODUCTION

A substantial population of human infants and young children has been exposed gestationally to cocaine, with recent reports indicating that approximately 4.5 to 18% of women have used cocaine during pregnancy in the United States (Frank et al., 1988; Neerhof et al., 1989). From initial clinical reports, infants born to cocaine-using mothers appear to differ from control infants on a number of measures. For instance, these infants have been reported to exhibit tremulousness, increased muscle tone, and irritability, along with poor state organization and a depression of interactive behavior (Chasnoff et al., 1985, 1989; Doberczak et al., 1987). Such clinical studies, although limited to the extent to which behavioral/cognitive capacities have been examined, suggest that prenatal cocaine exposure may influence neurological development of exposed offspring.

A number of laboratories have recently begun to examine the neurobehavioral consequences of early cocaine exposure in laboratory animals such as rodents and sheep (see Dow-Edwards, 1991, for a concise review). It is likely that data obtained with animal models may produce findings of eventual clinical relevance based on findings such as those presented at the Workshop on the Qualitative and Quantitative Comparability of Human and Animal Developmental Neurotoxicity sponsored by the Environmental Protection Agency and the National Institute on Drug Abuse. This Workshop examined a number of compounds including anticonvulsant drugs, ethanol, methylmercury, lead, polychlorinated biphenyls, and ionizing radiation to assess across-species comparability of results between humans and laboratory animals, pre-

Maternal Substance Abuse and the Developing Nervous System
155

dominantly rodents. From the data presented at this workshop it was concluded that, at the level of functional category (sensory, motivational, cognitive, motor function, and social behavior), "close agreement was found across species for all neurotoxic agents reviewed at this Workshop. If a particular agent produced, for example, cognitive or motor deficits in humans, corresponding deficits were also evident in laboratory animals. This was true even though the specific end points used to assess these functions often varied substantially across species" (Stanton and Spear, 1990, p. 265). Evidence for quantitative comparability across species, however, was most apparent when an internal measure of dose equivalency (e.g., blood level) was used, rather than administered doses per se (Rees et al., 1990). From such analyses, it would appear that the rodent model is appropriate for assessing the neurobehavioral teratogenicity of drugs like cocaine that have possible adverse effects on human development.

If animal models of developmental toxicants merely mirror effects obtained in human clinical studies, it perhaps could be argued that studies in laboratory animals are unnecessary. This conclusion is unwarranted for a number of reasons. First, data are accumulated more rapidly in work with laboratory animals such as rodents, particularly when conducting life-span studies. It will be many years before there is a substantial human study population of school-aged children, adolescents, and adults that have been exposed gestationally to cocaine. Rodent studies can be used to predict the kinds of alterations that might be expected in the current population of cocaine-exposed children as they begin to enter the school system, and eventually reach puberty and adulthood. Also, animal studies are useful in that dose and treatment duration can be carefully controlled, and substantial control can be exerted over environmental variables such as nutritional status and exposure to other drugs that typically complicate interpretations of human clinical data (e.g., see Farrar and Kearns, 1989). In addition, animal models are particularly useful for assessing underlying neural and physiological alterations produced by developmental toxicants; assessing the neural substrates affected by developmental toxicants is critical for predicting long-term outcome and for developing appropriate intervention strategies.

For these reasons, there has been substantial interest in examining the neurobehavioral consequences of early cocaine exposure in laboratory animal populations. The focus of this chapter will be to summarize the efforts of our laboratory in this area to date.

ANIMAL MODEL USED

The animal model that we have used involves the subcutaneous administration to Sprague–Dawley rat dams of 10, 20, or 40 mg/kg cocaine hydrochloride daily from around the time of neural tube closure to shortly before term—gestational days 8–20 (E8–20). Because of our interest in neurobehavioral consequences, the timing of drug administration was chosen to encompass a period of rapid development of the central nervous system, being roughly equivalent to end of the first trimester and most of the second trimester exposure in humans. This choice of exposure may have been fortuitous given evidence that women may decrease their cocaine use substantially during the course of pregnancy (Richardson and Day, 1991) (it should be noted, however, that this work was conducted in light of moderate users and needs confirmation in heavier users).

The dose range and injection route that we chose produces maternal plasma concentrations of cocaine in the range of, or slightly above, those reported in human cocaine users (see Spear et al., 1989a, for further discussion). In addition, as can be seen in Fig. 1, dose-related increases in plasma and brain levels of cocaine in the dams and fetuses were observed (Spear et al., 1989a). Thus, this animal model meets the dual criteria (dose-dependent and clinically relevant plasma levels in the dams) that have been suggested as requirements for an appropriate animal model for human exposed offspring (e.g., Kutscher and Nelson, 1985; see also Rees et al., 1990).

There are advantages and disadvantages associated with any model system. Although a discussion of the pros and cons of various other cocaine-administration protocols that led to our choice of the subcutaneous route is beyond the scope of this chapter, a potential disadvantage of the route we chose should be acknowledged. With subcutaneous cocaine administration, skin necrosis is produced at the site of injection, which can be minimized but not eliminated by increasing injection volume (3 cc/kg) and hence diluting the concentration of cocaine administered. These sites do not become infected and quickly develop into discrete scabs. If injection sites are varied to avoid these sites, the dams show no apparent adverse reaction to the injection process (and indeed are typically at least as complacent about the injection process as are control animals receiving saline). Although it is nevertheless possible that this necrosis may somehow be stressful to the dams, there is no evidence at present to support this possibility. We feel that this potential disadvantage of the subcutaneous route is outweighed by the advantages of this route (that are not necessarily seen using other administration routes); no maternal toxicity even when using doses of cocaine that produce clinically relevant plasma levels and low variability in the plasma and brain levels of cocaine produced in the dams and fetuses.

In much of our work, we have focused only on offspring of dams given the 40 mg/kg dose (C40), a dose well above threshold for producing neurobehavioral alterations. Control dams included in these experiments consisted of dams allowed ad lib access to lab chow (LC), and often one of two types of nutritional control groups designed to control for the transient reduction in food intake that is typically seen in C40 dams at the onset of drug treatment (see the following). The first nutritional control group (PF) consisted of dams pair-fed (and in later studies, also pair-watered) to dams in the 40 mg/kg dose group beginning at the onset of drug treatment on gestational day 8 (E8). Pair-feeding, however, is in certain respects more an experimental treatment than a control, in that this method of food restriction influences behavior and physiology of the dams as well as producing some long-term effects on their offspring (e.g., Gallo and Weinburg, 1981, 1982; Weinberg and Gallo, 1982). Consequently, we have recently begun to explore the utility of another type of nutritional control (NC) that matches chow intake with that of cocaine-exposed dams without the explicit food restriction associated with pair-feeding. Dams in this NC

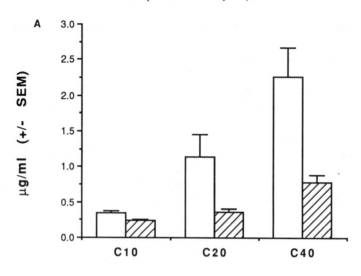

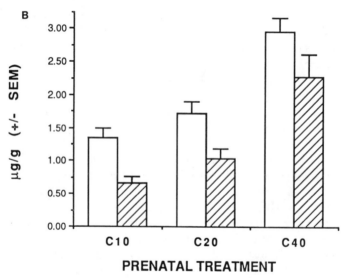

PRENATAL TREATMENT

FIGURE 1   (A) Plasma and (B) brain concentrations of cocaine in maternal and fetal samples 2.0 hr postinjection on gestational day 20 following chronic subcutaneous injections of 10 (C10), 20 (C20), or 40 (C40) mg/kg cocaine hydrochloride beginning on gestational day 8. Error bars reflect SEMs. Maternal (□); fetal (▨).

group are given ad lib access to a diet consisting of 60% lab chow and 40% cellulose (a nondigestable fiber) beginning on E8; the chow intake of these dams closely matches that of C40 dams, being transiently reduced for several days at the onset of the treatments on E8 before regaining control levels of intake (Moody et al., 1991).

In all but our initial study (Spear et al., 1989b), each litter of offspring was fostered to a surrogate dam on postnatal day 1 (P1), and maternal dams were sacrified for determination of number of implantation sites. Litter was used as the unit of analysis in all of our work, with no more than one pup per litter being placed into a given treatment condition. All litters were coded such that the experimenters conducting the testing or assays were unaware of the prenatal treatment condition of the offspring; postnatal testing conditions (drug versus control solutions, conditioning assignment group, etc.) were also coded whenever possible.

MATERNAL/LITTER DATA

We have used this cocaine dosing regimen in 11 series of studies to date and have observed few notable effects on the progress of gestation or fetal physical development (see Table 1). No maternal deaths have been observed in these cocaine-exposed dams. A slight but significant (6–9%) reduction in maternal weight gain was often seen, associated with a transient (2- to 5-day) reduction in food and water intake at the onset of drug treatment in the C40 dams. However, no significant differences in offspring body weights from P1 to adulthood have been observed in any of our work. Likewise, no differences have been found in the ratio of male:female pups in the litter or in litter size, although nonsignificant trends for an increase in resorption rates have

TABLE 1   Maternal/Litter Data with Gestational Cocaine Exposure

Measure	Effect seen
Maternal deaths	None
Maternal weight gain	Slight reduction in C40 dams (about 6–9%)
Food and water intake	Transient (2–5 days) reduction typically observed at onset of drug treatment in C40 dams
Gestational length	No effect
Reabsorption rates	No effect
Litter size	No effect
Male : female ratios	No effect
Offspring body weights	No effect
Reflex development	No effect
Physical maturation	No effect

been infrequently noted at the 40 mg/kg dose. Cocaine-exposed offspring do not vary from controls in terms of reflex maturation or the development of various physical landmarks (see Spear et al., 1989b). In other respects, these offspring do differ from control age-mates, as discussed in the following sections.

COGNITIVE EFFECTS IN OFFSPRING

Initially our focus was on cognitive assessment during the neonatal to weanling age period, although recently we have begun to examine exposed offspring in adulthood as well. In this work we have observed that offspring exposed gestationally to 40 mg/kg cocaine exhibited conditioning deficits in a variety of testing situations. For instance, when normal infant (P7–8) rat pups are given intraoral milk infusions intermittently in the presence of a particular odor, they develop a preference for that odor; C40 offspring did not exhibit conditioning in this simple appetitive classical conditioning task (Spear et al., 1989b). Similarly, as can be seen in Fig. 2, whereas P8 LC control pups learned to avoid an odor when that odor was paired with footshock, C40 offspring at this age did not show evidence of conditioning (Spear et al., 1989c; Heyser et al., 1990). This deficit in aversive classical conditioning was not related to a decreased sensitivity to the footshock unconditioned stimulus (US), as C40 offspring did not differ from control offspring in shock-sensitivity thresholds (Spear et al., 1989b).

There are two possible explanations of these classical conditioning deficits seen in C40 offspring. Although these offspring do not exhibit general maturational delays as indexed by reflex development, the development of physical landmarks, and body-weight measures (Spear et al., 1989b; see preceding sections), it is nevertheless possible that the conditioning deficits observed in these offspring early in life may reflect a subtle delay in cognitive development. Alternatively, exposed offspring may exhibit conditioning deficits owing to a fundamental alteration in cognitive function-ing/information processing. These two alternative possibilities are difficult to address in typical learning situations in which task performance improves with age. In a number of conditioning situations, however, young animals have been observed to learn more readily than older animals (e.g., see Spear and Kucharski, 1984). One such task is sensory preconditioning, a task that is readily learned by young rat pups in the age range of P7 to 18, but not by weanling (P21) animals (Chen et al., 1991). Given that there is normally a decline in performance on this task with age, performance of cocaine-exposed offspring at weaning would be expected to vary in opposite directions depending on the nature of the cognitive deficit. If there is a delay in cognitive development, cocaine-exposed weanlings should perform like younger animals, and hence exhibit better performance on this task than weanling control offspring. In contrast, if there is a fundamental alteration in cognitive performance, they should have difficulty performing this task regardless of age.

To test these alternatives, we assessed sensory preconditioning in C40 and LC offspring at three ages—P8 and P12 (ages when normal animals exhibit sensory

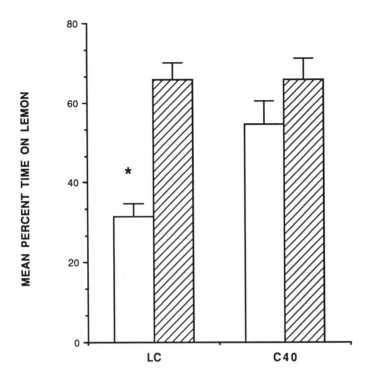

PRENATAL TREATMENT

FIGURE 2   Percentage of time spent by 8-day-old rat pups in the location of the lemon odor during preference testing. Conditioning was defined to occur if paired pups (□) (who received prior footshock exposure in the presence of the lemon odor) spent significantly less time in the presence of the lemon odor than did unpaired controls (▨) (whose prior footshock exposure was explicited unpaired with lemon exposure) (*, $p < 0.05$ for these comparisons). LC, saline control; C40, prenatally cocaine-exposed. Error bars reflect SEMs.

preconditioning) as well as P21 (an age when normal animals no longer exhibit significant sensory preconditioning) (Heyser et al., 1990). The procedures are outlined in Table 2. In Phase 1, paired animals received a 3-min simultaneous exposure to two odors—banana and lemon. In Phase 2, one of these odors, lemon, was paired with footshock. During the test, each animal was given a 3-min preference test between banana and a novel odor (orange). To the extent that the paired animals formed an association between banana and lemon in Phase 1, when lemon is paired with footshock in Phase 2, they should display an aversion to banana during the preference test. Two control groups were used: one that received unpaired exposures to banana and lemon in Phase 1, and one that received unpaired exposure to lemon and footshock in Phase 2. As can be seen in Fig. 3, LC animals showed significant sensory preconditioning at P8 and P12, but not P21, confirming the ontogenetic decline that

TABLE 2   Procedure for Sensory Preconditioning

	Phase I	Phase II	Test
Paired	CS1 + CS2 Simultaneous	CS- / CS2 + US Conditioning	CS1 vs neutral
Control phase I	CS1/CS2 (Unpaired)	CS- / CS2 + US Conditioning	CS1 vs neutral
Control phase II	CS1 + CS2 Simultaneous	US/CS-/ CS2 (Unpaired)	CS1 vs neutral

has been previously reported in this task (Chen et al., 1991). In contrast, C40 offspring did not show sensory preconditioning at any age. These data are consistent with the hypothesis that cocaine-exposed animals exhibit a fundamental alteration in cognitive performance rather than a delay in cognitive development.

In view of these findings, it would be expected that cognitive deficits may also be evident in adulthood. Indeed, Smith et al. (1989) have reported that adult offspring exposed gestationally to cocaine exhibit deficits in water maze performance and alterations in the acquisition of an operant response on a DRL-20 schedule (where reinforcement is obtained only if a lever press follows the preceding response by 20 sec. or more). To determine whether cognitive deficits are evident in adult exposed offspring using our treatment regimen, we examined C40 and control offspring on the acquisition and reversal of an operant conditional discrimination task in adulthood (Heyser et al., In press, a). Once shaped to bar-press for food, animals were gradually exposed to a decreasing rate of reinforcement until animals responded consistently on an FR-10 (one reinforcement for every 10 lever presses) schedule of reward. Animals remained on the FR-10 schedule throughout training on the conditional discrimination task, during which the odor present in the apparatus on a given day indicated which lever was to be reinforced on that day, with one odor indicating the left bar and another indicating the right bar. Once criterion was reached on the conditional discrimination task, the contingencies of the discrimination were reversed—that is, the odor that previously served as the cue for the left lever in the initial discrimination was switched to indicate the right lever and vice versa. The results indicated that C40 animals did not differ from PF, NC, and LC controls in rate of acquisition of the FR-10 response or on initial acquisition of the conditional discrimination. However, C40 animals required more sessions than all control groups to reach criterion on the reversal task, as can be seen in Fig. 4A. Moreover, once C40 offspring reached criterion for the reversal task, they still exhibited more errors on this task than control animals (see Fig. 4B). Thus, cocaine-exposed offspring examined in adulthood do exhibit cognitive deficits in terms of acquisition of a reversal task, although their performance does not differ from controls on the initial acquisition of the conditional discrimination. Indeed, reversal performance may be a particularly sensitive measure of cognitive disruption. For instance, a number of studies have demonstrated that, whereas initial ac-

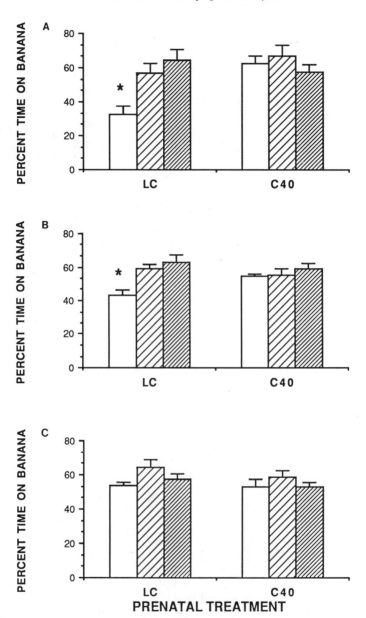

FIGURE 3 Assessment of sensory preconditioning for (A) 8-day-old, (B) 12-day-old and (C) 21-day-old rat pups as measured by percentage of time spent during the preference test in the location of the banana odor. Sensory preconditioning was defined to occur if the paired pups (□) spent significantly less time in the presence of the banana odor than did the unpaired controls (*, $p < 0.05$ for these comparisons). LC, saline control; C40, prenatally cocaine exposed; paired (□), conditioning procedure; CS1/CS2 UP (▨), Phase 1 unpaired odor-odor control; CS2/US UP (▨), Phase 2 unpaired odor-footshock control. Error bars reflect SEMs.

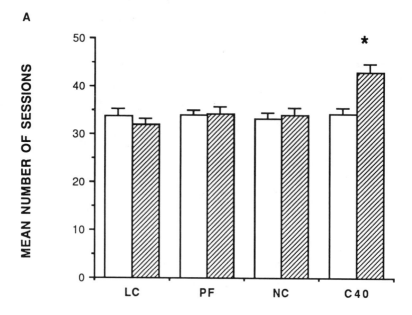

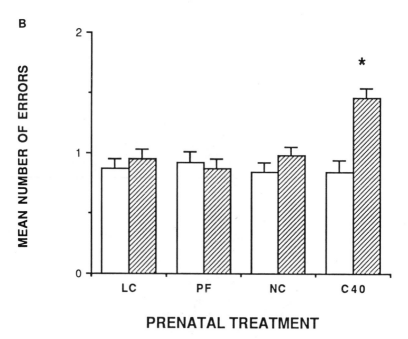

PRENATAL TREATMENT

FIGURE 4   (A) Mean number of sessions to criterion and (B) number of errors before first reward once criterion was reached for the initial acquisition (□) and reversal (▨) of a conditional discrimination task in adult offspring of dams exposed to cocaine during pregnancy (C40) or of various control dams (LC, nontreated control; PF, pair-fed; NC, nutritional control). Error bars reflect SEMs (*, $p < 0.05$ for these comparisons.)

quisition of a discrimination task was unaffected, reversal performance was altered by a variety of experimental manipulations, including lesions of specific brain regions (Eichenbaum et al., 1986; Fagan and Olton, 1986), and gestational exposure to stress (Weller et al., 1988) or adrenocorticotropic hormone (McGivern et al., 1987).

Taken together, the results of our conditioning experiments to date indicate that cocaine-exposed offspring exhibit deficits early in life in a variety of conditioning situations, deficits that do not appear to be related to any alteration in sensitivity to the unconditioned stimulus (e.g., footshock) or to the use of any particular sensory modality as a CS. These deficits do not appear to be the result of a delay in cognitive development, but rather appear to reflect a fundamental alteration in cognitive function/information processing. Consistent with this hypothesis, cognitive deficits appear to persist into adulthood, as reflected by deficits in acquisition and performance on a reversal task in adult C40 offspring.

NEURAL ALTERATIONS IN OFFSPRING

Effects on the Dopamine System

The work conducted in our laboratory has resulted in a number of lines of evidence to suggest that gestational exposure to the dopamine (DA) uptake inhibitor cocaine alters later functioning in the DA system. However, the ultimate impact that such exposure has on DA functioning still remains to be determined.

In our early work in this area, we observed that cocaine-exposed offspring exhibited a reduction in footshock-precipitated wall climbing at P12; given that wall climbing has been previously shown to be strongly related to levels of catecholamine activity at this age, we concluded from this finding that gestational cocaine exposure may result in an attenuation in catecholaminergic function (Spear et al., 1989b). Further data consistent with this hypothesis were provided by the observations that C40 offspring exhibit an attenuated psychopharmacological response to the DA agonist apomorphine (P12 test) along with a conversely accentuated behavioral response to the DA antagonist haloperidol (P8 test) (Spear et al., 1989c). We interpreted these data to suggest that gestational cocaine exposure may result in a down-regulation of the DA system (see Spear et al., 1989c). In retrospect, we were probably premature in drawing this conclusion, for in other respects, C40 offspring exhibit an attenuated response to haloperidol rather than the accentuated haloperidol response that was obtained psychopharmacologically. Measures from which these findings were obtained include haloperidol-induced increases in the ratio of the DA metabolites dihydroxyindoleacetic acid (DOPAC) and homovanillic acid (HVA) to DA in striatal/nucleus accumbens homogenates, as well as haloperidol-induced increases in plasma prolactin levels in P11 offspring (see Spear et al., 1990).

Although various hypotheses could undoubtedly be generated as to how these opposing haloperidol results can be reconciled, these data serve to remind us that it may be premature to do so. There are a multitude of possible alterations in the DA system that potentially could be induced by a developmental insult such as chronic

TABLE 3 D_1 and D_2 Dopamine Receptor Binding

	Striatum		Nucleus accumbems	
Prenatal treatment[a]	D_1	D_2[b]	D_1	D_2
Saline control	365.3	139.0	142.5	80.3
$n = 8$	±26.3	±7.9	±18.7	±6.0
Cocaine (40 mg/kg)	370.8	163.6*	150.3	82.4
$n = 8$	±9.8	±7.9	±17.9	±7.5

[a]Membranes from nucleus accumbens and striatum were incubated with either 1.0 nM [³H]SCH-23390 (D_1) or [³H]spiroperidol (D_2) with or without 1 μM (+)butaclamol. Binding is expressed as mean fmoles/mg protein ± SEM. n, number of rats per group.
[b]*, $p < 0.05$, significantly different from saline controls.

cocaine exposure. Some of these alterations may be induced by the drug exposure per se, whereas others may be compensatory to initial drug-induced modifications in the development of other portions of the system. Consequently, it may not be appropriate to infer from data limited to but a few indices of a neurotransmitter system how the system as a whole is altered by gestational drug exposure. Rather, to determine the ultimate impact of gestational cocaine treatment on DA function, it would appear to be necessary to derive a more complete profile of alterations in the DA system, including examinations of density of presynaptic DA terminal ending, number of DA uptake sites, D_1- and D_2-receptor expression, binding, and coupling to second messenger systems, and so on.

We are now just beginning to use this more multifaceted approach. In initial homogenate assays of D_1- and D_2-receptor binding in striatum and nucleus accumbens of weanling (P21) pups, we observed that C40 offspring exhibited an 18% increase in D_2-receptor binding in striatum (see Table 3). Skatchard analyses revealed that this effect was related to an increase in receptor affinity (Scalzo et al., 1990). Although this increase in D_2 binding is relatively modest, it does appear to be of functional significance, in that C40 animals at this age exhibit an increased responsiveness to the D_2-receptor agonist quinpirole. This is reflected by a shift to the left in the quinpirole dose–response curve, as illustrated in Fig. 5 for rearing behavior (Moody et al., in press). These complementary neurochemical and psychopharmacological data provide additional evidence that early cocaine exposure alters later DA functioning.

Currently we are examining other measures of DA function. It is hoped that the results of our research in conjunction with those from other laboratories working in the area will eventually lead to an elucidation of the potentially complex profile of alterations in the DA system induced by early cocaine exposure.

Effects on other Neural and Non-Neural Systems

Neural alterations induced by gestational cocaine exposure are not restricted to the DA system. For instance, we examined regional density of opiate receptors labelled

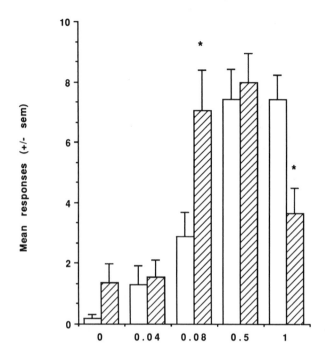

Dose of Quinpirole

FIGURE 5   Mean number of supported and unsupported rears by weanling offspring exposed gestationally to cocaine (C40; ▨) and weanling control offspring (LC; □) following various doses of the D_2-receptor agonist, quinpirole. Note the shift to the left in the dose–response curve to quinpirole in C40 offspring, with the number of supported/unsupported rears in C40 weanlings being significantly greater at the 0.08 mg/kg dose and significantly fewer at the 1.0 mg/kg dose than LC controls (*, $p < 0.05$ for these comparisons). Error bars reflect SEMs.

with [³H]naloxone in autoradiograms of P21 male offspring from cocaine-exposed or control dams. Significant increases in labeling were seen in numerous brain regions of C40 (and occasionally C20) offspring, with increases being observed not only in DA terminal brain regions but also other brain regions (Clow et al., 1991). A few of these findings are illustrated in Fig. 6. Given that opiate binding is increased in C40 offspring, it might be expected that these animals would display an increased psychopharmacological sensitivity to opiate agonists such as morphine. Indeed, consistent with this expectation, we have data showing that C40 offspring at P11 are more sensitive to morphine than are LC, PF, or NC offspring, in terms of a morphine-induced decrease in isolation-induced ultrasounds (Goodwin et al., in preparation).

Gestational cocaine exposure also produces transient alterations in whole brain levels of gangliosides and glycolipids, complex carbohydrates that play important roles in numerous developmental events such as cell contact responses and regulation of cellular growth. We observed that whole brain levels of both gangliosides and

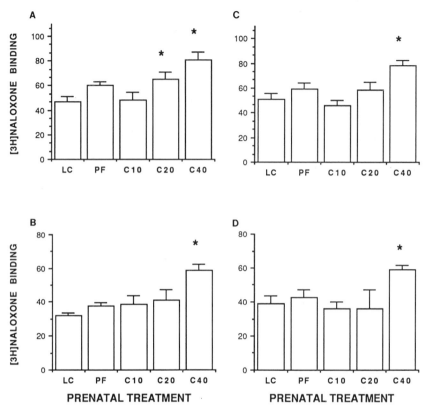

FIGURE 6   Mean (fmol/mg) ± SEM obtained from autoradiographs of 2.5 nM [³H]naloxone binding in selected brain regions of 21-day-old male offspring of dams exposed to 10 (C10), 20 (C20), 40 (C40) mg/kg cocaine during pregnancy or of pair-fed (PF) or untreated (LC) control dams (*, $p < 0.05$ for these comparisons). (A) rostral olfactory tubercle; (B) somatosensory cortex; (C) medial prefrontal cortex; (D) entorhinal cortex.

glycolipids were elevated in C40 offspring at P1, with levels being normalized by P11 (Leskawa et al., 1989). The significance of these transient elevations in gangliosides and glycolipids early in life in cocaine-exposed offspring has yet to be determined.

Alterations in cocaine-exposed offspring may not even be restricted to the nervous system. For instance, we observed that thymus weights of C40 offspring were lower than those of LC controls at P1, not different from those of control at P11, and significantly increased over LC levels at P21 (Spear et al., 1990). These age-related alterations in thymus weight may reflect potential alterations in immune function induced by gestational exposure. Indeed, Sobrian et al. (1990) has reported alterations in humoral immune responsivity in adult animals following early chronic cocaine exposure.

ALTERATIONS IN THE LATER REWARDING EFFICACY OF COCAINE: INCREASED DRUG ABUSE LIABILITY IN COCAINE-EXPOSED OFFSPRING?

We have completed two projects showing that cocaine-exposed offspring are less likely to develop a preference for stimuli previously associated with cocaine. In the first study, adult offspring were tested for conditioned place preference (CPP) for cocaine. As can be seen in Fig. 7, whereas significant place conditioning was obtained in LC and PF offspring when either 2.0 or 5.0 mg/kg of cocaine was paired with the designated place, C40 offspring did not exhibit place conditioning at either training dose (Heyser et al., 1992). The effects of gestational cocaine exposure on the rewarding properties of cocaine were further assessed in a second study, which examined cocaine-induced odor preferences in infant (P6–8) offspring from C40, NC, and LC dams (Heyser et al., in press b). In this study, LC and NC offspring were observed to exhibit a preference for the odor that had been paired with 2.0, 5.0, or 10.0 mg/kg cocaine. In contrast, C40 exhibited a significant odor preference only when the odor had been previously paired with 5.0 or 10.0 mg/kg cocaine.

There are a number of possible explanations of these data. One possibility is that the data may reflect potential alterations in cocaine pharmacokinetics in C40 offspring. For instance, if C40 offspring metabolized cocaine more rapidly than did control animals, higher delivered doses of cocaine might be necessary to support a cocaine-induced odor or place preference in these animals. Yet, results from the CPP study do not support a simple pharmacokinetics explanation. When chamber entries were analyzed, animals from all prenatal treatment groups that were exposed to 5 mg/kg cocaine during conditioning exhibited fewer compartment entries (i.e., less general activity) on the test day when compared with littermates who received saline during conditioning (and cocaine later in the day in their home cages) (Heyser et al., 1992). These data suggest that offspring exposed gestationally to cocaine are influenced on the test day by administration of 5 mg/kg cocaine during conditioning, although this influence is not manifest with respect to CPP. Thus, it is unlikely that the attenuated preference for stimuli previously associated with cocaine in C40 offspring is related simply to an alteration in the pharmacokinetics of cocaine.

Another possible explanation for these alterations in C40 offspring is that these findings reflect a general impairment of the learning process in these animals; indeed, as discussed previously, gestational cocaine exposure results in cognitive deficits that are evident both during development and in adulthood. It could perhaps be argued that because infant C40 offspring show conditioning at higher doses of cocaine (5.0 and 10.0 mg/kg), the lack of conditioning at the 2.0 mg/kg dose is unlikely to reflect a simple conditioning effect. Yet, US intensity influences the magnitude of conditioning (Rescorla and Wagner, 1972); hence, it is possible that higher US intensities (i.e., higher doses of cocaine) may have been sufficient to overcome deficits in associative learning evident in C40 offspring at a weaker US intensity (the 2.0 mg/kg dose). However, with regard to the deficits in adult CPP data, it is unlikely that the data

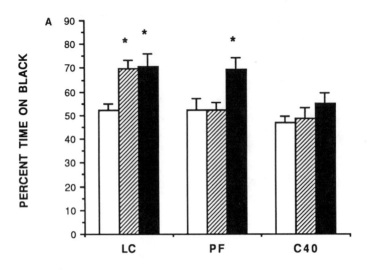

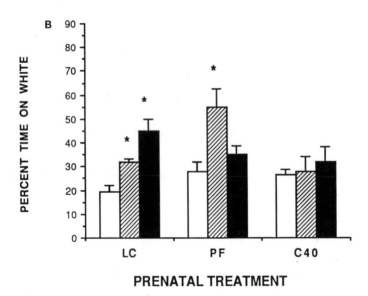

PRENATAL TREATMENT

FIGURE 7   Percentage of time spent on the test day in the previously drug-paired compartment when adult offspring from each of the prenatal treatment groups were conditioned in black (A) and conditioned in white (B). Place conditioning was defined to occur if the animals conditioned with cocaine spent significantly more time in the drug-paired chamber than animals receiving saline ($*$, $p < 0.05$ for these comparisons). C40, prenatally cocaine-exposed; PF, pair-fed control; LC, nontreated control. Error bars reflect SEMs. Saline (□); 2.0 mg/kg (▨); 5.0 mg/kg (■).

simply reflect a cognitive impairment. In adulthood, C40 animals have been observed to exhibit normal acquisition of a conditional discrimination response (Heyser, et al., In press, a; see preceding sections); hence, it would be expected that they should be at least as capable of acquiring the basic discrimination response necessary for CPP conditioning.

The final, and perhaps most viable, potential explanation for these data is that early cocaine exposure may alter the development of brain reward systems, resulting in a long-term attenuation in the rewarding properties of cocaine. According to this view, C40 offspring may not show conditioned preferences for stimuli associated with modest doses of cocaine because cocaine is a less effective reinforcer in these offspring. This reduction in reward efficacy induced by gestational cocaine exposure may be related to long-term alterations in the dopamine system (see preceding sections), given that mesolimbic DA regions have been implicated in the reward mechanisms of cocaine (e.g., Wise, 1987; Koob and Goeders, 1989).

Further work is needed to test the hypothesis that gestational cocaine exposure may result in a long-term attenuation in the rewarding properties of cocaine. The potential implications of these studies may be of importance. At least with respect to ethanol, animals that exhibit greater ethanol ingestion owing to genetic background (e.g., Randall and Lester, 1974; Lumeng et al., 1982) or chronic heavy metal exposure (Grover et al., 1991) show a reduced sensitivity to ethanol, including the reinforcing efficacy of ethanol (e.g., see Grover et al., 1991). Thus, a reduction in the rewarding properties of a substance may be associated with greater self-administration of that compound. This raises the possibility that exposure to cocaine during gestation may result in higher rates of self-administration in adulthood, a hypothesis that we plan to test in the near future in our laboratory. Taken together, this line of research may eventually have important implications with regard to the later drug-abuse liability of offspring exposed early in life to cocaine.

SUMMARY AND CONCLUSIONS

In our research to date we have observed that gestational cocaine exposure results in a number of neurobehavioral alterations in the offspring. Notable effects of early cocaine exposure include cognitive deficits that are evident early in life as well as in adulthood, as well as alterations in the DA system in young offspring. Observed physiological effects do not appear to be restricted, however, to the DA system or even to the nervous system.

We still have far to go in characterizing the pattern of neurobehavioral alterations induced by gestational cocaine exposure. For instance, more work is obviously necessary to characterize the effects of cocaine on the developing DA system. In addition, further research is needed to establish the profile of cognitive alterations seen in these animals, and to determine whether particular pharmacological therapies are effective in at least partially ameliorating these deficits. The intriguing possibility that early cocaine exposure may lead to an increase in subsequent self-administration of cocaine and possibly other drugs of abuse needs to be carefully evaluated in future work. We have also yet to determine critical periods for perinatal exposure in producing observed neurobehavioral alterations, and more dose–response studies are needed. Finally, it should be recognized that the focus of our initial studies was on assessment early in life to obtain data most relevant to the current human population of exposed infants and young children; we have yet to determine whether these neurobehavioral alterations are evident also in older animals.

Although we still have much work to do, it is clear from our research to date that cocaine is a neurobehavioral teratogen. Indeed, numerous other laboratories, using a variety of rodent model systems differing in dose, timing, and route of administration, have also reported that early cocaine exposure induces neural (Anderson-Brown et al., 1990; Church and Overbeck, 1990b; Dow-Edwards et al., 1988, 1990; Wiggins and Ruiz, 1990) and behavioral (Hutchings et al., 1989; Smith et al., 1989; Church and Overbeck, 1990a; Henderson and McMillen, 1990; Raum et al., 1990; Sobrian et al., 1990) alterations (although occasional negative findings have also been reported— Fung et al., 1989; Giordano et al., 1990; Riley & Foss, 1991a,b). The combined efforts from both basic research and clinical laboratories should result in a rapidly accumulating data base of replicable findings regarding the profile of neurobehavioral teratogenicity induced by cocaine. Given the large current population of young children who have been exposed prenatally to this drug of abuse, such rapid advancements of knowledge are clearly needed at this time.

ACKNOWLEDGMENTS

This research was supported by National Institute on Drug Abuse Grants R01 DA04478, K02 DA00140 and F31 DA05511.

REFERENCES

Anderson-Brown, T., Slotkin, T. A., and Seidler, F. J. (1990). Cocaine acutely inhibits DNA synthesis in developing rat brain regions: Evidence for direct actions. *Brain Res.* **537**, 197–202.

Chasnoff, I. J., Burns, W. J., Schnoll, S. H., and Burns, K. A. (1985). Cocaine use in pregnancy. *N. Engl. J. Med.* **313**, 666–669.

Chasnoff, I. J., Griffith, D. R., MacGregor, S., Dirkes, K., and Burns, K. A. (1989). Temporal patterns of cocaine use in pregnancy. *J. A. M. A.* **261**, 1741–1744.

Chen, W.-J., Lariviere, N. A., Heyser, C. J., Spear, L. P., and Spear, N. E. (1991). Age-related differences in sensory conditioning in rats. *Dev. Psychobiol.* **24**, 307–325.

Church, M. W., and Overbeck, G. W. (1990a). Prenatal cocaine exposure in the Long–Evans rat: II. Dose-dependent effects on offspring behavior. *Neurotoxicol. Teratol.* **12**, 335–343.

Church, M. W., and Overbeck, G. W. (1990b). Prenatal cocaine exposure in the Long–Evans rat: III. Developmental effects on the brainstem auditory-evoked potential. *Neurotoxicol. Teratol.* **12**, 345–351.

Clow, D. W., Hammer, R. P., Kirstein, C. L., and Spear, L. P. (1991). Gestational cocaine exposure increases opiate receptor binding in weanling offspring. *Dev. Brain Res.* **59**, 179–185.

Doberczak, T. M., Shanzer, S. R., and Kandall, S. R. (1987). Neonatal effects of cocaine abuse in pregnancy. *Pediatr. Res.* **21**, 359A.

Dow-Edwards, D. L. (1991). Cocaine effects on fetal development: A comparison of clinical and animal research findings. *Neurotoxicol. Teratol.* **13,** 347–352.

Dow-Edwards, D. L., Freed, L. A., and Milhorat, T. H. (1988). Stimulation of brain metabolism by perinatal cocaine exposure. *Dev. Brain Res.* **42,** 137–141.

Dow-Edwards, D. L., Freed, L. A., and Fico, T. A. (1990). Structural and functional effects of prenatal cocaine exposure in adult rat brain. *Dev. Brain Res.* **57,** 263–268.

Eichenbaum, H., Fagan, A., and Cohen, N. J. (1986). Normal olfactory discrimination learning set and facilitation of reversal learning after medial-temporal damage in rats: Implications for an account of preserved learning abilities in amnesia. *J. Neurosci.* **6,** 1876–1884.

Fagan, A. M., and Olton, D. S. (1986). Learning sets, discrimination reversal, and hippocampal function. *Behav. Brain Res.* **21,** 13–20.

Farrar, H. C., and Kearns, G. L. (1989). Cocaine: Clinical pharmacology and toxicology. *J. Pediatr.* **115,** 665–675.

Frank, D. A., Zuckerman, B. S., Amaro, H., Aboagye, K., Bauchner, H., Cabral, H., Fried, L., Hingson, R., Kayne, K., Levenson, S. M., Parker, S., Reese, H., and Vinci, R. (1988). Cocaine use during pregnancy; Prevalence and correlates. *Pediatrics* **82,** 888–895.

Fung, Y. K., Reed, J. A., and Lau, Y.-S. (1989). Prenatal cocaine exposure fails to modify neurobehavioral responses and the striatal dopaminergic system in newborn rats. *Gen. Pharmacol.* **20,** 689–693.

Gallo, P. V., and Weinberg, J. (1981). Corticosterone rhythmicity in the rat: Interactive effects of dietary restriction and schedule of feeding. *J. Nutr.* **111,** 208–218.

Gallo, P. V., and Weinberg, J. (1982). Neuromotor development and response inhibition following prenatal ethanol exposure. *Neurobehav. Toxicol. Teratol.* **4,** 505–513.

Giordano, M., Moody, C. A., Zubrycki, E. M., Dreshfield, L., Norman, A. B., and Sanberg, P. R. (1990). Prenatal exposure to cocaine in rats: Lack of long-term effects on locomotion and stereotypy. *Bull. Psychonom. Soc.* **28,** 51–54.

Grover, C. A., Nation, J. R., Reynolds, K. M., Benzick, A. E., Bratton, G. R., and Rowe, L. D. (1991). The effects of cadmium on ethanol self-administration using a sucrose-fading procedure. *Neurotoxicology* **12,** 235–244.

Henderson, M. G., and McMillen, B. A. (1990). Effects of prenatal exposure to cocaine or related drugs on rat developmental and neurological indices. *Brain Res. Bull.* **24,** 207–212.

Heyser, C. J., Spear, N. E., and Spear, L. P. The effects of prenatal exposure to cocaine on conditional discrimination learning in adult rats. *Behav. Neurosci.* In press (a).

Heyser, C. J., Chen, S.-J., Miller, J., Spear, N. E., and Spear, L. P. (1990). Prenatal cocaine exposure induces deficits in Pavlovian conditioning and sensory preconditioning among infant rat pups. *Behav. Neurosci.* **104,** 955–963.

Heyser, C. J., Miller, J. S., Spear, N. E., and Spear, L. P. (1992). Prenatal exposure to cocaine disrupts cocaine-induced conditioned place preference in rats. *Neurotox. Teratol.*, **14,** 57–64.

Heyser, C. J., Goodwin, G. A., Moody, C. A., and Spear, L. P. Prenatal cocaine exposure attenuates cocaine-induced odor preference in infant rats. *Pharmacol. Biochem. and Behav.,* in press (b).

Hutchings, D. E., Fico, T. A., and Dow-Edwards, D. L. (1989). Prenatal cocaine: Maternal toxicity, fetal effects, and locomotor activity in rat offspring. *Neurotoxicol. Teratol.* **11,** 65–69.

Koob, G. R., and Goeders, N. E. (1989). Neuroanatomical substrates of drug self-administration. *In* "The Neuropharmacological Basis of Reward" (J. M. Liebman and S. J. Cooper, eds.), pp. 214–263. Oxford University Press, New York.

Kutscher, C. L., and Nelson, B. K. (1985). Dosing considerations in behavioral teratology testing. *Neurobehav. Toxicol. Teratol.* **7,** 663–664.

Leskawa, K. C., Jackson, G. H., Moody, C. A., and Spear, L. P. (1989). Effect of prenatal cocaine exposure on neonatal and maternal brain glycosphingolipids. *Glycoconjugate J.* **5,** 447.

Lumeng, L., Waller, M. B., McBride, W. J., and Li, T.-K. (1982). Different sensitivities to ethanol in alcohol-preferring and -nonpreferring rats. *Pharmacol. Biochem. Behav.* **16,** 125–130.

McGivern, R. F., Rose, G., Berka, C., Clancy, A. N., Sandman, C. A., and Beckwith, B. E. (1987). Neonatal exposure to a high level of ACTH impairs adult learning performance. *Pharmacol. Biochem. Behav.* **27,** 133–142.

Moody, C. A., Frambes, N. A., and Spear, L. P. Psychopharmacological responsiveness to the dopamine agonist quinpirole in normal weanlings and in weanling offspring exposed gestationally to cocaine. *Psychopharmacology,* in press.

Moody, C., Goodwin, G., Heyser, C., and Spear, L. (1991). A novel nutritional control for transient drug-induced anorexia during pregnancy. *Teratol.* **43,** 495.

Neerhof, M. G., MacGregor, S. N., and Sullivan, T. P. (1989). Cocaine abuse during pregnancy: Peripartum prevalence and perinatal outcome. *Am. J. Obstet. Gynecol.* **161,** 633–638.

Randall, C. L., and Lester, D. (1974). Differential effects of ethanol and pentobarbital on sleep time in C57BL and BALB mice. *J. Pharmacol. Exp. Ther.* **188,** 27–33.

Raum, W. J., McGivern, F. R., Peterson, M. A., Shryne, J. F., and Gorski, R. A. (1990). Prenatal inhibition of hypothalamic sex steroid uptake by cocaine: Effects on neurobehavioral sexual differentiation in male rats. *Dev. Brain Res.* **53,** 230–236.

Rees, D. C., Francis, E. Z., and Kimmel, C. A. (1990). Qualitative and quantitative comparability of human and animal developmental neurotoxicants: A workshop summary. *Neurotoxicology* **11,** 257–270.

Rescorla, R. A., and Wagner, A. R. (1972). A theory of Pavlovian conditioning: Variations in the effectiveness of reinforcement and nonreinforcement. *In* "Classical Conditioning II: Current Research and Theory" (A. H. Black and W. F. Prokasy, eds.), pp. 64–99. Appleton-Century-Crofts, New York.

Richardson, G. A., and Day, N. L. (1991). Maternal and neonatal effects of moderate cocaine use during pregnancy. *Neurotoxicol. Teratol.* **13,** 455–460.

Riley, E. P. and Foss, J. A. (1991a). Exploratory behavior and locomotor activity: a failure to find effects in animals prenatally exposed to cocaine. *Neurotox. Teratol.* **13,** 553–558.

Riley, E. P. and Foss, J. A. (1991b). The acquisition of passive avoidance, active avoidance, and spatial navigation tasks by animals prenatally exposed to cocaine. *Neurotox. Teratol.* **13,** 559–564.

Scalzo, F. M., Ali, S. F., Frambes, N. A., and Spear, L. P. (1990). Weanling rats exposed prenatally to cocaine exhibit an increase in striatal D_2-dopamine binding associated with an increase in ligand affinity. *Pharmacol. Biochem. Behav.* **37,** 371–373.

Smith, R. F., Mattran, K. M., Kurkjian, M. F., and Kurtz, S. L. (1989). Alterations in offspring behavior induced by chronic prenatal cocaine dosing. *Neurotoxicol. Teratol.* **11,** 35–38.

Sobrian, S. K., Burton, L. E., Robinson, N. L., Ashe, W. K., James, H., Stokes, D. L., and Turner, L. M. (1990). Neurobehavioral and immunological effects of prenatal cocaine exposure in rat. *Pharmacol. Biochem. Behav.* **35,** 617–629.

Spear, L. P., Frambes, N. A., and Kirstein, C. L. (1989a). Fetal and maternal brain and plasma levels of cocaine and benzoylecgonine following chronic subcutaneous administration of cocaine during gestation in rats. *Psychopharmacology* **97,** 427–431.

Spear, L. P., Kirstein, C. L., Bell, J., Yoottanasumpun, V., Greenbaum, R., O'Shea, J., Hoffmann, H., and Spear, N. E. (1989b). Effects of prenatal cocaine exposure on behavior during the early postnatal period. *Neurotoxicol. Teratol.* **11,** 57–63.

Spear, L. P., Kirstein, C. L., and Frambes, N. A. (1989c). Cocaine effects on the developing central nervous system: Behavioral, psychopharmacological, and neurochemical studies. *Ann. N.Y. Acad. Sci.* **562,** 290–307.

Spear, L. P., Kirstein, C. L., Frambes, N. A., and Moody, C. A. (1990). Neurobehavioral teratogenicity of gestational cocaine exposure. *NIDA Research Monograph (Problems of Drug Dependence, 1989)* **95,** 232–238.

Spear, N. E., and Kucharski, D. (1984). Ontogenetic differences in the processing of multi-element stimuli. *In* "Animal Cognition" (H. Roitblat and H. Terrace, eds.), pp. 545–567. Erlbaum, Hillsdale, New Jersey.

Stanton, M. E., and Spear, L. P. (1990). Workshop on the qualitative and quantitative comparability of human and animal developmental neurotoxicity, work group I report: Comparability of measures of developmental neurotoxicity in humans and laboratory animals. *Neurotoxicol. Teratol.* **12,** 261–267.

Weinberg, J., and Gallo, P. V. (1982). Prenatal ethanol exposure: Pituitary–adrenal activity in pregnant dams and offspring. *Neurobehav. Toxicol. Teratol.* **4,** 515–520.

Weller, A., Glaubman, H., Yehuda, S., Caspy, T., and Ben-Uria, Y. (1988). Acute and repeated gestational stress effect offspring learning and activity in rats. *Physiol. Behav.* **43,** 139–143.

Wiggins, R. C., and Ruiz, B. (1990). Development under the influence of cocaine. II. Comparison of the effects of maternal cocaine and associated undernutrition on brain myelin development in the offspring. *Metabol. Brain Dis.* **5,** 101–109.

Wise, R. A. (1987). The role of reward pathways in the development of drug dependence. *Pharmacol. Ther.* **35,** 227–263.

— 9 —

Maternal Opioid Drug Use
and Child Development

§▲

Sydney L. Hans

Department of Psychiatry
The University of Chicago
Chicago, Illinois

NATURE OF OPIOID DRUG USE

The use of opioid drugs by women during pregnancy dates back thousands of years in those parts of the world in which the opium poppy was grown. Recognition that the use of opioid drugs during pregnancy has potentially harmful effects for the fetus also extends back to antiquity. Hippocrates noted that "uterine suffocation" often occurred in conjunction with maternal opium use (see Zagon and McLaughlin, 1984).

Widespread use of opioid drugs by women has a long history in American society as well. In the nineteenth and early twentieth centuries, patent medicines containing opium were widely used by middle-class women, and physicians often prescribed opium to their female patients. During this same period, medical journals carried numerous reports of infants of opium-using women who were born addicted (see Ghodse et al., 1977; Zagon and McLaughlin, 1984; Cobrinik et al., 1959).

Since the middle of the twentieth century, the opioid drug most widely used by Americans, including pregnant women, is intravenously injected heroin. This drug is obtained illegally and at great expense. A large proportion of heroin addicts are from lower socioeconomic strata, and their drug abuse further depletes their limited physical, medical, and social–emotional resources. Heroin users must structure their life styles around drug procurement, and usually resort to illegal activities to support their habits.

Since the mid-1970s, chronic heroin addicts who seek treatment in the United States have often been admitted to government-funded methadone maintenance programs, where they receive free daily oral doses of methadone, a synthetic opioid (Hutchings, 1985). Methadone has effects similar to but longer lasting than those of

Maternal Substance Abuse and the Developing Nervous System

heroin, thus making methadone better suited for treatment programs in which medication is administered on a once-daily basis. Methadone is generally dispensed in a clinic setting that may also provide a variety of social services and medical supervision to its clients. In general, people maintained on methadone lead more stable lives than they did as heroin addicts and are assured of better pharmacological control of their habit.

Whereas use of opioid drugs waned somewhat during the 1980s as cocaine became the illegal drug of choice, the National Institute on Drug Abuse projects that heroin use will be likely to increase again in the 1990s as purer, smokable forms are introduced into the United States (Isikoff, 1989).

IMPORTANCE OF RESEARCH ON HUMANS

A sizable literature exists on the effects of prenatal exposure to opioid drugs in laboratory animals, particularly rats, mice, and chicks (cf. Zagon and McLaughlin, 1984; Hutchings and Fifer, 1986). Although complicated by methodological problems, these studies nevertheless have documented early growth and behavioral effects that can plausibly be attributed to the teratological or toxicological effects of opioid drugs.

Scientists studying the development of children exposed to drugs prenatally typically have adopted a teratological perspective. They have tried to adopt quasi-experimental research designs that control for all nonpharmacological risk factors, and they have chosen to assess key growth and neurobehavioral outcomes that might be most vulnerable to early disruption in central nervous system development. The studies on the effects of prenatal exposure to opioids in humans have been limited in number and scope. Conducting studies with drug-using families presents enormous logistical challenges to researchers that limit the number of such studies, the size of their samples, and the length of longitudinal follow-up. Consequently, most human studies are confined to the neonatal period or have relatively small sample sizes.

Scientifically, working with human populations also presents enormous challenges. The scientist has no control over the primary variable under study—maternal drug use—and has only very poor access to information about women's individual patterns of drug use. Mothers are self-medicating without knowledge of the quantities or purities of the drugs they take; they take numerous drugs other than opioids; their patterns of drug use vary from day to day and month to month during pregnancy; and because of the social stigma attached to their use of illegal drugs, women are likely not to provide accurate information to examiners about their patterns of use, even if they could accurately recall such a complex pattern of behavior. In addition, maternal use of drugs is confounded with numerous other factors that are likely to affect development in utero and after birth. These include, but are not limited to, maternal infection associated with needle use and sexual promiscuity, maternal and infant undernutrition, poor maternal and infant health care, infant exposure to lead and environmental toxins, passive exposure of infants to drugs postnatally, and suboptimal child-rearing environments in drug-using families.

Despite these confounds in research design, human studies are essential for several reasons. First, species differences in response to behavioral teratogens (Vorhees, 1986) make human research a necessity. Nonprimate laboratory animals differ greatly from humans in their gestation, brain development, and behavioral repertoire. In fact, there are no animal models for the types of behavior—particularly higher-order cognitive processes—whose disruption is of greatest concern in human development.

Second, even if it is not possible to determine whether problems in opioid-exposed children have a teratologic origin, it remains important from public policy and mental health perspectives to document the nature of development in such children.

Third, human research is essential because only human work can document the interactions between teratology and environment in child development. Documenting such interactions is essential for understanding potential intervention strategies for exposed children and their families.

The purpose of this chapter is, first, to briefly review and summarize what is known about the development of infants and children who were exposed in utero to opioid drugs, and, second, to explore different interpretations of these findings that may be elucidated in further research.

PHYSICAL GROWTH

Fetal Growth

Numerous studies of infants born to heroin-addicted mothers suggest that heroin exposure in utero decreases birth weight (Finnegan, 1976; Fricker and Segal, 1978; Kandall et al., 1976; Lifschitz et al., 1985; Reddy et al., 1971; Stone et al., 1971; Wilson et al., 1973; Wilson et al., 1981; Zelson et al., 1973). Whereas they are larger than heroin-exposed infants, methadone-exposed infants also have lower birth weights than unexposed infants (Chasnoff et al., 1982; Finnegan, 1976; Harper et al., 1974; Jeremy and Hans, 1985; Kaltenbach and Finnegan, 1987; Kandall et al., 1976; Newman et al., 1975; Stimmel et al., 1982–1983; Wilson et al., 1981; Zelson et al., 1973). The larger birth weights of methadone-exposed compared to heroin-exposed infants is probably owing to better prenatal care and medical supervision of maternal drug use for women enrolled in methadone treatment. Within-groups analyses of methadone-exposed infants support this conclusion, by showing that higher birth weight is related to better prenatal care (Finnegan, 1976; Green et al., 1979; Stimmel et al., 1982–1983), supervision of use of nonmethadone drugs (Stimmel et al., 1982–1983), and, in some studies, higher methadone dosages (Doberczak et al., 1987; Kandall et al., 1976).

Tables 1 and 2 summarize data from studies reporting birth weights. The proportion of low-birth-weight infants averages about 45% for heroin-exposed infants, 25% for methadone-exposed infants, and 15% for unexposed infants.

Whereas there are some reports of differences between heroin-exposed and unexposed infants in rates of premature birth, prematurity does not seem to be a major factor in the differences in birth weight between methadone-exposed and unexposed

TABLE 1   Percentage of Opioid-Exposed Infants Weighing Less than 2500 Grams at Birth

Authors	Heroin (n)	Heroin (%)	Methadone (n)	Methadone (%)	Clinic population/ drug-free sample (n)	(%)
Fricker and Segal (1978)	149	37			210	7
Reddy et al. (1971)	40	62				
Stone et al. (1971)	384	49			35270	15
Stimmel et al. (1982–1983)	49	29	79[a]	24		
			78[b]	18		
Finnegan (1976)	63	47	88[c]	36	1586	15
			152[d]	17		
Zelson et al. (1973)	45	44	46	47		
Jeremy and Hans (1985)			38	24	37	3
Harper et al. (1974)			51	33	4916	15
Newman et al. (1975)			313	35		

[a]Supervised methadone treatment.
[b]Unsupervised methadone use.
[c]Poor quality prenatal care.
[d]Adequate prenatal care.

infants (Doberczak et al., 1987; Kandall et al., 1977). Thus the lowered birth weights of methadone-exposed infants are typically associated with normal gestational age and appear to be the result of fetal growth retardation.

The few studies that have looked at parameters of fetal growth other than birth weight have also reported head circumference reductions of between 0.5 to 2.0 cm in

TABLE 2   Mean Birth Weights of Opioid-Exposed Infants

Authors	Heroin (n)	Heroin (grams)	Methadone (n)	Methadone (grams)	Clinic population/ drug-free sample (n)	(grams)
Fricker and Segal (1978)	149	2710			210	3420
Zelson et al. (1973)	45	2461	46	2625		
Wilson et al. (1973)	15	2675				
Lifschitz et al. (1985)	25	2759	26	2910	41	3289
Kandall et al. (1976)	61	2490	59[a]	2535	40	3282
			106[b]	2961		
Wilson et al. (1981)	30	2745	39	2857	58	3286
Olofsson et al. (1983)	48	2661	41	2973		
Chasnoff et al. (1982)			39	2815	27	3492
Kaltenbach and Finnegan (1987)			141	2963	127	3210
Strauss et al. (1974)			72	2898	36	3003
Doberczak et al. (1987)			150	2800	150	3248
Newman et al. (1975)			313	2738		

[a]Mothers mixing methadone with other drugs.
[b]Nonmixers.

TABLE 3   Mean Head Circumferences of Opioid-Exposed Infants

Authors	Heroin		Methadone		Drug-free	
	(n)	(cm)	(n)	(cm)	(n)	(cm)
Wilson et al. (1973)	10	32.7				
Wilson et al. (1981)	30	32.8	39	32.9	58	34.3
Lifschitz et al. (1985)	25	33.0	26	33.2	41	34.5
Doberczak et al. (1987)			150	32.6	150	33.8
Chasnoff et al. (1982)			39	32.5	27	34.6
Kaltenbach and Finnegan (1987)			141	33.3	127	33.9

opioid-exposed neonates compared to unexposed neonates (Doberczak et al., 1987; Chasnoff et al., 1982; Kaltenbach and Finnegan, 1987; Lifschitz et al., 1985; Rosen and Johnson, 1982; Wilson et al., 1981; Wilson et al., 1973). Table 3 summarizes the available data on head circumference at birth. Doberczak et al. (1987) have argued that small head circumference and low birth weight in opioid-exposed infants suggest symmetric growth retardation that is the result of an insult in early pregnancy, resulting in the reduction of organ cell number. This conclusion is supported by the work of Naeye, Blanc, Leblanc, and Khatamee (1973) that found reduced body and brain weight due to reduced cell number in heroin-exposed fetuses studied at a mean gestational age of 30 weeks.

Postnatal Growth

Data on physical growth after birth in opioid-exposed children present a more mixed pattern of findings. Several investigators report reduced physical growth in opioid-exposed infants and children. Wilson and associates (1973) compared 27 infants who were exposed to heroin in utero to pediatric norms. At the time of the children's last pediatric follow-up—when they ranged in age between 2 and 24 months—37% of the heroin-exposed children were below the tenth percentile for weight, 44% were below the tenth percentile for height, and 30% were below the tenth percentile for head circumference. In a different sample of 22 3- and 4-year-old children exposed to heroin prenatally, Wilson and colleagues (1979) found an increased incidence of subnormal stature and head circumference (but not weight) compared to both healthy low-socioeconomic-status controls and a medically high-risk comparison group.

Ting, Keller, Berman, and Finnegan (1974) studied 25 methadone-exposed toddlers between the ages of 6 and 41 months and an unexposed comparison group. Of the methadone-exposed toddlers, 26% were below the third percentile in height, in comparison to none of the unexposed group.

Strauss and colleagues (1976) compared the physical growth of a sample of 21 methadone-exposed and 24 unexposed infants at 12 months of age. A higher proportion of methadone-exposed infants were below the tenth percentile for height than were comparison infants (52 versus 27%), and a higher proportion of methadone-exposed infants were below the tenth percentile for weight (14 versus 4%), although these differences in growth were not statistically significant. There was no difference between the groups in head circumference.

Hans (1989) reported that at 24 months of age, methadone-exposed infants showed poorer physical growth than unexposed infants, particularly head circumferences that averaged 1 cm smaller than those of comparison infants and linear height that averaged 2 cm shorter than that of comparison infants. These effects remained even after controlling for differences in birth weight. There were no differences in weight. The means for both groups were within the normal range for height and head circumference.

Johnson, Diano, and Rosen (1984) recruited during pregnancy a sample of 59 infants exposed in utero to methadone and a control group of 31 unexposed infants matched for birth weight. In the somewhat reduced sample available at follow-up, the authors observed no differences in growth parameters of methadone-exposed infants at 12 and 24 months, except a higher incidence of head circumferences below the third percentile in the methadone-exposed sample, compared to an unexposed sample at both ages. In follow-up of the same sample at 36 months of age, Johnson and associates (1987) reported a greater proportion of methadone-exposed children (34%) than unexposed children (13%) had head circumference below the third percentile.

Other samples suggested little or minimal reduction of growth in opioid-exposed children after birth. Wilson, Desmond, and Wait (1981) followed 29 heroin-exposed, 35 methadone-exposed, and 55 drug-free infants through their first birthdays. There were no statistically significant differences between the groups in weight, height, or head circumference, although 16% of the opioid-exposed infants had heights less than the tenth percentile, compared to 4% of the drug-free group. A follow-up of the same sample at preschool age (Lifschitz et al., 1983) also revealed no differences in physical growth.

Chasnoff, Hatcher, and Burns (1980) followed 15 infants exposed to methadone prenatally and observed that whereas mean birth weight and length were at the tenth percentile, and mean head circumference was at the fifth percentile at birth, an accelerated pattern of growth after 4 months enabled infants to catch up. By 7.5 months, the mean length and head circumferences were above the twenty-fifth percentile.

Summary of Physical Growth Data

Studies of opioid exposure on human fetal growth have replicated the basic finding that maternal use of opioid drugs—even medically supervised use of methadone—is related to lower weight at birth and probably to reduced head circumference. Whereas in these human studies, maternal opioid use has generally been confounded with exposure to other drugs (including cigarettes and alcohol), poor-quality prenatal care, poor maternal nutrition, and maternal infection, the number of replications of the basic findings and the similar lowered birth weight in opioid-exposed rat pups (reviewed by Zagon and McLaughlin, 1984), suggest that the opioids themselves are at least one contributor to lowered birth weight in exposed children. Currently, the reduced birth weight and head circumference are believed to be the result of intrauterine growth retardation rather than premature birth. The biological mechanism for the growth-retarding effects of the opioids are not yet clear, although Naeye et al. (1973a) have speculated that they may be related to deficient growth hormone or to reduced uterine or placental blood flow during brief periods of maternal narcotics withdrawal. The seemingly paradoxical findings of improved growth with higher doses of methadone also remain poorly understood. Doberczak, Kandall, and colleagues (Doberczak et al., 1987) have speculated that improved growth with higher methadone doses may be related to the fetal steroidal environment.

A smaller number of studies point also to growth retardation after birth—particularly slower growth in height and head circumference—although the mechanism for such postnatal growth retardation remains unclear. Animal studies provide little to elucidate the possible mechanism for postnatal growth retardation, although Slotkin, Seidler, and Whitmore (1980) have suggested that opioids have direct effects on growth, independent of nutritional status, possibly related to a disturbance of the ornithine decarboxylase/polyamine system that regulates nucleic acid, protein synthesis, and tissue growth during development.

SUDDEN INFANT DEATH SYNDROME

Sudden infant death syndrome (SIDS) is the leading cause of infant deaths after the neonatal period (Beckwith, 1979). Whereas a host of factors have been identified that contribute to risk for SIDS (cf. Anderson and Rosenblith, 1971; Naeye, 1980; Lewak et al., 1979; Shannon and Kelly, 1982), no single risk factor is more powerful than in utero exposure to opioid drugs.

Data on this issue can be gathered only from longitudinal, prospective studies of infants. Epidemiological studies based on autopsies, parent interviews, and perinatal records have access to little information on drug exposure and will underestimate the incidence of opioid exposure in SIDS cases. Virtually every prospective study of opioid-exposed children followed from pregnancy has reported an increased incidence of SIDS in children exposed to opioids prenatally (Pierson et al., 1972; Harper et al., 1973; Rajegowda et al., 1978; Finnegan, 1979; Chavez, et al., 1979; Wilson et al., 1981; Hans and Snook, 1986; Rosen and Johnson, 1988). These studies are summarized in Table 4. The general incidence of SIDS is between 2.5 and 3 infants per 1000 live births (Valdes-Dapena, 1980). The incidence of SIDS among offspring of opioid-addicted women can conservatively be estimated at 2 per 100—about an eightfold increase.

Although no clear theory of causation has been developed linking prenatal opioid exposure and SIDS, the plausibility of a direct link between the two is bolstered by experimental evidence showing that infants exposed to opioid drugs in utero have abnormal respiratory patterns after birth, including decreased ventilatory response to carbon dioxide (Olsen and Lees, 1980) and abnormal sleeping ventilatory patterns

TABLE 4   Incidence of Sudden Infant Death Syndrome
in Opioid-Exposed Infants

Authors	Drug-exposed infants (n)	SIDS cases (n)	Incidence (%)
Pierson et al. (1972)	14	3	21
Harper et al. (1973)	244	4	1.5
Rajegowda et al. (1978)	383	8	2
Finnegan (1979)	389	5	1
Chavez et al. (1979)	688	17	2.5
Wilson et al. (1981)	69	1	1
Hans and Snook (1986)	42	3	7
Rosen and Johnson (1988)	25	1	4
Total	1843	42	2.3

(Ward et al., 1986). These findings are consistent with theories that relate SIDS to chronic postnatal hypoventilation or recurrent apneic spells (Shannon and Kelly, 1982; Hoppenbrouwers and Hodgman, 1982; Schulte et al., 1982), although there are no data demonstrating postnatal hypoxia in drug-exposed infants.

NEONATAL BEHAVIOR

Clinical Assessments

Children of opioid-using mothers are born passively addicted to the drug and experience withdrawal effects when they no longer are exposed to the drug after birth. Within 1 to 3 days of birth, most infants born to opioid-addicted women show clear clinical signs of narcotics abstinence or withdrawal signs (Desmond and Wilson, 1975; Pierog, 1977; Zelson et al., 1971). These signs include a large number of behavioral and physiological symptoms that are indicators of both nonspecific central and autonomic nervous system dysfunctioning. Among the more frequently observed signs of neonatal abstinence are high-pitched crying, poor sleeping, hyperactive reflexes, tremors, hypertonicity, convulsions, frantic sucking of fists, poor feeding, regurgitation, diarrhea, dehydration, yawning, sneezing, nasal stuffiness, sweating, skin mottling, fever, rapid respiration, and excoriation of skin.

Despite the large number of clinical reports of abstinence symptoms, data systematically comparing the behavior of opioid-exposed and unexposed neonates are more sparse, coming from only seven different samples: (1) Soule et al. (1974); (2) Strauss et al. (1975), Strauss et al. (1976); (3) Kron et al. (1975), Kron et al. (1977); (4) Lodge et al. (1975); (5) Chasnoff et al. (1980, 1982); (6) Marcus et al. (1982a), Jeremy and Hans (1985); and (7) Lesser-Katz (1982).

Whereas these samples of infants exposed prenatally to opioid drugs were quite heterogeneous in terms of the patterns of maternal drug use and infant pharmacologic treatment, they all employed the same measure of neonatal behavior, and they all showed similar patterns of results. In each of these samples, infant behavior during the first week of life was evaluated using the Neonatal Behavioral Assessment Scale (NBAS) (Brazelton, 1973). The NBAS is a widely used instrument in clinical research that was designed to assess a set of typical neonatal behaviors, particularly those behaviors that are likely to have consequences for caregiving exchanges and for the establishment of an affectionate bond between the infant and his or her caregivers. Although the formal examination requires a highly skilled clinical observer to administer, behaviors assessed are ones that can be observed in newborns by individuals without special equipment or training.

The core of the NBAS consists of 26 items measured on nine-point scales. Analyses of the structure of the examination from a variety of samples (Sameroff, 1978) including at least one sample of methadone-exposed neonates (Jeremy and Hans, 1985) have suggested eight types of items: arousal, quieting, hand-to-mouth, motor control, tone, defensive movements, alertness, and response decrement. The results of the seven studies will be summarized by these behavioral categories.

1. *Arousal* items measure the infant's highest observed state of arousal, how quickly into the examination the infant becomes upset, the number of situations in which the infant becomes upset, the amount of spontaneous activity shown by the infant, and the number of swings in arousal state shown by the infant during the examination. All seven studies reported that opioid-exposed neonates were more aroused at the time of testing than comparison-group infants. Opioid-exposed newborns were more quickly aroused to higher degrees of upset by less obnoxious stimuli.

2. *Quieting* items measure the success with which the infant self-quiets and the ease with which she or he can be consoled by the examiner. Three of the seven studies (Kron et al., 1975, 1977; Chasnoff et al., 1980, 1982; Lesser-Katz, 1982) reported that opioid-group neonates required higher levels of intervention to console than did comparison-group newborns; in contradiction, two groups of investigators (Soule et al., 1974; Jeremy and Hans, 1985) commented on the good consolability of most opioid newborns and the quickness of their responsiveness to soothing by handling, use of a pacifier, or swaddling. Reported mean scores from the NBAS consolability items across samples ranged from a high of 5.9 (Jeremy and Hans, 1985) to a low of 4.5 (Chasnoff et al., 1980, 1982) for the opioid-group neonates. These scores suggest that most of the opioid-exposed neonates who needed consolation could be consoled simply by being held, picked up, or rocked.

3. The *hand-to-mouth* item measures the success with which the infant can bring his or her hand to the mouth. In most samples, but not those composed of methadone-exposed infants, this ability is highly associated with self-quieting skill. Three of the studies (Strauss et al., 1975, 1976; Lodge et al., 1975; Jeremy and Hans, 1985) reported more bringing of hands to mouth in opioid-exposed newborns. This finding is consistent with clinical reports of frantic fist sucking in infants going through narcotics abstinence. Fist sucking in the opioid infants appears not to be an organized attempt at self-regulation, but rather a frenzied, automatic response.

4. *Motor control* items measure the jerkiness and tremulousness of the infant's

movements. In all but one sample (Lodge et al., 1975), it was reported that opioid-exposed newborns have poorer motor control: their limb movements tend to be jerky and tremulous.

5. *Tone* items measure the rigidity of the infant's whole body, strength of neck and shoulder, and resistance to molding while being held. Four of the seven studies (Soule et al., 1974; Lodge et al., 1975; Jeremy and Hans, 1985; Lesser-Katz, 1982) reported differences between opioid-exposed and comparison neonates in muscle tone, with the general body tone of opioid-exposed newborns tending toward hypertonicity.

6. *Defensive movement* items measure the infant's success at removing a cloth placed over the face. None of the studies reported any differences between opioid and comparison newborns in ability to defend themselves against obstruction of the face.

7. *Alertness* items measure the degree to which the infant will turn his or her head to various auditory and visual stimuli and assessed length and quality of sustained period of alertness. On the NBAS, most of the information from which judgments of alertness are made comes from a series of orientation tasks in which the infant is expected to turn his or her head in the direction of auditory or visual stimuli. Four of the studies (Soule et al., 1974; Strauss et al., 1975, 1976; Lodge et al., 1975; Chasnoff et al., 1980, 1982) reported poorer mean levels of performance by opioid-exposed newborns on these items. One study (Jeremy and Hans, 1985) found no mean differences between opioid and comparison neonates in their overall levels of alertness and responsiveness to stimuli, but did find much higher rates of missing data on the orientation items for the opioid-group newborns. This indicates that they were less likely to be in awake, noncrying states than were comparison infants. The only other study to report sample sizes for items (Strauss et al., 1975, 1976) also reported more missing data on the orientation items in the opioid-exposed group (almost 50%).

Four research groups (Soule et al., 1974; Strauss et al., 1975, 1976; Lodge et al., 1975; Lesser-Katz, 1982) reported that opioid-group newborns are less responsive primarily to items requiring orientation to visual stimuli and that they show normal response patterns to items that use only auditory stimulation.

8. *Response decrement* items assess the speed with which the sleeping infant habituates to repetitions of light, sound, and tactile stimuli; it is assumed that rapid habituation to repetition is adaptive. Three studies (Soule et al., 1974; Strauss et al., 1975, 1976; Lesser-Katz, 1982) reported that opioid-exposed infants were slower than comparison infants to show a response decrement to light, but not to other stimuli. On the light item, the examiner records the sleeping infant's magnitude of response as a flashlight beam is passed across the eyes.

Table 5 summarizes the pattern of results across the seven studies using the NBAS.

There are few systematically collected data on the diminution of withdrawal symptoms during the course of the first month of life. At the time of release from the hospital, opioid-exposed infants are typically no longer showing acute withdrawal symptoms, but not all symptoms have disappeared. Only two of the NBAS studies described repeated the examination at the end of the neonatal period. Strauss et al. (1976) reassessed a third of their sample at 1 month of age using the NBAS. They reported that the opioid-exposed infants differed from a comparison group in only one

TABLE 5 Effects of Opioid Exposure on Neonatal Behavior

Authors	Opioid (n)	Comparison (n)	Arousal	Quieting	Hand-to-mouth	Motor control	Tone	Defensive movements	Alertness/orientation	Response decrement
Soule et al. (1974)	19	42	Higher			Poorer	Higher		Poorer	Slower
Strauss et al. (1975, 1976)	46	46	Higher		Better	Poorer			Poorer	Slower
Kron et al. (1975, 1977)	23	10	Higher	Poorer		Poorer				
Lodge et al. (1975)	29	10	Higher		Better		Higher		Poorer	
Chasnoff et al. (1980, 1982)	39	27	Higher	Poorer	Better	Poorer			Poorer	
Jeremy and Hans (1985)	27	44	Higher	Poorer	Better	Poorer	Higher			
Lesser-Katz (1982)	10	14	Higher	Poorer		Poorer	Higher		Poorer	Slower

area: motor control. Opioid-group infants remained more tremulous at 1 month. Jeremy and Hans (1985) found no strong group differences between opioid and comparison infants at 1 month. They report that, in all areas of the NBAS, the two groups converged by 1 month of age, with opioid-group infants becoming more alert, less stiff, and smoother in their movements, although there was still a strong trend for opioid-group infants to show greater body tonus.

Laboratory Assessments

A number of neonatal differences in behavior between opioid-exposed and comparison infants have been investigated using laboratory, usually electrophysiological, assessment procedures.

Cry Characteristics

Blinick and his colleagues (Blinick et al., 1971) made sound spectrograms of cries produced in the delivery room by 31 infants born to drug-addicted mothers and a large comparison group. Over half of the infants born to drug-addicted mothers had abnormal cry spectrograms compared to only 11% of the comparison infants. Drug-exposed infants, even most of those whose cries were considered normal, tended to have high-pitched cries. These results are consistent with clinical reports of high-pitched cries in withdrawing neonates. The authors suggest that group differences in pitch are most likely owing to differences in the tonus of the laryngeal musculature.

Huntington, Hans, and Zeskind (1990) compared the cries of eight infants exposed prenatally to methadone with twelve infants of similar socioeconomic status. Cries were recorded between 2 to 3 days of age in a hospital nursery using a rubberband snap to the heel as an eliciting stimulus. Methadone-exposed infants showed significantly shorter duration first cry expiration. Whereas there were no differences in average or peak fundamental frequency between the two groups, the trend was in the direction of higher fundamental frequency for the methadone-exposed infants.

Sucking

Kron and his colleagues (Kaplan et al., 1975; Kron et al., 1975; Kron et al., 1977; Kron et al., 1976) have measured sucking behavior in opioid-exposed infants using a special nursing instrument that monitors sucking rate, sucking pressure, amount of nutrient consumed, and percentage of time during the experiment that the infant is actively sucking. Forty-three infants born to opioid-dependent women were monitored twice daily just before scheduled feedings for 3 days beginning at 24 to 36 hr of age. The infants born to opioid-dependent women showed depressed sucking rates (22 sucks per min) in contrast to a comparison group of 10 infants (40 sucks per min). The opioid-group infants also sucked with less pressure and consumed less nutrient during the trials than comparison infants. Although most of the infants were tested while pharmacologically treated, those infants tested in the absence of drug treatment showed comparably depressed rates of sucking.

Muscle Tone

Marcus and Hans (1982a) investigated the muscle tone of opioid-exposed infants using a quantified electromyography (EMG) procedure. From the limbs of 18 opioid-exposed neonates, EMG was recorded during periods of rest and of passive movement of the limbs. Compared to 26 comparison infants, the opioid-exposed infants showed higher electrical output from both arms and legs, both while at rest and during limb manipulation. This is congruent with clinical and NBAS reports of hypertonicity in withdrawing infants.

Sleep Patterns

Infant sleep patterns typically show cycles of active sleep [characterized by rapid eye movements (REM); body movement; variable heart rate; low-voltage, electroencephalogram fast wave (EEG); variable respiration) followed by periods of quiet sleep (characterized by high-voltage, slow-wave EEG activity; regular heart rate and respiration; infrequent body movement; and absence of REM). Schulman (1969) monitored sleep rates for 45 min in eight neonates born to heroin-addicted women and eight normal comparison infants. During active sleep, heroin-group infants showed greater REM and body movement activity than comparison infants. Unlike comparison infants, heroin-group infants rarely entered a quiet sleep phase during the testing session. The author suggested that this pattern of sleep disturbance is one observed among newborns at high risk for central nervous system (CNS) impairment.

Dinges, David, and Glass (1980) monitored sleep in 28 neonates born to women maintained on methadone (some with concomitant heroin use) and 30 comparison neonates. Drug-exposed newborns averaged less quiet sleep and more active sleep than did the comparison infants. This effect seemed to be dose related, with infants exposed to higher doses of methadone having less quiet sleep, more active sleep, and being more likely to awaken during the testing period.

Brain Electrophysiology

Lodge et al. (1975) examined a sample of 29 opioid-exposed and 10 comparison infants in an evoked potential paradigm. Occipital and temporal lobe EEG activity were monitored during exposure to repeated visual (light and checkerboard) and auditory (click) stimuli and averaged across 50 trials for each type of stimulus. The auditory data were measured mostly during sleep state; most of the visual data were collected with infants in awake states. Both auditory and visual evoked potentials from the opioid-exposed group were "more irregular and more unreliable" (page 247) than those of the comparison group. Desynchronized high-frequency activity was often present in the opioid-exposed infants' EEGs even during seemingly quiet sleep. Both visual and auditory evoked potentials were characterized by early, sharp, high-amplitude components in the opioid group. The investigators felt that these characteristics most likely reflected greater CNS irritability in the opioid-exposed neonates. Differences in response frequencies were found only in the vertex region responses and only to visual stimuli; opioid infants showed lowered vertex but normal occipital arousal responses to light stimulation. The authors concluded that the opioid-exposed infants showed adequate auditory processing, but abnormal visual processing that, given the regional response pattern, was most likely a reflection of poor modulation of arousal rather than deficits in sensory abilities.

Summary of Neonatal Findings

Both clinical and laboratory assessments of infants who were exposed in utero to opioid drugs have shown a clear pattern of abnormal neonatal behavior. Opioid-exposed infants are much more easily aroused and aroused to higher levels than other newborns. This pattern of behavior is similar to the increased irritability reported by adults withdrawing from opioid drugs. The infants' state-control problems are also reflected in a relatively small proportion of quiet compared to active sleep. In addition, opioid-exposed newborns show poorer motor control (tremulousness and jerkiness) and higher muscle tonus—both signs of increased CNS irritability and similar to motoric symptoms reported in adults experiencing narcotics abstinence. Their hypertonicity has been confirmed by electromyographic assessments of muscle tone and suggested in high-pitched cries that could be caused by hypertonic laryngeal musculature. Drug-exposed neonates probably also are less likely to be in alert states during testing, possibly owing to higher states of arousal when nonmedicated or drowsiness when medicated, but it is not clear that when they are alert, the quality of their perception or responsiveness is any poorer. They orient better and habituate better to auditory stimuli than to visual stimuli, which require more highly organized states of attentiveness to process effectively. Electroencephalographic studies showed a pattern of normal auditory but abnormal visual processing in drug-exposed infants that was probably related to modulation of arousal rather than to sensory abilities.

Symptoms of neonatal abstinence diminish quickly over the first month of life to the point at which they are usually not clinically significant. The important empirical question that remains is whether more subtle developmental problems can be observed after the neonatal period, either effects of lingering mild withdrawal signs or more permanent signs of neurobehavioral damage. Whereas information about the behavioral effects of in utero exposure to opioid drugs on the neonate is important for the medical management of the child during the immediate postpartum period, it is usually assumed that the dramatic behavioral abnormalities of this period are primarily transitory symptoms of withdrawal rather than signs of permanent neurological dysfunctioning. Teratological effects for which long-term intervention might be required can be assessed only at later ages, after withdrawal symptoms have presumably subsided.

INFANCY AFTER THE NEONATAL PERIOD

Neurobehavioral Development in Infancy

The most frequently used instrument for assessing the mental and motor skills of infants past the neonatal period is the Bayley Scales of Infant Development (Bayley, 1969). Its three parts are the Mental Record, the Psychomotor Record, and the Infant Behavior Record (IBR). The first two consist of specific skill items arranged in order of difficulty and scored as pass or fail, with summative scores that can be converted to normalized scores (analogous to IQ): the Mental Development Index (MDI) and the Psychomotor Development Index (PDI). These scales measure the age-appropriate attainment of developmental milestones. In contrast, the IBR makes global, qualitative judgments of various aspects of the infant's functioning rather than assessing particular skills.

A number of samples of opioid-exposed infants have been assessed on the Bayley scales, but only six have reported comparisons to a sample of drug-free infants matched for socioeconomic status: (1) Strauss et al. (1976), (2) Kaltenbach and Finnegan (1986), (3) Kaltenbach and Finnegan (1987), (4) Wilson et al. (1981), Wilson (1989), (5) Rosen and Johnson (1982), Johnson et al. (1984), and (6) Hans (1989), Hans et al. (1984), Hans and Jeremy (1984).

Tables 6 and 7 list mean MDI and PDI scores for infants assessed in the six samples on the Bayley scales. In the samples with longitudinal data, both opioid and comparison infants show a trend for decreasing scores, especially MDI scores during the first 2 years of life. Across the samples, there were very few reports of significant differences between exposed and unexposed infants. Yet, across the studies, there was a general trend for opioid-exposed infants to perform more poorly than comparison-

TABLE 6   Mean Bayley Mental Development Indices for Opioid-Exposed and Nonexposed Infants

Study	Age in Months				
	3–4	6–9	12	18	24
Strauss et al. (1976)					
Opioid (n = 25)	112	116	113		
Comparison (n = 26)	115	114	115		
Kaltenbach et al (1986)					
Opioid (n = 34)		105	103		99
Comparison (n = 22)		106	109		104
Kaltenbach and Finnegan (1987)					
Opioid (n = 100)			103		
Comparison (n = 60)			105		
Wilson (1989)					
Heroin (n = 29)		101		87	84
Methadone (n = 39)		98		92	89
Comparison (n = 57		106		97	90
Rosen and Johnson (1982); Johnson et al (1984)					
Opioid (n = 45)		95	98	96	90
Comparison (n = 25)		101	107	106	97
Hans (1989); Hans and Jeremy (1984)					
Opioid (n = 30)	111	118	106	96	92
Comparison (n = 44)	113	122	109	105	96

TABLE 7   Mean Bayley Psychomotor Development Indices for Opioid-Exposed and Nonexposed Infants

	Age in Months				
Study	3–4	6–9	12	18	24
Strauss et al. (1976)					
Opioid (n = 25)	119	109	103		
Comparison (n = 26)	117	112	110		
Wilson et al. (1981)					
Opioid (n = 64)		91			
Comparison (n = 55)		99			
Rosen and Johnson (1982);					
Johnson et al (1984)					
Opiod (n = 45)		101	95	93	99
Comparison (n = 25)		105	103	105	108
Hans (1989);					
Hans and Jeremy (1984)					
Opioid (n = 19)	117	112	108	106	101
Comparison (n = 23)	122	112	110	110	109

group infants. In the more than 30 comparisons of opioid-exposed and unexposed infants on the MDI and PDI reported in the six samples, exposed infants had higher mean scores on only two of the comparisons, and these were not close to being statistically significant.

The MDI and PDI scales of the Bayley scales are measures of skill acquisition; they measure *whether* certain skills are performed at a particular age. The Bayley IBR, on the other hand, measures more clinical aspects of behavior: *how well* the skills are performed. Only two of these studies have reported Bayley IBR data on opioid-exposed and comparison infants.

In the first of these samples, Wilson et al. (1981) found that their 9-month-old opioid-exposed infants showed poorer fine-motor coordination and shorter span of attention than comparison infants. In the other sample, Marcus, Hans, and colleagues (Marcus et al., 1982b; Hans and Marcus, 1983; Hans et al., 1984) compared opioid-exposed and unexposed infants on sums of IBR items in the areas of activity, motor coordination, and attention at 4, 8, 12, and 18 months of age. The activity level of all infants increased greatly from 4 to 18 months. At 4 months, methadone-exposed infants were more active than unexposed infants, but this difference disappeared at the later assessment ages. Whereas motor coordination improved with age for all infants, at all ages the mean motor coordination scores were poorer for methadone-group infants, although significantly so only at 4 months. All infants showed dramatic increases in attention between 4 and 8 months. At all ages, methadone-group infants had lower mean levels of attention; this was statistically significant only at 12 months.

To supplement standardized developmental testing, Rosen and colleagues administered a neurological assessment to methadone-exposed and unexposed infants at 12 and 24 months of age (Johnson et al., 1984). At both ages, there was a greater incidence of abnormal neurological findings in the methadone group. These included nystagmus and/or strabismus, tone, and coordination abnormalities.

Infant Temperament

Considerable developmental theory and research has been devoted to documenting the presumably stable, biologically determined individual differences in children referred to as temperament (Thomas and Chess, 1977). The variations in temperament in infants have often led to certain children being described as easy infants and others as "difficult" infants. Difficult infants adapt slowly to new situations and events, are irregular in their biological patterns, are active, show frequent negative mood, and often react with intensity. Whereas opioid-exposed infants would certainly be described as difficult infants during their withdrawal, it remains an open question as to whether they differ from other children in their temperamental characteristics after the neonatal period.

Johnson and Rosen (1990) videotaped 75 infants who included both drug-free and opioid-exposed infants. Tapes made at 2, 4, and 6 months of age were coded for seven infant temperament characteristics described by Thomas and Chess: activity, adaptability, approach, intensity, mood, distractibility, and persistence. When the infants were 9 months of age, their mothers completed the Carey Infant Temperament Questionnaire (Carey, 1970). There was little agreement between mothers' reports at 9 months and the observations made by the researchers at 2, 4, and 6 months. The only dimensions of temperament that showed moderate and significant correlations between researcher and mother ratings were activity level and persistence. There was no relation between observers' ratings of the children's temperament and the children's methadone exposure. There were, however, correlations between maternal reports of temperament and prenatal exposure to drugs. Maternal drug use related to reports of more difficult infant characteristics. Specifically, drug-using mothers were more likely to report that their children were negative in mood, not approaching, and not adaptable. Thus, intensity of maternal drug use was related to increased reports by the mother of the infant's difficult temperament.

Kaltenbach, Nathanson, and Finnegan (1989) asked 65 methadone-maintained mothers to complete the Carey McDevitt Infant Temperament Questionnaire (Carey and McDevitt, 1978) when their infants were 6 months of age. The investigators found no differences in the reports of drug-using and nonusing mothers on any of the nine infant temperament dimensions or in the categorization of easy and difficult temperament patterns. Both their drug-using and the low-socioeconomic-status control mothers, however, reported a much higher incidence of infants with difficult temperament (43 and 31% respectively) than were found with Carey's standardization sample (9.4%).

The data from these two studies provide no evidence for temperament characteristics associated with maternal opioid use during pregnancy and raise a concern that has been voiced by others (e.g., Vaughn et al., 1987) that maternal reports of temperament are basically measures of maternal characteristics rather than reliable assessments of infant characteristics.

Infant Socioemotional Development

The hallmark variable in the study of early social–emotional development has been the infant's attachment to the mother at the end of the first year of life. Examination of 1-year-old children's behavior when separated from the mother and during reunion after separation has pointed to individual differences among infants that are markers for later social competence (cf. Bretherton and Waters, 1985).

Only one study has examined attachment behavior in infants exposed in utero to methadone. Using a modified Strange Situation (Ainsworth et al., 1978) with only one separation and reunion (a procedure that would tend to underestimate the incidence of insecure attachments), Goodman (1990) examined mother–infant attachment in a sample of 35 1-year-old infants who had been exposed in utero to methadone and 46 unexposed comparison infants. There were no differences between methadone-exposed and comparison infants in their security of attachment as assessed using the standard Ainsworth coding categories. Using the alternative categories developed by Main and Solomon (1986) and by Crittenden (1985), Goodman found that 20% of methadone-exposed infants showed disorganized or mixed insecure patterns of attachment, compared to only 4% of the unexposed comparison group. Analyses suggested that these classifications related to both infant characteristics (increased arousal at 1 month, possibly associated with prolonged withdrawal) and maternal characteristics (mothers' possibly intrusive labeling of objects during free play at 12 months).

Summary of Infant Findings

In summary, all studies of infants born to opioid-using women have repeatedly shown small, but typically not statistically significant, lags in the acquisition of developmental skills as measured by standardized tests in opioid-exposed infants. The replication of this finding across studies suggest that opioid exposure may play a role in these developmental lags, but they could be related to confounds in experimental design, such as a failure to match carefully for social class. Differences between opioid-exposed and unexposed infants have been more clearly observed in qualitative measures of infant behavior such as those assessed on the Infant Behavior Record and neurological examinations. These data suggest that infants exposed in utero to opioids have poor motor coordination, high activity level, and poor attention. The differences in motor coordination and activity observed with the IBR seem to be especially pronounced early in the first year of life and may be subtle signs of continuing withdrawal from opioid drugs. Differences in attentional functioning seem to emerge later and may be the early signs of a more permanent neurobehavioral syndrome in some of the children, perhaps a type of attention deficit hyperactivity disorder (APA, 1987).

There are few reports on opioid-exposed infants outside the domain of neurobehavioral development. These reports suggest that opioid exposure is not related to early markers of temperament, but may be related to risk for disorganized attachment behavior.

DEVELOPMENT IN EARLY CHILDHOOD

Cognitive and Neurobehavioral Development in Early Childhood

Past the period of infancy, follow-up data on children who were exposed to opioids in utero become scarce. Wilson et al. (1979) reported on the behavior of 22 3- to 6-year-old children whose mothers were addicted to heroin during their pregnancies. Contrasted with comparison-group children, the opioid-exposed children showed poorer performance on the auditory memory portion of the Illinois Test of Psycholinguistic Abilities (1968), three of the subscales of the McCarthy Scales of Children's Abilities (Perceptual-Performance, Quantitative, and Memory) (McCarthy, 1972), and on several perceptual tasks. Subjective evaluations of the children's activity level during physical examination rated the heroin-exposed group as more active. Similar ratings did not detect differences in attention or alertness.

Strauss, Lessen-Firestone, Chavez, and Stryker (1979) examined 33 children of drug-dependent women and a comparison group on the McCarthy Scales at 5 years of age. They concluded that there were no differences between the groups on any of the subscales. However, examiners' clinical ratings of the children showed that the children of the drug-dependent women were more active and energetic during testing, showed more task-irrelevant activity, and tended to show poorer fine-motor coordination.

Johnson et al. (1987) reported a higher incidence of suspect or abnormal neurological findings at 36 months in 39 methadone-exposed infants (32%) compared to 23 unexposed infants (13%). The Merrill-Palmer Scale of Mental Tests (Stutsman, 1931) was also administered at this age. There were no statistically significant group differences between the two groups in their scores, although there was a trend for higher scores in the drug-free group (51st percentile for methadone, and 58th percentile for drug-free). In a further follow-up of this sample at age six, Rosen and Johnson (unpublished) found that 45% of methadone-exposed children showed neurological abnormalities in tone, gross and fine motor coordination, balance, and hyperactivity, compared to only 20% of unexposed children. The exposed children had significantly poorer McCarthy scores on the perceptual, quantitative, and motor subscales. The General Cognitive Index (GCI) for methadone-exposed children was 89 and for unexposed children, 94.5.

Kaltenbach and Finnegan (1987) found that a sample of twenty-seven 3½ to 4½ year-old methadone-exposed children did not differ from unexposed preschool-age children on the McCarthy General Cognitive Index of any of the six subscales of the test.

Lifschitz et al. (1985) found no differences between heroin-exposed, methadone-exposed, and socioeconomic-matched controls on intellectual performance as assessed by the McCarthy Scale of Children's abilities in a sample of 3-year-old children. They noted, however, an increase of General Cognitive Index (GCI) scores less than 84 in the heroin group (56%) than in the comparison group (22%). Methadone-group children were intermediate, with 35% having scores less than 84. Predictors of intellectual performance included amount of prenatal care, prenatal risk score, the home environment, but not head circumference.

Table 8 summarizes the findings of the early childhood studies that used the McCarthy Scales.

TABLE 8   Mean McCarthy Scales of Children's Abilities Scores
for Opioid-Exposed and Nonexposed Preschool

Study	Age (yrs)	GCI	Verbal	Perceptual	Quantitative	Memory	Motor
Wilson et al. (1979)	3						
Opioid (n = 22)		89	—	43.5	41.2	44.3	—
Comparison (n = 20)		97	—	48.8	51.7	51.2	—
Strauss et al. (1979)	5						
Opioid (n = 33)		87	44.3	40.7	42.4	43.6	44.5
Comparison (n = 30)		86	44.1	40.7	40.2	44.0	46.0
Lifschitz et al. (1985)	3–6						
Heroin (n = 25)		90	—	—	—	—	—
Methadone (n = 26)		89	—	—	—	—	—
Comparison (n = 41)		89	—	—	—	—	—
Rosen and Johnson, (1986)	6						
Opioid (n = 31)		89	46.8	43.9	40.6	44.0	46.3
Comparison (n = 15)		95	45.1	50.8	46.1	47.1	53.5
Kaltenbach and Finnegan (1987)	4						
Opioid (n = 45)		107	53.4	55.5	51.3	49.5	52.3
Comparison (n = 18)		106	54.3	53.0	53.4	52.3	52.3

Social and Adaptive Functioning in Early Childhood

Even fewer studies have examined the social and adaptive functioning of preschool-age children exposed in utero to opioid drugs. Wilson et al. (1979) studied a group of heroin-exposed 3- to 6-year-old children. Their parents rated the children as having greater difficulty in areas of self-adjustment, social adjustment, and physical adjustment than did the parents of children not exposed to drugs. Parents also reported that their heroin-exposed children had more difficulties with uncontrollable temper, impulsiveness, poor self-confidence, aggressiveness, and difficulty making and keeping friends.

Strauss et al. (1979) observed methadone-exposed and unexposed 5-year-olds for 15 min in a waiting room before their appointment. They coded the incidence of playing and talking, relating to caregiver, relating to other children, and wandering. There were no differences between the two groups in these behaviors, although the authors point out that the relatively unstructured setting was not one likely to elicit signs of behavior problems.

In their follow-up of 6-year-old methadone-exposed children, Rosen and Johnson (1986) gathered data from teachers and mothers. Methadone-exposed children differed from unexposed controls on the school behavior checklist in their need for achievement, aggressiveness, and school disturbance. The methadone-exposed children had more referrals for social problems and for child developmental and emotional needs.

Olofsson, Buckley, Andersen, and Friis-Hansen (1983) followed a sample of 89 Danish children of opioid-dependent mothers. The sample contained children who ranged in age from 1 to 10 years at time of follow-up, but the mean age was 3.5 years, and most children appeared to be between the ages of 2 and 5. There was no unexposed comparison group, and the assessments do not appear to have been made by individuals blind to the purpose of the study. The investigators reported that only 25% of the opioid-exposed children were physically, mentally, and behaviorally normal. Fifty-six percent were hyperactive and aggressive, with lack of concentration and social inhibition.

Among the more provocative reports of behavior in young children prenatally exposed to opioid are those of altered sex-dimorphic behavior. Sandberg, Meyer-Bahlburg, Rosen, and Johnson (1990) studied 30 methadone-exposed children at ages 6 through 8. Based on standardized questionnaires completed by children's caregivers, they concluded that methadone-exposed boys showed more stereotypically feminine behavior than unexposed male control subjects. They reported no differences between exposed and unexposed girls. The investigators did not investigate possible confounds with respect to male role models available to the male children.

Similarly, Ward, Kopertowski, Finnegan, and Sandberg (1989) assessed a sample of 48 opioid-exposed children 5 to 7 years old using two projective tests—the Draw-A-Person Test (Machover, 1949) and the IT Scale for Children (Brown, 1957). Compared to unexposed children, drug-exposed boys more likely to draw a female figure on Draw-A-Person Test and to view the neuter "It" figure as having feminine characteristics. In free play, opioid-exposed boys generally chose masculine toys, but were more likely to at least briefly play with feminine toys than were comparison boys. The investigators report that this tendency to play with feminine toys was unrelated to the presence of a father figure in the boys' lives. There were no differences in test behavior between opioid-exposed and unexposed girls.

Both studies interpret their findings as consistent with studies of behavioral demasculinization in male rats exposed to opioid drugs in utero (Ward et al., 1983).

Summary of Early Childhood Findings

All in all, the data from studies of opioid-exposed children during early childhood show little difference between opioid-exposed children and unexposed children as assessed on standardized tests of cognitive ability. As was seen in the infant data, however, the older offspring of drug-using women also have a number of the characteristics of attention deficit disorder: high activity, impulsivity, poor self-control, poor motor coordination, and poor performance on cognitive tests requiring focused attention.

More provocative patterns of behavior are emerging in the domain of early sex-role development. The two studies of very small samples have reported feminized behavior in male children exposed to opioid in utero, consistent with data on sexual differentiation in opioid-exposed rats.

DEVELOPMENT DURING MIDDLE CHILDHOOD AND ADOLESCENCE

Studies of the effects of opioid exposure on child development past the years of early childhood are all flawed by design problems—they lack either a prospective design or a comparison group. There is only one published report of opioid-exposed children after the age of 6 in which the sample has been followed prospectively. Wilson (1989) was able to follow 40 heroin-exposed children from previous samples (Wilson et al., 1973, 1979) to elementary school ages. All children were between first and fifth grades at the time of middle-childhood assessment. Sixty-five percent of the opioid-exposed children required placement in special educational services or had repeated one or more grades at the time of the study. School and parent reports, supplemented by psychologist and pediatrician observations, suggested that two thirds of the sample had problem behavior. School reports indicated that half of the children had problems related to inattention and poor self-discipline, and a third of them were described as having low self-confidence, poor peer relations, and failure to participate.

Whereas prospective methodology is important for having adequate measures of drug use during pregnancy and for having available data on antecedent functioning that might elucidate causal mechanisms, cross-sectional studies can be valuable for generating hypotheses. Two cross-sectional studies of school-age children and adolescents living with drug-using parents—although not necessarily exposed to drugs in utero—provide interesting information.

Herjanic, Barredo, Herjanic, and Tomelleri (1979) assessed 32 6- to 17-year-old children from 14 African-American families in which the father resided with his wife and children and was receiving treatment for opiate addiction. Compared to other children, the children with drug-using fathers functioned less well on cognitive tests. Of the 6- to 11-year-olds in drug-using families, 44% were of borderline intelligence or were mildly retarded compared to 14% of the children without drug-using fathers. Of the 12- through 17-year-olds living in drug-using families, 43% were of borderline intelligence or were mildly retarded compared to 20% of the other children. Adolescent children in drug-using families showed earlier and stronger antisocial trends; 43% of them had a conduct disorder compared to none of the children in the control group.

Sowder and Burt (1980) reported on 126 8- to 17-year-old children whose parents were present or recovering heroin addicts at the time of the study and a matched group of 126 children whose parents were not using drugs. Many, but far from all, of the children of drug-using parents would have been exposed to heroin prenatally. The investigators collected data from parent and child interviews, from schools, and from community agencies. Results indicated that children of drug-using parents were at greater risk than the comparison group in two areas: school adjustment problems and delinquency/behavioral problems. Specifically, in the area of school problems, they had more teacher-reported behavioral problems in the classroom; they were more likely to have received tutoring; their parents were more likely to have been contacted by school for child's misbehavior; the children were more likely to have had school failure; and they were more likely to have missed 6 or more days of school. In the area of behavioral problems, children in drug-using families were more likely to have appeared before juvenile court; they had had more encounters with police and for more serious offenses; they were more likely to get into a lot of fights; they were more destructive of school property; they were more likely to be runaways; and they were more likely to use drugs and/or alcohol themselves.

Bauman and Levine (1986) compared preschool-age children of methadone-maintained mothers with a matched control group. They found that the children of opioid-using parents performed less well on measures of intelligence and adaptive behavior, were more impulsive, immature, irresponsible, and lacking in empathy than children of non-drug-using parents. Children of drug-using mothers who experienced withdrawal symptoms at the time of birth had more developmental delays, lower IQ scores, and lower heights and weights than those not experiencing withdrawal.

Johnson, Boney, and Brown (1991) studied a group of 35 children between the ages of 8 and 14 years with parents who either were enrolled in methadone-maintenance programs or were active crack-cocaine users, and 37 whose parents did not use drugs. Children of substance-abusing parents were found to score less well on measures of depression, trait anxiety, and standardized tests of arithmetic. The investigators did not make comparisons between the children of opioid-using parents and crack cocaine-using parents.

In summary, although we know relatively little about drug-exposed children at school-age and adolescence, the available data suggest that this is a period in which school and behavior problems may emerge strongly in this group.

STAGES IN HUMAN RESEARCH ON PRENATAL SUBSTANCE EXPOSURE

In attempting to understand the effects of a drug on child development, research typically is conducted in two stages (Hans et al., in press). The first of these stages is to document whether there are developmental differences between children exposed to a substance and those not exposed. Most research on opioid-exposed children to date has been conducted with this goal. The second of these stages is to explore various hypotheses for the source of differences between substance-exposed and other children.

The first stage of research conducted on opioid-exposed children has documented that opioid-exposed and unexposed children do differ in some regards. Opioid-exposed infants as a group are born at lower birth weights and they exhibit a dramatically disturbed pattern of behavior during the first week of life. Opioid-exposed infants are at risk for sudden infant death. Past the neonatal period, however, the offspring of opioid-using women may be more likely than unexposed children to have neurological signs including hyperactivity, motor dyscoordination, and attention problems. They may also be of shorter stature and have reduced head circumferences. As children of opioid-using mothers enter school, they seem to be at risk for school and behavior problems, although the precise nature of these problems has not been well studied.

Much remains to be learned, however, about differences between opioid-exposed and unexposed children. Our present body of knowledge on differences between exposed and unexposed infants is limited by the small number of follow-up studies and by choice of measures for these studies. The largest amount of information gathered on opioid-exposed infants and children has come from standardized developmental and intelligence tests. These tests have suggested the possibility of problems in some opioid-exposed children but have often have failed to detect overall differences between groups of exposed and unexposed children. Whereas this failure to detect group differences could be a reflection of the normality of the children being studied or of the inadequate statistical power provided by very small samples, it could also be related to the inadequacy of global tests—such as tests of intelligence—as assessments of subtle developmental and cognitive processes. Beyond the neonatal period, no studies of opioid-exposed children have used measures that tap into specific behavioral and cognitive deficits. The few studies of opioid-exposed children that have supplemented developmental test data with more subjective and qualitative ratings of behavior made by blind observers have often detected differences between exposed and unexposed children and lead one to believe that sensitive measures of psychological processes might reveal differences between these groups of children. Measures of specific attentional and motoric processes would seem to be good candidates for future studies.

The second stage in research on the effects of drugs in child development involves the exploration of plausible hypotheses for the source of differences between opioid-exposed and other children. Human studies of opioid-exposed children have only begun to explore such hypotheses. There are many different ways through which maternal opioid use might affect the development of children (cf. Aylward, 1982; Marcus and Hans, 1982b). These include

- *teratological effects* related to dysmorphogenesis early in embryonic/fetal life,
- *toxicological effects* related to the presence of circulating drugs in the system at the time of birth,
- *genetic effects* related to familial behavioral and learning disorders that may be transmitted genetically,
- *poor nutrition* both before and after birth,
- *pregnancy and delivery complications* that may be related to maternal drug use, and
- *rearing environment,* including the direct interaction with the primary caregiver.

Unfortunately, most studies of infants exposed to opioids in utero have simply assumed that any findings are related to the direct teratological or toxicological effect of the drug. The role of the scientist is to better understand the possible alternative explanations, and if not to rule out all alternative explanations, at least to explore which are the most plausible. The remainder of this chapter will explore what is currently known about routes through which maternal opioid use may affect the development of children.

EVIDENCE FOR TERATOGENIC OR TOXICOLOGICAL EFFECTS

The signs of neonatal narcotics abstinence are so dramatic, so specific to opioid-exposed infants, so time limited, and so similar to the signs of abstinence in human adults that they almost certainly are caused by withdrawal from the drug. None of the other differences between opioid-exposed and unexposed children, however, can be so easily related to the direct teratological or toxicological effect of drug exposure. Classic types of evidence for teratogenic effects—documentation of dose–response relationships, identification of physical birth defects, and first-trimester exposure as a critical period—are lacking.

The evidence for dose–response relationships is weak in human studies. Although a number of investigators have explored the correlations between methadone dose and birth weight, in most of these reports there is no consistent relationship between methadone dose and birth weight (e.g., Newman, 1974; Kaltenbach and Finnegan, 1987; Thakur et al., 1990), or higher doses were associated with higher birth weight (Doberczak et al., 1987; Hagopian et al., 1991). Similarly there is not strong evidence that methadone dose relates to severity of neonatal withdrawal symptoms. A number of studies have shown a correlation between dose and severity of withdrawal (Rosen and Pippenger, 1976; Connaughton et al., 1977; Kandall et al., 1977; Newman et al., 1975), but others have not (Harper et al., 1977; Stone et al., 1971; Zelson et al., 1971; Ostrea et al., 1976; Hagopian et al., 1991). No data suggest a relationship between maternal dose and child growth or behavior past the neonatal period. Yet existing studies do not rule out the possibility of dose–response relationships for opioid exposure. Those studying heroin-using mothers have no access to reliable information about the amount of drug used by the women. In studies of methadone-treated women, samples are typically drawn from one geographic area in which methadone doses fall within a very constricted range, with different treatment communities generally having a policy of low-dose (mean around 20 mg) or high-dose (mean greater than 40 mg) methadone maintenance. Any individual sample probably has such a narrow range of doses as to render correlations with dose meaningless. Also, doses at all methadone programs fall far short of the amount that human beings can tolerate, that would perhaps lead to floor effects. Another problem is that methadone dosage is probably a poor approximation of actual amount of opioid drug ingested since a high proportion of methadone-maintained women self-medicate with other opioids and are most likely to do so if their methadone doses are relatively low (e.g., McLothlin and Anglin, 1981), thus tending to equalize the differences between low- and high-dose women.

The evidence for physical birth defects in opioid-exposed children is also almost nonexistent. Whereas there are isolated case reports in the literature of opioid-exposed children with physical birth defects, there is no evidence of a consistent pattern of either major physical birth defects or of minor physical anomalies in the offspring of opioid-using women.

Evidence for the importance of first-trimester exposure is also lacking in studies to date. For some drugs of abuse, one can compare the outcome of infants for whom exposure to drugs was primarily during the first trimester with those whose exposure was primarily at other stages of gestation or throughout gestation. This model has been well used in the cocaine literature, in which comparisons have been made between infants whose mothers successfully completed drug treatment before the birth of the child (Chasnoff and Griffith, 1989). With respect to the differential effects of opioid exposure in different trimesters of pregnancy, the data simply are not there. Opioid drugs are so addictive that women controlling their own use tend not to vary their patterns of use dramatically during different stages of pregnancy. Because maternal detoxification from opioids is believed to endanger the fetus (Rementeria and Nunag, 1973; Connaughton et al., 1975), women who enter treatment programs during pregnancy generally are maintained on methadone throughout their pregnancies and encouraged to detoxify only after the birth of the infant.

EVIDENCE FOR OTHER BIOLOGICAL EFFECTS

In the studies of human offspring exposed to opioid prenatally, a number of biological factors that could potentially account for differences between opioid-exposed and unexposed infants have not yet been adequately investigated. These alternative explanations include the teratological or toxicological effects of drugs other than the opioids, nutritional factors, genetic factors, and pregnancy and birth complications.

The effects of nonopioid drugs on the children of opioid-using mothers have hardly been studied to date. Opioid-using women—even those on methadone maintenance—tend to be simultaneous users of nicotine, alcohol, marijuana, barbiturates, benzodiazepines, and beginning in the late 1980s, cocaine. Yet there is virtually no information on the extent to which long-term growth or behavior in opioid-exposed children is related to exposure to these drugs.

Opioid addiction is likely to affect maternal appetite and eating patterns. Whereas clearly maternal nutrition during pregnancy is related to intrauterine growth retardation and possibly to other infant outcomes (e.g., Naeye et al., 1973b), there has been little investigation into whether maternal opioid use affects development through poorer nutrition. Naeye et al. (1973b) have reported that fetal growth retardation in offspring of opioid-using women is unrelated to signs of maternal undernutrition. They suggest that organ size reductions in heroin-exposed infants are associated with a subnormal number of cells rather than a combination of subnormal cell number and cell size observed in undernutrition uncomplicated by drug addiction. On the other hand, Lifschitz et al. (1985) has reported that head circumference at birth is related to maternal nutritional status in heroin- and methadone-exposed infants.

Similarly, there has been little attention to the possibility of genetic factors accounting for differences between offspring of drug-exposed and unexposed children. It is clear that there is a high incidence of mental disorders in drug-using populations (e.g., Ross et al., 1988; Rounsaville et al., 1982). In fact, there is an enduring theory that some individuals who become involved in drug abuse do so to self-medicate their behavioral problems (Khantzian, 1985). Many of the types of behavioral problems most commonly observed in drug-using individuals—depression, antisocial behavior, attentional disorders—are known to have a familial, possibly genetic basis (Cantwell, 1975; Beardslee et al., 1983; Moffitt, 1987). No research to date has seriously explored genetic hypotheses in relation to the neurobehavioral problems observed in opioid-exposed children. Such research would require collection of complete family behavioral histories of parents and siblings and direct attentional and neurological assessments of family members, particularly fathers. Interpretation of such assessments would be complicated by parents' concurrent drug use.

Finally, pregnancy and birth complications may play a role in accounting for differences between opioid-exposed and unexposed children. The pregnancies of opioid-dependent women are often complicated by problems such as toxemia, maternal infection, placental problems, abnormal presentation, and fetal distress (Finnegan, 1975; Rementeria and Lotongkum, 1977; Perlmutter, 1974; Naeye et al., 1973a; Kandall et al., 1977; Stone et al., 1971; Ostrea and Chavez, 1979; Wilson et al., 1981). Some of these conditions may be directly related to action of the drug (such as fetal distress accompanying a period of fetal withdrawal), and others are associated with drug use but are independent of the pharmacology of the drug (such as hepatitis associated with needle use or syphilis related to prostitution).

Whereas it remains a plausible hypothesis that the effects of maternal drug use on child outcome are mediated through these pregnancy and birth complications, there has been little investigation into whether known prenatal events or chronic conditions are related to later developmental outcome in exposed infants. Lifschitz et al. (1985) showed a relationship between the Hobel prenatal risk score and intellectual performance at preschool measured on the McCarthy GCI at preschool. Hans (1989) reported that methadone-exposed 2-year-olds with few documented pregnancy and birth complications still lagged behind comparison infants in a variety of aspects of motoric functions, suggesting that complications may not be the primary cause of postnatal growth retardation in opioid-exposed children. Rosen and Johnson (Rosen and Johnson, 1982; Johnson et al., 1984) reported no simple correlation between complications and Bayley developmental scores in toddlers or neurological scores. In a path analysis of a larger set of variables, labor and delivery complications combined with neonatal outcome were mediating variables between extent of maternal drug abuse and developmental outcome at 36 months of age.

EVIDENCE FOR ENVIRONMENTAL EFFECTS

In every study of the effects of maternal drug use on human infants, there are likely to be confounds between drug-abuse variables and the children's rearing environments. After all, it seems only plausible that there be major life-style differences between families in which there is drug use and other families that do not use drugs. Certainly, mothers' patterns of drug procurement—be they legal or illegal—have an effect on the day-to-day activities of a family. Children whose mothers actively use street drugs are at risk for witnessing drug-using and drug-procuring activities and are probably at risk for inadequate care while their mothers are using and procuring. When mothers are actively involved in drug treatment, the demands of the treatment on parents' time will structure parents' daily routines and are likely to affect the experiences of children.

Only a very limited number of studies have tried to document the differences between the rearing environments of children living in drug-using families and non-drug-using families. Marcus et al. (1984a) report that opioid-using women are at high risk for a variety of factors that adversely affect the development of children, but particularly psychiatric disturbance. Rounsaville, Weissman, Kleber, and Wilber (1982) found that almost three quarters of methadone-treatment patients met DSM-III criteria for a major affective disorder, and over a third met criteria for an anxiety disorder. Thus, opioid-exposed children are at high risk for being reared by a caregiver with a major mental disorder. Numerous studies on non-drug-using populations provide evidence that maternal psychiatric disorder is a major environmental risk factor for child development (e.g., Orvaschel et al., 1988; Rolf et al., 1990).

Data also suggest that opioid-exposed children are more likely than children in non-drug-using families to experience frequent changes in custodial care. In a sample of heroin-exposed children followed through school age, Wilson (1989) reported that only 12% lived with their biological mother, while 60% lived with extended family or friends, 25% had been legally adopted, and one profoundly retarded child had been institutionalized. In a sample of heroin-exposed and methadone-exposed children followed through preschool age, Wilson (1989) found that by their first birthdays 48% of infants of heroin-using women were living away from their mothers, and by preschool year only 9% of the heroin-exposed infants lived with their biological mothers. In contrast, almost half of the methadone-treated women still cared for their infants by the preschool years. In this same study, Lawson and Wilson (1980) reported that, other than the differences in custody arrangements, the environments of methadone-exposed, heroin-exposed, and drug-free controls did not differ in terms of educational or occupational classifications of the head of the household, characteristics that promote child development as assessed on the Caldwell HOME scale (Caldwell and Bradley, 1978), or in terms of family stability (measured in terms of factors such as regularity of income and involvement in illegal activities).

Data also indicate that opioid-using women caring for their children are more likely to experience difficulties in caring for their children. Fiks, Johnson, and Rosen (1985) reported on a sample of 57 methadone-maintained mothers and 31 drug-free control mothers who had been followed from pregnancy until their children were aged 3. Methadone mothers required more referrals to day-care services, preschool nurseries, and special education programs (comparison mothers were more likely to have independently sought such services). Methadone mothers had greater financial problems and more emotional problems. They were less likely to be living with a male partner and had more ambivalent relationships with their partners. They were less likely to have pursued vocational and educational activities while their children were young.

Limited data exist on the types of parenting attitudes espoused by drug-using parents. Using an attitude questionnaire, Wellisch and Steinberg (1980) found that addicted women were especially high on the "authoritarian overinvolvement" factor typical of parents who try to overcontrol their children.

Bauman and Dougherty (1983) found no differences in parental attitudes toward preschool-age children, between 15 drug-using mothers and comparison mothers. However, they did find that in actual interaction with their children, drug-using mothers were more likely to use a threatening disciplinarian approach and less likely to employ positive reinforcement.

Studies of parent–infant interaction have also reported differences between drug-using and drug-free mothers. Householder (1980) observed the interaction between opioid-using mothers and their 3-month-old infants. The opioid-using mothers exhibited more physical activity and less emotional involvement in communicating with their infants. They seemed to enjoy the mothering role less and gazed into their infants' eyes less often. Opioid-using mothers appeared either unresponsive, distant, and uninvolved; or intrusive and unable to give their infants time alone.

Bernstein, Jeremy, and colleagues (Bernstein et al., 1984; Bernstein et al., 1986; Jeremy and Bernstein, 1984) observed 4-month-old opioid-exposed and unexposed infants in interaction with their mothers. Opioid-using mothers responded less often and encouraged less their infants' communicative behavior. This difference was attributable to differences in psychosocial resources between the drug-using groups.

Fitzgerald, Kaltenbach, and Finnegan (1990) videotaped 21 exposed infants born to opioid-using women and 28 unexposed infants at birth and at 4 months of age. Using the Greenspan–Lieberman Observational System to evaluate the interactions (Greenspan et al., 1983), drug-using dyads received lower global ratings of dyadic interaction quality when the infants were neonates (and presumably showing signs of narcotics abstinence). Mothers of drug-exposed infants were less socially engaged and showed less positive affect. By the time the infants were 4 months old, the groups did not differ.

Johnson and Rosen (1990) reported on maternal interaction in a sample of 75 multirisk infants for whom approximately half were methadone maintained. At 2, 4, and 6 months of age the infants were videotaped playing with their mothers. Maternal behavior was coded on five dimensions—interacting, negative participation, apathy, vocalization, and holding. There was no relation between severity of maternal drug abuse and maternal responsiveness toward the infant, although the investigators reported concern that the frequency of the target behaviors was so low in the comparison group to make it impossible to observe differences.

In summary, there is some evidence that the rearing environments of drug-exposed children may differ from those of children from non-drug-using families. The ultimate reason for understanding the environments of drug-exposed children, however, is to see how they affect child development. Only a few studies have examined the effects of these environmental factors on the neurobehavioral development of children in drug-using families. Wilson et al. (1979) reported on the development of preschool children of heroin-addicted mothers, comparing children exposed in utero to heroin (many of whom had been adopted or were in foster care) to unexposed children being reared in a drug-using environment to unexposed children being reared in drug-free environments. In most of the measures of physical, cognitive, and behavioral development, heroin-exposed infants did the worst; unexposed children in drug-free environments did the best; and unexposed children in heroin environments performed intermediate to the other two groups.

Marcus et al. (1984b) report that differences between exposed and unexposed infants in motor functioning disappear after 4 months of age in infants from families with higher psychosocial resources, but not in families with lower psychosocial resources. Differences in attentional functioning remain between the two groups regardless of the level of psychosocial resource. Hans (1989) reports that methadone-exposed infants living in extremely impoverished circumstances developed less well mentally at age 2 than do exposed infants in more adequate environmental circumstances or than unexposed infants living in extreme poverty. The author interprets the finding that the methadone may not cause a behavioral deficit but may create a vulnerability in the children that then makes them more susceptible to impoverished environments. Hans, Bernstein, and Henson (1990) report very poor developmental outcomes in drug-exposed children whose mothers have mental disorders, especially personality disorders.

Johnson et al. (1987) found that 3-year outcome in a sample of children that included many opioid-exposed children was related to adverse maternal drug-use practices, but that these practices were mediated through labor and delivery complications and neonatal condition in one path and maternal functioning during pregnancy and subsequent social disorganization in the family environment in another path.

MODELS FOR THE FUTURE

As the data reviewed indicate, studies of children exposed in utero to opioid drugs have only begun the second stage of research on fetal drug exposure: to look at possible alternative explanations to teratology and toxicology as sources of differences between drug-exposed and unexposed children. Whereas the data suggest that there are developmental differences between opioid-exposed and unexposed children, they do not clearly implicate the teratological or toxicological effects of opioids as the source of these differences, except in the case of neonatal abstinence syndrome. There are simply too many other potential biological and environmental sources of these differences that have not yet been adequately investigated.

Clearly variability in the rearing environments of drug-exposed children is tremendous and has the potential to be a major factor in determining developmental outcome of these children. Understanding the role of environment in the development of drug-exposed children is a topic of study that should have a high priority. Aside from its basic scientific relevance, information of this type is critical for informing those who are developing interventions for drug-addicted parents and making policy decisions related to child custody issues. Research in the future needs to provide much more detailed documentation of the home environments provided in drug-using families.

Even research with a primarily teratological focus needs to adopt a different perspective with respect to issues of child environment. In the past, most research on drug-exposed children has either ignored environmental factors or has tended to view environmental factors as confounds that need to be either matched for in experimental design or controlled for in statistical analyses. Yet it is a mistake to view environment as a nuisance variable or to separate the environmental effects of drugs from teratological/toxicological effects. Children are biological organisms, and they always develop within an environmental context; to study their development without taking this context into account can be misleading. Biological and environmental factors operate as a system to affect child development. One of the basic principles of behavioral teratology outlined by Vorhees (1986, p. 36) relates to the necessary coexistence of teratological and environmental influences: "The type and magnitude of a behavioral teratogenic effect depend on the environmental influences on the organism, including both prenatal and postnatal environmental factors." Given the relatively great role environment plays in determining developmental outcome in the human species, this principle is particularly important to keep in mind with research in clinical populations. Little research to date on opioid-exposed children has tried to look at biological and environmental variables as a system or even tried to determine how they interact with one another. Yet, those studies that have attempted such an approach (Hans, 1989; Johnson et al., 1987) provide suggestive evidence that environmental variables may play key roles in either mediating or modulating teratological effects of opioids. Clearly the next stage of research in this field needs to address the joint effects of environment and teratological action in the development of opioid-exposed children.

ACKNOWLEDGMENT

While preparing this paper, the author was supported by Grant RO1 DA05396-01A1 from the National Institute on Drug Abuse.

REFERENCES

Ainsworth, M. D. C., Blehar, M. C., Waters, E., and Wall, S. (1978). "Patterns of Attachment: A Psychological Study of the Strange Situation." Erlbaum, Hillsdale, New Jersey.

Anderson, R. B., and Rosenblith, J. F. (1971). Sudden unexpected death syndrome: Early indicators. *Biol. Neonate* **18**, 395–406.

American Psychiatric Association (1987). "Diagnostic and Statistical Manual of Mental Disorders-III-Revised." American Psychiatric Association, Washington, D. C.

Aylward, G. P. (1982). Methadone outcome studies: Is it more than the methadone? *J. Pediatr.* **10**, 214–215.

Bauman, P. S., and Dougherty, F. E. (1983). Drug-addicted mothers' parenting and their children's development. *Int. J. Addict.* **18**, 291–302.

Bauman, P., and Levine, S. A. (1986). The development of children of drug addicts. *Int. J. Addict.* **21**, 849–863.

Bayley, N. (1969). "Manual for the Bayley Scales of Infant Development." Psychological Corporation, New York.

Beardslee, W. R., Bemporad, J., Keller, M., and Klerman, G. L. (1983). Children of parents with major affective disorder: A review. *Am. J. Psychiatr.* **140**, 825–832.

Beckwith, J. B. (1979). The sudden infant death syndrome. *Curr. Probl. Pediatr.* **3**, 1–36.

Bernstein, V. J., Jeremy, R. J., Hans, S. L., and Marcus, J. (1984). A longitudinal study of offspring born to methadone-maintained women: II. Dyadic interaction and infant behavior at four months. *Am. J. Alcohol Drug Abuse* **10**, 161–193.

Bernstein, V. J., Jeremy, R. J., and Marcus, J. (1986). Mother–infant interaction in multiproblem families: Finding those at risk. *J. Am. Acad. Child Psychiatry* **25**, 631–640.

Blinick, G., Tavolga, W. N., and Antopol, W. (1971). Variations in birth cries of newborn infants from narcotic-addicted and normal mothers. *Am. J. Obstet. Gynecol.* **110**, 948–958.

Brazelton, T. B. (1973). Neonatal Behavioral Assessment Scale. "Clinics in Developmental Medicine," vol. 50. Lippincott, Philadelphia, Pennsylvania.

Bretherton, I., and Waters, E. (eds.) (1985). Growing points of attachment theory and research. *Monogr. Soc. Res. Child Dev.* **50** (1–2, Serial No. 209).

Brown, D. G. (1957). Masculinity–femininity development in children. *J. Consult. Psychol.* **21**, 197–202.

Caldwell, B. M., and Bradley, R. H. (1978). "Home Observation for Measurement of the Environment." University of Arkansas, Little Rock, Arkansas.

Cantwell, D. P. (1975). Genetic studies of hyperactive children: Psychiatric illness in biological and adopting parents. *In* "Genetic research in psychiatry" (R. R. Fieve, D. Rosenthal, and H. Brill eds.), pp. 259–272. Johns Hopkins University Press, Baltimore, Maryland.

Carey, W. C. (1970). A simplified method for measuring infant temperament. *J. Pediatr.* **77**, 188–195.

Carey, W. C., and McDevitt, S. C. (1978). Revision of the Infant Temperament Questionnaire. *Pediatr.* **61**, 735–739.

Chasnoff, I. J., and Griffith, D. R. (1989). Cocaine: Clinical studies of pregnancy and the newborn. *Ann. N.Y. Acad. Sci.* **562**, 260–266.

Chasnoff, I. J., Hatcher, R., and Burns, W. J. (1980). Early growth patterns of methadone-addicted infants. *Am. J. Dis. Child.* **134**, 1049–1051.

Chasnoff, I. J., Hatcher, R., and Burns, W. J. (1982). Polydrug- and methadone-addicted newborns: A continuum of impairment? *Pediatrics* **70**, 210–213.

Chavez, C. J., Ostrea, E. M., Jr., Stryker, J. C., and Smialek, Z. (1979). Sudden infant death syndrome among infants of drug-dependent mothers. *J. Pediatr.* **70**, 210–213.

Cobrinik, R. W., Hood, T., and Chusid, E. (1959). The effect of maternal narcotic addiction on the newborn infant. *Pediatrics* **24**, 288–304.

Connaughton, J. F., Finnegan, L. P., Schut, J., and Emich, J. P. (1975). Current concepts in the management of the pregnant opiate addict. *Addict. Dis. Int. J.* **2**, 21–35.

Connaughton, J. F., Reese, D., Schut, J., and Finnegan, L. P. (1977). Perinatal addiction: Outcome and management. *Am. J. Obstet. Gynecol.* **129**, 79–85.

Crittenden, P. M. (1985). Maltreated infants: Vulnerability and resilience. *J. Child Psychol. Psychiatr.* **26**, 85–96.

Desmond, M. M., and Wilson, G. S. (1975). Neonatal abstinence syndrome: Recognition and diagnosis. *Addict. Dis.* **2**, 113–121.

Dinges, D. F., Davis, M. M., and Glass, P. (1980). Fetal exposure to narcotics: Neonatal sleep as a measure of nervous system disturbance. *Science* **209**, 619–621.

Doberczak, T. M., Thorton, J. C., Berstein, J., and Kandall, S. R. (1987). Impact of maternal drug use dependency on birth weight and head circumference of offspring. *Am. J. Dis. Child.* **141**, 1163–1167.

Fiks, K. B., Johnson, H. L., and Rosen, T. S. (1985). Methadone-maintained mothers: Three-year follow-up of parental functioning. *Int. J. Addict.* **20**, 651–660.

Finnegan, L. P. (1975). Narcotics dependence in pregnancy. *J. Psychedel. Drugs* **7**, 299–311.

Finnegan, L. P. (1976). Clinical effects of pharmacologic agents on pregnancy, the fetus, and the neonate. *Ann. N.Y. Acad. Sci.* **281**, 74–89.

Finnegan, L. P. (1979). In utero opiate dependence and sudden infant death syndrome. *Clin. Perinatol.* **6**, 163–180.

Fitzgerald, E., Kaltenbach, K., and Finnegan, L. (1990). Patterns of interaction among drug dependent women and their infants, *Pediatr. Res.* **27**, 10A.

Fricker, H. S., and Segal, A. (1978). Narcotic addiction, pregnancy, and the newborn. *Am. J. Dis. Child.* **132**, 360–366.

Ghodse, A. H., Reed, J. L., and Mack, J. W. (1977). The effect of maternal narcotic addiction on the newborn infant. *Psychol. Med.* **7**, 667–675.

Goodman, G. (1990). "Identifying Attachment Patterns and Their Antecedents among Opioid-exposed

12-month-old Infants." Unpublished doctoral dissertation, Northwestern University Medical School, Chicago, Illinois.

Green, M., Silverman, I., Suffet, F., Taleporos, E., and Turkel, W. (1979). Outcomes of pregnancy for addicts receiving comprehensive care. *Am. J. Drug Alcohol Abuse* **6**, 413–429.

Greenspan, S., Lieberman, A., and Poisson, S. (1983). "Greenspan–Lieberman Observation System for Assessment of Caregiver–Infant Interaction during Semistructured Play (GLOS)." Division of Maternal and Child Health, HRSA, DHHS, Rockville, Maryland.

Hans, S. L. (1989). Developmental consequences of prenatal exposure to methadone. *Ann. N.Y. Acad. Sci.* **562**, 195–207.

Hans, S. L., and Jeremy, R. J. (1984). Post-neonatal motoric signs in infants exposed in utero to methadone. *Inf. Behav. Dev.* **7**, 158.

Hans, S. L., and Marcus, J. (1983). Motor and attentional behavior in infants of methadone maintained women. *Nat. Inst. Drug Abuse Res. Monogr.* **43**, 287–293.

Hans, S. L., and Snook, S. S. (1986). Sudden infant death in infants exposed to opioid drugs in utero. *Inf. Behav. Dev.* **9**, 161.

Hans, S. L., Marcus, J., Jeremy, R. J., and Auerbach, J. G. (1984). Neurobehavioral development of children exposed in utero to opioid drugs. "Neurobehavioral Teratology" (J. Yanai ed.), pp. 249–273. Elsevier, New York.

Hans, S. L., Bernstein, V. J., and Henson, L. G. (1990). Interaction between drug-using mothers and their toddlers. *Inf. Behav. Dev.* **13**(Special), 190.

Hans, S. L., Henson, L. G., and Jeremy, R. J. The development of infants exposed in utero to opioid drugs. In "Longitudinal Studies of Children at Risk: Cross National Perspectives." (C. W. Greenbaum and J. G. Auerbach, eds.), Ablex, Norwood, New Jersey. In press.

Harper, R. G., Sia, C. G., and Blenman, S. (1973). Observations on the sudden death of infants born to addicted mothers. *Proceedings of the Fifth National Conference on Methadone Treatment*, 1122–1127.

Harper, R. G., Solish, G. I., Purow, H. M., Sang, E., and Panepinto, W. C. (1974). The effect of a methadone treatment program upon pregnant heroin addicts and their newborn infants. *Pediatrics* **54**, 300–305.

Harper, R. G., Solish, G., Feingold, E., Gersten-Woolf, N. B., and Sokal, M. M. (1977). Maternal ingested methadone, body fluid methadone, and the neonatal withdrawal syndrome, *Am. J. Obstet. Gynecol.* **129**, 417–424.

Herjanic, B. M., Barredo, V. H., Herjanic, M., and Tomelleri, C. J. (1979). Children of heroin addicts. *Int. J. Addict.* **14**, 919–931.

Hoppenbrouwers, T., and Hodgman, J. E. (1982). Sudden infant death syndrome (SIDS): An integration of ontogenetic pathologic, physiologic and epidemiologic factors. *Neuropediatrics* **13**(Suppl.), 36–51.

Householder, J. (1980). An investigation of mother–infant interaction in a narcotic-addicted population. Unpublished doctoral dissertation, Northwestern University.

Huntington, L., Hans, S. L., and Zeskind, P. S. (1990). The relations among cry characteristics, demographic variables, and developmental test scores in infants prenatally exposed to methadone. *Inf. Behav. Dev.* **13**, 533–538.

Hutchings, D. E. (1985). "*Methadone: A Treatment for Drug Addiction.*" Chelsea House, New York.

Hutchings, D. E., and Fifer, W. P. (1986). Neurobehavioral effects in human and animal offspring following prenatal exposure to methadone. In "Handbook of Behavioral Teratology" (E. P. Riley and C. V. Vorhees, eds.), pp. 141–160. Plenum Press, New York.

Illinois Test of Psycholinguistic Abilities. (1968). University of Illinois Press, Urbana.

Isikoff, M. (1989, August 2). New drug crisis feared as purity of heroin rises. *The Washington Post*, pp. A1, A11.

Jaffe, J. H. (1980). Drug addiction and drug abuse. In "Goodman and Gilman's The Pharmacological Basis of Therapeutics" (L. S. Goodman and A. G. Gilman, eds.), 6th Ed., pp. 535–584. MacMillan, New York.

Jaffe, J. H. and Martin, W. R. (1980). Opioid analgesic and antagonists. In "Goodman and Gilman's The Pharmacological Basis of Therapeutics" 6th Ed., (L. S. Goodman and A. G. Gilman, eds.), pp. 494–534. MacMillan, New York.

Jeremy, R. J. and Bernstein, V. J. (1984). Dyads at risk: Methadone-maintained women and their four-month-old infants. *Child Dev.* **55**, 1141–1154.

Jeremy, R. J., and Hans, S. L. (1985). Behavior of neonates exposed *in utero* to methadone as assessed on the Brazelton scale. *Inf. Behav. Dev.* **8**, 323–336.

Johnson, H. L. and Rosen, T. S. (1990). Difficult mothers of difficult babies: Mother–infant interaction in a multi-risk population. *Am. J. Orthopsychiatry* **60**, 281–288.

Johnson, H. L., Diano, A., and Rosen, T. S. (1984). 24-month neurobehavioral follow-up of children of methadone-maintained mothers. *Inf. Behav. Dev.* **7**, 115–123.

Johnson, H. L., Glassman, M. B., Fiks, K. B., and Rosen, T. (1987). Path analysis of variables affecting 36-month outcome in a population of multi-risk children. *Inf. Behav. Dev.* **10**, 451–465.

Johnson, J. L., Boney, T. Y., and Brown, B. S. (1991). Evidence of depressive symptoms in children of substance abusers. *Int. J. Addict.* **25**, 465–479.

Kaltenbach, K. and Finnegan, L. P. (1986). Developmental outcome of infants exposed to methadone *in utero*: A longitudinal study. *Pediatric Res.* **20**, 162A.

Kaltenbach, K. and Finnegan, L. P. (1987). Perinatal and developmental outcome of infants exposed to methadone *in utero*. *Neurotoxicol. Teratol.* **9**, 311–313.

Kaltenbach, K. K., Nathanson, L., and Finnegan, L. P. (1989). Temperament characteristics of infants born to drug-dependent women. *Pediatric Res.* **25**, 15A.

Kandall, S. R., Albin, S., Lowinson, J., Berle, B., Eidelman, A. I., and Gartner, L. M. (1976). Differential effects of maternal heroin and methadone use on birth weight. *Pediatrics* **58**, 681–685.

Kandall, S. R., Albin, S., Gartner, L. M., Lee, K. S., Eidelman, A., and Lowinson, J. (1977). The narcotic dependent mother: Fetal and neonatal consequences. *Early Hum. Dev.* **1**, 159–169.

Kaplan, S. O., Kron, R. E., Litt, M., Finnegan, L. P., and Phoenix, M. D. (1975). Correlations between scores on the Brazelton Neonatal Behavioral Assessment Scale, measures of newborn sucking behavior, and birthweight in infants born to narcotic addicted mothers. *In* "Aberrant Development in Infancy: Human and Animal Studies."(N. R. Ellis, ed.), pp. 139–148. Erlbaum, Hillsdale, New Jersey.

Khantzian, E. J. (1985). The self-medication hypothesis of addictive disorders: Focus on heroin and cocaine dependence. *Am. J. Psychiatry* **142**, 1259–1264.

Kron, R. E., Kaplan, S. L., Finnegan, L. P., Litt, M., and Phoenix, M. D. (1975). The assessment of behavioral change in infants undergoing narcotic withdrawal: Comparative data from clinical and objective methods. *Addict. Dis.* **2**, 257–275.

Kron, R. E., Litt, M., Phoenix, M. D., and Finnegan, L. P. (1976). Neonatal narcotic abstinence: Effects of pharmacotherapeutic agents and maternal drug usage on nutritive sucking behavior. *J. Pediatr.* **88**, 637–641.

Kron, R. E., Kaplan, S. L., Phoenix, M. D., and Finnegan, L. P. (1977). Behavior of infants born to drug-dependent mothers: Effects of prenatal and postnatal drugs. *In* "Drug Abuse in Pregnancy and Neonatal Effects" (J. L. Rementeria, ed.), pp. 129–144. Moseley, St. Louis, Missouri.

Lawson, M. S. and Wilson, G. W. (1980). Parenting among women addicted to narcotics. *Child Welfare* **59**, 67–79.

Lesser-Katz, M. (1982). Some effects of maternal drug addiction on the neonate. *Int. J. Addict.* **17**, 887–896.

Lewak, N., van den Berg, B. J., and Beckwith, J. B. (1979). Sudden infant death syndrome risk factors. *Clin. Pediatr.* **18**, 404–411.

Lifschitz, M. H., Wilson, G. S., Smith, E. O., and Desmond, E. (1983). Fetal and postnatal growth of children born to narcotic dependent women. *J. Pediatr.* **102**, 686–691.

Lifschitz, M. H., Wilson, G. S., Smith, E. O., and Desmond, M. M. (1985). Factors affecting head growth and intellectual function in children of drug addicts. *Pediatrics* **75**, 269–274.

Lodge, A., Marcus, M. M., and Ramer, C. M. (1975). Behavioral and electrophysiological characteristics of the addicted neonate. *Addict. Dis.* **2**, 235–255.

Machover, K. (1949). Personality projection in the drawing of the human figure: A method of personality investigation. Thomas, Springfield, Illinois.

Main, M., and Solomon, J. (1986). Discovery of an insecure-disorganized/disoriented attachment pattern.

In "Affective Development in Infancy" (T. B. Brazelton and M. W. Yogman, eds.), pp. 95–124. Ablex, Norwood, New Jersey.

Marcus, J. and Hans, S. L. (1982a). Electromyographic assessment of neonatal muscle tone. *Psychiatr. Res.* **6,** 31–40.

Marcus, J., and Hans, S. L. (1982b). A methodological model to study the effects of toxins on child development. *Neurobehav. Toxicol. Teratol.* **4,** 483–487.

Marcus, J., Hans, S. L., and Jeremy, R. J. (1982a). Differential motor and state functioning in newborns of women on methadone. *Neurobehav. Toxicol. Teratol.* **4,** 459–462.

Marcus, J., Hans, S. L., and Jeremy, R. J. (1982b). Patterns of 1-day and 4-month motor functioning in infants of women on methadone. *Neurobehav. Toxicol. Teratol.* **4,** 473–476.

Marcus, J., Hans, S. L., Patterson, C. B., and Morris, A. J. (1984a). A longitudinal study of offspring born to methadone-maintained women: I. Design, methodology and description of women's resources for functioning. *Am. J. Drug Alcohol Abuse* **10,** 135–160.

Marcus, J., Hans, S. L., and Jeremy, R. J. (1984b). A longitudinal study of offspring born to methadone-maintained women: III. Effects of multiple risk factors on development at four, eight, and twelve months. *Am. J. Alcohol Drug Abuse* **10,** 195–207.

McCarthy, D. (1972). "Manual for the McCarthy Scales of Children's Abilities." Psychological Corporation, New York.

McGlothlin, W. H., and Anglin, M. D. (1981). Long-term follow-up of clients of high- and low-dose methadone programs. *Arch. Gen. Psychiatry* **38,** 1055–1063.

Moffitt, T. E. (1987). Parental mental disorder and offspring criminal behavior: An adoption study. *Psychiatry: Interpersonal Biol. Processes* **50,** 346–360.

Naeye, R. L. (1980). Sudden infant death. *Sci. Am.* **242**(4), 56–62.

Naeye, R. L., Blanc, W., Leblanc, W., and Khatamee, M. A. (1973a). Fetal complications of maternal heroin addiction: Abnormal growth, infections, and episodes of stress. *J. Pediatr.* **83,** 1055–1061.

Naeye, R. L., Blanc, W., and Paul, C. (1973b). Effects of maternal nutrition on the human fetus. *Pediatrics* **52,** 494–503.

Newman, R. G. (1974). Pregnancies of methadone patients: findings in the New York City methadone maintenance treatment program. *N.Y. State J. Med.* **74,** 52–54.

Newman, R. G. Bashkow, S., and Calko, D. (1975). Results of 313 consecutive live births of infants delivered to patients in the New York City Methadone Maintenance Treatment Program. *Am. J. Obstet. Gynecol.* **121,** 233–237.

Olofsson, M., Buckley, W., Andersen, G. E., and Friis-Hansen, B. (1983). Investigation of 89 children born by drug-dependent mothers: Follow-up 1-19 years after birth. *Acta Paediatr. Scand.* **72,** 407–410.

Olsen, G. D., and Lees, M. H. (1980). Ventilatory response to carbon dioxide of infants following chronic prenatal methadone exposure. *J. Pediatr.* **96,** 983–989.

Orvaschel, H., Walsh-Allis, G., and Ye, W. (1988). Psychopathology in children of parents with recurrent depression. *J. Abnorm. Child Psychol.* **16,** 17–28.

Ostrea, E. M., Jr., and Chavez, C. J. (1979). Perinatal problems (excluding neonatal withdrawal) in maternal drug addiction: A study of 830 cases. *J. Pediatr.* **94,** 292–295.

Ostrea, E. M., Jr., Chavez, C. J., and Strauss, M. E. (1976). A study of factors that influence the severity of neonatal narcotic withdrawal. *J. Pediatr.* **88,** 642–645.

Perlmutter, J. F. (1974). Heroin addiction and pregnancy. *Obstet. Gynecol. Surv.* **29,** 439–446.

Pierog, S. (1977). The infant in narcotic withdrawal: Clinical picture. *In* "Drug Abuse in Pregnancy and Neonatal Effects" (J. L. Rementeria, ed.), pp. 95–102. Moseley, St. Louis, Missouri.

Pierson, P. S., Howard, P., and Kleber, H. D. (1972). Sudden deaths in infants born to methadone-maintained addicts. *J. A. M. A.* **220,** 1733–1734.

Rajegowda, B. K., Kandall, S. R., and Falciglia, H. (1978). Sudden unexpected death in infants of narcotic-dependent mothers. *Early Hum. Dev.* **2/3,** 219–225.

Reddy, A. M., Harper, R. G., and Stern, G. (1971). Observations on heroin and methadone withdrawal in the newborn. *Pediatrics* **48,** 353–358.

Rementeria, J. L., and Lotongkhum, K. (1977). The fetus of the drug-addicted woman: Conception, fetal wastage, and complications. *In* "Drug Abuse in Pregnancy and Neonatal Effects" (J. L. Rementeria, ed.), pp. 1–18. Moseley, St. Louis, Missouri.

Rementeria, J. L., and Nunag, N. N. (1973). Narcotic withdrawal in pregnancy: Stillbirth incidence with a case report. *Am. J. Obstet. Gynecol.* **116,** 1152–1156.

Rolf, J., Masten, A. S., Cicchetti, D., Nuechterlein, K. H., and Weintraub, S. (eds.) (1990). "Risk and Protective Factors in the Development of Psychopathology." Cambridge University Press, Cambridge, Massachusetts.

Rosen, T. S. and Johnson, H. L. (1982). Children of methadone-maintained mothers: Follow-up to 18 months of age. *J. Pediatr.* **101,** 192–196.

Rosen, T. S. and Johnson, H. L. (1988). Drug-addicted mothers, their infants, and SIDS. *Ann. N.Y. Acad. Sci.* **533,** 89–95.

Rosen, T. S. and Johnson, H. L. "Children of Mothers on Methadone Maintenance: 6-Year Follow-Up." 1986.

Rosen, T. S., and Pippenger, C. E. (1976). Pharmacologic observation of the neonatal withdrawal syndrome. *J. Pediatr.* **88,** 1044–1048.

Ross, H. E., Glaser, F. B., and Germanson, T. (1988). The prevalence of psychiatric disorders in patients with alcohol and other drug problems. *Arch. Gen. Psychiatry* **45,** 1023–1031.

Rounsaville, B. J., Weissman, M. M., Kleber, H. D., and Wilber, C. H. (1982). The heterogeneity of psychiatric disorders in treated opiate addicts. *Arch. Gen. Psychiatry* **39,** 161–166.

Sameroff, A. J. (ed.) (1978). Organization and stability of newborn behavior: A commentary on the Brazelton Neonatal Behavioral Assessment Scale. *Monogr. Soc. Res. Child Dev.* **43** (5–6).

Sandberg, D. E., Meyer-Bahlburg, H. F. L., Rosen, T. S., and Johnson, H. L. (1990). Effects of prenatal methadone exposure on sex-dimorphic behavior in early school-age children. *Psychoneuroendocrinology* **15,** 77–82.

Schulman, C. A. (1969). Alterations of the sleep cycle in heroin-addicted and "suspect" newborns. *Neuropaediatrie* **1,** 89–100.

Schulte, F. J., Albani, M., Schnizer, H., Bentele, K., and Klingsporn, R. (1982). Neuronal control of neonatal respiration—Sleep apnea and the sudden infant death syndrome. *Neuropediatrics* **13** (Supple), 3–14.

Shannon, D. C., and Kelly, D. H. (1982). SIDS and near-SIDS. *N. Engl. J. Med.* **306,** 959–965, 1022–1028.

Slotkin, T. A., Seidler, F. J., and Whitmore, W. L. (1980). Effects of maternal methadone administration on ornithine decarboxylase in brain and heart of the offspring: Relationships of enzyme activity to dose and to growth impairment in the rat. *Life Sci.* **26,** 861–867.

Soule, A. B., III, Standley, K., Copans, S. A., and Davis, M. (1974). Clinical uses of the Brazelton Neonatal Scale. *Pediatrics* **54,** 583–586.

Sowder, B. J., and Burt, M. R. (1980). "Children of Heroin Addicts: An Assessment of Health, Learning, Behavioral and Adjustment Problems." Praeger, New York.

Stimmel, B., Goldberg, J., Reisman, A., Murphy, R. J., and Teets, K. (1982–1983). Fetal outcome in narcotic-dependent women: The importance of the type of maternal narcotic used. *Am. J. Drug Alcohol Abuse* **9,** 383–395.

Stone, M. L., Salerno, L. J., Green, M., and Zelson, C. (1971). Narcotic addiction in pregnancy. *Am. J. Obstet. Gynecol.* **109,** 716–723.

Strauss, M. E., Andresko, M., Stryker, J. C., Wardell, J. N., and Dunkel, L. D. (1974). Methadone maintenance during pregnancy: Pregnancy, birth, and neonate characteristics. *Am. J. Obstet. Gynecol.* **120,** 895–900.

Strauss, M. E., Lessen-Firestone, J. K., Starr, R. H., and Ostrea, E. M., Jr. (1975). Behavior of narcotics addicted newborns. *Child Dev.* **46,** 887–893.

Strauss, M. E., Starr, R. H., Ostrea, E. M., Jr., Chavez, C. J., and Stryker, J. C. (1976). Behavioral concomitants of prenatal addiction to narcotics. *J. Pediatr.* **89,** 842–846.

Strauss, M. E., Lessen-Firestone, J. K., Chavez, C. J., and Stryker, J. C. (1979). Children of methadone-treated women at five years of age. *Pharmacol. Biochem. Behav.* **11,** 3–6.

Stutsman, R. (1931). "Mental Measurement of Preschool Children." World Book, Yonkers-on-Hudson, New York.

Thakur, N., Kaltenbach, K., Peacock, J., Weiner, S., and Finnegan, L. (1990). The relationship between maternal methadone dose during pregnancy and infant outcome. *Pediatr. Res.* 227A.

Thomas, A., and Chess, S. (1977). "Temperament and Development." Brunner/Mazel, New York.

Ting, R., Keller, A., Berman, P., and Finnegan, L. (1974). Follow-up studies of infants born to methadone-addicted mothers. *Pediatr. Res.* **8,** 346.

Valdes-Dapena, M. A. (1980). Sudden infant death syndrome: A review of the medical literature, 1974–1979. *Pediatrics* **66,** 597–614.

Vaughn, B. E., Bradley, C. F., Joffe, L. S., Seifer, R., and Barglow, P. (1987). Maternal characteristics measured prenatally are predictive of ratings of temperamental "difficulty" on the Carey Infant Temperament Questionnaire. *Dev. Psychol.* **23,** 152–161.

Vorhees, C. V. (1986). Principles of behavioral teratology. *In* "Handbook of Behavioral Teratology." (E. P. Riley and C. V. Vorhees, eds.), pp. 23–48. Plenum Press, New York.

Wang, C., Pasulka, P., Perry, B., and Pizzi, W. J. (1986). Effect of perinatal exposure to methadone on brain opioid and alpha Z-adrenergic receptors. *Neurobehav. Toxicol. Teratol.* **8,** 399–402.

Ward, O. B., Orth, T. M., and Weisz, J. (1983). A possible role of opiates in modifying sexual differentiation. *In* "Monographs in Neural Sciences" (M. Schlumpf and W. Lichtensteiger, eds.), vol. 9, pp. 194–200. S. Karger, Basel, Switzerland.

Ward, S. L. D., Schuetz, S., Krishna, V., Bean, X., Wingert, W., Wachsman, L., and Keens, T. G. (1986). Abnormal sleeping ventilatory pattern in infants of substance-abusing mothers. *Am. J. Dis. Child.* **140,** 1015–1020.

Ward, O. B., Kopertowski, D. M., Finnegan, L. P., and Sandberg, D. E. (1989). Gender-identity variations in boys prenatally exposed to opiates. *Ann. N.Y. Acad. Sci.* **562,** 365–366.

Wellisch, D. K., and Steinberg, M. R. (1980). Parenting attitudes of addicted mothers. *Int. J. Addict.* **15,** 809–819.

Wilson, G. S. (1989). Clinical studies of infants and children exposed prenatally to heroin. *Ann. N.Y. Acad. Sci.* **562,** 183–194.

Wilson, G. S., Desmond, M. M., and Verniaud, W. M. (1973). Early development of infants of heroin-addicted mothers. *Am. J. Dis. Child.* **126,** 457–462.

Wilson, G. S., McCreary, R., Kean, J., and Baxter, J. C. (1979). The development of preschool children of heroin-addicted mothers: A controlled study. *Pediatrics* **63,** 135–141.

Wilson, G. S., Desmond, M. M., and Wait, R. B. (1981). Follow-up of methadone-treated and untreated narcotic-dependent women and their infants: Health, developmental, and social implications. *J. Pediatr.* **98,** 716–722.

Zagon, I. S., and McLaughlin, P. J. (1984). An overview of the neurobehavioral sequelae of perinatal opioid exposure. *In* "Neurobehavioral Teratology." (J. Yanai, ed.), pp. 197–233. Elsevier, Amsterdam.

Zelson, C., Rubio, E., and Wasserman, E. (1971). Neonatal narcotic addiction: 10-year observation. *Pediatrics* **48,** 178–189.

Zelson, C., Ja Lee, S., and Casalino, M. (1973). Neonatal narcotic addiction: Comparative effects of maternal intake of heroin and methadone. *N. Engl. J. Med.* **289,** 1216–1220.

— 10 —

Effects of Opiates on the Physiology of the Fetal Nervous System

৯৯

Hazel H. Szeto and Peter Y. Cheng

Department of Pharmacology
Cornell University Medical College
New York, New York

INTRODUCTION

There is ample clinical evidence demonstrating that prenatal opiate exposure is associated with a variety of behavioral disturbances, many of which persist into early childhood. These infants have abnormal sleep patterns, with a decrease in quiet sleep (QS) and an increase in rapid eye movement sleep (REMS) and wakefulness (Dinges et al., 1980). Many children have been found to be hyperactive, with shorter attention spans and decreased fine motor coordination (Wilson et al., 1981). These persistent neurological and behavioral effects are often considered a consequence of the neonatal withdrawal insult. However, it is also possible that repeated intrauterine exposure to opiates may have direct effects on the development of the nervous system. Pharmacokinetic studies have shown that the opiate drugs are readily distributed across the placenta to the fetus (Szeto et al., 1978, 1981, 1982).

Our laboratory has been actively involved in investigating the direct effects of opiate exposure on the fetal nervous system. In this chapter, we shall present our findings on the effects of opiates on cardiorespiratory control, electroencephalogram (EEG) and sleep–wake regulation in the fetus. We shall then compare the effects observed in the fetus with effects reported in adults. Attempts at studying the sites and mechanisms of action of the opiates will also be presented. In the course of studying the effects of exogenously administered opiates, we have also examined the role of the endogenous opiate peptides in the control of fetal breathing. Finally, the clinical implications of these findings will be discussed.

215

ANIMAL MODEL

Because of both technical and ethical constraints, research of this type relies largely on experimental animals. The pregnant sheep and its fetal lamb provide a close model for human pregnancy and have been studied extensively to provide insight into conditions affecting human development before birth (Hecker, 1983). The use of pregnant sheep for study of fetal physiology was originated by Barcroft and his co-workers in Cambridge in the 1930s (Barcroft, 1946). In contrast to more common experimental animals, the sheep normally has only one or two lambs with birth weights similar to that of the human baby, and a relatively long gestational period (145 days). The fetal lamb in the early third trimester is large enough to permit the implantation of catheters in the fetal blood vessels, making it possible to obtain fetal blood samples, as well as monitoring fetal arterial pressure and heart rate. Experimental approaches have been developed for investigating cardiorespiratory control and neurobehavioral regulation in the fetus. Initially, most studies were carried out with the ewe under anesthesia. However, increasing recognition that maternal anesthesia may affect placental perfusion and fetal cardiovascular function has led to the use of the chronic preparation. All the investigations reported in this chapter were carried out using the unanesthetized, chronically instrumented fetal lamb model.

EFFECTS OF EXOGENOUSLY ADMINISTERED OPIATES ON THE FETAL NERVOUS SYSTEM

Respiratory Control

Despite considerable advances that have been made in the past decade, our understanding of respiratory control during fetal and neonatal life remains incomplete. Spontaneous breathing movements have been observed in the fetus in many mammalian species, including humans, baboons, sheep, and guinea pigs. In humans, fetal breathing movements (FBM) can be observed as chest and diaphragm movements by ultrasound techniques. In animal models, FBM can be detected by electromyographic (EMG) bursts in the diaphragm as well as negative changes in intrathoracic pressure (Dawes et al., 1972). These FBM have been shown to be correlated with phrenic nerve activity in fetal sheep (Bahoric and Chernick, 1975). In all species, FBM have been found to occur intermittently and do not become continuous until after birth. In both the primate and the sheep, FBM have been shown to occur in aggregates interspersed with periods of apnea (Dawes et al., 1972; Martin et al., 1974; Patrick et al., 1980). This has prompted most investigators to characterize fetal breathing by quantitating the incidence of FBM, which is defined as the percentage of time with FBM, and this has been reported to range between 30 and 60% in both the fetal sheep and the fetal baboon (Dawes et al., 1972; Rey et al., 1989). Another unique characteristic of FBM is the large variability in instantaneous breathing rates, with rates ranging up to 300 breaths/min (Szeto et al., 1991).

The precise nature of the control mechanism in fetal breathing remains unknown.

The fetus is largely unresponsive to the usual chemical stimuli associated with enhanced breathing after birth (for review, see Jansen and Chernick, 1983). For instance, FBM do not appear to be affected by normal fluctuations in pO_2 or pCO_2. In contrast to that of the adult, fetal hypoxemia results in a decrease in the incidence of FBM, although hypercarbia has been shown to increase the incidence and depth of the FBM (Dawes et al., 1982). Spontaneous FBM are diminished during fetal hypoglycemia, and are increased during glucose infusion (Richardson et al., 1982). Many pharmacological agents have also been shown to significantly affect the incidence of FBM, including muscarinic agonists (Brown et al., 1981; Hinman and Szeto, 1988), prostaglandins (PG) and PG synthetase inhibitors (Kitterman et al., 1979), and ethanol (Patrick et al., 1988). In the past few years, we and others have also found significant changes in the breathing pattern of the fetal lamb with opiate administration.

Dose—Response and Time—Action Relationships

Early studies in our laboratory revealed that maternal administration of methadone, at doses that result in maternal plasma methadone concentrations similar to those reported for pregnant patients on a methadone maintenance program, resulted in continuous breathing movements in the fetus (Szeto, 1983). The FBM after methadone administration were deeper and more regular than in control conditions. When 10% of the methadone dose was administered directly to the fetus, the same response was observed (Szeto, 1983). Subsequently, we demonstrated a similar response to both maternal and fetal administration of morphine (Umans and Szeto, 1983). This apparent stimulation of FBM by opiates has since been confirmed by other investigators after either constant rate infusion of morphine to the fetus (Olsen et al., 1983) or intravenous (i.v.) bolus of morphine to the fetus (Sheldon and Toubas, 1984; Toubas et al., 1985), although there appeared to be a delay in onset of action (28 ± 7.3 min) after bolus administration.

However, there have also been reports that morphine suppresses FBM. Olsen and Dawes (1983) first reported a decrease in FBM after prolonged constant rate infusion of morphine to the fetus. We later confirmed that FBM were indeed suppressed when a high dose of morphine was infused directly to the fetus (Umans and Szeto, 1985).

To obtain a better understanding of the effects of opiates on fetal breathing, we examined the complete dose—response relationship for morphine on the incidence of FBM. The quantal dose—response curve shown in Fig. 1A illustrates the increase in the percentage of animals responding by stimulation of FBM with increasing doses of morphine from 0.15 to 2.5 mg/hr. At doses greater than 2.5 mg/hr, an increasing number of animals responded by a decrease in the incidence of FBM (Szeto et al., 1988b). Probit analysis revealed that there is a 30-fold difference between the 50% effective dose (ED_{50}) for stimulation (0.15 mg/hr) and for suppression (4.43 mg/hr).

The graded dose—response curve in Fig. 1B shows that the incidence of FBM increased as a function of dose up to 2.5 mg/hr. There was subsequently a rapid decrease in the incidence of FBM with further increases in dose. This biphasic dose—response relationship may account for the initial apnea followed by a delay in onset of

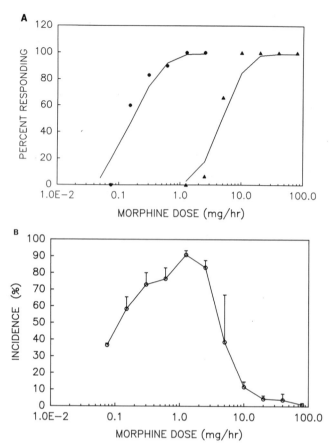

FIGURE 1   (A) Quantal dose–response curve for stimulation (●; ED, 0.15 mg/hr) and suppression (▲; ED, 4.43 mg/hr) of FBM due to morphine administration. Reprinted with permission from Szeto et al. (1988b). (B) Effects of morphine on incidence of FBM. Reprinted with permission from Szeto et al. (1988b). Dual action of morphine on fetal breathing movements. *J. Pharmacol. Exp. Ther.* **245**, 537–542.

a prolonged breathing period when morphine was administered as an i.v. bolus to the fetus (Sheldon and Toubas, 1984; Toubas et al., 1985). When morphine is given as an i.v. bolus, initial plasma drug levels are very high, and suppression of FBM would be expected. As plasma levels decline exponentially, stimulation of FBM would be observed. The time lag would be a function of the dose of morphine and the distribution and elimination of morphine from the fetus.

Besides increasing the incidence of FBM, the lower doses of morphine also significantly increased the continuity and stability of the fetal breathing pattern. Doses between 0.6 and 2.5 mg/hr resulted in a more continuous breathing pattern, with a significant increase in the number of breaths/hr and a decrease in the number of apneas (Szeto et al., 1991). The stability of the breathing rate was also enhanced by these doses of morphine (Szeto et al., 1991). Following 1-hr infusion of morphine, significant effects on FBM can be observed for at least 3 hr after infusion.

Comparison with Effects in the Adult

Although it is often thought that opiates depress respiration in the adult (Martin, 1983), there is some evidence that they may also have a stimulatory effect on respiration. Morphine has been reported to increase ventilatory rate in the cat when administered into the third ventricle (Florez et al., 1968). Rapid, shallow breathing (panting) has been observed in dogs after i.v. morphine, although the effect was not dose related (Breckenridge and Hoff, 1952). More recently, morphine has been reported to decrease tidal volume but increase ventilatory rate in conscious unanesthetized dogs (Haddad et al., 1984). There is general agreement, however, that opiates decrease both tidal volume and ventilatory rate in anesthetized animals (Martin, 1983).

A biphasic effect of morphine on respiration has also been reported in the adult cat (Willette and Sapru, 1982). The administration of morphine into the right atrium of decerebrate cats resulted in an initial apneic period followed by rapid shallow breathing. A brief respiratory pause was also observed before the increase in respiratory frequency and minute ventilation after intra-arterial injection of high doses of morphine in conscious adult dogs (Haddad et al., 1984). Thus the effects of opiates on fetal breathing do not appear to be that different from their effects in conscious unanesthetized adult animals.

Sites and Mechanisms of Action

Involvement of Specific Opiate Receptors   The biphasic dose–response relationship suggested that the observed effects of morphine on FBM may be mediated by two different receptor–effector systems (Szeto et al., 1988b). Bennet et al. (1986) reported that the suppression of FBM by morphine can be reversed by naloxone, suggesting that it is mediated by specific opiate receptors. We subsequently showed that the stimulation of FBM by low doses of morphine could also be blocked by naloxone pretreatment, indicating that it is also a specific opiate response (see Fig. 2) (Szeto et al., 1988b).

Central versus Peripheral Sites of Action   The dual action of morphine may be mediated by opiate receptors at two different sites of action. Morphine-induced respiratory depression in the adult is thought to be the result of an inhibition of pontine respiratory centers (Florez and Borison, 1969; Hassen et al., 1976). However, morphine-induced stimulation is not so well understood. There is some evidence that the stimulation is mediated by peripheral mechanisms since it was not observed in vagotomized animals (Willette and Sapru, 1982), and it has been suggested that it is the result of a direct excitatory effect of morphine on spinal cord neurons (Millhorn et al., 1985). A peripheral action has also been ascribed to the effect of Met-enkephalin on ventilation control in adult dogs (Evanich et al., 1985). Pretreatment with naltrexone methylbromide abolished all enkephalin-induced increases in minute ventilation. In

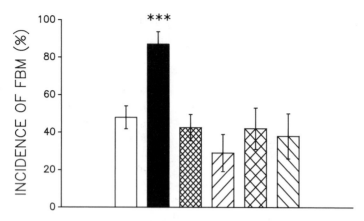

FIGURE 2   Effect of morphine (2.5 mg/hr) alone (■■■) and in the presence of naloxone (▨▨▨ , 6 mg/hr i.v.), methylnaloxone (▨▨ , 300 μg/hr i.c.v.), naloxonazine (▨▨ , 34 mg i.v.) and methyl-atropine (▨▨▨ , 100 μg/hr i.c.v.) on the incidence of FBM. ***, $p < 0.001$ when compared to the incidence of FBM before drug infusion (▭).

the fetal lamb, however, we found that the respiratory stimulation by morphine was effectively blocked by intracerebroventricular (i.c.v.) administration of methylnalox-one (see Fig. 2), thus suggesting a central rather than peripheral site of action (Szeto et al., 1991). It is still possible, however, that the action of morphine is site specific within the CNS.

Roles of Different Opiate Receptor Subtypes   The dual action of morphine may also be explained by the involvement of two or more subtypes of the opiate receptor. Naloxone unfortunately lacks specificity for the various receptor subtypes (Kosterlitz et al., 1980). Although morphine is generally considered a mu agonist, there is actually only a fivefold to tenfold selectivity for the mu receptor as compared to the delta-receptor, as determined by in vitro bioassays (Kosterlitz et al., 1980; Mosberg et al., 1983). It can therefore be expected that the effects of morphine on respiration after systemic administration would be the sum of both mu and delta response, and the contribution of the two would be a function of the dose of morphine. At low doses, we might expect more mu than delta contribution. It is important to point out that both mu and delta binding have been demonstrated in fetal lamb brains of comparable gestational ages to those animals used in our study, with development of the mu site preceding that of the delta site (Dunlap et al., 1985). Thus the contribution of the mu response may be even greater early in development.

Current pharmacologic and receptor-binding studies support the presence of two subtypes of mu receptors: the mu_1-receptor, which binds both opiates and most en-kephalins with similar very high affinity, and the mu_2 site, which binds morphine preferentially (Wolozin and Pasternak, 1981). There is currently no available informa-tion on the developmental profile of the two mu binding sites. Naloxonazine has been demonstrated to bind irreversibly to the mu_1-receptor (Hahn and Pasternak, 1982).

With the use of naloxonazine selectivity, the mu_1-receptor has been implicated in opiate-induced analgesia (Ling et al., 1985) and locomotor and respiratory stimulation (Paakkari et al., 1990). Studies in our laboratory showed that pretreatment with naloxonazine effectively blocked the respiratory stimulation effect of morphine in the fetal lamb (see Fig. 2), suggesting that this low-dose response is mediated by mu_1-receptors (Szeto et al., 1991). As proposed by Ling et al. (1985) for the adult rat, the respiratory depression seen with higher doses of morphine in the fetal lamb may be mediated by mu_2-receptors. There is unfortunately no selective mu_2 antagonist available to directly test this hypothesis.

The role of the delta receptor in control of respiration in the adult is not clear. Our understanding has been greatly hampered, until recently, by the lack of highly selective delta agonists and antagonists. We have attempted to investigate the role of the delta receptor in respiratory control in the fetal lamb by using the highly selective agonist, [D-Pen^2,D-Pen^5]-enkephalin (DPDPE) (Mosberg et al., 1983). Our results showed that i.c.v. infusion of DPDPE to the fetus caused respiratory stimulation, resulting in an almost continuous breathing pattern (see Fig. 3A). These findings are quite different from those reported in adult animals with DPDPE. Kiristy-Roy et al. (1989) reported that DPDPE caused respiratory depression in adult rats. However, their conclusion was based only on a decrease in arterial pH, even though there was no change in pO_2 or pCO_2. Conversely, May et al. (1989) and Yeadon and Kitchen (1989) reported no effect of DPDPE on respiratory rate or blood gases. The reason behind the differences between fetal sheep and adult rats is not known. The DPDPE response in the fetal lamb appeared to be a specific opiate response, in that it was attenuated by naloxone pretreatment, although it was not affected by the delta-selective antagonist, naltrindole (Portoghese et al., 1988), or the mu_1 antagonist, naloxonazine (see Fig. 3B). These results raise the possibility that DPDPE may not be so selective for the delta receptor as had been thought, or that there may be different subtypes of the delta receptor. The idea of multiple subtypes of the delta receptor was recently suggested by Sofuoglu et al. (1991), based on the antinociceptive effects of different delta agonists.

Put together, results to date suggest that opiate-induced respiratory stimulation in the fetal lamb may be mediated by either mu_1- or delta-receptors in the CNS, whereas the respiratory depression at higher doses is probably mediated by mu_2 receptors.

Involvement of Other Neurotransmitters   There is evidence that other neurotransmitters are involved in the respiratory response to low doses of morphine. We had been impressed by the similarity between the respiratory response pattern to low doses of morphine and that to pilocarpine, a muscarinic agonist. Systemic administration of pilocarpine has been shown to result in a dose-dependent increase in the incidence of FBM as well as a more continuous and regular breathing pattern (Brown et al., 1981; Hinman and Szeto, 1985). This pilocarpine response was effectively blocked by i.v. atropine, but not i.v. methylatropine, thus implying activation of central muscarinic pathways (Hinman and Szeto, 1985). We have now demonstrated that the respiratory response to low doses of morphine can also be blocked by i.c.v. administration of

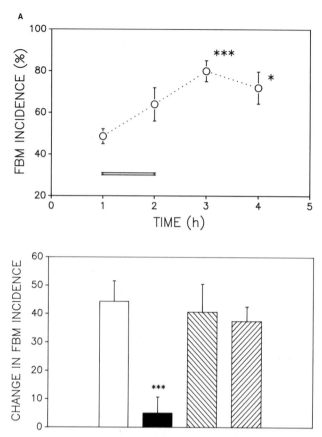

FIGURE 3   (A) Time course for changes in the incidence of FBM before, during, and after infusion of DPDPE (30 μg/hr i.c.v.) to the fetal lamb. Horizontal bar indicates duration of DPDPE infusion. *, $p < 0.05$; ***, $p < 0.005$ when compared to control value (ANOVA). (B) Effect of DPDPE alone (☐) and in the presence of naloxone (■, 6 mg/hr i.v.), naloxonazine (▨ -, 34 mg i.v.) and naltrindole (300 μg/hr ▨ i.c.v.) on the incidence of FBM. *, $p < 0.05$ when compared to DPDPE alone (ANOVA).

methylatropine (see Fig. 2) (Szeto et al., 1991). This potential link between mu_1 receptors and central muscarinic pathways has not been suggested before.

Development of Tolerance with Prolonged Exposure

In the adult, prolonged exposure to opiates is associated with the development of tolerance and dependence. To determine whether tolerance develops to the respiratory response to low doses of morphine, fetal lambs were exposed to constant-rate i.v. infusion of morphine via miniosmotic pumps for a period of 7 days (Szeto et al., 1988a). There was a significant increase in the incidence of FBM on day 1 of morphine exposure. The response was absent by day 2, and the incidence of FBM remained at control levels for the remaining 5 days. Thus, tolerance appears to develop very rapidly to this response in the fetus. This may have significant clinical implications, since continuous breathing activity would greatly increase metabolic demands, which in the presence of constant energy supply, may result in compromised fetal growth (see Clinical Implications). The rapid adaptation of the fetus to this response may therefore be very important in maintaining fetal growth.

Cardiovascular Control

Opiates are known to be important modulators of cardiovascular function in the adult. However, until recently, the roles of opiates in the modulation of fetal cardiovascular function have been rather limited. Early studies in the fetal lamb showed that methadone administration to the mother resulted in an increase in fetal heart rate (FHR) with no significant change in fetal blood pressure (FBP) (Szeto, 1983). The same response was observed when either methadone or morphine was administered directly to the fetus (Umans and Szeto, 1983). This was rather unexpected since opiates have traditionally been thought to produce bradycardia and hypotension in the adult, and prompted a systematic investigation of the cardiovascular actions of opiates in the fetus.

Dose–Response and Time–/Action Relationships

As in the early studies, direct administration of morphine to the fetal lamb resulted in an increase in FHR without significant change in FBP. The increase in FHR was found to be dose dependent, with peak effect observed after 2.5 mg/hr (see Fig. 4). Further increases in dose resulted in an attenuation of the response until no significant change in FHR could be seen after 40 mg/hr. This resulted in a rather unusual bell-shaped dose–response curve (Zhu and Szeto, 1989). This increase in FHR persisted for more than 3 hr after termination of the morphine infusion.

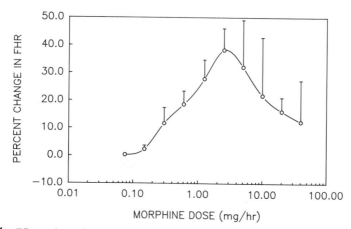

FIGURE 4 Effects of morphine on fetal heart rate. Data are presented as percentage change from control. Reprinted with permission from Zhu and Szeto (1989). Morphine-induced tachycardia in fetal lamb: A bell shaped dose–response curve. *J. Pharmacol. Exp. Ther.* **249**, 78–82.

Comparison with Effects in the Adult

Morphine has traditionally been thought to produce bradycardia and hypotension in the adult. More recently, morphine-induced tachycardia has also been reported by several investigators. These conflicting results cannot be explained simply by species differences or the presence of anesthetic agents since such tachycardia has been found in both anesthetized (Faden and Feuerstein, 1983) and unanesthetized (Holaday, 1982) rats, anesthetized cats (Feldberg and Wei, 1986), and conscious dogs (Given et al., 1986). Thus the response in the fetal lamb is not unlike the responses that have been observed in adult animals. It is of interest, however, that no matter how high the dose was increased, bradycardia was never observed in the fetus.

Sites and Mechanisms of Action

Involvement of Specific Opiate Receptors The mechanism behind the morphine-induced tachycardia in the adult is not understood. Although Faden and Feuerstein (1983) reported that the tachycardia could be reversed by naloxone, other investigators have reported that this effect is not a specific opiate effect (Given et al., 1986; Feldberg and Wei, 1986). We have evidence, however, that the fetal tachycardia is clearly mediated by specific opiate receptors, as it was abolished by naloxone treatment (Zhu and Szeto, 1989).

The bell-shaped dose–response curve suggests that the complex action of morphine may be owing to its interaction with two receptor–effector systems, causing effects that counteract each other. In this case, one would predict that if the tachycardia could be blocked, a bradycardic response would then be unmasked. However, when the morphine-induced tachycardia was blocked with propranolol, no significant decrease in FHR was observed (Zhu and Szeto, 1989). This led to another proposal that perhaps the binding of morphine to the second receptor system altered the affinity or intrinsic activity of the binding to the first receptor system (i.e., autoinhibition) (Zhu and Szeto, 1989). Binding to the second receptor may not result in any change by itself. This may be a reasonable hypothesis if the two receptor systems represent two different subtypes of the opiate receptor. Cowan et al. (1977) have proposed a similar mechanism to explain the bell-shaped dose–response curve of buprenorphine in an antinociceptive assay. Furthermore, interaction between mu- and delta-opiate binding sites has been suggested (Rothman and Westfall, 1983).

Role of Different Opiate-Receptor Subtypes There is accumulating evidence that the effects of opioids on blood pressure and heart rate in the adult are receptor specific. A number of studies have demonstrated that mu- and kappa-opioid agonists elicit qualitatively different cardiovascular responses (Feuerstein and Faden, 1982; Hassen et al., 1983; Pfeiffer et al., 1983a). Highly selective mu agonists seem to consistently increase blood pressure, whereas kappa agonists tend to produce a depressor effect and decrease in heart rate. On the other hand, initial studies with highly selective delta agonists suggest that delta-receptors probably do not modulate cardiovascular function (Kiritsy-Roy et al., 1989). The bell-shaped dose–response curve may be explained by the interaction between the binding of morphine to the mu and delta sites.

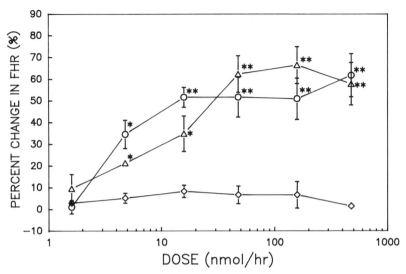

FIGURE 5   Comparison of effects of DAGO (O), DPDPE (◇), and DADLE (△) on fetal heart rate (FHR). Data are presented as percentage change from control. There were significant differences between both DAGO and DADLE when compared with DPDPE at corresponding dose (*, $p < 0.01$; **, $p < 0.001$). There was no difference between DAGO and DADLE at any dose. Reprinted with permission from Szeto et al. (1990).

To investigate the role of mu- and delta-receptors in the modulation of fetal cardiovascular function, we compared the effects of a highly selective mu agonist [DAGO; (D-Ala^2,N-Me-Phe^4,Gly^5-ol)-enkephalin] and a highly selective delta agonist (DPDPE) (Szeto et al., 1990). Administration of DAGO to the fetus (i.c.v.) resulted in a dose-dependent increase in FHR with no significant changes in FBP (see Fig. 5). In contrast, DPDPE failed to elicit any significant changes in either FHR or FBP. The lack of a pressor response to DAGO is similar to that of morphine (Zhu and Szeto, 1989), and is different form the response in adult animals (Pfeiffer et al., 1983a; Kiritsy-Roy et al., 1986, 1989). Since the pressor response in adults appear to be mediated by vasopressin, the lack of a pressor response to DAGO or morphine may indicate that mu-receptors are not linked to AVP pathways in the fetus (Szeto et al., 1990). The lack of effect of DPDPE may indicate either the absence of delta-receptors at this stage of development or that delta-receptors do not modulate cardiovascular function in the fetal lamb. Since specific binding has been demonstrated using both mu- and delta-ligands in fetal lambs of comparable gestational ages (Dunlap et al., 1985), these data suggest that opioid receptors regulating cardiovascular function in the fetus are of the mu subtype. A difference between the response to DAGO and morphine is that despite the large range of doses studied, there was no evidence of any attenuation in the FHR response with higher doses of DAGO (Szeto et al., 1990). Since the relative selectivity of DAGO for mu- versus delta-receptors is much better than that of morphine, the attenuation of tachycardia with higher doses of morphine may indeed be owing to its binding to the delta site, which may affect its action at the mu site.

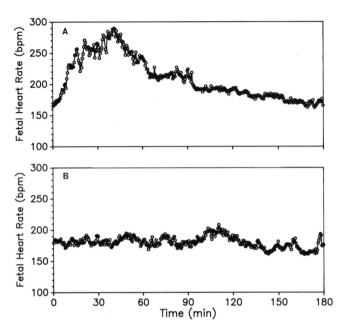

FIGURE 6   Effects of DADLE infusion on fetal heart rate before (A) and 24 hr after (B) naloxonazine treatment. The duration of DADLE infusion (158 nmol/hr) was 90 min.

There is some evidence suggesting that the mu-receptors involved in the tachycardic response are of the mu_1 subtype. Although DADLE [(D-Ala^2,D-Leu^5)-enkephalin] was originally considered to be a putative delta-ligand, Fig. 5 shows that its dose–response curve was similar to that of DAGO, and quite unlike that of DPDPE. It has now been shown that DADLE labels mu_1 sites with affinities equal to or greater than those for delta sites (Wolozin and Pasternak, 1981), whereas the affinity of DPDPE to delta sites is 14-fold greater than to mu_1 sites (Clark et al., 1986). This led us to propose that the tachycardic response to DAGO and DADLE are mediated by mu_1-receptors. We now have evidence that the tachycardic response to DADLE can be effectively abolished by pretreatment with naloxonazine, the irreversible mu_1 antagonist (see Fig. 6). The mu_1-receptor has not been implicated in adult cardiovascular control. In fact, a previous study reported that naloxazone, also a mu_1 antagonist, did not have any significant effect on morphine-induced bradycardia in the adult (Holaday et al., 1983).

Involvement of Other Neurotransmitters   There is evidence that other neurotransmitters are involved in the cardiovascular actions of morphine in the fetal lamb. Previous studies showed that the cardioacceleratory effect of morphine can be completely abolished by propranolol, suggesting that it is mediated via the beta-adrenergic system (Zhu and Szeto, 1989). These findings are consistent with those reported in adult animals (Simon et al., 1978; Giles and Sander, 1983; Pfeiffer et al., 1983b;

Feldberg and Wei, 1986). The administration of morphine and other opioid peptides has been shown to increase both brain and plasma levels of epinephrine, norepinephrine, and dopamine (Moore et al., 1965; Van Loon et al., 1981; Appel and Loon, 1986). There is also evidence that opiates increase central sympathetic outflow to adrenal medulla and peripheral sympathetic nerve endings (Van Loon et al., 1981; Pfeiffer et al., 1983b; Feldberg and Wei, 1986). In addition, we have evidence that the morphine tachycardia can be blocked by i.c.v. methylatropine, suggesting that central muscarinic pathways may be involved (unpublished data, 1992). Activation of central muscarinic sites has been reported to result in a dose-dependent tachycardia in the fetal lamb (Szeto and Hinman, 1990).

In summary, evidence to date suggests that the morphine-induced fetal tachycardia is mediated by specific opiate receptors, most likely of the mu_1 subtype, and involves activation of central muscarinic pathways and subsequent release of catecholamines.

Electroencephalograms

Opiates are known to cause profound changes in the adult EEG. Although attempts have been made to record EEG activity in the human fetus, it is sufficiently difficult to preclude any systematic pharmacological studies in the human fetus. Because EEG activity can be monitored relatively easily in the unanesthetized fetal lamb, much is known about the normal maturation pattern in that species. By early third trimester, two distinct EEG patterns become apparent visually, a high-voltage slow-activity (HVSA) pattern and a low-voltage fast-activity (LVFA) pattern (Dawes et al., 1972). There is also a small amount of a transitional (TRANS) pattern, which is of intermediate voltage and frequency. Over the past 5 years, our laboratory has carried out a systematic study of the maturational profile of fetal lamb EEG using power spectral analysis (Szeto, 1990, 1991; Szeto et al., 1985; Vo et al., 1986; McNerney and Szeto, 1990). The results confirmed the establishment of HVSA and LVFA by 115 days. The frequency characteristics of the three EEG states in a 131-day fetal lamb are summarized in Table 1.

Figure 7A shows a representative example of the transition pattern between the three EEG states. Under normal physiological conditions, the incidence of LVFA,

TABLE 1 Spectral Parameters of the Three EEG States in the Fetal Lamb

State[a]	Total power (μV^2)	% Total Power in				
		1–4 Hz	5–7 Hz	8–13 Hz	15–32 Hz	SEF
LVFA	3272 ± 38	33.0 ± 1	10.9 ± 0.4	18.0 ± 0.2	20.6 ± 0.4	18.4 ± 0.1
TRANS	5505 ± 103	66.5 ± 0.5	10.7 ± 0.3	9.2 ± 0.1	10.9 ± 0.2	13.5 ± 0.1
HVSA	13,157 ± 11	85.8 ± 0	7.0 ± 0.1	4.2 ± 0.1	2.5 ± 0	5.2 ± 0.1

[a]LVFA, low voltage fast activity; TRANS, transitional; HVSA, high voltage slow activity; SEF, spectral edge frequency.

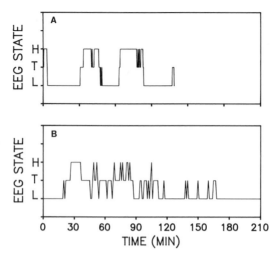

FIGURE 7 Electroencephalogram (EEG) state transition (A) before and (B) during morphine infusion (5 mg/hr). L, low voltage, fast activity (LVFA); T, transitional; H, high voltage, slow activity (HVSA). Reprinted with permission from Szeto (1991).

TRANS, and HVSA in the third trimester was found to be 45.6 ± 2.2%, 14.8 ± 1.4%, and 39.6 ± 2.2%, respectively. With maturation, there is a progressive increase in power in the delta band (1–4 Hz) of the HVSA pattern, and an acquisition of faster frequencies in the LVFA pattern (Szeto et al., 1985). These developmental changes can be quantitated by changes in the spectral-edge frequency of the two EEG patterns (Szeto, 1990).

Dose–Response and Time–Action Relationships

Early investigations on the effects of opiates on fetal EEG relied only on visual analysis (Szeto, 1983; Umans and Szeto, 1983; Olsen et al., 1983; Toubas et al., 1985; Bennet et al., 1986). All reported that both methadone and morphine resulted in a low-voltage EEG pattern. Recent studies in our laboratory using power spectral analysis revealed that morphine significantly altered fetal EEG in three different ways (Szeto, 1991).

First, morphine altered the relative incidence of the three EEG states. Specifically, morphine administration resulted in a dose-dependent reduction in HVSA. This reduction in HVSA was compensated by a reciprocal increase in TRANS at lower doses and an increase in LVFA at higher doses. Taken together, these data suggest that morphine results in a dose-dependent increase in faster frequencies. This is supported by the dose-dependent increase in mean spectral-edge frequency during the morphine infusion period.

Second, morphine significantly affected the temporal pattern in EEG state transition. At lower doses, morphine resulted in a much more unstable, fragmented EEG pattern, with a significantly larger number of state-to-state transitions (Fig. 7B).

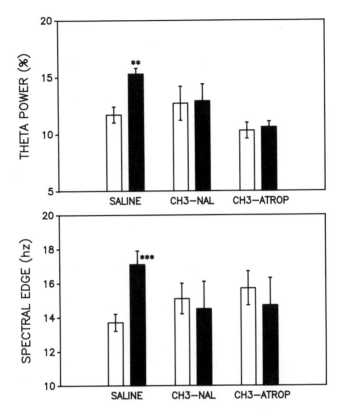

FIGURE 8   Effects of morphine (2.5 mg/hr) on theta power and spectral-edge frequency in the presence of i.c.v. saline, methylnaloxone (CH₃-NAL), or methyl atropine (CH₃-ATROP) infusion. $**p < 0.01$; $***p < 0.005$ compared with predrug control values in each group (paired t-test). Control (□); morphine (■).

Third, morphine also had a significant effect on EEG waveform characteristics. Of particular interest is the dose-dependent increase in the distribution of power in the theta band (see Fig. 8).

Comparison with Effects in the Adult

It has been well established in both human and animals studies that systemic administration of opiates results in an increase in spindle bursts and synchronized high-voltage, slow-activity EEG (for review, see Martin and Kay, 1977). Spectral analysis of the EEG revealed that the HVSA bursts seen after acute morphine and methadone administration to the rat were associated with dose-related increases in spectral power in the 0 to 10-Hz range (Young and Khazan, 1984). Few attempts have been made to quantify EEG changes in response to opiates in humans, but both slowing of alpha activity and increased delta activity have been reported after morphine administration (Matejeck et al., 1988). However, there have also been reports of EEG activation after opiate administration. Intravenous or i.c.v. administration of low doses of morphine in the dog or rabbit resulted in a period of EEG desynchronization (Albus and Herz, 1972; Novack et al., 1976). Higher doses resulted in EEG synchronization. Furthermore, Horita et al. (1989) reported an increase in theta activity in the hippocampus of rabbits treated with morphine or fentanyl. Thus the response of the fetal lamb EEG to low doses of morphine appears to be similar to that in adult animals.

Sites and Mechanisms of Action

Involvement of Specific Opiate Receptors   This EEG activation by morphine in the fetal lamb appears to be mediated via specific opiate receptors in the CNS (Szeto, 1991). The effects on incidence of HVSA, spectral-edge frequency, and theta activity were all abolished by concurrent i.c.v. administration of methylnaloxone (see Fig. 8).

Involvement of Other Neurotransmitters   This opiate-mediated EEG activation also appears to involve central muscarinic pathways (Szeto, 1991). These effects were successfully blocked by i.c.v. administration of methylatropine alone (see Fig. 8). This is consistent with previous studies in our laboratory, in which we found similar EEG activation with systemic administration of pilocarpine, a muscarinic agonist, to the fetal lamb (Hinman and Szeto, 1988). The effects of pilocarpine were blocked by i.v. administration of atropine, but not methylatropine, thus suggesting that the site of action is in the CNS. Interestingly, Horita et al. (1989) also reported that the morphine-induced EEG activation in the rabbit involved central cholinergic mechanisms, but apparently was not mediated by opiate receptors. The discrepancy between the findings in the fetal lamb and those in the adult rabbit is not understood.

Sleep–Wake Control

Disturbances in behavior and sleep–wake patterns have been reported in children who were exposed to opiates in utero (Dinges et al., 1980; Wilson et al., 1981). Since the opiates have been found to alter EEG in the fetal lamb, it was reasonable to propose that prenatal exposure to opiates may have direct effects on fetal behavioral activity or sleep–wake patterns.

Although it is possible to monitor human fetal activity using ultrasound techniques, it is technically very difficult to obtain long recordings without imposing a fair amount of stress on both mother and fetus. We have therefore addressed these questions in the fetal lamb. Even though fetal behavioral activity cannot be observed directly, the concurrent recording of fetal EEG, nuchal muscle EMG, and eye movements (electro-oculogram) has permitted us to utilize established polygraphic criteria to infer the behavioral state of the fetus (see Table 2) (Szeto and Hinman, 1985). With these polygraphic techniques, and scoring criteria, we have been able to monitor the ontogenetic development of organized sleep–wake patterns in the fetal lamb throughout the third trimester (Szeto and Hinman, 1985).

Organized sleep–wake states are established before 120 days of gestation. As in

TABLE 2   Criteria for Sleep-State Scoring
in the Fetal Lamb[a]

State	EEG	EMG	EOG
Arousal	LVFA	+++	Random movements
QS	HVSA	Tone	Slow/absent
REMS	LVFA	Atonia	REMS

[a]QS, quiet sleep; REMS, rapid eye movements; EOG, electrooculogram; EMG, electromyogram.

the human and other species, there is a predominance of REMS ($>$ 50%) in the fetal lamb as well as a tendency to enter REMS without an episode of QS. The sequence from arousal to REMS is the reverse of that in adults. A direct transition from the arousal state to REMS is also frequently seen in the human infant until 6 months of age (Roffwarg et al., 1966). With maturation, there is also a progressive increase in the time spent in arousal, which is accompanied by a significant decrease in REMS throughout the third trimester. Even so, the incidence of arousal periods is very low before birth, ranging only from 5 to 15%; these are generally of very short duration (2–3 min). In addition to the maturational changes in overall distribution of behavioral states, the duration of both QS and REMS increases with age until a stable level is achieved by 130 days of gestation in the fetal lamb (Szeto and Hinman, 1985). The durations of both QS and REMS average 5–15 min throughout the third trimester. In summary, studies to date have established that the transition from disorganized to organized behavioral state patterns occurs at approximately 115 to 120 days of gestation. However, quantitative changes in the distribution of these behavioral states, and in cycle duration, continue throughout the rest of gestation.

Dose–Response and Time–Action Relationships

Infusion of methadone to the mother, at a dose that has been shown to result in maternal plasma methadone levels comparable to those reported in methadone maintenance patients, resulted in a behavioral pattern that would be scored as arousal according to the criteria stated in Table 2 (Szeto, 1983). Within 30 min of the infusion, there was a transition to a LVFA EEG pattern, accompanied by enhanced tone and activity in the nuchal muscle, as well as random eye movements. This enhanced activity pattern persisted even after the methadone infusion was terminated, and the first QS cycle was not detected until 1.5 hr later. Alternating QS–REMS cycles of normal duration were not apparent until 3 hr or more after termination of methadone infusion. Thus methadone appears to induce a state of persistent arousal in the fetus, with a suppression of both QS and REMS.

This methadone-induced behavioral activation appears to be the result of a direct action of methadone on the fetus, since similar results were obtained when methadone was administered directly to the fetus (Szeto, 1983). Subsequent studies revealed that both maternal and fetal administration of morphine resulted in the same behavioral activity pattern (Umans and Szeto, 1983). However, higher doses of morphine resulted in increased synchronization of the fetal EEG accompanied by decreases in body, eye, and breathing movements and heart rate variability (Umans and Szeto, 1985). Thus it appears that low doses of opiates cause behavioral excitation in the fetus, whereas higher doses result in CNS depression. Both responses were reversible on termination of drug exposure.

Comparison with Effects in the Adults

Although it is generally thought that opiates produce CNS depression in the adult, a number of studies have actually reported behavioral excitation after opiate administration in adult animals. After either i.v. or i.c.v. administration of morphine in the dog or rabbit, a period of EEG desynchronization was frequently observed, associated with behavioral excitation, increased respiratory rate, and bradycardia (Albus and Herz, 1972; Novack et al., 1976). This was then followed by behavioral sedation and a predominantly synchronized EEG. These observations suggest that low doses of morphine may have excitatory actions in the adult as well. This is supported by more recent studies, which showed that low doses of morphine or opioid peptides can increase locomotor activity and respiratory activity (Brady and Holtzman, 1981; Locke and Holtzman, 1986; Paakkari et al., 1990). In addition, Horita et al. (1983, 1989) reported that both morphine and fentanyl can arouse rabbits from pentobarbital-induced narcosis.

Sites and Mechanisms of Action

Involvement of Specific Opiate Receptors   The mechanism behind the excitatory action of opiates in the adult is not understood, and both delta- (Locke and Holtzman, 1986) and mu_1-receptors (Paakkari et al., 1990) have been implicated. Our own evidence indicates that the excitatory response in the fetus is clearly mediated by specific opiate receptors in the CNS (Szeto, 1991), and these appear to be of the mu_1 subtype (unpublished data, 1992).

Involvement of other neurotransmitters   Horita et al. (1989) reported that the analeptic effect of morphine could be disrupted by i.v. atropine (but not methylatropine), suggesting that the opiates activated a central cholinergic arousal system. Interestingly, recent studies in our laboratory suggest that central muscarinic pathways are also involved in the behavioral excitatory effects of morphine in the fetus (Szeto, 1991). The only difference is that this appears to be an opiate receptor-mediated response in the fetal lamb but not in adult rabbits (Horita et al., 1989). A central cholinergic arousal system has been demonstrated in the fetal lamb (Hinman and Szeto, 1988). Administration of pilocarpine to the fetal lamb results in a behavioral excitation that can be blocked by i.v. atropine, but not methylatropine. The mechanism by which opiates activate central muscarinic pathways in the fetal lamb is not known. Morphine has been shown to stimulate cholinergic neuronal activity in the rat hippocampus (Vallano and McIntosh, 1980). It is possible that morphine may increase cholinergic neuronal activity in the hippocampus, and this may result in an arousal response.

Development of Tolerance with Prolonged Exposure

It is important to determine whether these neurobehavioral changes in the fetus persist with continuous intrauterine exposure to opiates. Persistent alterations in fetal behavioral activity may affect the normal maturation of sleep–wake behavior, as well as modify metabolic demands. When a low dose of morphine was administered i.v. to the fetus at a constant rate for a total of 7 days, we found that the excitatory effects of morphine declined rapidly within the first 24 hr, and sleep–wake patterns by day 3 of morphine exposure were not significantly different from control patterns (Szeto et al., 1988a). These results demonstrate that, as with the respiratory response, tolerance develops rapidly to the effects of morphine on sleep–wake control in the fetus. In fact, there does not appear to be any difference in the rate of tolerance development to the different effects of morphine. In contrast, during long-term subcutaneous administration of morphine to the adult, arousal and REM suppression persisted, though greatly attenuated, whereas complete tolerance developed to the decrease in delta sleep (Kay, 1975; Kay et al., 1981). However, tolerance did develop to most of the effects of methadone on sleep, following chronic exposure in methadone maintenance patients (Kay, 1975) and to the effects of morphine in the rat (Khazan et al., 1967). It should also be mentioned that this tolerance in the fetal lamb appeared to be a functional tolerance, rather than dispositional tolerance, as it was not associated with a decline in plasma morphine levels (Szeto et al., 1988a). Thus it appears that these neurobehavioral changes do not persist with continuous intrauterine drug exposure, and the development of tolerance may serve to protect the fetus from irreversible changes in the development of sleep–wake control. However, it should be pointed out that in these studies, morphine exposure did not begin until after organized sleep–wake patterns had been established. In the human, however, fetal drug exposure would have started before the development of organized behavioral patterns. It is not known whether responses in the fetal lamb would be different if drug exposure were started earlier in gestation.

INTRAUTERINE WITHDRAWAL

As in the adult, the development of tolerance in the fetus is accompanied by the development of physical dependence. Physical dependence to morphine was manifested by a mild abstinence syndrome characterized by disturbances in sleep–wake regulation and rapid, deep breathing activity (Szeto et al., 1988a). This abstinence syndrome is not so severe as that following naloxone precipitation, which consisted of intense total body movements, neck tone, and eye movements; continuous, rapid, deep breathing movements; and an immediate bradycardia associated with transient increase in systolic, diastolic, and pulse pressures (Umans and Szeto, 1985). The changes in sleep–wake regulation included a decrease in REMS, increase in arousal time, and shortened sleep–wake cycle time. This resulted in a rapid alternation pattern between arousal and QS. This syndrome resembles the one described in neonates born to opiate-dependent mothers, in that there is increased body and eye movements, diarrhea, tachypnea, and hyperpnea (Glass et al., 1972; Schulman, 1969).

ROLE OF ENDOGENOUS OPIATE PEPTIDES IN CONTROL OF FETAL BREATHING

It has been common belief that the endogenous opiate peptides play an important role in the control of breathing in adults. Based on the profound effects that morphine has on fetal breathing, it would not be unreasonable to think that the endogenous opiate peptides may also play a role in the control of fetal breathing. However, administration of naloxone generally failed to affect breathing patterns of respiratory responses to hypoxia and hypercapnia in adult animals and humans (Fleetham et al., 1980; Weinberger et al., 1985), except when employed in very high doses (Cohen et al., 1982). Thus the role of physiological levels of endorphins on ventilatory control in the adult remains unclear. Likewise, previous studies on the effects of naloxone in the fetal sheep also did not show any significant effect on the incidence of FBM (Adamson et al., 1984; Joseph et al., 1987).

Although naloxone appears to be ideal for investigating the role of endogenous opiate peptides, since it freely crosses the blood–brain barrier and has no apparent agonist actions of its own, a growing concern with the use of naloxone is its lack of selectivity for the various opiate receptor subtypes. Whereas naloxone is considered to have some selectivity for the mu-receptor, its affinity for the mu site is only fourfold to tenfold greater than for the delta site and slightly more than for the kappa site (Paterson et al., 1983). As we have already demonstrated the complex actions of the various opiate receptor subtypes in control of fetal breathing, nonselective antagonism may block opposing actions and thus show no net effect. Thus more selective opiate antagonists are necessary in order to study the role of the endogenous opiate peptides at each receptor subtype.

When naloxonazine was administered to the fetal lamb under normal physiological conditions, it significantly increased the number of apneic periods and respiratory pauses, resulting in a more discontinuous or fragmented FBM pattern (Cheng et al., 1991). In addition, naloxonazine significantly decreased the overall stability or regularity of the breathing rate. In contrast, naloxone had no significant effect on any of the breathing parameters (Cheng et al., 1991). These data suggest that the endogenous opiate peptides play a tonic role at the mu_1-receptor to maintain both the stability and the continuity of the breathing pattern.

CLINICAL IMPLICATIONS

Adverse Effects of Prenatal Opiate Exposure on Fetal Growth and Development

These studies have clearly demonstrated that opiate exposure can profoundly affect the fetal nervous system. Furthermore, they demonstrate that two very different response clusters can result, depending on the extent of fetal drug exposure. Low doses of morphine and methadone result in CNS excitation, with desynchronization of the EEG, increased body movements and eye movements, continuous breathing movements, and increased heart rate and heart rate variability. This response profile is associated with increased blood flow to many regions of the fetal brain, as well as increased blood flow to the heart and the diaphragm (Olsen et al., 1983). There is also an increase in oxygen consumption by the fetus (unpublished data, 1992), indicating increased metabolic demand. Higher doses result in CNS depression, with increased synchronization of the EEG accompanied by decreases in body, eye and breathing movements, and heart rate variability. Metabolic demand is most likely to be decreased under these conditions. Because of this dual nature of opioid effects in the fetus, long-term outcome may depend greatly on the magnitude of fetal drug exposure. Persistent alterations in fetal behavioral states may affect the growth, development, and function of the brain. We have found that cerebral metabolism in the fetus is very different during desynchronized versus synchronized EEG activity (Clapp et al., 1980). Long-term alterations in cerebral metabolism may have an effect on the development of the CNS. Increase in metabolic demands with low doses of opiates may also deprive the fetus of energy substrates for growth and result in intrauterine growth retardation.

Long-term outcome may also depend on the particular opiate drug that is used, since we have previously shown that the extent of fetal exposure to methadone is three times greater than that to morphine (Szeto et al., 1982). The lower exposure to morphine is predominantly owing to a more restricted placental clearance because of its relative polarity compared to methadone (Szeto et al., 1982).

The frequency of opiate exposure may be another important factor in determining the long-term outcome of prenatal opiate exposure. Our findings indicate that tolerance develops relatively quickly if drug consumption is constant (Szeto et al., 1988a). In this case, the fetus adapts rapidly, and there is no apparent effect on the nervous system, and consequently no changes in metabolic demand. However, with less frequent drug exposure, tolerance may develop more slowly and to a lesser extent, and the effect on the fetus would be more sustained. On the other hand, less frequent drug exposure may also be accompanied by intermittent fetal withdrawal, which is clearly also a highly stressful situation (Szeto et al., 1988a). Thus, the consequences of opiate abuse on development of the fetus is highly complicated, and dependent on many variables.

Use of Opiate Antagonists in the Treatment of Infant Apnea

It has been proposed that elevated levels of endogenous opiate peptides may be associated with increased incidence of apnea in early development (Orlowski, 1982; Myer et al., 1990). Several investigators have reported increased plasma and cerebrospinal fluid levels of immunoreactive β-endorphin in infants with sleep apnea and apnea of prematurity (Orlowski, 1982, 1986; Sankaran et al., 1984; MacDonald et al., 1986; Myer et al., 1990). Furthermore, administration of naloxone has been reported to reduce the number of apneas in infants with sleep apnea who had high cerebrospinal fluid endorphin levels (Orlowski, 1986). Similar success has been reported in a subsequent study with naltrexone (Myer et al., 1990). However, other studies reported no significant change in respiratory pattern or occurrence of apnea following opiate antagonist treatment in infants with apnea of prematurity (MacDonald et al., 1986) or infants at risk for sudden infant death syndrome (Haidmayer et al., 1986).

Based on our studies of the role of different opiate receptor subtypes in the control of fetal breathing, these conflicting results may result from the nonselectivity of naloxone, and more consistent results might be obtained if more selective antagonists were used rather than naloxone or naltrexone. Our results may even suggest that the use of a selective mu_1 agonist may be beneficial. It is unfortunate that there are currently no selective mu_1 agonists available to test this idea in an animal model.

REFERENCES

Adamson, S. L., Patrick, J. E., and Challis, J. R. G. (1984). Effects of naloxone on the breathing, electrocortical, heart rate, glucose, and cortisol responses to hypoxia in the sheep fetus. *J. Dev. Physiol.* **6,** 495–507.

Albus, K., and Herz, A. (1972). Inhibition of behavioral and EEG activation induced by morphine acting on lower brain stem structures. *Electro-encephalogr. Clin. Neurophysiol.* **33,** 579–590.

Appel, N. M., and Van Loon, G. R. (1986). β-Endorphin-induced stimulation of central sympathetic outflow: Inhibitory modulation by central noradrenergic neurons. *J. Pharm. Exp. Ther.* **237,** 695–701.

Bahoric, A., and Chernick, V. (1975). Electrical activity of phrenic nerve and diaphragm in utero. *J. Appl. Physiol.* **39,** 513–518.

Barcroft, J. (1946). "Researches in Pre-Natal Life." Blackwell Scientific Publications, Oxford.

Bennet, L., Johnston, B. M., and Gluckman, P. D. (1986). The central effects of morphine on fetal breathing movements in the fetal sheep. *J. Dev. Physiol.* **8,** 297–305.

Brady, L. S., and Holtzman, S. G. (1981). Effects of intraventricular morphine and enkephalins on locomotor activity in nondependent, morphine-dependent and post-dependent rats. *J. Pharmacol. Exp. Ther.* **218,** 613–620.

Breckenridge, C. G., and Hoff, H. E. (1952). Influence of morphine on respiratory patterns. *J. Neurophysiology* **15,** 57–74.

Brown, E. R., Lawson, E. E., Jansen, A., Chernick, V., and Taeusch, H. W. (1981). Regular fetal breathing induced by pilocarpine infusion in the near-term fetal lamb. *J. Appl. Physiol.* **50,** 1348–1352.

Cheng, P. Y., Decena, J. A., Wu, D. L., Cheng, Y., and Szeto, H. H. (1991). The effects of selective mu_1-opiate receptor blockade on breathing patterns in the fetal lamb. *Pediatr. Res.* **30,** 202–206.

Clapp, J. F., Szeto, H. H., Abrams, R., Larrow, R., and Mann, L. I. (1980). Physiologic variability and fetal electrocortical activity. *Am. J. Obstet. Gynecol.* **136,** 1045–1050.

Clark, J. A., Itzhak, Y., Hruby, V. J., Yamamura, H. I., and Pasternak, G. W. (1986). [D-Pen^2,D-Pen^5]en-kephalin (DPDPE): a δ-selective enkephalin with low affinity for μ_1-opiate binding sites. *Eur. J. Pharmacol.* **128**, 303–304.

Cohen, M. R., Cohen, R. M., Pickar, D., Murphy, D. L., and Bunney, W. E. (1982). Physiological effects of high-dose naloxone administration to normal adults. *Life Sci.* **30**, 2025–2031.

Cowan, A., Lewis, J. W., and MacFarlane, I. R. (1977). Agonist and antagonist properties of buprenorphine, a new antinociceptive agent. *Br. J. Pharmacol.* **60**, 537–545.

Dawes, G. S., Fox, H. E., Leduc, B. M., Liggins, G. S., and Richards, R. T. (1972). Respiratory movements and rapid-eye-movement sleep in the fetal lamb. *J. Physiol. (London).* **220**, 119–143.

Dawes, G. S., Gardner, W. N., Johnston, B. M., and Walker, D. W. (1982). Effects of hypercapnia on tracheal pressure, diaphragm and intercostal electromyograms in unanesthetized fetal lambs. *J. Physiol. (London)* **326**, 461–474.

Dinges, D. F., Davis, M. M., and Glass, P. (1980). Fetal exposure to narcotics: Neonatal sleep as a measure of nervous system disturbance. *Science* **209**, 619–621.

Dunlap, C. E., Christ, G. J., and Rose, J. C. (1985). Characterization of opioid receptor binding in adult and fetal sheep brain regions. *Dev. Brain Res.* **24**, 279–285.

Evanich, M. J., Sander, G. E., Rice, J. C., and Giles, T. D. (1985). Ventilatory response to intraveneous methionine enkephalin in awake dogs. *J. Pharmacol. Exp. Ther.* **234**, 677–680.

Feldberg, W., and Wei, E. (1986). Analysis of cardiovascular effects of morphine in the cat. *Neuroscience* **17**, 495–506.

Feuerstein, G. and Faden, A. I. (1982). Differential cardiovascular effects of mu and kappa opiate agonists at discrete hypothalamic sites in the anesthetized rat. *Life Sci.* **31**, 2197–2200.

Fleetham, J. A., Clarke, H., Dhingra, S., Chernick, V., and Anthonisen, N. R. (1980). Endogenous opiates and chemical control of breathing in humans. *Am. Rev. Resp. Dis.* **121**, 1045–1049.

Florez, J., and Borison, H. L. (1969). Effects of central depressant drugs on respiratory regulation in the decerebrate cat. *Resp. Physiol.* **6**, 318–329.

Florez, J., McCarthy, L. E., and Borison, H. L. (1968). A comparative study in the cat of the respiratory effects of morphine injected intravenously and into the cerebrospinal fluid. *J. Pharmacol. Exp. Ther.* **163**, 448–455.

Giles, F. D., and Sander, G. E. (1983). Mechanism of the cardiovascular response to systemic intravenous administration of Leu-enkephaline in the conscious dog. *Peptides* **4**, 171–175.

Given, M. B., Sander, G. E., and Giles, T. D. (1986). Non-opiate and peripheral-opiate cardiovascular effects of morphine in conscious dogs. *Life Sci.* **38**, 1299–1303.

Glass, L., Rajegowda, B. K., and Evans, H. E. (1972). Effects of heroin withdrawal on respiratory rate and acid–base status in the newborn. *N. Engl. J. Med.* **186**, 746–748.

Haddad, G. G., Schaeffer, J. I., and Chang, K. J. (1984). Opposite effects of the delta- and mu-receptor agonists on ventilation in conscious adult dogs. *Brain Res.* **323**, 73–82.

Hahn, E. F., and Pasternak, G. W. (1982). Naloxonazine, a potent, long-lasting inhibitor of opiate-binding sites. *Life Sci.* **31**, 1385–1388.

Haidmayer, R., Kerbl, R., Meyer, U., Kershhaggl, P., Kurz, R., and Kenner, T. (1986). Effects of naloxone on apnoea duration during sleep in infants at risk for SIDS. *Eur. J. Pediatr.* **145**, 357–360.

Hassen, A. H., St. John, W., and Wang, S. C. (1976). Selective respiratory depressant action of morphine compared to meperidine in the cat. *Eur. J. Pharmacol.* **39**, 61–70.

Hassen, A. H., Feuerstein, G., and Faden, A. I. (1983). Differential cardiovascular effects mediated by mu- and kappa-opiate receptors in hindbrain nuclei. *Peptides* **4**, 621–625.

Hecker, J. F. (1983). "The Sheep as an Experimental Model." Academic Press, London.

Hinman, D. J., and Szeto, H. H. (1988). Cholinergic influences on sleep–wake patterns and breathing movements in the fetus. *J. Pharmacol. Exp. Ther.* **247**, 372–378.

Holaday, J. W. (1982). Cardiorespiratory effects of μ- and δ-opiate agonists following third or fourth ventricle injections. *Peptides* **3**, 1023–1029.

Holaday, J. W., Pasternak, G. W., and Faden, A. I. (1983). Naloxazone pretreatment modifies cardiorespiratory, temperature, and behavioral effects of morphine. *Neurosci. Lett.* **37**, 199–204.

Horita, A., Carino, M. A., and Yamawaki, S. (1983). Morphine antagonizes pentobarbital-induced anesthesia. *Neuropharmacology* **22**, 1183–1186.

Horita, A., Carino, M. A., and Chinn, C. (1989). Fentanyl produces cholinergically mediated analeptic and EEG arousal effects in rats. *Neuropharmacology* **28**, 481–486.

Jansen, A. H., and Chernick, V. (1983). Development of respiratory control. *Physiol. Rev.* **63**, 437–483.

Joseph, S. A., McMillen, I. C., and Walker, D. W. (1987). Effects of naloxone on breathing movements during hypercapnia in the fetal lamb. *J. Appl. Physiol.* **62**, 673–678.

Kay, D. C. (1975). Human sleep during chronic morphine intoxication. *Psychopharmacologia* **44**, 117–124.

Kay, D. C., Pickworth, W. B. and Neidert, G. L. (1981). Morphine-like insomnia from heroin in nondependent human addicts. *Br. J. Clin. Pharmacol.* **11**, 159–69.

Khazan, N., Weeks, J. R., and Schroeder, L. A. (1967). Electroencephalographic, electromyographic, and behavioral correlates during a cycle of self-maintaining morphine addiction in the rat. *J. Pharmacol. Exp. Ther.* **155**, 521–531.

Kiritsy-Roy, J. A., Appel, N. M., Bobbitt, F. G., and Van Loon, G. R. (1986). Effects of mu-opioid receptor stimulation in the hypothalamic paraventricular nucleus on basal and stress-induced catecholamine secretion and cardiovascular responses. *J. Pharmacol. Exp. Ther.* **239**, 814–822.

Kiritsy-Roy, J. A., Marson, L., and Van Loon, G. R. (1989). Sympathoadrenal, cardiovascular, and blood gas responses to highly selective mu- and delta-opioid peptides. *J. Pharmacol. Exp. Ther.* **251**, 1096–1103.

Kitterman, J. A., Liggins, G. C., Clements, J. A., and Tooley, W. H. (1979). Stimulation of breathing movements in fetal sheep by inhibitors of prostaglandin synthesis. *J. Dev. Physiol.* **1**, 453–466.

Kosterlitz, H. W., Lord, J. A. H., Paterson, S. J., and Waterfield, A. A. (1980). Effects of changes in the structure of enkephalins and of narcotic analgesic drugs on their interactions with μ- and δ-receptors. *Br. J. Pharmacol.* **68**, 333–342.

Ling, G. S. F., Spiegel, K., Lockhart, S. H., and Pasternak, G. W. (1985). Separation of opioid analgesia from respiratory depression: Evidence for different receptor mechanisms. *J. Pharmacol. Exp. Ther.* **232**, 149–155.

Locke, K. W., and Holtzman, S. G. (1986). Behavioral effects of opioid peptides selective for mu- or delta-receptors. Locomotor activity in nondependent and morphine-dependent rats. *J. Pharmacol. Exp. Ther.* **238**, 997–1003.

MacDonald, M. G., Moss, I. R., Kefale, G. G., Ginzburg, H. M., Fink, R. J., and Chin, L. (1986). Effect of naltrexone on apnea of prematurity and on plasma beta-endorphinlike immunoreactivity. *Dev. Pharmacol. Ther.* **9**, 301–309.

Martin, C. B., Murata, V., Petrie, R. H., and Parer, J. T. (1974). Respiratory movements in fetal rhesus monkey. *Am. J. Obstet. Gynecol.* **119**, 939–948.

Martin, W. R. (1983). Pharmacology of opioids. *Pharmacol. Rev.* **35**, 283–323.

Martin, W. R., and Kay, D. C. (1977). Effects of opioid analgesics and antagonists on the EEG. In "Handbook of Electroencephalography and Clinical Neurophysiology" (V. G. Longo, ed.), Vol. 7, pp. 7C-97 to 7C-132. Elsevier/North Holland, Amsterdam.

Matejcek, M., Pokorny, R., Ferber, G., and Klee, H. (1988). Effect of morphine on the electroencephalogram and other physiological and behavioral parameters. *Neuropsychobiology* **19**, 202–211.

May, C. N., Dashwood, M. R., Whitehead, C. J., and Mathias, C. J. (1989). Differential cardiovascular and respiratory responses to central administration of selective opioid agonists in conscious rabbits: Correlation with receptor distribution. *Br. J. Pharmacol.* **98**, 903–913.

McNerney, M. E., and Szeto, H. H. (1990). Automated identification and quantitation of four patterns of electrocortical activity in the near-term fetal lamb. *Pediatr. Res.* **28**, 106–110.

Millhorn, D. E., Eldridge, F. L., Kiley, J. P., and Waldrop, T. G. (1985). Excitatory and inhibitory effects of morphine on the intercostal-to-phrenic respiratory reflex. *Respir. Physiol.* **62**, 79–84.

Moore, K. E., McCarthy, L. E., and Borison, H. L. (1965). Blood glucose and brain catecholamine levels in the cat following the injection of morphine into the cerebrospinal fluid. *J. Pharm. Exp. Ther.* **148**, 169–175.

Mosberg, H. I., Hurst, R., Hruby, V. J., Gee, K., Yamamura, H. I., Galligan, J. J., and Burks, T. F. (1983).

Bis-penicillamine enkephalins possess highly improved specificity toward opioid receptors. *Proc. Natl. Acad. Sci. U.S.A.* **80**, 5871–5874.

Myer, E. C., Morris, D. L., Brase, D. A., Dewey, W. L., and Zimmerman, A. W. (1990). Naltrexone therapy of apnea in children with elevated cerebrospinal beta-endorphin. *Ann. Neurol.* **27**, 75–80.

Novack, G. D., Winter, W. D., and Nakamura, J. (1976). EEG and behavioral effects of morphine in dogs. *Proc. West. Pharmacol. Soc.* **19**, 239–242.

Olsen, G. D., and Dawes, G. S. (1983). Morphine effects on fetal lambs. *Fed. Proc.* **42**, 1251.

Olsen, G. D., Hohimer, A. R., and Mathis, M. D. (1983). Cerebral blood flow and metabolism during morphine-induced stimulation of breathing movements in fetal lambs. *Life Sci.* **33**, 751–754.

Orlowski, J. P. (1982). Endorphins in infant apnea. *N. Engl. J. Med.* **307**, 186–187.

Orlowski, J. P. (1986). Cerebrospinal fluid endorphins and the infant apnea syndrome. *Pediatrics* **78**, 233–237.

Paakkari, P., Paakkari, I., Siren, A. L., and Feuerstein, G. (1990). Respiratory and locomotor stimulation by low doses of dermorphin, a mu$_1$-receptor-mediated effect. *J. Pharmacol. Exp. Ther.* **252**, 235–240.

Paterson, S. J., Robson, L. E., and Kosterlitz, H. W. (1983). Classification of opioid receptors. *Br. Med. Bull.* **39**, 31–36.

Patrick, J., Campbell, K., Carmichael, L., Natale, R., and Richardson, B. (1980). Patterns of human fetal breathing during the last 10 weeks of pregnancy. *Obstet. Gynecol.* **56**, 24–30.

Patrick, J., Carmichael, L., Richardson, B., Smith, G., Homan, J., and Brien, J. (1988) Effects of multiple-dose maternal ethanol infusion on fetal cardiovascular and brain activity in lambs. *Am. J. Obstet. Gynecol.* **159**, 1424–1429.

Pfeiffer, A., Feuerstein, G., Kopin, I. J., and Faden, A. I. (1983a). Cardiovascular and respiratory effects of mu-, delta-, and kappa-opiate agonists microinjected into the anterior hypothalamic brain area of awake rats. *J. Pharmacol. Exp. Ther.* **225**, 735–741.

Pfeiffer, A., Feuerstein, G., Zerbe, R. L., Faden, A., and Kopin, I. J. (1983b). μ-receptors mediate opioid cardiovascular effects at anterior hypothalamic sites through sympatho-adrenomedullary and parasympathetic pathways. *Endocrinology* **113**, 929–38.

Portoghese, P. S., Sultana, M., and Takemori, A. E. (1988). Naltrindole, a highly selective non-peptide delta opioid receptor antagonist. *Eur. J. Pharmacol.* **146**, 185–186.

Rey, H. R., Stark, R. I., Kim, Y. I., Daniel, S. S., MacGregor, G., and James, L. S. (1989). Method of processing of fetal breathing epoch analysis: Studies in the primate. *IEEE Eng. Med. Biol.* **8**, 30–42.

Richardson, B., Hohimer, A. R., Mueggler, P., and Bissonnette, J. (1982). Effects of glucose concentration on fetal breathing movements and electrocortical activity in fetal lambs. *Am. J. Obstet. Gynecol.* **142**, 678–683.

Roffwarg, H. P., Muzio, J. N., and Dement, W. C. (1966). Ontogenetic development of the human sleep–dream cycle. *Science* **152**, 604–619.

Rothman, R. B., and Westfall, T. C. (1983). Further evidence for an opioid receptor complex. *J. Neurobiol.* **14**, 341–351.

Sankaran, K., Hindmarsh, K. W., and Watson, V. G. (1984). Plasma beta-endorphin concentration in infants with apneic spells. *Am. J. Perinatol.* **1**, 331–334.

Schulman, C. A. (1969). Alterations in the sleep cycle in heroin-addicted and "suspect" newborns. *Neuropediatrics* **1**, 89–100.

Sheldon, R. E., and Toubas, P. L. (1984). Morphine stimulates rapid, regular, deep, and sustained breathing efforts in fetal lambs. *J. Appl. Physiol.* **57**, 40–43.

Simon, W., Schaz, K., Ganten, U., Stock, G., Schlor, K. H., and Ganten, D. (1978). Effects of enkephalins on arterial blood pressure are reduced by propranolol. *Clin. Sci. Mol. Med.* **55**, 237s–241s.

Sofuglu, M., Portoghese, P. S., and Takemori, A. E. (1991). Differential antagonism of delta opioid agonists by naltrindole and its benzofuran analog (NTB) in mice: Evidence for delta opioid receptor subtypes. *J. Pharmacol. Exp. Ther.* **257**, 676–680.

Szeto, H. H. (1983). Effects of narcotic drugs on fetal behavioral activity: Acute methadone exposure. *Am. J. Obstet. Gynecol.* **146**, 211–217.

Szeto, H. H. (1990). Spectral edge frequency as a simple quantitative measure of the maturation of electrocortical activity. *Pediatr. Res.* **27**, 289–292.

Szeto, H. H. (1991). Morphine-induced activation of fetal EEG is mediated via central muscarinic pathways. *Am. J. Physiol.* **260**, R509–R517.

Szeto, H. H., and Hinman, D. J. (1985). Prenatal development of sleep–wake patterns in sheep. *Sleep* **8**, 347–355.

Szeto, H. H., and Hinman, D. J. (1990). Central muscarinic modulation of fetal blood pressure and heart rate. *J. Dev. Physiol.* **13**, 17–23.

Szeto, H. H., Mann, L. I., Bhakthavathsalan, A., Liu, M., and Inturrisi, C. E. (1978). Meperidine pharmacokinetics in the maternal–fetal unit. *J. Pharmacol. Exp. Ther.* **206**, 448–459.

Szeto, H. H., Clapp, J. F., Larrow, R. W., Hewitt, J., Inturrisi, C. E., and Mann, L. I. (1981). Disposition of methadone in the ovine maternal–fetal unit. *Life Sci.* **28**, 2111–2117.

Szeto, H. H., Umans, J. G., and McFarland, J. W. (1982). A comparison of morphine and methadone disposition in the maternal–fetal unit. *Am. J. Obstet. Gynecol.* **143**, 700–706.

Szeto, H. H., Vo, T. D., Dwyer, G., Dogramajian, M. E., Cox, M., and Senger, G. (1985). The ontogeny of fetal lamb electrocortical activity: A power spectral analysis. *Am. J. Obstet. Gynecol.* **153**, 462–468.

Szeto, H. H., Zhu, Y. S., Amione, J., and Clare, S. (1988a). Prenatal morphine exposure and sleep–wake disturbances in the fetus. *Sleep* **11**, 121–130.

Szeto, H. H., Zhu, Y. S., Umans, J. G., Dwyer, G., Clare, S., and Amione, J. (1988b). Dual action of morphine on fetal breathing movements. *J. Pharmacol. Exp. Ther.* **245**, 537–542.

Szeto, H. H., Zhu, Y. S., and Cai, L. Q. (1990). Central opioid modulation of fetal cardiovascular function: Role of μ- and δ-receptors. *Am. J. Physiol.* **258**, R1453–R1458.

Szeto, H. H., Cheng, P. Y., Dwyer, G., Decena, J. A., Wu, D. L., and Cheng, Y. (1991). Morphine-induced stimulation of fetal breathing: Role of mu_1-receptors and central muscarinic pathways. *Am. J. Physiol.* **261**, R344–R350.

Toubas, P. L., Pryor, A. L., and Sheldon, R. E. (1985). Effect of morphine on fetal electrocortical activity and breathing movements in fetal sheep. *Dev. Pharmacol. Ther.* **8**, 115–128.

Umans, J. G., and Szeto, H. H. (1983). Effects of opiates on fetal behavioral activity *in utero*. *Life Sci.* **33**, 639–642.

Umans, J. G., and Szeto, H. H. (1985). Precipitated opiate abstinence *in utero*. *Am. J. Obstet. Gynecol.* **151**, 441–444.

Vallano, M. L., and McIntosh, T. K. (1980). Morphine stimulates cholinergic neuronal activity in the rat hippocampus. *Neuropharmacology* **19**, 851–853.

Van Loon, G. R., Appel, N. M., and Ho, D. (1981). β-Endorphin-induced stimulation of central sympathetic outflow: β-Endorphin increases plasma concentrations of epinephrine, norepinephrine, and dopamine in rats. *Endocrinology* **109**, 46–53.

Vo, T. D. H., Dwyer, G., and Szeto, H. H. (1986). A distributed microcomputer-controlled system for data acquisition and power spectral analysis of EEG. *J. Neurosci. Methods* **16**, 141–148.

Weinberger, S. E., Steinbrook, R. A., Carr, D. B., von Gal, E. R., Fisher, J. E., Leith, D. E., Fencl, V., and Rosenblatt, M. (1985). Endogenous opioids and ventilatory response to hypercapnia in normal humans. *J. Appl. Physiol.* **58**, 1415–1420.

Willette, R. N., and Sapru, H. N. (1982). Peripheral versus central cardiorespiratory effects of morphine. *Neuropharmacology* **21**, 1019–1026.

Wilson, G. S., Desmond, M. M., and Wait, R. B. (1981). Follow-up of methadone-treated and untreated narcotic-dependent women and their infants: Health, developmental, and social implications. *J. Pediatr.* **98**, 716–722.

Wolozin, B. J., and Pasternak, G. W. (1981). Classification of multiple morphine- and enkephalin-binding sites in the central nervous system. *Proc. Natl. Acad. Sci. U.S.A.* **78**, 6181–6185.

Yeadon, M., and Kitchen, I. (1989). Opioids and respiration. *Prog. Neurobiol.* **33**, 1–16.

Young, G. A., and Khazan, N. (1984). Differential neuropharmacological effects of mu-, kappa-, and sigma-opiate agonists on cortical EEG power spectra in the rat. Stereospecificity and naloxone antagonism. *Neuropharmacology* **23**, 1161–1165.

Zhu, Y. S., and Szeto, H. H. (1989). Morphine-induced tachycardia in fetal lambs: A bell-shaped dose–response curve. *J. Pharmacol. Exp. Ther.* **249**, 78–82.

— 11 —

Maternal Exposure to Opioids and the Developing Nervous System: Laboratory Findings

ે**

Ian S. Zagon and Patricia J. McLaughlin

Department of Neuroscience and Anatomy
Pennsylvania State University
M. S. Hershey Medical Center
Hershey, Pennsylvania

INTRODUCTION

Human consumption of opium for religious, social, medicinal, and/or personal reasons dates back over 6000 years (Blum, 1970; Terry and Pellens, 1970; Musto, 1973). An important corollary to opioid (the term is used in the generic sense and refers to exogenous and endogenous substrates, natural or synthetic in origin, that possess opium- or morphinelike properties; Jaffe and Martin, 1980; Wikler, 1980) consumption is the passive dependence of the fetus, neonate, and infant when this drug is consumed by pregnant or nursing women. Precisely how long the relationship between opium and dependence of the fetus/neonate/infant has been perceived is not clear, but Hippocrates (see Martin, 1893) mentions opium in connection with "uterine suffocation"; this may be a reference to the toxicity of opium on the embryo and/or fetus. Clearly, by the end of the nineteenth century, clinical reports evidenced the unusual behavior of the fetus and the neonate passively addicted to opium (see Zagon et al., 1982 for references). Féré (1883) and Fürst (1889) first noted clinical information about opioids and human development, making the observation that violent kicking of the fetus occurred when the mother was denied opioids. Infants with "chronic opium intoxication," "congenital morphinism," or "congenital narcotic addiction" were reported to exhibit excessive nervousness, rapid breathing, and convulsive movements right after birth, with death occurring within the first few days of life (Terry and Pellens, 1970). A review of the early literature reveals an impressive amount of information and insight into the field of perinatal opioid dependence. For example, in 1895, Bureau found that morphine could pass through the human placenta. Other workers recognized the possibilities that paternal drug addiction might influence the fetus (Sainsbury, 1909), and that a sufficient amount of opium passed into breast milk to alleviate withdrawal (Laase, 1919; Langstein, 1930; Petty, 1912, 1913; Van Kleek, 1920) or even to establish drug dependency in a normal infant breast fed by a woman who became dependent on opioids after parturition (Lichenstein, 1915). The importance of therapeutic intervention by administration of morphine, heroin, or paregoric to prevent withdrawal was also well recognized by the beginning of the twentieth century (Menninger-Lerchenthal, 1934; Terry and Pellens, 1970). Even the long-term implications of perinatal opioid exposure were noted by these early investigators. For example, in a discussion following the presentation of a paper by Earle (1888) concerning the adverse effects of fetal opioid exposure, a number of participants reported the association of neurobiological abnormalities in offspring (including individuals reaching adulthood) maternally exposed to opium.

The incidence of perinatal exposure to opioids is difficult to gauge. Perlstein (1947) and Petty (1913) refer to the "rare" occurrence of such situations in earlier periods (up to the 1940s). However, Graham-Mulhall (1926) recorded over 800 pregnant women consuming opioids in New York during a 1-year period in the early 1900s. The type of opioid abused has changed over the years. Until the 1950s, morphine appeared to be the drug of choice, with the 1956 report by Goodfriend et al. signaling a change to heroin. Following Dole and Nyswander's (1965) advocation of the methadone maintenance treatment program as an alternative to heroin dependency, numerous reports have documented offspring dependent on methadone. Paralleling the change from morphine to the utilization of heroin and methadone in the 1950s, a marked increase in the number of births to opioid-dependent women was recorded. Estimates by Carr (1975) place the births to heroin- and methadone-addicted mothers at approximately 3000/year in New York City alone. Assuming that New York City has one third to one half of the total number of chronic heroin and methadone users in the United States (Carr, 1975; Salerno, 1977), one could extrapolate 6000 to 9000 births/year to opium-consuming women. Placed within the context of roughly 3.3 million births every year in the United States, at least 1 in 1000 births involves a mother using heroin or methadone. The incidence of the perinatal opioid syndrome, as it has been termed (see Zagon, 1985), rivals estimates for many well-known and highly publicized problems of early life, including Down's syndrome, neural tube defects, and the fetal alcohol syndrome, and it far exceeds the incidence of cancer in children aged 1 to 15 years. Thus, over the last 30 to 40 years, it is quite possible that over 250,000 infants, children, and adults, or 1 in 1000 people, have been exposed to opium in early life in the United States. By themselves these numbers are significant. However, unreported use of opioids by pregnant women and the possible influence exerted by paternal opioid consumption may indicate an error of underestimation.

CLINICAL OBSERVATIONS

The clinical findings regarding the fetus, neonate, and infant offspring born of mothers consuming opioid substances have been reviewed in another chapter in this book, as well as in earlier reports (see the bibliographies of Zagon et al., 1982, 1984, 1989 for citations; also reviews by Zagon and McLaughlin, 1984b; Wilson, 1992; Kaltenbach and Finnegan, 1992). However, a brief review of clinical observations is included in order to place the laboratory findings in perspective.

Even before birth, the offspring of opioid-dependent women are subjected to a host of potential problems. During the course of pregnancy, the mothers of these children often encounter complications, including infectious diseases, nutritional deficits, and an abnormal incidence of venereal disease. Obstetrical complications include toxemia of pregnancy and intrauterine growth retardation. Additionally, spontaneous miscarriage, abortion, and stillbirths have been suspected of being higher than normal among opioid-dependent women (Salerno, 1977). Many of these coassociated medical and obstetrical complications appear to be secondary to the life style and habits of the pregnant addict (Perlmutter, 1974), with prenatal care often being neglected. The fact that enrollment in the methadone program places women in touch with health professionals who encourage prenatal and postnatal care has been an extremely positive feature of this program.

A decrease in birth weight of infants born to heroin-dependent mothers has been recorded (Finnegan et al., 1972; Wilson et al., 1979), with low-birth-weight infants often being small for gestational age. These growth delays associated with prenatal heroin exposure do not appear to be related to inadequate nutrition or prenatal care (e.g., Naeye et al., 1973). In contrast to heroin-exposed offspring, methadone-dependent neonates have higher birth weights and are of greater gestational ages, although mean weight is often less than that of nondrug-dependent controls (Kandall et al., 1974).

Conclusive evidence that maternal opioid dependency is related to congenital malformations has not been documented, although some incidences (e.g., congenital heart defects, inguinal hernias) have been reported (e.g., Ostrea and Chavez, 1979). Abrams (1975) and Amarose and Norusis (1976) recorded increased chromosomal aberrations in infants exposed to heroin, but not necessarily in those exposed to methadone (Abrams, 1975), whereas Chavez et al. (1979) noted an increased incidence of sudden infant death syndrome in opioid-exposed infants. A lower incidence and severity of neonatal jaundice is known to occur for heroin-exposed infants (Zelson et al., 1973), and premature infants of heroin-dependent mothers often demonstrate a lower incidence of respiratory distress syndrome (Glass et al., 1971). Van Baar and colleagues (1989a) found that more infants of drug-dependent mothers had electroencephalograms rated as suspect or abnormal than did control children.

The proportion of stillbirths and mortality in the opioid-exposed population generally appears to be increased in comparison to those in control data (Ostrea and Chavez, 1979; Perlmutter, 1967), but the higher mortality rate may not be necessarily related to the neonatal withdrawal syndrome (Ostrea and Chavez, 1979). Few studies have pursued details of postmortem examination of drug-dependent infants, although evidence from neuropathological studies (e.g., Rorke et al., 1977) suggests that, in addition to a host of nonspecific secondary gestational complications, primary and specific effects of addictive drugs on the developing nervous system occur.

Perhaps the most well recognized sign of fetal exposure to opioids is the neonatal withdrawal syndrome, with 60 to 90% of opioid-exposed infants undergoing some degree of abstinence (Perlmutter, 1974). Symptoms of withdrawal include irritability, tremors, high-pitched cry, hyperactivity, wakefulness, diarrhea, disorganized sucking reflex, respiratory alkalosis, and lacrimation, as well as hiccups, sneezing, twitching, myoclonic jerks, or seizures. Opioids reported to cause these symptoms in the neonate are heroin, methadone, meperidine, morphine, codeine, and pentazocine. The onset of symptoms may occur at birth or may begin up to 14 to 28 days after birth. This depends on the drug the infant was exposed to in utero and the profile of pharmacokinetic excretion. Subacute symptoms may last for 4 to 6 months after birth (Desmond and Wilson, 1975).

Depending on the severity of the withdrawal syndrome, treatment with such agents as paregoric, phenobarbital, diazepam, and chlorpromazine may be required. The period of treatment may last from a few days to several months (Zelson et al., 1970). The question of advisability of breast feeding by women on opioids has been raised. Opioids are known to be present in breast milk (Blinick et al., 1975), and numerous reports describe "addictive" effects in infants consuming opioid-containing milk (Menninger-Lerchenthal, 1934). Given the permeability of the blood–brain barrier during perinatal life, and the fact that only nanogram or picogram quantities of drug are required to occupy opioid receptors, it could easily be envisioned that opioids in milk would be capable of influencing the infant.

The neurobehavioral sequelae associated with perinatal opioid exposure have been extensively reviewed by Zagon and McLaughlin (1984b), Householder et al. (1982), Wilson (1992), and Kaltenbach and Finnegan (1992). In summary, a number of highlights should be mentioned. "Classical" neurological tests (Davis and Shanks, 1975) have shown opioid-exposed infants to exhibit central nervous system hyperactivity, irritability, tremors and hypertonicity, impaired nutritive sucking, vomiting, severe sleep deficit, autonomic dysfunction (vasomotor lability and diarrhea), and a lability of states manifested by frequent shifts between sleep and wakefulness. Hyperactivity was a persistent finding that endured through the early school years (Davis and Shanks, 1975). Utilizing the Brazelton Neonatal Behavioral Assessment Scale (BNBAS), a number of investigators have tried to quantitate infant response to external stimuli, motor organization, and ability to regulate state of consciousness (Kaplan et al., 1978; Strauss et al., 1976, 1979a; Lodge, 1977; Lodge et al., 1975; Marcus et al., 1982). These studies show an impaired ability of opioid-dependent babies to adequately organize their responses to the environment, with a lessened capacity to attend and react to noxious stimuli and to habituate to disturbing events (Davis and Shanks, 1975). Opioid-exposed infants exhibited a depressed response decrement to light, with problems in visual and auditory orientation evident (Chasnoff et al., 1982). Using the BNBAS to decipher the specific effects of methadone on the neonate, Marcus et al. (1982) have suggested that behavioral problems in methadone-exposed infants may be owing not to generalized central nervous system irritability, but actually to neuromuscular dysfunction. Other investigators have suggested that the orientation responsiveness and excitability recorded in the BNBAS have substantial impact on caregivers' perceptions of infants, which may lead to long-term consequences for caregiver–infant interactions.

During the period of infancy, a significant decrease in quiet sleep (Schulman, 1969) along with an increase (Dinges et al., 1980) or decrease (Sisson et al., 1974) in rapid eye movement has been noted in children subjected to opioids. Kron and colleagues (1976) have reported an uncoordinated and ineffective sucking reflex as a manifestation of the opioid abstinence syndrome, whereas Lodge and colleagues (1975) found that the withdrawal period was characterized by heightened auditory responsiveness and orientation, lowered overall alertness, and poor attentiveness to visual stimuli.

Differences in neurological status between opioid-dependent and control groups have not been noted by some workers (Kaltenbach et al., 1979). However, Rosen and Johnson (1982a) reported tone discrepancies, developmental delays, and eye problems in children up to 18 months of age delivered by methadone-dependent mothers. Between 6 and 12 months, no real language development was detected. Van Baar and co-workers (1989a) reported that opioid-dependent infants had jerky movements, were easily excitable, and quickly irritated; these aspects of behavior were thought to be related to drug withdrawal.

Beyond the withdrawal period, Wilson et al. (1973) have found adaptive behavior, language performance, and personal–social development of heroin-exposed children to be normal, but gross motor coordination was more advanced than fine motor coordination. Disturbances of activity levels and/or attention span, along with sleep disturbances, temper tantrums, and low tolerance to frustration were recorded by Wilson at 1 year, whereas the onset of hyperactivity began at 12 to 18 months. Using the Bayley Scale of Infant Behavior as a monitor for mental and motor development, a number of investigators have found that children subjected perinatally to opioids are within the normal range (Blatman and Lipsitz, 1972; Chasnoff et al., 1980; van Baar et al., 1989b). However, Wilson et al. (1981) have noted the less attentive nature of both methadone- and heroin-exposed infants, whereas Kaltenbach et al. (1979) recorded a higher failure rate on three specific items: naming of two objects, pointing to five pictures, and naming of three objects. Kaltenbach suggested that subtle differences in cognitive behavior occur in opioid-exposed children that may be occluded by summary scores. Rosen and Johnson (1982a) reported a lower than normal score for methadone-exposed infants on both the mental and motor development indices at 12 and 18 months of age, and Strauss et al. (1976) also noted delays in the physical developmental index.

Opioid-subjected infants tested on the Gesell scales (Finnegan et al., 1977; Wilson et al., 1973), Merrill-Palmer test, and Peabody Picture Vocabulary test (Blatman and Lipsitz, 1972), as well as the Object Permanence Scales (Johnson and Rosen, 1982), appeared normal. A number of workers have reported disorders in speech and/or language development (Blatman and Lipsitz, 1972; Nichtern, 1973; Sardemann et al., 1976) in children subjected to opioids, although Lodge et al. (1975) reported strength in the realm of language for methadone-exposed offspring.

Few studies that extend beyond the first few years of childhood have been conducted on children maternally subjected to opioids. Nichtern (1973), examining children up to 15 years of age born of heroin-dependent mothers, recorded a number of problems related to their capacity for human relationships, demonstration of excessive adult–peer interactions, poor socialization, and the use of withdrawal to deal with difficult situations. Rosen and Johnson (1982b), in a study of children exposed in utero to methadone, recorded signs of impaired development, including smaller head circumferences, strabismus and/or nystagmus, and abnormalities in muscle tone, coordination, and language. Strauss et al. (1979a,b) found a cluster of behavioral differences related to greater task-irrelevant activity in methadone-exposed children during testing situations. These authors suggest that in structured and demanding situations, qualities of attentiveness and motor inhibition may be a domain in which opioid-exposed children might reveal particular vulnerabilities.

The neurobehavioral implications of exposure to heroin during early life have been reported by Wilson and colleagues (1979). Wilson did not find problems with speech and language function, but on a battery of perceptual measures, visual, tactile, and auditory modalities were less normal. Moreover, the parents of these children noted difficulties for their heroin-exposed children in the areas of self-adjustment, social adjustment, and physical adjustment. Items included were uncontrolled temper, impulsiveness, poor self-confidence, aggressiveness, and difficulty in making and keeping friends. These children were very active, although ratings of attention, cooperation, and alertness were not abnormal. Wilson concludes that heroin-exposed children may have a problem common to the general process of perception rather than a specific sensory deficit, and feels that behavioral problems in these children may be manifestations of impaired attention and organizational abilities.

In addition to the neurobehavioral sequelae described for children born of opioid-dependent women, a number of studies have monitored growth. Some studies (Blatman and Lipsitz, 1972) reveal no abnormalities, whereas others (Strauss et al., 1976; Wilson et al., 1981; Finnegan, 1975, 1976) have recorded lower (but not significantly reliable) values for offspring exposed to opioids. A substantial number of reports (Wilson et al., 1973; Wilson, 1975; Ting et al., 1974, 1978; Rosen and Johnson, 1982a,b; Chasnoff et al., 1982) document delayed growth properties associated with perinatal opioid exposure. These delays may take the form of noticeably lower body weights, shorter body lengths (height), and/or smaller head circumferences. Few studies have pursued details of postmortem examination of drug-dependent infants. In the most complete study to date, Rorke and colleagues (1977) performed neuropathological examinations on 10 infants exposed to heroin or methadone. Gliosis and brain developmental retardation were noted in some of these infants; these findings were thought to bear some relationship to maternal drug dependence.

As in most areas of human research, data collection and interpretation may be fraught with difficulties. The potential for problems is only magnified by the behavior of the population (i.e., opioid addicts) under consideration. The licit and illicit use of opioids so alters psychological, behavioral, and physiological processes that it elevates the drug to a paramount position in the life of any user. These individuals consume opioids during pregnancy despite ramifications of such consumption on the health of their children. Moreover, in actuality, two individuals—the mother and her offspring—are involved [and conceivably other individuals (e.g., caregivers)], and all must be considered in evaluating all of the repercussions associated with maternal opioid abuse. Thus, the interaction of these forces serves only to confuse essential issues further. Fortunately, in the case of clinical studies on opioids and development, recognition of and discussion about problems in experimental design and interpretation have not been overlooked (e.g., Householder et al., 1982; Strauss et al., 1979a; Wilson et al., 1981; Aylward, 1982). Some of these confounding variables include polydrug abuse, poor prenatal and/or postnatal care, demographics, socioeconomic status, length of hospitalization, neonatal withdrawal, breast feeding, mother/infant/child interpretations, paternal influences, sample selection, types of tests used for investigation, structuring of appropriate comparison/control groups, "dropout" rate of patients in the study, and statistical analysis.

Rather than being overwhelmed by these confounding issues and simply dismissing reports about the repercussions of perinatal opioid exposure, one needs to be aware that although the goal of clinical research may be to determine a relationship between cause and effect, this type of distinction may be unattainable. Searching for the etiology of opioid-related problems (much less defining all specific abnormalities) could be frustrating and unrewarding. Given all of the possible confounding variables enumerated above, it is therefore no wonder that defining pathognomonic features of perinatal opioid exposure is a very difficult task. An example of the problems in attempting to ascribe specific characteristics to perinatal opioid exposure concerns the neonatal abstinence syndrome. Neonatal withdrawal has been considered a hallmark of maternal opioid consumption; yet, a variety of nonopioid drugs consumed by the mother can elicit somewhat similar effects. Some of these agents include alcohol, barbiturates, chlorodiazepoxides, ethylchlorvynol, glutethimide, hydroxyzine, and meprobamate (World Health Organization Committee on Drugs, 1973). Thus, at least in the clinical realm, we may have to be satisfied with the knowledge that opioids are part of a cumulative effect from an overall potentially dangerous milieu that contributes to any damaging sequelae encountered.

LABORATORY FINDINGS: OPIOID EXPOSURE AND NEURAL DEVELOPMENT

Laboratory studies offer another perspective as to the influences of opioids on developmental processes. Through laboratory investigations, the goal of achieving information in regard to opioids and biological development can be realized. Laboratory models permit us to address such questions with a variety of methodologies that include maternal models of opioid consumption, short- and long-term studies on postnatal animals, regenerating tissues, tissues and cells in culture, and other in vitro preparations. Thus, laboratory studies allow investigators a more unrestricted view in research design in order to address the important questions of whether opioids influence biological development, and, if they do, what are the mechanisms involved and what are some possible strategies for intervention and treatment. A number of reviews (e.g., Zagon, 1983a; 1985; Zagon and McLaughlin, 1984b) and bibliographies (Zagon et al., 1982, 1984, 1989) have been published regarding laboratory studies and the perinatal opioid syndrome, and the reader is referred to this earlier literature.

The methodology employed in studies of perinatal opioid exposure, and discussion of some potentially confounding influences introduced by these techniques, have recently been reviewed (Zagon and McLaughlin, 1990a) and need not be reiterated herein. In Table 1 we present protocols of experiments involved with assessing the effects of opioids on the developing nervous system. As can be noted, routes of administration, drug dosages, and schedule of treatment are variable. Maternal exposure to opioids does not appear to have an effect on the estrous cycle, fertility, length of gestation, or parturition (e.g., Zagon and McLaughlin, 1977a,b,c), although difficulties with conception and a protracted period of gestation (Buchenauer et al., 1974), as well as positional malformations of the fetus (Chandler et al., 1975) have been reported. Fujinaga and Mazze (1988), implanting osmotic minipumps containing morphine into pregnant rats, found a significantly lower pregnancy rate in animals receiving 35 or 70 mg/kg/day, but not in animals treated with 10 mg/kg/day. A reduction in maternal body weight during pregnancy appears to be a common finding in terms of the effects of maternal opioid consumption (e.g., Ford and Rhines, 1979; Markham et al., 1971; Middaugh and Simpson, 1980; McGinty and Ford, 1976; White et al., 1978; Seidler et al., 1982; Zagon and McLaughlin, 1977c), but these weight deficits do not appear to reflect poor nutritional status (e.g., Ford and Rhines, 1979; White et al., 1978). Maternal exposure to morphine or methadone, in general, does not have a detrimental influence on litter size, although some decreases in litter size with higher doses of methadone have been cited (Middaugh and Simpson, 1980). Teratogenicity appears to be associated with high drug dosages administered acutely or over a short period (Geber and Schramm, 1975; Harpel and Gautierei, 1968; Arcuri and Gautieri, 1973; Jurand, 1973, 1980, 1985), but not with drugs administered chronically (e.g., Fujinaga and Mazze, 1988). An increase in stillborns has been observed with high drug dosages by Freeman (1980) and Sobrian (1977), whereas other investigations reveal little problem in this area (Davis and Lin, 1972; Fujinaga and Mazze, 1988).

The effect of transplacental exposure to opioids on postnatal viability seems to be determined by drug dosage, and whether the neonate continues to receive opioids postnatally (e.g., breast milk, direct injection). Neonates that do not continue to receive opioids after birth often may be hypersensitive to stimuli and exhibit tremors at birth, with substantial neonatal mortality found in the first few days of life (e.g., Davis and Lin, 1972; Zagon and McLaughlin, 1977c; Freeman, 1980; Fujinaga and Mazze, 1988).

The effects of opioid exposure in early life on the structure (Table 2), physiology (Table 3), behavior (Tables 4–9), and chemistry (Table 10) of the developing nervous system have been summarized. In general, perinatal opioid exposure often appears to be associated with a retardation in neural development that covers a number of aspects; this altered pattern of nervous system ontogeny may have ramifications beyond the developmental period. In some cases, these changes may persist into adult life. Another generality emerging from studies on exposure to opioids in early life is that drug exposure of the embryo and fetus is more detrimental if the opioid is withdrawn at birth. Laboratory animals exposed to opioids during gestation, and given continued exposure during lactation, are often not so adversely affected. This might

TABLE 1   Drug Protocols in Laboratory Studies of Exposure to Exogenous Opioids in Early Life[a]

Authors	Species	Dosage	Route	Treatment schedule
Methadone				
Choi and Visekul (1988)	Mouse	1 μM–mM	tc	Cultures from days 14–17 gestation
Darmani et al. (1991)	Rat	9 mg/kg/day	mp	Days 7–20 gestation
DeMontis et al. (1983)	Rat	60–80 mg/kg	po	Before mating to 10 days lactation
Field et al. (1977)	Rat	1.9–2.9 mg/kg/day	po	Before mating to weaning
Ford and Rhines (1979)	Rat	5 mg/kg	sc	Day 13 gestation to parturition
Freeman (1980)	Rat	4, 16 mg/kg	sc	Days 8–22 gestation
Friedler (1977)	Mouse	20 mg/kg	sc	Exposure for 5 days at least 6 days before mating (twice daily)
Gibson and Vernadakis (1983a)	Chick	20–500 mg	io	Days 3–20 embryo
Grode and Murray (1973)	Mouse	2×10^{-3} M	tc	Day 19 gestation, exposed up to 55 days
Grove et al. (1979)	Rat	0.125 mg/ml	po	Day 1 gestation to weaning
Hammer et al. (1989)	Rat	10 mg/kg/day	mp	Day 12 gestation to lactation day 5
Hovious and Peters (1984)	Rat	5 mg/kg	ip	7 days before mating to weaning
Hovious and Peters (1985)		see Hovious and Peters (1984)		
Hui et al. (1978)	Mouse	2 mg/kg	sc	Birth to day 28
Hutchings et al. (1979a)	Rat	10, 15 mg/kg	po	Days 8–22 gestation
Hutchings et al. (1979b)	Rat	10 mg/kg	po	Days 8–22 gestation
Hutchings et al. (1980)	Rat	5, 7.5, 10 mg/kg	po	Days 8–22 gestation
Jakubovic et al. (1978)	Chick	0.4, 0.8, 4 mg/kg	io	Days 2–12 embryo
Lau et al. (1977)		see Slotkin et al. (1976)		
McGinty and Ford (1980)	Rat	0.5 mg/ml	po	Day 15 gestation to weaning
McGinty and Ford (1980)	Rat	6 mg/kg	sc	Day 16 gestation to parturition, twice daily
Middaugh and Simpson (1980)	Mouse	2.5, 5, 10 mg/kg	sc	Days 15–22 gestation
Pertshuk et al. (1977)		8 mg/kg	sc/im	Days 15–21 gestation
Peters (1977)	Rat	5 mg/kg	ip	7 days before mating to weaning
Peters (1978)		sees Peters (1977)		

(*continued*)

TABLE 1 (*Continued*)

Authors	Species	Dosage	Route	Treatment schedule
Rech et al. (1980)	Rat	10 mg/kg	sc	Day 5 gestation to weaning
Ricalde and Hammer (1990)	see Hammer et al. (1989)			
Sakellaridis et al. (1986)	Chick	10^{-5}–10^{-6} M	tc	Day 6 embryo
Seidler et al. (1982)	see Slotkin et al. (1980)			
Singh et al. (1980)	Rat	5, 7.5, 10, 20 mg/kg	sc	Days 14–19 gestation
Slotkin et al (1976)	Rat	5 mg/kg	sc	Day 10 gestation to weaning
		5 mg/kg	sc	[b]Birth to day 37
Slotkin et al. (1979)	Rat	5 mg/kg	sc	Day 10 gestation to weaning
		5 mg/kg	sc	[b]Birth to day 35
Slotkin et al. (1980a)	Rat	5 mg/kg	sc	Day 10 gestation to postnatal day 14
Slotkin et al. (1980b)	Rat	5 mg/kg	sc	Day 10 gestation to weaning
Slotkin et al. (1981)	Rat	5 mg/kg	sc	Day 10 gestation to weaning
		5 mg/kg	sc	[b]Days 1–10
Slotkin et al. (1982)	Rat	5 mg/kg	sc	Birth to day 21
Soyka et al. (1978)	Rat	10 mg/kg	sc	12 days before mating
Thompson and Zagon (1981)	see Zagon and McLaughlin (1977a)			
Thompson et al. (1979)	see Zagon and McLaughlin (1977a)			
Tiong and Olley (1988)	Rat	8 mg/kg	sc	Before mating to gestation
Tsang et al. (1986)	Rat	10 mg/kg	sc	7 days before mating to weaning
Van Wagoner et al. (1980)	see Zagon and McLaughlin (1977a)			
Vernadakis et al. (1982)	Chick	10^{-5}–10^{-6} M	tc	Days 8 and 15
Walz et al. (1983)	Rat	15 mg/kg	po	60 days before mating to weaning
Wang et al. (1986)	Rat	9 mg/kg/day	mp	Day 8 gestation to lactation day 10
Willson et al. (1976)	Rat	5 mg/kg	sc	3 months to 3 days lactation
		1×10^{-4} M	tc	Days 1–3
Zagon and McLaughlin (1977a)	Rat	5 mg/kg	ip	5 days before mating to weaning
Zagon and McLaughlin (1977b)	see Zagon and McLaughlin (1977a)			
Zagon and McLaughlin (1977c)	see Zagon and McLaughlin (1977a)			
Zagon and McLaughlin (1978a)	see Zagon and McLaughlin (1977a)			

(*continued*)

TABLE 1   (*Continued*)

Authors	Species	Dosage	Route	Treatment schedule
Zagon and McLaughlin (1978b)	see Zagon and McLaughlin (1977a)			
Zagon and McLaughlin (1980)	see Zagon and McLaughlin (1977a)			
Zagon and McLaughlin (1981a)	see Zagon and McLaughlin (1977a)			
Zagon and McLaughlin (1981b)	see Zagon and McLaughlin (1977a)			
Zagon and McLaughlin (1982a)	see Zagon and McLaughlin (1977a)			
Zagon and McLaughlin (1982b)	see Zagon and McLaughlin (1977a)			
Zagon and McLaughlin (1982c)	see Zagon and McLaughlin (1977a)			
Zagon and McLaughlin (1984a)	see Zagon and McLaughlin (1977a)			
Zagon et al. (1979a)	see Zagon and McLaughlin (1977a)			
Zagon et al. (1979b)	see Zagon and McLaughlin (1977a)			
LAAM				
Gibson and Vernadakis (1983a)	Chick	20–300 mg	io	Days 3–20 embryo
Gibson and Vernadakis (1983b)	Chick	10^{-6} M	tc	Day 7 embryo
Kuwahara and Sparber (1981)	Chick	1.5, 7.5, 15 mg/kg	io	Day 3 embryo
Kuwahara and Sparber (1982)	Chick	2.5, 5, 10 mg/kg 0.15, 0.3, 0.5 mg	io ip	Day 3 embryo [b]Days 1–2
Kuwahara and Sparber (1983)	Chick	10 mg/kg	io	Day 3, embryo
Lichtblau and Sparber (1982)	Rat	0.2 mg/kg	po	3 weeks before mating to parturition
Lichtblau and Sparber (1983)	Rat	0.2 mg/kg	po	3 weeks before mating to parturition
Buprenorphine				
Tiong and Olley (1988)	Rat	2 mg/kg	sc	Before mating to weaning
Pentazocine				
Andres et al. (1988)	Rat	23.4 mg/kg	im	Days 1–19 gestation
Heroin				
Laskey et al. (1977)	Rat	5 mg/kg	ip	5 days before mating to weaning (5 mg/kg once daily or 2.5 mg/kg twice daily)
Morphine				
Anderson and Slotkin (1975)	Rat	40 mg/kg	sc	Day 8 gestation to weaning (twice daily)
Arjune and Rodnar (1989)	Rat	1 or 20 μg	sc	[b]Days 1–7
Bardo et al. (1982)	Rat	5 mg/kg	sc	[b]Days 1–21
Butler and Schanberg (1975)	Rat	40 mg/kg	sc	Day 13 gestation to weaning
Castellano and Ammassari-Teule (1984)	Mouse	40 mg/kg	sc	5 days before mating to day 15 lactation

(*continued*)

TABLE 1 (Continued)

Authors	Species	Dosage	Route	Treatment schedule
Caza and Spear (1980)	Rat	0.1, 0.5, 1, 5 mg/kg	sc	[b]Days 10, 17, or 24
Coyle and Pert (1976)	Rat	10 mg/kg	ip	Days 15–20 gestation (three times/day)
Crain et al. (1979)	Mouse	1 μM	tc	Day 14 gestation, 2–3 days exposure
Crain et al. (1982)	see Crain et al. (1979)			
Davis and Lin (1972)	Rat	45 mg/kg	sc	Days 5–18 gestation
Di Guilio et al. (1988)	Rat	0.40 mg/ml	po	Day 1 gestation to weaning
		0.03 mg/ml	po	Postweaning
Friedler (1977)	Mouse	240 mg/kg	sc	5-day exposure occurring at least 6 days before mating
Ghadrian (1969)	Rabbit, dog	2.4–6 × 10•−4 M	tc	Neonatal
Gibson and Vernadakis (1983a)	Chick	20–200 mg	io	Days 3–20 embryo
Glick et al. (1977)	Rat	0.5 g/liter	po	Day 1 gestation to day 15 lactation
Hammer et al. (1989)	Rat	10 mg/kg/day	mp	Day 12 gestation to day 5 postnatal
Handelmann and Dow-Edwards (1985)	Rat	1 μg	sc	[b]Days 1–7
Handelmann and Quirion	Rat	1 μg	sc	[b]Days 1–7
Hansson and Rönnbäck (1983)	Rat	10^{-5}–10^{-6} M	tc	[b]Days 1–7
Hansson and Rönnbäck (1985)	see Rönnbäck and Hansson (1983)			
Hovious and Peters (1985)	Rat	5 mg/kg	ip	7 days before mating to weaning
Jakubovic et al. (1978)	Chick	0.8 mg/kg	io	Days 2–12, embryo
Johannesson and Becker (1972)	Rat	20 mg/kg	sc	Days 2–5 gestation Days 7–9 gestation Days 11–23 gestation
Johannesson et al. (1972)	Rat	5 mg/kg	iv	Days 21–22 gestation
Kirby (1979)	Rat	1.25, 5, 10, 20, 40 mg/kg	sc	Day 18 gestation
		10 mg/kg	sc	Days 7–18 gestation (twice daily)
Kirby (1980)	Rat	5 mg/kg	sc	Days 12–18 gestation (four times/day)
Kirby (1981)	Rat	10 mg/kg	sc	Acute injection, days 15–21
Kirby (1983)	Rat	5 mg/kg	sc	Day 12 gestation to parturition (four times/day)

(continued)

TABLE 1 (*Continued*)

Authors	Species	Dosage	Route	Treatment schedule
Kirby and Aronstam (1983)		see Kirby (1983)		
Kirby and Holtzman (1982)	Rat	20 mg/kg	sc	Days 12–21 gestation (5 mg/kg every 6 hr or 10 mg/kg every 12 hr)
Kirby et al. (1982)		see Kirby and Holtzman (1982)		
LaPointe and Nosal (1982)	Rat	56 mg/kg	ip	5 days before mating to day 16 gestation
Mattio and Kirby (1982)	Rat	5 mg/kg	sc	[b]Days 5–18 (three times/day)
McGinty and Ford (1976)	Rat	0.3 mg/ml	po	Day 15 gestation to weaning
Meriney et al. (1985)	Chick	20–200 μM/day	io	Days 7–14, embryo
Newby-Schmidt and Norton (1981a)	Chick	20 mg/kg	io	Days 12–19 or 16–19, embryo
Newby-Schmidt and Norton (1981b)	Chick	20 mg/kg	io	Days 12–19, embryo
Newby-Schmidt and Norton (1983)	Chick	1 mg/kg	ip	[b]Days 1–2 or 4–5
Nosal (1979a)	Rat	0.5 mg/kg 0.5 mg/ml	ip po	3 weeks before mating and days 1–21 gestation
Nosal (1979b)		see Nosal (1979b)		
O'Callaghan and Holtzman (1976)	Rat	10 mg/kg	sc	Days 5–12 gestation (twice daily)
Peters	Rat	7.5 mg/kg	ip	7 days before mating to weaning
Peterson et al. (1974)	Chick	2.5×10^{-4} M	tc	Day 7 embryo
Plishka and Neale (1984)	Mouse	75 μM	tc	Day 14.5 gestation, 6 days in culture
Ricalde and Hammer (1990)		see Hammer et al. (1989)		
Rönnbäck and Hansson (1985)	Rat	10^{-5}–10^{-7} M	tc	Neonate, 14 days in culture
Sakellaridis et al. (1986)	Chick	10^{-5}–10^{-6} M	tc	Day 6 embryo
Shuster et al. (1975)	Mouse	25 mg/kg	sc	[b]Days 10–14, 10–15, 12–16, 16–20, 19–23, 20–24
Sobrian (1977)	Rat	40 mg/kg	sc	5 days before mating to day 15 lactation
Sonderegger and Zimmermann (1976)	Rat	4, 8, 16 mg/kg	sc	[b]Days 0–7, 8–14, or 15–22 (twice daily)
Sonderegger and Zimmermann (1978)	Rat	8 mg/kg	sc	Days 3–12 or 12–21
Sonderegger et al. (1977)	Rat	75 mg	imp	[b]Day 5 or 11
Sonderegger et al. (1979)		see Sonderegger et al. (1979)		

(*continued*)

TABLE 1 (*Continued*)

Authors	Species	Dosage	Route	Treatment schedule
Steele and Johannesson (1975a)	Rat	5 mg/kg/hr	iv	Day 21 or 22 gestation
Steele and Johannesson (1975b)	Rat	20 mg/kg	sc	Days 17–20 gestation
		5 mg/kg	iv	Days 21 or 22 gestation
Steele and Johannesson (1975c)	same as Steele and Johanesson (1975a)			
Tempel et al. (1988)	Rat	75 mg/kg	imp	Beginning day 16 gestation
		5 mg/kg	sc	[b]Days 1–28
Tsang and Ng (1980)	Rat	10 mg/kg	ip	1 week before mating to parturition (twice daily)
Zagon and McLaughlin (1977d)	Rat	40 mg/kg	ip	5 days before mating to day 21 or 60

[a]Unless otherwise stated doses are postnatal. Tc, tissue culture; mp, minipump; po, oral; sc, subcutaneous; io, in ovo; im, intramuscular; ip, intraperitoneal; imp, implant.
[b]Direct injection to offspring.

suggest that the embryo/fetus adjusts to drug exposure, at least to some degree, and that drug withdrawal at birth introduces yet another destabilizing influence during a critical window of development.

In evaluating the laboratory studies on perinatal opioid exposure, interpretation of the results in light of experimental paradigms is important. As mentioned above, an earlier report has examined many of the methodological aspects that might introduce confounding influences in studies examining the effects of drugs in the fetus and neonate (see Zagon and McLaughlin, 1990a). Without being exhaustive, it is instructive to consider some of the issues about the literature on perinatal opioid exposure that might have a bearing on the credibility and importance of the findings. Many of these issues have been raised and answered earlier (see Zagon, 1983b). One such issue revolves around the premise that opioid action is dependent on the presence of drug in the plasma, and that protocols in which plasma levels are maintained would be advisable. However, selective drug uptake and drug persistence may occur in tissues and organs, and the absence of drug in the plasma does not preclude the presence of drug in body compartments, explaining why drugs having very short plasma half-lives may possess prolonged action. The presence of even small amounts of drug in an area (e.g., brain) essential to drug response is of utmost importance. Thus, plasma concentrations may be low or negligible, but withdrawal reactions may not occur because tissue (e.g., brain) concentrations are sufficiently high. Perhaps the best example of this concerns endogenous opioid activity in which only nanogram or picogram quantities are necessary to bind with opioid receptors in order to effect a response. In regard to methadone, this drug has been reported to have a half-life in rodents ranging from a few hours (Misra et al., 1973) to over 24 hr (Ziring et al., 1981). It appears that

TABLE 2   Effects of Opioids during Development on the Structure of Neural Cells and Tissues

Authors	Observations[a]
Morphine	
Ghadrian (1969)	↑ mitosis and cell proliferation
Hammer et al. (1989)	↓ neuronal packing density in primary somatosensory cortex and preoptic areas of hypothalamus, day 5 ↑ glial packing in hypothalamus, day 5
Kirby (1980)	↓ volume of 1st thoracic spinal cord segment, day 18 of gestation NS length of 1st thoracic spinal cord segment, day 18 of gestation
Kirby (1983)	↓ volume of 1st thoracic spinal cord segment, day 6 ↓ gray/white ratio, day 15 ↑ length of segment, day 15 NS adult volume, gray/white ratio, length of segment
Meriney et al. (1985)	Delayed normal cell death in ciliary ganglion
Nosal, G. (1979a)	Altered morphology cerebellar Purkinje cells, neonate
Nosal, G. (1979b)	Altered morphology of cerebellar Purkinje cells, day 21
Ricalde and Hammer (1990)	↓ dendritic length in primary somatosensory cortex, day 5
Zagon and McLaughlin (1977c)	↓ brain dimensions, developing and adult
Methadone	
Choi and Visekul (1988)	↓ neurotoxicity by N-methyl-D-aspartate
Ford and Rhines (1979)	↓ cortical thickness up to day 14 ↓ cortical cells up to day 14
Grode and Murray (1973)	Altered pattern of neuronal density during first 4 weeks
Pertschuk et al. (1977)	Immunocytochemical localization of methadone in neonatal retina
Willson et al. (1976)	↓ cerebellar explant outgrowth
Zagon and McLaughliin (1977a)	↓ brain dimensions
Zagon and McLaughlin (1977d)	↓ brain dimensions
Zagon and McLaughlin (1982c)	↓ cerebellar area and internal granule neurons, day 21

[a]Arrows represent direction of change from controls. Unless otherwise indicated, all ages are postnatal. NS, not significant.

plasma concentrations of methadone have no direct correlation with the rate of metabolism or with brain concentrations of this drug (Liu and Wang, 1975). In fact, methadone is known to bind firmly to tissue proteins (Sung et al., 1953) and to accumulate in body tissues (Dole and Kreek, 1973; Misra et al., 1973), and methadone has been reported to persist in several organs including the brain for up to 10 weeks (Harte et al., 1976). It is also known that immature animals accumulate much larger amounts of methadone than do adults (Peters et al., 1972; Shah et al., 1976;

TABLE 3   Effects of Opioids during Development on Physiological Parameters

Authors	Observations[a]
Morphine	
Arjune and Bodnar (1988)	Response to morphine in adults ↑ opioid-mediated cold water swimming ↓ nonopioid-mediated cold water swimming ↓ analgesia, males ↑ analgesia, females
Castellano and Ammassari-Teule (1984)	Response to morphine in adults ↑ activity ↑ passive avoidance ↑ (enhanced) analgesia
Crain et al. (1979)	↓ sensory-evoked dorsal-horn network responses to morphine
Crain et al. (1982)	Tolerance of sensory-evoked dorsal-horn network responses to morphine
Handelmann and Dow-Edwards (1985)	↓ metabolic activity in adult brain
Hovious and Peters (1984)	↓ latencies on hot plate and tail-flick tests in response to methadone or morphine, days 25 and 120
Johannesson and Becker (1972)	↓ nociceptive response to morphine, days 12–13 and days 21–22
Johannesson et al. (1972)	↑ analgesic response to morphine, day 12
Kirby et al. (1982)	↑ (enhanced) analgesia, 20 mg/kg (4 injections, 5 mg/kg) of morphine in adult NS analgesia, 20 mg/kg (2 × 10 mg/kg) of morphine in adult
O'Callaghan and Holtzman (1976)	↓ analgesic response to morphine, 3, 5, and 11 weeks
Shuster et al. (1975)	↑ running, response to morphine in adult
Sonderegger and Zimmerman (1978)	↓ (reduced) analgesic sensitivity to morphine in adults
Sonderegger et al. (1977)	↓ (reduced) analgesic sensitivity to morphine
Sonderegger et al. (1979)	NS analgesic response to morphine in adults
Methadone	
Hovious and Peters (1984)	↓ latencies on hot plate and tail-flick tests in response to methadone or morphine, days 25 or 120
Singh et al. (1980)	Fetal blood testosterone and Δ⁴-androstenedione ↓ males NS females
Thompson and Zagon (1981)	Altered thermal regulation, young and adult rats
Thompson et al. (1979)	Altered thermal regulation, juvenile rats
Zagon and McLaughlin (1980)	↑ analgesia at 21 → adult
Zagon and McLaughlin (1981a)	↓ body temperature, response to methadone in adults ↑ (enhanced) analgesia, response to methadone in adults
Zagon and McLaughlin (1981b)	↓ analgesia, day 45 ↓ body temperature, day 30 ↑ head and body shakes, days 30–120 ↑ withdrawal with naloxone administration

(continued)

TABLE 3   (*Continued*)

Authors	Observations[a]
Zagon and McLaughlin (1982a)	↑ analgesia at 21 to 60 days
Zagon and McLaughlin (1984)	Response in adults Tolerance to thermoregulatory actions of chlorpromazine and amphetamine Sensitivity to thermoregulatory actions of morphine ↑ (enhanced) analgesic response to amphetamine, cocaine, morphine

[a]Arrows represent direction of change from controls. Unless otherwise indicated, all ages are postnatal. NS, not significant.

Shah and Donald, 1979). In particular, the brains of fetal (Peters et al., 1972) and preweaning (Shah and Donald, 1979) rats have high concentrations of methadone following acute injections, presumably because developing animals have an increased permeability due to the gradual formation of the blood–brain barrier postnatally. It probably is worth noting that immature organisms also may not be able to cope with drugs as well as adults since drug metabolism has not fully matured (Yeh and Krebs, 1980), and fetal/neonatal storage and delayed metabolism and/or excretion could produce pharmacological effects that are even more pronounced than those in adults (e.g., Kandall, 1977). Finally, it is interesting to note that close observations of pregnant animals undergoing treatment with at least one opioid, methadone, revealed no signs of abstinence over a 24-hr period (Zagon and McLaughlin, 1980), suggesting that an episodic pattern of drug exposure whereby animals have alternating periods of

TABLE 4   Effects of Opioids during Development on Behavior Monitored in Ovo or in Utero

Authors	Age examined	Observation[a]
Morphine		
Kirby (1979)	18 days in utero	↓ spontaneous activity
Kirby (1981)	15–21 days in utero	↓ spontaneous activity
Kirby and Holtzman (1982)	18–20 days in utero	↓ spontaneous activity (twice daily injection group)
Newby-Schmidt and Norton (1981b)	16 days in ovo	NS spontaneous activity
Newby-Schmidt and Norton (1983)	18 days in ovo	NS spontaneous activity
LAAM		
Kuwahara and Sparber (1982)	19 days in ovo	↓ spontaneous activity

[a]Arrows indicate the direction of significant change from controls for all opioid-treated groups studied or those within parentheses; NS, not significant.

TABLE 5   The Effects of Opioids during Development on Behavior in the
Preweaning Period (Days 0–21)

Authors	Age examined (days)	Observation[a]
Morphine		
Arjune and Bodnar (1984)	0–21	NS eye opening
Castellano and Ammassari-Teule (1984)	2–20 8–16	↓ (delayed) motor control NS reflexes ↓ spontaneous activity
Caza and Spear (1980)	10	↓ open field, locomotion (0.5, 1, 5 mg/kg groups) ↑ catalepsy (5 mg/kg group) NS gnawing NS grooming, sniffing, rearing
	17	↑ open field, locomotion (0.5 mg/kg group) ↓ open field, locomotion (5 mg/kg group) ↑ catalepsy (5 mg/kg group) NS gnawing, grooming, sniffing, rearing
	24	NS open field, locomotion ↑ gnawing (5 mg/kg group) NS grooming, sniffing, rearing
Friedler (1977)	0–21	↓ (delayed) ear and eye opening
La Pointe and Nosal (1982)	0–21	NS battery of sensorimotor reflexes except: ↓ (delayed) eye opening (prenatal) ↓ frequency of pivoting (prenatal/postnatal) ↓ respiratory rate (0–3 days) ↑ frequency of crying (0–3 days) ↑ restlessness (0–3 days)
McGinty and Ford (1976)	0–21	↓ open field, ambulation ↓ (delayed) reflex and motor skill development ↓ (delayed) ear, eye, vaginal opening, hair covering
Newby-Schmidt and Norton (1981a)	1, 2 7	↑ neuromuscular weakness (16–19 day group) ↑ disruption of stride (16–19 day group)
Newby-Schmidt and Norton (1981b)	1, 2 4, 5	NS distress vocalization ↑ distress vocalization
Newby-Schmidt and Norton (1983)	1	Altered distress vocalization in response to morphine
Sobrian (1977)	1, 5, 10, 15, 20, 25, 30	NS activity monitor ↑ activity monitor
Methadone		
Friedler (1977)	0–21	↓ (delayed) ear and eye opening

(continued)

TABLE 5   (*Continued*)

Authors	Age examined (days)	Observation[a]
Hutchings et al (1979a)	17, 22	↑ activity monitor, 3 littermates/group (10 mg/kg group) ↑ number of rest–activity fluctuations, 3 littermates/group (10 mg/kg group)
Hutchings et al. (1979b)	0–21	NS time of eye opening and hair covering
Hutchings et al. (1980)	2	↓ activity monitor, 8–10 littermates/group (10 mg/kg group)
	5	↑ activity monitor, 8–10 littermates/group (5 mg/kg group)
	14–32	NS activity monitor, 8–10 littermates/group
McGinty and Ford (1976)	11	↓ open field, ambulation
	0–21	↓ (delayed) reflex and motor skill development ↓ (delayed) ear, eye, and vaginal opening, hair covering
Walz et al. (1983)	0–21	↓ (delayed) reflex ontogeny
	0–14	↓ open field, exploration
	14–21	↑ open field, exploration
	10–21	↓ open field, defecation
Zagon and McLaughlin (1978b)	2–19	↓ (delayed) reflex and spontaneous motor development, physical characteristics
LAAM		
Kuwahara and Sparber (1982)	3, 4	↓ open field, escape jumps, ambulation (prenatal, postnatal groups) ↓ (retarded) acquisition of detour performance (mg/kg prenatal group) ↑ frequency of distress vocalizations (5 mg/kg prenatal group)

[a]Arrows represent direction of significant change from controls for either all opioid-treated groups studied or those specified within parentheses. NS, not significant.

drug exposure is not a participant in understanding the etiology of the problems seen in animals subjected to opioids in perinatal life.

Another important methodological principle in studies on perinatal opioid exposure is the necessity of adjusting drug dosage for body weight, rather than administering the same amount of drug to each animal. The practice of adjusting drug dosage for body weight is predicated on the basis of dosage/volume distribution. Furthermore, such procedures serve to control the pharmacological regimen in order to provide meaningful data for within- and between-group comparisons. In essence, keeping a dosage constant without adjusting to body weight makes each animal different from all others, since body weight gain (e.g., during pregnancy) may vary. An added variable introduced by this procedure is that as the animal becomes heavier, drug withdrawal

TABLE 6   Summary of Sensorimotor and Physical Parameters in Preweaning
Rats Maternally Subjected to Methadone

Parameter	Age[a] (days)	Treatment Group		
		Prenatal	Postnatal	Prenatal and postnatal
Spontaneous motor				
Unilateral head turn with no return	2	NS	↓	↓
Unilateral head turn to left and return	5	↓	NS	NS
Head raise	8	↓	NS	NS
Pivoting 360°	8	↓	↓	NS
Foreleg and hindleg movement	11	↓	NS	NS
Head, foreleg, hindleg movement	12	NS	↓	NS
Walking	13	↓	↓	NS
Reflex tests				
Startle	3	↓	↓	↓
Righting reflex	3	NS	↓	NS
Pain—paw withdrawal	5	↓	NS	NS
Bar grasping	11	↓	NS	NS
Tail hanging	12	↓	↓	↓
Edge aversion	12	NS	↓	NS
Visual orientation	15	↓	↓	↓
Auditory reflex	16	NS	NS	↓
Physical development				
Incisor eruption	9	↓	↓	NS
Hair covering	11	↓	↓	NS
Ear opening	13	↓	↓	↓
Eye opening	14	↓	NS	NS

[a]The day when 50% of all animals (pups from both control and experimental groups) displayed a particular behavior. Arrows indicate the direction of significant change from controls; NS, not significant. Modified from data presented in Zagon and McLaughlin (1978b).

may be exhibited because of insufficient quantities of drug, making interpretation of the results even more difficult. It is of interest to note that in the clinical setting, dosage adjustments for volume of distribution must also be considered. Rementeria and Lotongkum (1977) suggest that an increase in methadone dosage during later stages of pregnancy may need to be considered in order to compensate for the methadone utilized by the rapidly growing fetus. Furthermore, failure to adjust the dosage near term can lead not only to increased drug craving, but also to fetal distress (e.g., Zuspan et al., 1975).

Another aspect in the design of laboratory studies of perinatal opioid exposure is the issue of achieving a level of tolerance and physical dependence in the animals under investigation. Such paradigms would have the greatest similarity to the clinical setting and eliminate consideration of acute effects. For example, opioids are known to cause respiratory depression, and a schedule of drug administration alleviating this phenomenon would minimize confounding influences resulting from anoxia or hypoxia. Thus, beginning injections several days or even weeks before conception would be

TABLE 7   Effects of Opioids during Development on Behavior in the Weanling Period (Days 22–44)

Authors	Observation[a]
Morphine	
Castellano and Ammassari-Teule (1984)	NS spontaneous activity
Davis and Lin (1972)	↑ open field, ambulation, rearing ↓ open field, defecation NS open field, face washing NS activity cage
Friedler (1977)	↓ activity monitor, running, rearing (males) ↓ open field, exploration
Sonderegger and Zimmerman (1978)	NS open field
Sonderegger et al. (1979)	NS open field, ambulation, defecation
Heroin	
Laskey et al. (1977)	NS open field, ambulation NS activity cage NS activity wheel ↓ latency to step down from an elevated platform
Methadone	
Freeman (1980)	↓ activity cage, ambulation (16 mg/kg group) NS activity cage, defecation
Friedler (1977)	↓ activity monitor, running, rearing (males) ↓ open field, exploration (males)
Grove et al. (1979)	↓ activity cage ↓ open field, ambulation, rearing (prenatal and prenatal–postnatal groups NS open field, defecation
Middaugh and Simpson (1980)	↓ open field, ambulation ↓ reactivity to stimulus presentation (5 mg/kg females)
Peters (1977)	↑ shock avoidance, latency
Zagon et al. (1979a)	↓ open field, ambulation ↓ activity cage NS activity wheel ↑ latency to step down from an elevated platform (prenatal group)
Zagon and McLaughlin (1981b)	↑ withdrawal, head shakes (prenatal, postnatal groups)

[a] Arrows represent direction of significant change from controls for either all opioid-treated groups studied or those specified within parentheses. NS, not significant.

advantageous. Several studies have examined this issue closely. White and Zagon (1979) investigated the effects of an opioid (methadone) on arterial blood gases and pH. These authors found that tolerance to the respiratory depressive action of the dosages of methadone utilized occurred within 2 weeks of chronic, daily administration. Thus, opioid administration that begins well before mating would circumvent

TABLE 8   Effects of Opioids during Development on Behavior in Young Adult Animals (Days 45–89)

Authors	Observation[a]
Morphine	
Davis and Lin (1972)	↑ open field activity, ambulation, rearing NS open field, defecation, face washing ↑ activity cage NS audiogenic seizures
Peters (1978)	NS food maze ↑ treadmill (females) ↑ shock avoidance, latency
Sonderegger et al. (1979)	↑ errors in Lashley III maze
Methadone	
Grove et al. (1979)	↑ open field, ambulation (prenatal group, prenatal–postnatal group) NS open field, rearing, defecation ↑ activity cage
Hutchings et al. (1979b)	NS acquisition of lever pressing NS variable interval reinforcement NS auditory–visual discrimination NS response inhibition in a punishment paradigm
Peters (1978)	NS food maze NS treadmill (males) ↑ shock avoidance, latency ↑ treadmill (females)
Soyka et al. (1978)	↑ open field, ambulation ↓ open field, defecation
Zagon et al. (1979a)	↑ open field, ambulation (prenatal, postnatal groups) ↑ activity cage (postnatal, prenatal–postnatal group) ↑ activity wheel (postnatal group) ↓ latency to step down from an elevated platform (prenatal, prenatal–postnatal)
Zagon and McLaughlin (1981b)	↑ withdrawal, head shakes ↑ withdrawal, wet-dog shakes (postnatal, prenatal–postnatal groups)
LAAM	
Lichtblau and Sparber (1982)	NS neuromuscular development or exploratory activity ↑ pressure fixed-ratio procedure
Lichtblau and Sparber (1983)	↓ unconditioned exploratory activity

[a]Arrows represent direction of significant change from controls for either all opiod-treated groups studied or those specified within parentheses. NS, not significant.

insults related to acute exposure. Other evidence that offspring are physically dependent would also provide information that the dosages of opioid were indeed functionally relevant. Observations that neonates undergo distress would be an example similar to that occurring in the clinical situation. As mentioned earlier, 60–90% of the infants of mothers with a recent history of opioid abuse experience the withdrawal

TABLE 9   Effects of Opioids during Development on Behavior
in Mature Adult Animals

Authors	Observation[a]
Morphine	
Castellano and Ammassari-Teule (1984)	↑ passive avoidance learning ↑ activity monitor
Glick et al. (1977)	↑ (facilitation) self-administration behavior
Handelmann and Dow-Edwards (1985)	Altered gait Impaired motor coordination Altered open field activity
Sonderegger and Zimmermann (1976)	↓ conditioned suppression
Sonderegger and Zimmermann (1978)	NS conditioned emotional response, group injected days 12–21 ↓ conditioned emotional response, group injected days 3–12
Methadone	
Hovious and Peters (1985)	↑ oral self-administration of morphine NS oral self-administration of methadone
Middaugh and Simpson (1980)	NS acquisition of lever response (5 mg/kg group) ↑ fixed ratio reinforcement
Rech et al. (1980)	↑ active avoidance "massed" trials
Soyka et al. (1978)	NS open field, emotionality NS conditioned avoidance
Van Wagoner et al. (1980)	↓ discrimination learning (prenatal, postnatal groups) NS discrimination learning (prenatal–postnatal group)
Zagon and McLaughlin (1981b)	↑ withdrawal, head shakes
Zagon et al. (1979b)	↓ active avoidance "spaced" trials (prenatal, prenatal–postnatal groups) NS passive avoidance ↓ food-reward, light–dark discrimination maze (prenatal, postnatal groups)
LAAM	
Lichtblau and Sparber (1983)	↓ unconditioned exploratory activity

[a]Arrows represent direction of significant change from controls for either all opioid-treated groups studied or those specified within parentheses. NS, not significant.

syndrome (Blinick et al., 1976; Perlmutter, 1974) and, if left untreated, these infants may convulse and die. Neonatal withdrawal and mortality has been recorded in animals (e.g., Zagon and McLaughlin, 1977a,c; Lichtblau and Sparber, 1981), substantiating the clinical relevance of the laboratory findings.

Since opioids may affect the gastrointestinal tract (Burks, 1976; Jaffe and Martin, 1980), one must also consider that exposure of the mother and/or offspring may result in nutritional irregularities that could lead to alterations in developmental patterns. There is considerable literature from laboratory, as well as clinical, studies (Kirby, 1980; Naeye et al., 1973; Ford and Rhines, 1979; Seidler et al., 1982; White et al.,

TABLE 10   Effects of Opioids during Development on the Chemistry of the Nervous System

Authors	Observations[a]
Morphine	
Anderson and Slotkin (1975)	Altered pattern of tyrosine hydroxylase and dopamine β-hydroxylase, and the number of storage vesicles in the adrenal medulla
Bardo et al. (1982)	NS opioid receptors or monamine systems
Butler and Schanberg (1975)	↓ catecholamine levels ↓ β-hydroxylase activity
Coyle and Pert (1976)	NS opioid receptors
DiGiulio et al. (1988)	↑ [Met⁵] enkephalin NS Substance P
Gibson and Vernadakis (1983b)	NS ([³H]etorphine) opioid receptors (B_{max}) NS ([³H]etorphine) opioid receptors (K_d)
Hammer et al. (1989)	↓ (mu) opioid receptors
Handelmann and Quirion (1983)	↑ (mu) opioid receptors
Hansson and Rönnbäck (1983)	Altered protein synthesis
Hansson and Rönnbäck (1985)	Altered brain proteins
Jakubovic et al. (1978)	Altered brain proteins
Johannesson et al. (1972)	NS brain RNA, DNA, and protein
Kirby and Aronstam (1983)	↓ opioid receptors (B_{max}) NS catecholamine/norepinephrine
Mattio and Kirby (1982)	NS catecholamine/norepinephrine ↑ β-adrenergic receptors
Meriney et al. (1985)	↓ K⁺-stimulated synthesis of AChE
Peterson et al. (1974)	↑ choline acetyltransferase and acetylcholinesterase
Plishka and Neale (1984)	↑ clonidine-binding sites (B_{max}) NS clonidine-binding sites (K_d)
Rönnbäck and Hansson (1985)	Altered pattern of brain protein synthesis
Sakellaridis et al. (1986)	↓ choline acetyltransferase, days 4–6 NS choline acetyltransferase, days 6–8
Steele and Johannesson (1975a)	↓ monosomes and polysomes ↑ disome component of free ribosomes
Steele and Johannesson (1975b)	NS brain DNA and protein at birth
Steele and Johannesson (1975c)	↓ brain DNA, RNA, and protein in fetal brain NS RNA:DNA, and protein:DNA ratios
Tempel et al. (1988)	↓ (mu) opioid receptors (B_{max}) NS (mu) opioid receptors (K_d)
Tsang and Ng (1980)	Altered pattern of [Met⁵] enkephalin binding
Buprenorphine	
Tiong and Olley (1988)	NS [Met⁵] and [Leu⁵] enkephalin

(continued)

TABLE 10   (*Continued*)

Authors	Observations[a]
LAAM	
Gibson and Vernadakis (1983a)	↓ ([³H]etorphine) opioid receptors (B_{max}) ↑ ([³H]etorphine) opioid receptors (K_d)
Gibson and Vernadakis (1983b)	↓ ([³H]etorphine) opioid receptors (B_{max}) ↑ ([³H]etorphine) opioid receptors (K_d)
Kuwahara and Sparber (1981)	NS brain nucleic acids or protein content
Kuwahara and Sparber (1983)	NS ornithine decarboxylase, days 15, 17, and 19 ↑ ornithine decarboxylase at day 17 with naloxone administration
Lichtblau and Sparber (1983)	↓ brain protein concentration, day 19 of gestation ↓ brain RNA, DNA, and protein content postnatal day 1
Pentazocine	
Andres et al. (1988)	NS brain protein, phospholipid, and cholesterol contents ↑ inositol, phosphoglycerides, phosphatidyl-glycerol ↓ ethanolamine, serine phosphoglyceride
Methadone	
Darmani et al. (1991)	NS β_1 and β_2-adrenergic receptors
DeMontis et al. (1983)	NS muscarinic, serotonergic, opioid receptors ↓ β-adrenergic receptors ↓ imipramine-binding sites
Field et al. (1977)	↓ brain protein, neonate NS brain DNA, acetylocholinesterase, neonate ↓ brain protein, acetylcholinesterase, day 21 NS brain DNA, day 21
Gibson and Vernadakis (1983b)	NS ([³H]etorphine) opioid receptors (B_{max}) NS ([³H]etorphine) opioid receptors (K_d)
Hui et al. (1978)	↓ brain RNA and protein
Jakubovic et al. (1978)	↓ brain protein
Lau et al. (1977)	↓ brain tyrosine hydroxylase
McGinty and Ford (1980)	↓ norepinephrine and dopamine content and uptake, forebrain, days 1 and 20 NS norepinephrine and dopamine content and uptake, hindbrain, days 1, 20, and 40
Peters (1977)	↓ brain DNA, RNA, and protein content
Rech et al. (1980)	↓ monamines and metabolites
Sakellaridis et al. (1986)	↓ choline acetyltransferase, days 4–6
Seidler et al. (1982)	↓ brain ornithine decarboxylase, tyrosine hydroxylase
Slotkin et al. (1976)	↓ brain ornithine decarboxylase
Slotkin et al. (1979)	↓ 5-hydroxytryptamine, dopamine, norepinephrine
Slotkin et al. (1980a)	↓ brain ornithine decarboxylase
Slotkin et al. (1980b)	↑ (precocious) development of insulin-induced release of adrenal catecholamines

(*continued*)

TABLE 10   *(Continued)*

Authors	Observations[a]
Slotkin et al. (1981)	NS brain noradrenergic or dopaminergic systems
Slotkin et al. (1982)	↓ norepinephrine stimulation of ^{33}Pi
Tiong and Olley (1988)	↓ [Met^5] enkephalin (8 mg/kg) NS [Met^5] enkephalin (4 mg/kg)
Tsang et al. (1986)	↓ monoamine oxidase
Vernadakis et al. (1982)	↑ ornithine decarboxylase cyclic nucleotide phosphohydrolase
Wang et al. (1986)	↓ opioid receptors, α_2-adrenergic receptors
Zagon and McLaughlin (1977b)	↓ brain DNA
Zagon and McLaughlin (1978a)	↓ brain and cerebellar DNA, RNA, and protein
Zagon and McLaughlin (1982b)	↓ brain DNA, RNA, protein

[a]Arrows represent direction of change from controls. Unless otherwise indicated, all ages are postnatal. NS, not significant.

1978; McLaughlin et al., 1978; Slotkin et al., 1980; McGinty and Ford, 1980; Raye et al., 1977; Zagon and McLaughlin, 1977a,c, 1978a,b, 1982b) indicating that maternal undernutrition does not appear to be the etiologic basis for opioid-related problems. For example, White et al. (1978) found that food and water consumption in chronically treated mothers was not significantly reduced from control levels. Pair-feeding experiments also provide evidence that nutritional alterations are not the basis for the neurobiological effects observed with perinatal opioid exposure. Drug-treated litters have been noted to suckle frequently and to have milk in their stomachs as often as controls (Zagon and McLaughlin, 1977c). In addition, the significant decreases in litter size and birth weight that are often associated with mothers given restricted diets have not been observed with methadone-treated females (Zagon and McLaughlin, 1977c). Although cumulative undernutrition during gestation and lactation is known to be more detrimental than undernutrition during either gestation or lactation, animals exposed to opioids such as methadone during gestation and lactation do not demonstrate a cumulative response, but rather these pups often exhibit the fewest alterations (e.g., Zagon and McLaughlin, 1977a,c, 1978a,b, 1982b). Moreover, the ratios of brain weight to body weight indicate that the brains of methadone-treated pups were not "spared" as in the case of undernourished animals (Dobbing and Smart, 1973), but appeared to be a sensitive target to the actions of methadone (Zagon and McLaughlin, 1977a,b, 1978b, 1982b). Finally, the changes in organ weights as a result of perinatal opioid (i.e., methadone) exposure (McLaughlin and Zagon, 1980; McLaughlin et al., 1978), as well as the constellation of behavioral and neurochemical changes recorded in these animals (Zagon and McLaughlin, 1978a,b; Zagon et al.,

1979a,b; Seidler et al., 1982) are not compatible with those occurring with under-nutrition.

Another question that can be raised is whether postnatal undernutrition rather than prenatal undernutrition may be an important factor in understanding opioid studies. Zagon and McLaughlin (1982b) have published a report that addresses this issue. In one study, a direct comparison was made of the effects of postnatal exposure to methadone in animals raised in standard size litters of eight pups/mother to that of postnatal undernutrition in animals raised in litters of 18 pups/mother. The differences between these two treatments were striking. As anticipated from previous reports, 21-day-old undernourished rats weighed about half that of controls, but the under-nourished rats had brain weight deficits of only 12%; the brain:body weight ratios of these animals were considerably elevated over control levels. In contrast, the meth-adone-exposed pups had body-weight deficits of only 19%, but brain-weight deficits of 32%. Moreover, the methadone-treated rats had more severe deficits in the weight and total DNA content of the brain and cerebellum, as well as brain DNA concentra-tion, than did undernourished animals. Further evidence that the growth-inhibiting properties of methadone are not the result of reduced caloric intake was garnered by the inclusion of groups of rats subjected to postnatal methadone exposure and under-nutrition (control pups cross-fostered at birth to a mother receiving methadone and raised in litters of 18 pups/mother). Although these animals were expected to show the greatest changes in body weight and brain neurochemistry, deficiencies of smaller magnitude than those occurring with methadone alone were often recorded. The "protection" afforded by combined treatment may have been effected by the reduced milk consumption of the methadone-undernourished pups, thereby lowering their methadone intake, and hence, reducing the drug-related disturbances.

A central question that must be applied to the laboratory data is whether there are parallels between laboratory and clinical findings with regard to the perinatal opioid syndrome. A positive answer to this question would (1) provide affirmation for the basic biological actions of opioids, (2) support the use of animal studies to investigate the mechanisms of opioid action, and (3) endorse the utilization of laboratory systems as prognosticators of impending clinical problems, as well as for the exploration and evaluation of therapeutic protocols. Although clinical studies are often replete with confounding variables, and despite the difficulties of extrapolating laboratory results to the human situation, a number of striking parallels between clinical and laboratory findings do exist. These similarities include passive dependence of the fetus on opioids, occurrence of the neonatal abstinence syndrome, high rate of morbidity and mortality of the neonate when not given supportive therapy for withdrawal, sleep disturbances in the neonate, protracted/subacute withdrawal, delays in sensorimotor development, retardation in somatic growth, smaller head circumferences in hu-mans/smaller brain sizes in laboratory animals, delays in walking, problems in visual and/or auditory systems, abnormalities in neuropathological studies indicating altera-tions in neural ontogeny, diminished alertness, poor attention spans in early phases of development, hyperactivity in later phases of development, learning disabilities, and social maladjustments.

ETIOLOGICAL AND MECHANISTIC CONSIDERATIONS OF THE PERINATAL OPIOID SYNDROME

The results of laboratory experiments provide a compelling argument that opioids disturb the pattern of somatic and neurobiological development. However, it is difficult to ascribe any pathognomonic feature(s) or characteristic(s) that one could associate with the syndrome, or that can be utilized to distinguish opioid effects from the manifestations exerted by other detrimental agents and/or conditions. Of course, the same ends do not infer the same means. For example, smaller head size may be caused by a number of distinct situations such as cell death or diminishment of cell proliferation, with the final outcome showing a similarity. Rather, opioids appear to cause a variety of delays in ontogeny. These alterations occurring in early life may persist and have ramifications in later years. Opioids do appear to inhibit growth. It can be envisioned that maternal opioid consumption subjects the fetus and infant by way of the placenta and breast milk, respectively, to these substances and alters developmental events until tolerance is established. The magnitude of these effects depends on such things as drug dosage and the length of time needed to establish tolerance in the developing organism. Withdrawal from opioids at birth or in the fetus forces developing cells that have become tolerant to the opioid to readjust to the absence of drug. This withdrawal reaction could result in delayed growth, with the extent of these delays correlating to the extent of withdrawal. Amelioration of withdrawal by administration (breast feeding or direct administration) of opioids would serve to eliminate or minimize the withdrawal reaction and circumvent the need of the tissues and cells to readjust to a new environment and to continue in an established (albeit, opioid-containing) milieu. This hypothesis fits nicely with laboratory findings that prenatal exposure followed by postnatal withdrawal is often the most devastating to developmental processes, whereas prenatal exposure along with administration of opioids in the postnatal period usually exerts the least damage. It is understandable that elimination or substantial reduction of the opioid withdrawal syndrome probably attenuates many detrimental influences. Therefore, it is hardly any wonder that difficulties might exist in establishing firm evidence separating the effects of perinatal opioid exposure from other potentially damaging influences. Woven into the fabric of this intricate network, one must realize that perinatal opioid exposure may, directly or indirectly, involve a host of other factors such as hormonal imbalance, genetic and epigenetic influences, stress, altered nutrition status, and dysfunctional behavior of the mother and/or infant. In addition to the presence of toxicity, physical dependence, tolerance, and/or withdrawal, it should be kept in mind that the fetus and neonate are growing rapidly, with cell proliferation, cell migration, and cell/tissue/organ differentiation occurring. All of these factors may contribute to the perinatal opioid syndrome and affect the status and welfare of the offspring.

It is valuable to consider how opioids may damage the developing organism. As mentioned earlier, opioids appear to selectively accumulate in the brains and nervous tissues of fetal rats (Peters et al., 1972), monkeys (Davis and Fenimore, 1978), and preweaning rats (Shah and Donald, 1979), presumably because of an increased permeability of the blood–brain barrier in developing organisms (Shah and Donald, 1979). Opioids have been reported to exert a stereospecific effect on the growth of animals (Smith et al., 1977; Crofford and Smith, 1973), suggesting that opioid action is quite specific and involves opioid receptors. The effects of opioids can be blocked by concomitant administration of an opioid antagonist such as naloxone (Crofford and Smith, 1973; Hui et al., 1978; Meriney et al., 1985), indicating that opioid action may reside at the level of the opioid receptor. In this regard, opioid receptors have been found in body and brain tissues of developing organisms (e.g., Gibson and Vernadakis, 1982; Clendeninn et al., 1976), often appearing only, or at the greatest concentration, during the developmental period. Morphine administered to 1-day-old rats inhibits [^{3}H]thymidine incorporation into brain DNA in a naloxone-reversible manner, suggesting that opioids can depress cell proliferation (Kornblum et al., 1987). Tempel and colleagues (1988) have reported that chronic administration of morphine to pre- and postnatal rats produces a marked decrease in brain mu-opiate receptor density without a change in receptor affinity; no changes in delta- or kappa-opiate receptors were observed. Interestingly, Tempel observed a tolerance to the analgesic reactions of morphine in these animals, indicating that exposure to some types of opioids may have notable functional influences with respect to receptor regulation. Alterations in receptor profiles by perinatal opioid exposure were also found by Gibson and Vernadakis (1983a,b) and Vernadakis and Gibson (1985). These investigators reported that 1-alpha-acetylmethadol (LAAM), a long-acting opioid that has been considered a possible substitute for methadone, altered both the binding affinity and capacity of [^{3}H]-etorphine, in both in vivo and in vitro preparations. Tissue culture studies have confirmed the stereospecific and naloxone-reversible effects of opioids on cell growth, and also reveal that tolerance can develop to these influences (e.g., Zagon and McLaughlin, 1984c). Additionally, cultured cells that are physically dependent on opioids and removed from drugs go through a withdrawal. A principle sign of cellular withdrawal is a diminution in mitotic activity. Vernadakis and co-workers (1982), using cultures of neuronal and glial cells, reported that ornithine decarboxylase, thought to play a role in neural growth, was altered in methadone-subjected cultures. Sakellaridis et al. (1986) have shown that brain cell cultures derived from chick embryos may react in a neurotoxic fashion to morphine, but not to methadone. However, methadone may influence some cholinergic functions, since a decrease in choline acetyltranferase activity was noted in opiate-treated cultures. Finally, Choi and Visekul (1988) have reported a selective attenuation of N-methyl-D-aspartate neurotoxicity of cortical neurons in opioid-subjected cultures.

Much of the approach in early studies on the effects of opioids on the developing organism could best be labeled phenomenological. By a wide variety of strategies (see Table 1), the fetus or infant was subjected to opioids, either by the maternal route or by direct injection, and the offspring studied with regard to some biological aspect (see Tables 2–10). What has emerged from these studies is a view that opioids can exert selective effects on growth, many involving the nervous system. The very nature of these selective actions in the face of such a variety of methodologies does not appear to support a theory that opioids work in a nonspecific, random, toxic fashion. Another possible explanation for the effects of opioids is the theory that the constellation of somatic and neurobiological effects is epiphenomenal. This term is used in the sense that the manifestations of the perinatal opioid syndrome are the result of other phenomena elicited by drug action. For example, opioids can depress respiration, alter food and water intake, and the effects observed in offspring exposed to opioids could be the result of these processes. Experiments and comparisons to the effects of some of these indirect influences have been conducted with regard to opioids and growth (see earlier discussion). In summary, it appears to be difficult to ascribe all, or even many, of the effects of opioids on the developing organism to these potential epiphenomena. Additionally, evidence from tissue culture studies suggests that the effects of opioids on growth processes need not involve the entire organism, but rather work at a cellular level. However, in stating this, one must be aware that, to some degree, other phenomena may be involved with the perinatal opioid syndrome. For example, the effects of drug administration or withdrawal certainly must have ramifications on a wide variety of physiological processes that may in turn affect the welfare of the animal. However, once again, at this juncture no singular element stands out that permits us to associate the perinatal opioid syndrome with another phenomenon.

Another theory of opioid action on growth emanates from knowledge gained in the 1970s. In 1973, the laboratories of Snyder, Terenius, and Simon made the discovery of a receptor associated with nervous tissue that was involved with opioids. This historic finding was followed by the elegant work of Hughes, Kosterlitz, and coworkers (1975), which demonstrated endogenous opioid substances (termed endogenous opioids). Many of the activities of exogenous opioids appear to be related to the nervous system, and one could postulate that the opioids are altering development because they are interacting with opioid receptors. If these endogenous opioids act as neurotransmitters at the opioid receptor, the excess or deficit (in the case of withdrawal) of opioids from perinatal opioid exposure may invoke problems in the ontogeny of the nervous system. This hypothesis is supported by the literature showing that neurotransmitters as well as opioid receptors are altered by opioid administration. Additionally, the blockade of opioid effects by concomitant administration of an opioid antagonist (see earlier discussion) suggests involvement of opioid receptors.

Although the hypothesis that opioids such as heroin, methadone, and morphine administered during early life could have effects on the nervous system that would produce the perinatal opioid syndrome, it should be kept in mind that a number of opioid effects are not related to the nervous system. For example, body weight may be reduced, as well as organ weights, in animals exposed to opioids during early life. Moreover, one might expect that alterations in neurotransmitters or opioid receptors would have little influence on cell proliferation. Another hypothesis that might underlie the perinatal opioid syndrome is based on observations initially reported in 1983 (Zagon and McLaughlin, 1983). Investigators found that administration of naltrexone, an opioid antagonist, profoundly influenced somatic and neurobiological development. In essence, a blockade of opioid receptors resulted in animals that demonstrated marked increases in growth that were dependent on the duration of opioid receptor blockade (Zagon and McLaughlin, 1984d). A complete blockade of opioid receptors during embryogenesis caused animals to be developmentally accelerated with respect to their control counterparts. Animals given opioid antagonists to completely block receptors for the entire period of potential interaction exhibited a greater number of neurons and glia engaged in proliferation, an increase in dendritic elaboration and synaptogenesis, and acceleration in the acquisition of physical and behavioral characteristics (Zagon and McLaughlin, 1985, 1987, 1991; Hauser et al., 1987, 1989; Isayama et al., 1991). These results indicated that opioid antagonists blocked the interaction of endogenous opioid peptides from the opioid receptor, and that this interaction must be related to growth processes. Thus, endogenous opioids must be able to regulate growth by an inhibitory process, serving as a negative regulator of growth. Moreover, the endogenous opioid related to growth must be tonically active, since interference with opioid–receptor interfacing produced dramatic growth alterations. The effects on growth by opioid antagonists obeyed pharmacologic principles since the ($-$) isomer was active and the ($+$) isomer, inactive (Zagon and McLaughlin, 1989). Continuing research revealed that the control of cell proliferation appears to be the target for opioid action; endogenous opioids depress DNA synthesis and mitosis in proliferating neural cells destined to become neurons and glia (Zagon and McLaughlin, 1987, 1991). Extensive study has shown that the pentapeptide, [Met^5]enkephalin, is one of the most potent peptides in regard to regulating cell proliferation (Zagon and McLaughlin, 1991). Concentrations as low as 100 μg/kg of [Met^5]enkephalin result in a marked reduction in DNA synthesis of replicating brain cells; administration of naloxone blocked the effect of [Met^5]enkephalin. Opioids selective for other receptors such as mu, delta, kappa, epsilon, and sigma did not alter DNA synthesis. Using antibodies to [Met^5]enkephalin, this substance was found to be associated with developing neural and non-neural tissues, but not adult differentiated counterparts (Zagon et al., 1985, 1986). Moreover, immunoreactivity observed in immunocytochemical preparations showed that [Met^5]enkephalin was detected in the cortical cytoplasm of germinative neural cells but not differentiated cells (Zagon and McLaughlin, 1990b; Zagon et al., 1985). Immunoelectron microscopic studies confirmed these observations and revealed that opioid activity was associated with most organelles in proliferating cells (Zagon and McLaughlin, 1990b). Interestingly, a great deal of immunoreactivity was associated with macroneurons that are generated much earlier in ontogeny. A major question to be addressed was concerned with the source of the opioid peptide regulating growth. To examine this question further, in situ hybridization was utilized. Studies with the 1-day-old retina, which contains germinative retinal cells and ganglion neurons, demonstrated that gene expression of preproenkephalin messenger RNA (mRNA) (the prohormone that gives rise to [Met^5]enkephalin) was associated with both the neuroblasts and the ganglion cells (Isayama and Zagon, 1991). This would suggest that [Met^5]enkephalin may be produced from both autocrine (neuroblasts) and paracrine (ganglion cells) sources.

Another aspect of this work has been the identification and characterization of the opioid receptor involved with growth. To study this matter further, the strategy was to probe developing nervous tissue with a radiolabeled [Met^5]enkephalin using a receptor-binding assay (Zagon et al., 1991). Specific and saturable binding with high

272   Ian S. Zagon and Patricia J. McLaughlin

affinity was detected in homogenates of developing cerebellar tissues. The binding affinity (K_d) of radiolabeled [Met^5]enkephalin was 2.2 nM, and the binding capacity (B_{max}) was 22.3 fmol/mg protein. Competitive inhibition profiles showed that [Met^5]enkephalin was the most avid compound displacing the radiolabeled ligand. Moreover, the location of [Met^5]enkephalin binding was nuclear rather than cytoplasmic/membranous (the putative location of other opioid receptors). This new opioid receptor has been termed zeta, in reference to its association with the proliferation of life (Greek, zoe) (Zagon et al., 1991). Studies in the developing human brain also show the presence of the zeta receptor (Zagon et al., 1990).

Placed in the context of the perinatal opioid syndrome, the discovery of an endogenous opioid system that regulates somatic and neurobiological growth raises the question of potential interaction between exogenous opioids and native biological phenomena. If opioid exposure in early life alters developmental events, it is possible that opioids are somehow mistaken for endogenous opioids in the regulatory pathways, thereby leading to alterations in the biological profile of the organism. An excess of opioids such as heroin, morphine, or methadone as a result of perinatal opioid exposure may depress growth by acting like the overexpression of trophic factors controlling growth. This exciting possibility could well provide valuable insights into the mechanisms of how opioids can exert a detrimental influence on the developing organism, and may even hold a key for prophylactic measures or provide insight for treatment.

REFERENCES

Abrams, C. A. L. (1975). Cytogenetic risks to the offspring of pregnant addicts. *Addict. Dis.* 2, 63–77.
Amarose, A. P., and Norusis, J. J. (1976) Cytogenetics of methadone-managed and heroin-addicted pregnant women and their newborn infants. *Am. J. Obstet. Gynecol.* 123, 653–639.
Anderson, T. R., and Slotkin, T. A. (1975). Maturation of the adrenal medulla. IV. Effects of morphine. *Biochem. Pharmacol.* 24, 1469–1474.
Andres, R., Cabezas, J. A., and Llanillo, M. (1988). Changes in forebrain, cerebellum, and brain stem phospholipid patterns from mothers and newborn rats after chronic pentazocine treatment during the gestation period. *Ital. J. Biochem.* 37, 275–283.
Arcuri, P. A., and Gautieri, R. F. (1973). Morphine-induced fetal malformations. III. Possible mechanisms of action. *J. Pharm. Sci.* 62, 626–1634.
Arjune, D., and Bodnar, R. J. (1989). Post-natal morphine differentially affects opiate and stress analgesia in adult rats. *Psychopharmacology* 98, 512–517.
Aylward, G. P. (1982). Methadone outcome studies: Is it more than the methadone? *J. Pediatr.* 101, 214–215.
Bardo, M. T., Bhatnagar, R. K., and Gebhart, G. F. (1982). Differential effects of chronic morphine and naloxone on opiate receptors, monoamines, and morphine-induced behaviors in preweanling rats. *Brain Res.* 256, 139–147.
Blatman, S., and Lipsitz, P. J. (1972). Children of women maintained on methadone: Accidental methadone poisoning of children. *In* "Proceedings of the Fourth National Conference on Methadone Treatment," pp. 175–176. NAPAN, New York.
Blinick, G., Inturrisi, C. E., Jerez, E., and Wallach, R. C. (1975). Methadone assays in pregnant women and progeny. *Am. J. Obstet. Gynecol.* 121, 617–621.
Blinick, G., Wallach, R. C., Jerez, E., and Ackerman, B. D. (1976). Drug addiction in pregnancy and the neonate. *Am. J. Obstet. Gynecol.* 125, 135–142.

Blum, R. H. (1970). A history of opium. *In* "Society and Drugs. I. Social and Cultural Observations," pp. 45–58. Jossey-Bass, San Francisco, California.

Buchenauer, D., Turnbow, M., and Peters, M. A. (1974). Effect of chronic methadone administration on pregnant rats and their offspring. *J. Pharmacol. Exp. Ther.* 189, 66–71.

Bureau, A. (1895). Accouchement d'une morphinomane: Prevue chimique du passage de la morphine à travers le placenta: reflexions. *Bull. Mem. Soc. Obstet. Gynecol.* (Paris) 356–362.

Burks, T. F. (1976). Gastrointestinal pharmacology. *Rev. Pharmacol. Toxicol.* 16, 15–31.

Butler, S. R., and Schanberg, S. M. (1975). Effect of maternal morphine administration on neonatal rat brain ornithine decarboxylase (ODC). *Biochem. Pharmacol.* 24, 1915–1918.

Carr, J. N. (1975). Drug patterns among drug-addicted mothers: Incidence, variance in use, and effects on children. *Pediatr. Ann.* 4, 408–417.

Castellano, C., and Ammassari-Teule, M. (1984). Prenatal exposure to morphine in mice: Enhanced responsiveness to morphine and stress. *Pharmacol. Biochem. Behav.* 21, 103–108.

Caza, P. A., and Spear, L. P. (1980). Ontogenesis of morphine-induced behavior in the rat. *Pharmacol. Biochem. Behav.* 13, 45–50.

Chandler, J. M., Robie, P., Schoolar, J., and Desmond, M. M. (1975). The effects of methadone on maternal–fetal interactions in the rat. *J. Pharmacol. Exp. Ther.* 192, 549–554.

Chasnoff, I. J., Hatcher, R., and Burns, W. J. (1980). Early growth patterns of methadone-addicted infants. *Am. J. Dis. Child.* 134, 1049–1051.

Chasnoff, I. J., Hatcher, R., and Burns, W. J. (1982). Polydrug- and methadone-addicted newborns: A continuum of impairment? *Pediatrics* 70, 210–213.

Chavez, C. J., Ostrea, E. M., Stryker, J. C., and Smialek, Z. (1979). Sudden infant death syndrome among infants of drug-dependent mothers. *J. Pediatr.* 95, 407–409.

Choi, D. W., and Visekul, V. (1988). Opioids and nonopioid enantiomers selectively attenuate N-methyl-D-aspartate neurotoxicity on cortical neurons. *Eur. J. Pharmacol.* 155, 27–35.

Clendeninn, N. J., Petraitis, M., and Simon, E. J. (1976). Ontological development of opiate receptors in rodent brain. *Brain Res.* 118, 157–160.

Coyle, J. T., and Pert, C. B. (1976). Ontogenetic development of [³H]naloxone binding in rat brain. *Neuropharmacology* 15, 555–560.

Crain, S. M., Crain, B., Finnigan, T., and Simon, E. J. (1979). Development of tolerance to opiates and opioid peptides in organotypic cultures of mouse spinal cord. *Life Sci.* 25, 1797–1802.

Crain, S. M., Crain, B., and Peterson, E. R. (1982). Development of cross-tolerance to 5-hydroxytryptamine in organotypic cultures of mouse spinal cord ganglia during chronic exposure to morphine. *Life Sci.* 31, 241–247.

Crofford, M., and Smith, A. A. (1973). Growth retardation in young mice treated with *dl*-methadone. *Science* 181, 947–949.

Darmani, N. A., Schnoll, S. H., Fuchs, B., and Martin, B. R. (1991). Does chronic prenatal methadone exposure affect β-receptor subtypes in placental, fetal, and maternal brain homgenates? *Neurotoxicol. Teratol.* 13, 43–48.

Davis, C. M., and Fenimore, D. C. (1978). The placental transfer and maternofetal disposition of methadone in monkeys. *J. Pharmacol. Exp. Ther.* 205, 577–586.

Davis, M. M., and Shanks, B. (1975). Neurological aspects of perinatal narcotic addiction and methadone treatment. *Addict. Dis.* 2, 213–226.

Davis, W. M., and Lin, C. H. (1972). Prenatal morphine effects on survival and behavior of rat offspring. *Res. Commun. Chem. Pathol. Pharmacol.* 3, 205–214.

DeMontis, G. M., Devoto, P., Angioi, R. M., Curreli, V., and Tagliamonte, A. (1983). *In utero* exposure to methadone produces a stable decrease of the cortex 5-HT transport system in rats. *Eur. J. Pharmacol.* 90, 57–63.

Desmond, M. M., and Wilson, G. S. (1975). Neonatal abstinence syndrome: Recognition and diagnosis. *Addict. Dis.* 2, 112–121.

DiGiulio, A. M., Restani, P., Galli, C. L., Tenconi, B., LaCroix, R., and Gorio, R. (1988). Modified ontogenesis of enkephalin- and substance P-containing neurons after perinatal exposure to morphine. *Toxicology* 49, 197–201.

Dinges, D. F., Davis, M. M., and Glass, P. (1980). Fetal exposure to narcotics: Neonatal sleep as a measure of nervous system disturbance. *Science* **209**, 619–621.

Dobbing, J., and Smart, J. L. (1973). Early undernutrition, brain development, and behavior. *Clin. Dev. Med.* **47**, 16–36.

Dole, V. P., and Kreek, M. J. (1973). Methadone plasma level: Sustained by a reservoir of drug in tissue. *Proc. Natl. Acad. Med. U.S.A.* **70**, 10.

Dole, V. P., and Nyswander, M. (1965). A medical treatment for diacetylmorphine (heroin) addiction. *J.A.M.A.* **193**, 646–650.

Earle, F. B. (1888). Maternal opium habit and infant mortality. *M. Standard (Chicago)* **3**, 2.

Férè, Ch. (1883). De la morphinomanie au point de vue de la grossesse et de la né du foetus. *Comm. Soc. Biol. Sem. Méd.* 294.

Field, T., McNelly, A., and Sadava, D. (1977). Effect of maternal methadone addiction on offspring in rats. *Arch. Int. Pharmacodyn. Therap.* **228**, 300–303.

Finnegan, L. P. (1975). Narcotics dependence in pregnancy. *J. Psychodel. Drugs* **7**, 299–311.

Finnegan, L. P. (1976). Clinical effects of pharmacologic agents on pregnancy, the fetus, and the neonate. *Ann. N.Y. Acad. Sci.* **281**, 74–89.

Finnegan, L. P., Connaughton, J. F., Emich, J. P., and Wieland, W. (1972). Comprehensive care of the pregnant addict and its effect on maternal and infant outcome. *Committee on Problems of Drug Dependence (34th Annual Scientific Meeting)*, Ann Arbor, Michigan, pp. 372–390.

Finnegan, L. P., Reeser, D. S., Ting, R. Y., Rozenzwerg, M., and Keller, A. (1977). Growth and development of children born to women maintained on methadone during pregnancy. *Pediatr. Res.* **11**, 377.

Ford, D., and Rhines, R. (1979). Prenatal exposure to methadone HCl in relationship to body and brain growth in the rat. *Acta Neurol. Scand.* **59**, 248–262.

Freeman, P. R. (1980). Methadone exposure in utero: Effects on open-field activity in weanling rats. *Int. J. Neurosci.* **11**, 295–300.

Friedler, G. (1977). Effect of pregestational morphine and methadone administration to mice on the development of their offspring. *Fed. Proc.* **36**, 1001.

Fujinaga, M., and Mazze, R. I. (1988). Teratogenic and postnatal developmental studies of morphine in Sprague–Dawley rats. *Teratology* **38**, 401–410.

Fürst, C. (1889). Morphium bei schwangeren, gebärenden und Säugender. *Wien. Klin. Wschr.* **2**, 191–220.

Geber, W. F., and Schramm, L. C. (1975). Congenital malformations of the central nervous system produced by narcotic analgesics in the hamster. *Am. J. Obstet. Gynecol.* **123**, 705–713.

Ghadrian, A. (1969). A tissue culture study of morphine dependence on the mammalian CNS. *Can. Psychiatr. Assoc. J.* **14**, 607–615.

Gibson, D. A., and Vernadakis, A. (1982). [³H]etorphine binding activity in early chick embryos: Brain and body tissue. *Dev. Brain Res.* **4**, 23–29.

Gibson, D. A., and Vernadakis, A. (1983a). Critical period for LAAM in the chick embryo: Toxicity and altered opiate receptor binding. *Dev. Brain Res.* **8**, 61–69.

Gibson, D. A., and Vernadakis, A. (1983b). Effects of N-LAAM on [³H]etorphine binding in neuronal-enriched cell cultures. *Neurochem. Res.* **8**, 1197–1202.

Glass, L., Rajegowda, B. K., and Evans, H. E. (1971). Absence of respiratory distress syndrome in premature infants of heroin-addicted mothers. *Lancet*, **ii**, 685–686.

Glick, S. D., Strumpf, A. J., and Zimmerberg, B. (1977). Effect of in utero administration of morphine on the subsequent development of self-administration behavior. *Brain Res.* **132**, 194–196.

Goodfriend, M. J., Shey, I. A., and Klein, M. D. (1956). The effects of maternal narcotic addiction on the newborn. *Am. J. Obstet. Gynecol.* **71**, 29–36.

Graham-Mulhall, S. (1926). "Opium the Demon Flower." Harold Vinal, New York (reprinted in 1981 by Arno Press).

Grode, M. L., and Murray, M. R. (1973). Effects of methadone-HCl on dorsal root ganglia in organotypic culture. *Exp. Neurol.* **40**, 68–81.

Grove, L. V., Etkin, M. K., and Rosecrans, J. A. (1979). Behavioral effects of fetal and neonatal exposure to methadone in the rat. *Neurobehav. Toxicol.* **1**, 87–95.

Hammer, R. P., Ricalde, A. A., and Seatriz, J. V. (1989). Effects of opiates on brain development. *Neurotoxicology* **10**, 475–483.

Handelmann, G. E., and Dow-Edwards, D. (1985). Modulation of brain development by morphine: Effects on central motor systems and behavior. *Peptides* **6** (Suppl.) 29–34.

Handelmann, G. E., and Quirion, R. (1983). Neonatal exposure to morphine increases mu-opiate binding in the adult forebrain. *Eur. J. Pharmacol.* **94**, 357–358.

Hansson, E., and Rönnbäck, L. (1983). Incorporation of ^{3}H-valine into soluble protein of cultivated astroglial cells after morphine treatment. *J. Neurosci. Res.* **10**, 279–288.

Hansson, E., and Rönnbäck, L. (1985). Amino acid incorporation during morphine intoxication. II. Electrophoretic separation of extracellular proteins from cerebral hemisphere slices and astroglia-enriched primary cultures. *J. Neurosci. Res.* **14**, 479–490.

Harpel, H. S., and Gautieri, R. F. (1968). Morphine-induced fetal malformations. I. Exencephaly and axial skeletal fusions. *J. Pharmacol. Sci.* **57**, 1590–1597.

Harte, E. H., Guthar, C. L., and Kreek, M. J. (1976). Long-term persistence of *dl*-methadone in tissues. *Clin. Res.* **24**, 663a.

Hauser, K. F., McLaughlin, P. J., and Zagon, I. S. (1987). Endogenous opioids regulate dendritic growth and spine formation in developing rat brain. *Brain Res.* **416**, 157–161.

Hauser, K. F., McLaughlin, P. J., and Zagon, I. S. (1989). Endogenous opioid systems and the regulation of dendritic growth and spine formation. *J. Comp. Neurol.* **281**, 13–22.

Hess, G. D., and Zagon, I. S. (1988). Endogenous opioid systems and neural development. Ultrastructural studies in the cerebellar cortex of infant and weanling rats. *Brain Res. Bull.* **20**, 473–475.

Householder, J., Hatcher, R., Burns, W., and Chasnoff, I. (1982). Infants born to narcotic-addicted mothers. *Psychol. Bull.* **92**, 453–468.

Hovious, J. R., and Peters, M. A. (1984). Analgesic effect of opiates in offspring of opiate-treated female rats. *Pharmacol. Biochem. Behav.* **21**, 555–559.

Hovious, J. R., and Peters, M. A. (1985). Opiate self-administration in adult offspring of methadone-treated female rats. *Pharmacol. Biochem. Behav.* **22**, 949–953.

Hughes, J. A., Smith, T. W., Kosterlitz, H. W., Fothergill, L. A., Morgan, B. A., and Morris, H. R. (1975). Identification of the pentapeptides from the brain with potent opiate agonist activity. *Nature (London)* **258**, 577–579.

Hui, F. W., Krikum, E., and Smith, A. A. (1978). Inhibition of *d,l*-methadone of RNA and protein synthesis in neonatal mice: Antagonism by naloxone or naltrexone. *Eur. J. Pharmacol.* **49**, 87–93.

Hutchings, D. E., Feraru, E., Gorinson, H. S., and Golden, R. R. (1979a). Effects of prenatal methadone on the rest–activity cycle of the preweanling rat. *Neurobehav. Toxicol.* **1**, 33–40.

Hutchings, D. E., Towey, J. P., Gorinson, H. S., and Hunt, H. F. (1979b). Methadone during pregnancy: Assessment of behavioral effects in the rat offspring. *J. Pharmacol. Exp. Ther.* **208**, 106–112.

Hutchings, D. E., Towey, J. P., and Bodnarenko, S. R. (1980). Effects of prenatal methadone on the activity level in the preweanling rat. *Neurobehav. Toxicol.* **2**, 331–335.

Isayama, T., and Zagon, I. S. (1991). Localization of preproenkephalin A mRNA in the neonatal rat retina. *Brain Res. Bull.* **27**, 805–808.

Isayama, T., McLaughlin, P. J., and Zagon, I. S. (1991). Endogenous opioids regulate cell proliferation in the retina of developing rat. *Brain Res.* **544**, 79–85.

Jaffe, J. H., and Martin, W. R. (1980). Opioid analgesics and antagonists. In "The Pharmacological Basis of Therapeutics" (A. G. Gilman, L. S. Goodman, and A. Gilman, eds.), 6th Ed., pp. 494–534. Macmillan, New York.

Jakubovic, A., McGeer, E. G., and McGeer, P. L. (1978). Effects of *dl*-methadone and morphine on developing chick embryo. *Experientia* **34**, 1617–1618.

Johannesson, T., and Becker, B. A. (1972). The effects of maternally administered morphine on rat fetal development and resultant tolerance to the analgesic effect of morphine. *Acta Pharmacol. Toxicol.* **31**, 305–313.

Johannesson, T., Steele, W. J., and Becker, B. A. (1972). Infusion of morphine in maternal rats at near-term: Maternal and fetal distribution and effects on analgesia, brain DNA, RNA, and protein. *Acta Pharmacol. Toxicol.* **31**, 353–368.

Johnson, H. L., and Rosen, T. S. (1982). Prenatal methadone exposure: Effects on behavior in early infancy. *Pediatr. Pharmacol.* **2**, 113–120.

Jurand, A. (1973). Teratogenic activity of methadone hydrochloride in mouse and chick embryos. *J. Embryol. Exp. Morphol.* **30,** 449–458.

Jurand, A. (1980). Malformations of the central nervous system induced by neurotropic drugs in mouse embryos. *Dev. Growth Diff.* **22,** 61–78.

Jurand, A. (1985). The interference of naloxone hydrochloride in the teratogenic activity of opiates. *Teratology* **31,** 235–240.

Kaltenbach, K., and Finnegan, L. P. (1992). Methadone maintenance during pregnancy: Implications for perinatal and developmental outcome. In "Perinatal Substance Abuse: Research Findings and Clinical Implications" (T. B. Sonderegger, ed.), Johns Hopkins Press, Baltimore, Maryland. In press.

Kaltenbach, K., Graziani, L. T., and Finnegan, L. P. (1979). Methadone exposure *in utero*: Developmental state at one and two years of age. *Pharmacol. Biochem. Behav.* **11** (Suppl.), 15–17.

Kandall, S. R. (1977). Late complications in passively addicted infants. In "Drug Abuse in Pregnancy and Neonatal Effects" (J. L. Rementeria, ed.), pp. 116–128. C. V. Mosby, St. Louis, Missouri.

Kandall, S. R., Gartner, L. M., and Berle, B. B. (1974). Birthweights and maternal narcotic use. *Pediat. Res.* **8,** 364.

Kaplan, S. L., Kron, R. E., Phoenix, M. D., and Finnegan, L. P. (1978). Brazelton Neonatal Assessment at three and twenty-eight days of age: A study of passively addicted infants, high risk infants, and normal infants. In "Critical Concerns in the Field of Drug Abuse" (A. Schecter, H. Alksne, and E. Kaufman, eds.), pp. 726–730. Marcel Dekker, New York.

Kirby, M. L. (1979). Effects of morphine on spontaneous activity of 18-day rat fetus. *Dev. Neurosci.* **2,** 238–244.

Kirby, M. L. (1980). Reduction of fetal rat spinal cord volume following maternal morphine injection. *Brain Res.* **202,** 143–150.

Kirby, M. L. (1981). Effects of morphine and naloxone on spontaneous activity of fetal rats. *Exp. Neurol.* **73,** 430–439.

Kirby, M. L. (1983). Recovery of spinal cord volume in postnatal rats following prenatal exposure to morphine. *Brain Res.* **282,** 211–217.

Kirby, M. L., and Aronstam, R. S. (1983). Levorphanol-sensitive [³H]naloxone binding in developing brain stem following prenatal morphine exposure. *Neurosci. Lett.* **35,** 191–195.

Kirby, M. L., and Holtzman, S. G. (1982). Effects of chronic opiate administration on spontaneous activity of fetal rats. *Pharmacol. Biochem. Behav.* **16,** 263–269.

Kirby, M. L., DeRossett, S. E., and Holtzman, S. G. (1982). Enhanced analgesic response to morphine in adult rats exposed to morphine prenatally. *Pharmacol. Biochem. Behav.* **17,** 1161–1164.

Kornblum, H. I., Loughlin, S. E., and Leslie, F. M. (1987). Effects of morphine on DNA synthesis in neonatal rat brain. *Dev. Brain Res.* **31,** 45–52.

Kron, R. E., Litt, M., Phoenix, M. D., and Finnegan, L. P. (1976). Neonatal narcotic abstinence: Effects of pharmacotherapeutic agents and maternal drug usage on nutritive sucking behavior. *J. Pediatr.* **88,** 637–641.

Kuwahara, M. D., and Sparber, S. B. (1981). Continuous exposure of the chick embryo to l-alpha-noracetylmethadol does not alter protein or nucleic content. *Dev. Pharmacol. Therap.* **3,** 12–24.

Kuwahara, M. D., and Sparber, S. B. (1982). Behavioral consequences of embryonic or early postnatal exposure to 1-α-noracetyl methadol (NLAAM) in the domestic chicken. *Neurobehav. Toxicol. Teratol.* **4,** 323–329.

Kuwahara, M. D., and Sparber, S. B. (1983). Opiate withdrawal increases ornithine decarboxylase activity which is otherwise unaltered in brains of dependent chicken fetuses. *Life Sci.* **32,** 495–502.

Laase, C. F. J. (1919). Narcotic drug addiction in the newborn, report of a case. *Amer. Med.* **25,** 283–286.

Langstein, L. (1930). Uber das schicksal von morphiumsuchtigen frauen geborener sauglinge. *Med. Klin.* **26,** 500–501.

LaPointe, G., and Nosal, G. (1982). Morphine treatment during rat pregnancy: Neonatal and preweaning consequences. *Biol. Neonate* **42,** 22–30.

Lasky, D. I., Zagon, I. S., and McLaughlin, P. J. (1977). Effect of maternally administered heroin on the motor activity of rat offspring. *Pharmacol. Biochem. Behav.* **7,** 281–284.

Lau, C., Bartolome, M., and Slotkin, T. A. (1977). Development of central and peripheral catecholaminergic systems in rats addicted perinatally to methadone. *Neuropharmacology* **16,** 473–478.

Lichtblau, L., and Sparber, S. B. (1981). Outcome of pregnancy in rats chronically exposed to l-alpha-acetylmethadol (LAAM). *J. Pharmacol. Exp. Ther.* **218**, 303–308.

Lichtblau, L., and Sparber, S. B. (1982). Congenital behavioral effects in mature rats prenatally exposed to *levo*-alpha-acetylmethadol (LAAM). *Neurobehav. Toxicol. Teratol.* **4**, 557–565.

Lichtblau, L., and Sparber, S. B. (1983). Prenatal *levo*-alpha-acetylmethadol (LAMM) and/or naloxone: Effects on brain chemistry and postweaning behavior. *Neurobehav. Toxicol. Teratol.* **5**, 479–486.

Lichtenstein, P. M. (1915). Infant drug addiction. *N.Y. Med. J.* **15**, 905.

Liu, S. J., and Wang, R. I. J. (1975). Increased analgesia and alterations in distribution and metabolism and methadone by desipramine in the rat. *J. Pharmacol. Exp. Ther.* **195**, 94–104.

Lodge, A. (1977). Developmental findings with infants born to mothers on methadone maintenance: A preliminary report. *In* "NIDA Symposium on Comprehensive Health Care for Addicted Families and Their Children" (G. Beschner and R. Brotman, eds.), pp. 79–85. U.S. Government Printing Office, Washington, D.C.

Lodge, A., Marcus, M. M., and Ramer, C. M. (1975). Neonatal addiction: A two-year study. II. Behavioral and electrophysiological characteristics of the addicted neonate. *Addict. Dis.* **2**, 235–255.

Marcus, J., Hans, S. L., and Jeremy, R. J. (1982). Differential motor and state functioning in newborns of women on methadone. *Neurobehav. Toxicol. Teratol.* **4**, 459–462.

Markham, J. K., Emmerson, J., and Owen, N. W. (1971). Teratogenicity studies of methadone HCl in rats and rabbits. *Nature* **233**, 342–343.

Martin, E. (1893). "L'opium, ses abus, mangeurs et fumerus d'opium morphinomanes." Paris.

Mattio, T. G., and Kirby, M. L. (1982). Effects of chronic morphine administration on catecholamines and beta-adrenergic receptors of the superior cervical ganglion and iris of the rat. *Life Sci.* **30**, 1435–1442.

McGinty, J. F., and Ford, D. H. (1976). The effects of maternal morphine or methadone intake on the growth reflex development and maze behavior of rat offspring. *In* "Tissue Responses to Addictive Drugs" (D. H. Ford and D. H. Clouet, eds.), pp. 611–629. Spectrum Publications, New York.

McGinty, J. F., and Ford, D. H. (1980). Effects of prenatal methadone in rat brain catecholamines. *Dev. Neurosci.* **3**, 224–234.

McLaughlin, P. J., and Zagon, I. S. (1980). Body and organ development of young rats maternally exposed to methadone. *Biol. Neonate* **38**, 185–196.

McLaughlin, P. J., Zagon, I. S., and White, W. J. (1978). Perinatal methadone exposure in rats: Effects on body and organ development. *Biol. Neonate* **34**, 48–54.

Menninger-Lerchenthal, E. (1934). Die morphinkrankheit der neugeboren morphinstischer mutter. *Monatsschr. Kinderheilkd.* **60**, 182–193.

Meriney, S. D., Gray, D. B., and Pilar, G. (1985). Morphine-induced delay of normal cell death in the avian ciliary ganglion. *Science* **228**, 1451–1453.

Middaugh, L. D., and Simpson, L. W. (1980). Prenatal maternal methadone effects on pregnant C57BL/6 mice and their offspring. *Neurobehav. Toxicol.* **2**, 307–313.

Misra, A. L., Mule, S. J., Bloch, R., and Vadlaman, N. L. (1973). Physiological disposition and metabolism of levo-methadone-l-^{3}H in nontolerant and tolerant rats. *J. Pharmacol. Exp. Ther.* **185**, 287–299.

Musto, D. F. (1973). "The American Disease." Yale University Press, New Haven, Connecticut.

Naeye, R. L., Blanc, W., LeBlanc, W., and Khatamee, M. A. (1973). Fetal complications of maternal heroin addiction: Abnormal growth, infections, and episodes of stress. *J. Pediatr.* **83**, 1055–1061.

Newby-Schmidt, M. B., and Norton, S. (1981a). Alterations of chick locomotion produced by morphine treatment *in ovo.* *Neurotoxicology* **2**, 743–748.

Newby-Schmidt, M. B., and Norton, S. (1981b). Development of opiate tolerance in the chick embryo. *Pharmacol. Biochem. Behav.* **15**, 773–778.

Newby-Schmidt, M. B., and Norton, S. (1983). Drug withdrawal prior to hatching in the morphine-tolerant chick embryo. *Pharmacol. Biochem. Behav.* **18**, 817–820.

Nichtern, S. (1973). The children of drug users. *J. Am. Acad. Child Psychiatry* **12**, 24–31.

Nosal, G. (1979a). Influence exercée sur la progéniture de rat par l'exposition à la morphine maternelle. I. Exposition foetale et neuronogénèse dans le cervelet néonatal. *Acta Neurol. Latinoam.* **25**, 27–45.

Nosal, G. (1979b). Influence exercée sur la progéniture de rat par l'exposition à la morphine maternelle. II. Exposition pré- et postnatale et maturation cérébelleuse chez le raton. *Acta Neurol. Latinoam.* **25**, 151–165.

O'Callaghan, J. P., and Holtzman, S. G. (1976). Prenatal administration of morphine to the rat: Tolerance to the analgesic effect of morphine in the offspring. *J. Pharmacol. Exp. Ther.* **197,** 533–544.

Ostrea, E. M., and Chavez, D. J. (1979). Perinatal problems (excluding neonatal withdrawal) in maternal drug addiction. A study of 830 cases. *J. Pediatr.* **94,** 292–295.

Perlmutter, J. F. (1967). Drug addiction in pregnant women. *Am. J. Obstet. Gynecol.* **99,** 569–572.

Perlmutter, J. F. (1974). Heroin addiction and pregnancy. *Obstet. Gynecol. Surv.* **29,** 439–446.

Perlstein, M. A. (1947). Congenital morphinism. A rare cause of convulsions in the newborn, *J.A.M.A.* **135,** 633.

Pert, C. B., and Snyder, S. H. (1973). Opiate receptor: A demonstration in nervous tissue. *Science,* **179,** 1011–1014.

Pertschuk, L. P., Ford, D. H., and Rainford, E. A. (1977). Localization of methadone in the fetal rat eye by the immunofluorescence technique. *Exp. Eye Res.* **24,** 547–552.

Peters, M. A. (1977). The effect of maternally administered methadone on brain development in the offspring. *J. Pharmacol. Exp. Ther.* **203,** 340–346.

Peters, M. A. (1978). A comparative study on the behavioral response of offspring of female rats chronically treated with methadone and morphine. *Proc. West. Pharmacol. Soc.* **21,** 411–418.

Peters, M. A., Turnbow, M., and Buchenauer, D. (1972). The distribution of methadone in the nonpregnant, pregnant and fetal rat after acute methadone treatment. *J. Pharmacol. Exp. Ther.* **181,** 273–278.

Peterson, G. R., Webster, G. W., and Shuster, L. (1974). Effects of narcotics on enzymes of acetylcholine metabolism in cultured cells from embryonic chick brains. *Neuropharmacology* **13,** 365–376.

Petty, G. E. (1912). Congenital morphinism with report of cases: General treatment of morphinism. *Memphis M. Monthly* **32,** 37–63.

Petty, G. E. (1913). "Narcotic Drug Diseases and Allied Ailments." Davis, Memphis, Tennessee.

Plishka, R. J., and Neale, J. (1984). Morphine treatment increases clonidine binding in brain cell cultures. *Neurosci. Lett.* **51,** 281–186.

Raye, J. R., Dubin, J. W., and Blechner, J. N. (1977). Fetal growth retardation following maternal morphine administration: Nutritional or drug effect? *Biol. Neonate* **32,** 222–228.

Rech, R. H., Lomuscio, G., and Algeri, S. (1980). Methadone exposure *in utero:* Effects on brain biogenic amines and behavior. *Neurobehav. Toxicol.* **2,** 75–78.

Rementeria, J. L., and Lotongkhum, L. (1977). The fetus of the drug-addicted woman: Conception, fetal wastage, and complications. *In* "Drug Abuse in Pregnancy and Neonatal Effects" (J. L. Rementeria, ed.), pp. 3–18. C. V. Mosby, St. Louis, Missouri.

Ricalde, A. A., and Hammer, R. P. (1990). Perinatal opiate treatment delays growth of cortical dendrites. *Neurosci. Lett.* **115,** 132–143.

Rönnbäck, L., and Hansson, E. (1985). Amino acid incorporation during morphine intoxication. I. Dose and time effects of morphine on protein synthesis in specific regions of the rat brain and in astroglia-enriched primary cultures. *J. Neurosci. Res.* **14,** 461–477.

Rorke, L. B., Reeser, D. S., and Finnegan, L. P. (1977). Pathological findings in the nervous system of infants born to substance-abusing women. *Committee on Problems of Drug Dependence (39th Annual Scientific Meeting),* Cambridge, Massachusetts, pp. 551–571.

Rosen, T. S., and Johnson, H. L. (1982a). Children of methadone-maintained mothers: Follow-up to 18 months of age. *J. Pediatr.* **101,** 192–196.

Rosen, T. S., and Johnson, H. L. (1982b). *In utero* methadone exposure—three-year follow-up. *Pediatr. Res.* **16,** 130A.

Sainsbury, H. (1909). "Drugs and the Drug Habit." Methuen, London.

Sakellaridis, N., Mangoura, D., and Vernadakis, A. (1986). Effects of opiates on the growth of neuron-enriched cultures from chick embryonic brain. *Int. J. Dev. Neurosci.* **4,** 293–302.

Salerno, L. J. (1977). Prenatal care. *In* "Drug Abuse in Pregnancy and Neonatal Effects" (J. L. Rementeria, ed.), pp. 19–29. C. V. Mosby, St. Louis.

Sardemann, H., Madsen, K. S., and Friis-Hansen, B. (1976). Follow-up of children of drug-addicted mothers. *Arch. Dis. Child.* **51,** 131–134.

Schulman, C. A. (1969). Alterations of the sleep cycle in heroin-addicted and "suspect" newborns. *Neuropediatrie* **1,** 89–100.

Seidler, F. J., Whitmore, W. L., and Slotkin, T. A. (1982). Delays in growth and biochemical development of rat brain caused by maternal methadone administration: Are the alterations in synaptogenesis and cellular maturation independent of reduced maternal food intake? *Dev. Neurosci.* **5,** 13–18.

Shah, N. S., and Donald, A. G. (1979). Pharmacological effects and metabolic fate of levo-methadone during postnatal development in rat. *J. Pharmacol. Exp. Ther.* **208,** 491–497.

Shah, N. S., Donald, A. G., Bertolatus, J. A., and Hixson, B. (1976). Tissue distribution of levo-methadone in nonpregnant and pregnant female and male mice: Effect of SKF-525-A. *J. Pharmacol. Exp. Ther.* **199,** 103–116.

Shuster, L., Webster, G. W., and Yu, G. (1975). Perinatal narcotic addiction in mice: Sensitization to morphine stimulation. *Addict. Dis.* **2,** 277–292.

Simon, E. J., Miller, J. M., and Edelman, I. (1973). Stereospecific binding of the potent narcotic analgesic [³H]etorphine to rat brain homogenate. *Proc. Natl. Acad. Sci. U.S.A.* **70,** 1947–1949.

Singh, H. H., Purohit, V., and Ahluwalia, B. S. (1980). Effect on methadone treatment during pregnancy on the fetal testes and hypothalamus in rats. *Biol. Reprod.* **22,** 480–485.

Sisson, T. R. C., Wickler, M., Tsai, P., and Rao, I. P. (1974). Effect of narcotic withdrawal on neonatal sleep patterns. *Pediatr. Res.* **8,** 451.

Slotkin, T. A., Lau, C., and Bartolome, M. (1976). Effects of neonatal or maternal methadone administration on ornithine decarboxylase activity in brain and heart of developing rats. *J. Pharmacol. Exp. Ther.* **199,** 141–148.

Slotkin, T. A., Seidler, F. J., and Whitmore, W. L. (1980a). Effects of maternal methadone administration on ornithine decarboxylase in brain and heart of the offspring: Relationships of enzyme activity to dose and to growth impairment in the rat. *Life Sci.* **26,** 861–867.

Slotkin, T. A., Seidler, F. J., and Whitmore, W. L. (1980b). Precocious development of sympatho-adrenal function in rats whose mothers received methadone. *Life Sci.* **26,** 1657–1663.

Slotkin, T. A., Weigel, S. J., Barnes, G. A., Whitmore, W. L., and Seidler, F. J. (1981). Alterations in the development of catecholamine turnover induced by perinatal methadone: Differences in central vs. peripheral sympathetic nervous systems. *Life Sci.* **29,** 2519–2525.

Slotkin, T. A., Weigel, S. J., Whitmore, W. L., and Seidler, F. J. (1982). Maternal methadone administration: Deficit in development of α-noradrenergic responses in developing rat brain as assessed by norepinephrine stimulation of ³³Pi incorporation into phospholipids *in vivo*. *Biochem. Pharmacol.* **31,** 1899–1902.

Slotkin, J. A., Whitmore, J. L., Salvaggio, M., and Seidler, F. W. (1979). Perinatal methadone addiction affects brain synaptic development of biogenic amine systems in the rat. *Life Sci.* **24,** 1223–1230.

Smith, A. A., Hui, F. W., and Crofford, M. J. (1977). Inhibition of growth in young mice treated with *d,l*-methadone. *Eur. J. Pharmacol.* **43,** 307–314.

Sobrian, S. K. (1977). Prenatal morphine administration alters behavioral development in the rat. *Pharmacol. Biochem. Behav.* **7,** 285–288.

Sonderegger, T., and Zimmermann, E. (1976). Persistent effects of neonatal narcotic addiction in the rat. *In* "Tissue Responses to Addictive Drugs" (D. H. Ford and D. H. Clouet, eds.), pp. 589–609. Spectrum Publications, New York.

Sonderegger, T., and Zimmermann, E. (1978). Adult behavioral adrenocortical function following neonatal morphine treatment in rats. *Psychopharmacology* **56,** 103–109.

Sonderegger, T., Bromley, B., and Zimmermann, E. (1977). Effects of morphine pellet implantation in neonatal rats. *Proc. Soc. Exp. Biol. Med.* **154,** 435–438.

Sonderegger, T., O'Shea, S., and Zimmermann, E. (1979). Consequences in adult female rats of neonatal morphine pellet implantation. *Neurobehav. Toxicol.* **1,** 161–167.

Soyka, L. F., Peterson, J. M., and Joffe, J. M. (1978). Lethal and sublethal effects on the progeny of male rats treated with methadone. *Toxicol. Appl. Pharmacol.* **45,** 797–807.

Steele, W. J., and Johannesson, T. (1975a). Effects of morphine infusion in maternal rats at near-term on ribosome size distribution in fetal and maternal rat brain. *Act Pharmacol. Toxicol.* **36,** 236–242.

Steele, W. J., and Johannesson, T. (1975b). Effects of prenatally administered morphine on brain development and resultant tolerance to the analgesic effect of morphine in offspring of morphine-treated rats. *Acta Pharmacol. Toxicol.* **36,** 243–256.

Steele, W. J., and Johannesson, T. (1975c). Distribution of ^{14}C-morphine and macromolecules in the brain and liver and their nuclei in pregnant rats and their fetuses after infusion of morphine into pregnant rats at near-term. *Acta Pharmacol. Toxicol.* **37**, 265–273.

Strauss, M. E., Andresko, M., Stryker, J. C., and Wardell, J. N. (1976). Relationship of neonatal withdrawal to maternal methadone dose. *Am. J. Drug Alcohol Abuse* **3**, 339–345.

Strauss, M. E., Lessen-Firestone, J. K., Chavez, C. J., and Stryker, J. C. (1979a). "Psychological Characteristics and Development of Narcotic-Addicted Infants." Abstract presented at "Genetic, Perinatal, and Developmental Effects of Abused Substances," NIDA Conference, Arlie, Virginia.

Strauss, M. E., Lessen-Firestone, J. K., Chavez, C. J., and Stryker, J. C. (1979b). Children of methadone-treated women at five years of age. *Pharmacol. Biochem. Behav.* **11**, (Suppl.), 3–6.

Sung, C.-Y., Way, E. L., and Scott, K. G. (1953). Studies on the relationship of metabolic fate and hormonal effects of d,l-methadone to the development of drug tolerance. *J. Pharmacol. Exp. Ther.* **107**, 12–23.

Tempel, A., Habas, J., Paredes, W., and Barr, G. A. (1988). Morphine-induced down-regulation of mu-opioid receptors in neonatal rat brain. *Brain Res.* **469**, 129–133.

Terenius, L. (1973). Stereospecific interaction between narcotic analgesics and a synaptic plasma membrane fraction of rat cerebral cortex. *Acta Pharmacol. Toxicol.* **32**, 317–320.

Terry, C. E., and Pellens, M. (1970). "The Opium Problem." Patterson Smith, New Jersey (originally published in 1928 by the Bureau of Social Hygiene, Inc.).

Thompson, C. I., and Zagon, I. S. (1981). Long-term thermoregulatory changes following perinatal methadone exposure in rats. *Pharmacol. Biochem. Behav.* **14**, 653–659.

Thompson, C. I., Zagon, I. S., and McLaughlin, P. J. (1979). Impaired thermal regulation in juvenile rats following perinatal methadone exposure. *Pharmacol. Biochem. Behav.* **10**, 551–556.

Ting, R., Keller, A., Berman, P., and Finnegan, L. P. (1974). Follow-up studies of infants born to methadone-dependent mothers. *Pediatr. Res.* **8**, 346.

Ting, R. Y., Keller, A., and Finnegan, L. P. (1978). Physical, neurological, and developmental assessment of infants born to methadone-dependent mothers. In "Drug Abuse: Modern Trends, Issues, and Perspectives" (A. Schecter, H. Alksne, E. Kaufman, V. Shorty, A. Henderson, and J. H. Lowinson, eds.), pp. 632–641. Marcel Dekker, New York.

Tiong, G. K., and Olley, J. E. (1988). Effects of exposure *in utero* to methadone and buprenorphine on enkephalin levels in the developing rat brain. *Neurosci. Lett.* **31**, 101–106.

Tsang, D., and Ng, S. C. (1980). Effect of antenatal exposure to opiates on the development of opiate receptors in rat brain. *Brain Res.* **188**, 199–206.

Tsang, D., Ho, K. P., and Wen, H. L. (1986). Effect of maternal methadone administration on the development of multiple forms of monoamine oxidase in rat brain and liver. *Brain Res.* **391**, 187–192.

Van Baar, A. L., Fleury, P., Seopatmi, S., Ultee, C. A., and Wesselman, P. J. M. (1989a). Neonatal behaviour after drug-dependent pregnancy. *Arch. Dis. Child.* **64**, 235–240.

Van Baar, A. L., Fleury, P., and Ultee, C. A. (1989b). Behaviour in first year after drug-dependent pregnancy. *Arch. Dis. Child.* **64**, 241–245.

Van Kleek, L. A. (1920). Symptoms of morphine withdrawal in an infant. *Am. Med.* **15**, 51–52.

Van Wagoner, S., Risser, J., Moyer, M., and Lasky, D. (1980). Effect of maternally administered methadone on discrimination learning of rat offspring. *Percept. Mot. Skills* **50**, 1119–1124.

Vernadakis, A., and Gibson, A. (1985). Neurotoxicity of opiates during brain development: In vivo and in vitro studies. *Prog. Clin. Biol. Res.* **163**, 245–253.

Vernadakis, A., Estin, C., Gibson, D. A., and Amott, S. (1982). Effects of methadone on ornithine decarboxylase and cyclic nucleotide phosphohydrolase in neuronal and glial cell cultures. *J. Neurosci. Res.* **7**, 111–119.

Walz, M. A., Davis, W. M., and Pace, H. B. (1983). Parental methadone treatment: A multigenerational study of development and behavior in offspring. *Dev. Pharmacol. Ther.* **6**, 125–137.

Wang, C., Pasulka, P., Perry, B., Pizzi, W. J., and Schnoll, S. H. (1986). Effect of perinatal exposure to methadone on brain opioid and alpha$_2$-adrenergic receptors. *Neurobehav. Toxicol. Teratol.* **8**, 399–402.

White, W. J., and Zagon, I. S. (1979). Acute and chronic methadone exposure in adult rats: Studies on arterial blood gas concentrations and pH. *J. Pharmacol. Exp. Ther.* **209**, 451–455.

White, W. J., Zagon, I. S., and McLaughlin, P. J. (1978). Effects of chronic methadone treatment on maternal body weight and food and water consumption in rats. *Pharmacology* **17,** 227–232.

Wikler, A. (1980). "Opioid Dependence. Mechanisms and Treatment." Plenum Press, New York.

Willson, N. J., Schneider, J. F., Roizin, L., Fleiss, J. F., Rivers, W., and DeMartini, J. E. (1976). Effects of methadone hydrochloride on the growth of organotypic cerebellar cultures prepared from methadone-tolerant and control rats. *J. Pharmacol. Exp. Ther.* **199,** 368–374.

Wilson, G. S. (1975). Somatic growth effects of perinatal addiction. *Addict. Dis.* **2,** 233–345.

Wilson, G. S. (1992). Heroin/morphine. In "Perinatal Substance Abuse" (T. B. Sonderegger, ed.), pp. 207–223. Johns Hopkins University Press, Baltimore.

Wilson, G. S., Desmond, M. M., and Verniaud, W. M. (1973). Early development of infants of heroin-addicted mothers. *Am. J. Dis. Child.* **126,** 457–462.

Wilson, G. S., McGreary, R., Kean, J., and Baxter, J. C. (1979). The development of preschool children of heroin-addicted mothers: A controlled study. *Pediatrics* **63,** 135–141.

Wilson, G. S., Desmond, M. M., and Wait, R. B. (1981). Follow-up of methadone-treated and untreated narcotic-dependent women and their infants: Health, developmental, and social implications. *J. Pediatr.* **98,** 716–722.

World Health Organization Expert Committee on Drug Dependence. (1973). 20th Annual Report, WHO Tech. Report Series 516, Geneva, Switzerland.

Yeh, S. Y., and Krebs, H. A. (1980). Development of narcotic drug-metabolizing enzymes in the newborn rat. *J. Pharmacol. Exp. Ther.* **213,** 28–32.

Zagon, I. S. (1983a). Behavioral effects of prenatal exposure to opiates. *Monogr. Neural Sci.* **9** 159–168.

Zagon, I. S. (1983b). Preclinical, perinatal, and developmental effects of methadone: Behavioral and biochemical aspects. In "Research on the Treatment of Narcotic Addiction" (J. R. Cooper, F. Altman, B. S. Brown, and D. Czechowicz, eds.), pp. 376–390. U. S. Government Printing Office, Washington, D.C.

Zagon, I. S. (1985). Opioids and development: New lessons from old problems. *NIDA Research Monograph* **60,** 58–77.

Zagon, I. S., and McLaughlin, P. J. (1977a). The effects of different schedules of methadone treatment on rat brain development. *Exp. Neurol.* **56,** 538–552.

Zagon, I. S., and McLaughlin, P. J. (1977b). Methadone and brain development. *Experientia* **33,** 1486–1487.

Zagon, I. S., and McLaughlin, P. J. (1977c). Effect on chronic maternal methadone exposure on perinatal development. *Biol. Neonate* **31,** 271–282.

Zagon, I. S., and McLaughlin, P. J. (1977d). Morphine and brain growth retardation in the rat. *Pharmacology* **15,** 276–282.

Zagon, I. S., and McLaughlin, P. J. (1978a). Perinatal methadone exposure and brain development: A biochemical study. *J. Neurochem.* **31,** 49–54.

Zagon, I. S., and McLaughlin, P. J. (1978b). Perinatal methadone exposure and its influence on the behavioral ontogeny of rats. *Pharmacol. Biochem. Behav.* **9,** 665–672.

Zagon, I. S., and McLaughlin, P. J. (1980). Protracted analgesia in young and adult rats maternally exposed to methadone. *Experientia* **36,** 329–330.

Zagon, I. S., and McLaughlin, P. J. (1981a). Enhanced sensitivity to methadone in adult rats perinatally exposed to methadone. *Life Sci.* **29,** 1137–1142.

Zagon, I. S., and McLaughlin, P. J. (1981b). Withdrawal-like symptoms in young and adult rats maternally exposed to methadone. *Pharmacol. Biochem. Behav.* **15,** 887–894.

Zagon, I. S., and McLaughlin, P. J. (1982a). Analgesia in young and adult rats perinatally exposed to methadone. *Neurobehav. Toxicol. Teratol.* **4,** 455–457.

Zagon, I. S., and McLaughlin, P. J. (1982b). Comparative effects of postnatal undernutrition and methadone exposure on protein and nucleic acid contents of the brain and cerebellum in rats. *Dev. Neurosci.* **5,** 385–393.

Zagon, I. S., and McLaughlin, P. J. (1982c). Neuronal cell deficits following maternal exposure to methadone in rats. *Experientia* **38,** 1214–1216.

Zagon, I. S., and McLaughlin, P. J. (1983). Increased brain size and cellular content in infant rats treated with an opiate antagonist. *Science* **221,** 1179–1180.

Zagon, I. S., and McLaughlin, P. J. (1984a). Perinatal exposure to methadone alters sensitivity to drugs in adult rats. *Neurobehav. Toxicol. Teratol.* **6**, 319–323.

Zagon, I. S., and McLaughlin, P. J. (1984b). An overview of the neurobehavioral sequelae of perinatal opioid exposure. In "Neurobehavioral Teratology" (J. Yanai, ed.), pp. 197–234. Elsevier Science Publishers BV, Amsterdam.

Zagon, I. S., and McLaughlin, P. J. (1984c). Opiates alter tumor cell growth and differentiation in vitro. *NIDA Res. Monogr.* **49**, 344–350.

Zagon, I. S., and McLaughlin, P. J. (1984d). Naltrexone modulates body and brain development in rats: A role for endogenous opioids in growth. *Life Sci.* **35**, 2057–2064.

Zagon, I. S., and McLaughlin, P. J. (1985). Naltrexone's influence on neurobehavioral development. *Pharmacol. Biochem. Behav.* **22**, 441–448.

Zagon, I. S., and McLaughlin, P. J. (1987). Endogenous opioid systems regulate cell proliferation in the developing rat brain. *Brain Res.* **412**, 68–72.

Zagon, I. S., and McLaughlin, P. J. (1989). Naloxone modulates body and organ growth of rats: Dependency on the duration of opioid receptor blockade and stereospecificity. *Pharmacol. Biochem. Behav.* **33**, 325–328.

Zagon, I. S., and McLaughlin, P. J. (1990a). Drugs of abuse and the fetus and neonate: Testing and evaluation in animals. In "Modern Methods in Pharmacology. Testing and Evaluation of Drugs of Abuse," Vol. 6 (A. Cowan and M. Adler, eds.), pp. 241–254. Wiley-Liss, New York.

Zagon, I. S., and McLaughlin, P. J. (1990b). Ultrastructural localization of enkephalinlike immunoreactivity in developing rat cerebellum. *Neuroscience* **34**, 479–489.

Zagon, I. S., and McLaughlin, P. J. (1991). Identification of opioid peptides regulating proliferation of neurons and glia in the developing nervous system. *Brain Res.* **542**, 318–323.

Zagon, I. S., McLaughlin, P. J., and C. I. Thompson. (1979a). Development of motor activity in young rats following perinatal methadone exposure. *Pharmacol. Biochem. Behav.* **10**, 743–749.

Zagon, I. S., McLaughlin, P. J., and Thompson, C. I. (1979b). Learning ability in adult female rats perinatally exposed to methadone. *Pharmacol. Biochem. Behav.* **10**, 889–894.

Zagon, I. S., McLaughlin, P. J., Weaver, D. J., and Zagon, E. (1982). Opiates, endorphins, and the developing organism: A comprehensive bibliography. *Neurosci. Biobehav. Rev.* **6**, 439–479.

Zagon, I. S., McLaughlin, P. J., and Zagon, E. (1984). Opiates, endorphins, and the developing organism: A comprehensive bibliography, 1982–1983. *Neurosci. Biobehav. Rev.* **8**, 387–402.

Zagon, I. S., Rhodes, R. E., and McLaughlin, P. J. (1985). Distribution of enkephalin immunoreactivity in germinative cells of developing rat cerebellum. *Science* **227**, 1049–1051.

Zagon, I. S., Rhodes, R. E., and McLaughlin, P. J. (1986). Localization of enkephalin immunoreactivity in diverse tissues and cells of the developing and adult rat. *Cell Tissue Res.* **246**, 561–565.

Zagon, I. S., Zagon, E., and McLaughlin, P. J. (1989). Opioids and the developing organism: A comprehensive bibliography, 1984–1988. *Neurosci. Biobehav. Rev.* **13**, 207–235.

Zagon, I. S., Gibo, D. M., and McLaughlin, P. J. (1990). Adult and developing human cerebella exhibit different profiles of opioid-binding sites. *Brain Res.* **523**, 62–68.

Zagon, I. S., Gibo, D. M., and McLaughlin, P. J. (1991). Zeta (ζ), a growth-related opioid receptor in developing rat cerebellum: Identification and characterization. *Brain Res.* **551**, 28–35.

Zelson, C., Kahn, E. J., Neumann, L., and Polk, G. (1970). Heroin withdrawal syndrome. *J. Pediatr.* **76**, 483.

Zelson, C., Lee, S. T., and Casalino, M. (1973). Neonatal narcotic addiction. *N. Engl. J. Med.* **289**, 1216–1220.

Ziring, B. S., Kreek, M. J., and Brown, L. T. (1981). Methadone disposition following oral versus parenteral dose administration in rats during chronic treatment. *Drug Alcohol Depend.* **7**, 311–318.

Zuspan, F. P., Gumpel, J. A., Mejia-Zelaya, A., Madden, J., Davis, R., Filer, M., and Tiamson, A. (1975). Fetal stress from methadone withdrawal. *Am. J. Obstet. Gynecol.* **122**, 43–46.

— 12 —

Benzodiazepines and the Developing Nervous System: Laboratory Findings and Clinical Implications

ૢ

Carol K. Kellogg

Department of Psychology
University of Rochester
Rochester, New York

INTRODUCTION

The benzodiazepines were introduced in the early 1960s and, having a plethora of pharmacologically advantageous effects (antianxiety, sedative, hypnotic, muscle relaxant, autonomic system depressant, and anticonvulsant) as well as a large safety factor (Randall and Kappell, 1973), they quickly became some of the most commonly prescribed medications in the United States. These drugs were placed on the Drug Enforcement Administration controlled substance list in 1975. Whereas there is considerable evidence that dependence can develop following continued use of benzodiazepines (De Witt et al., 1984; Griffiths and Sannerud, 1987) and that a subset of patients on chronic benzodiazepine (BZD) treatment suffers severe withdrawal symptoms on discontinuation of treatment (Nutt, 1990), the actual potential of these drugs for misuse has not been clearly established (Woods et al., 1987; Cole and Charello, 1990). However, whereas it is not clear that BZDs are drugs of abuse, they are certainly drugs of widespread use. Prescriptions of these drugs peaked at 103 million in 1975, decreased to 67 million in 1981, and have increased moderately since then (Woods et al., 1987). In 1982, over 70 million prescriptions for BZDs were written in the United States (Mellinger and Balter, 1983). This converts to a prevalence rate for use of BZDs of approximately 11% of the adult population.

The implication of such widespread use of these drugs for the unborn organism, regardless of intent of exposure, is considerable. Research has clearly indicated that the final organization of the nervous system and the behavioral capacity of an organism involve the interaction during development among three major influences: the genome, the prenatal and postnatal internal chemical environment, and the external (for example, sensory, maternal, and peer) environment of the organism (Oppenheim and Haverkamp, 1986). Any drugs to which an organism in utero is exposed may become part of its internal chemical environment. It has been well accepted for some time that drugs exert their effects on organisms through interaction with specific receptive molecules (Langley, 1906). As a major target organ for psychoactive drugs, the developing central nervous system (CNS) is particularly vulnerable to exposures to drugs such as the BZDs. Plasticity is a major characteristic of the developing brain, so exposure to selective drugs over specific developmental periods could influence the continued course of neural development. The influence of a specific drug on the developing brain will depend on several factors: (1) the presence of specific recognition sites for the drug, (2) the maturational state of effector systems that translate the interaction of a drug with its binding site into a response, (3) the maturational state of the receptive cell, and (4) the maturational state of specific neural circuitry. The relationship between developmental events taking place in the rat brain during the period of exposure to diazepam (DZ) used in our laboratory is illustrated in Fig. 1.

In reviewing the effects of prenatal exposure to BZDs, it is the goal of this chapter to develop and support the hypothesis that occupancy of BZD binding sites in the brain during specific developmental stages alters the organization of neural mechanisms underlying behavioral responses to challenge, i.e., adaptive behaviors. Before reviewing the effects of early BZD exposure, however, the neural mechanism mediating the pharmacologic effects of BZDs will be presented. Following review of the behavioral and neural effects of prenatal exposure of experimental animals to BZDs, clinical implications of this information will be considered.

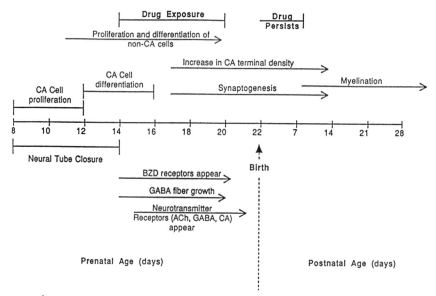

FIGURE 1   Relationship of the period of prenatal DZ administration to pregnant rats and persistence of the drug in the brains of the progeny to some developmental events in the fetal and neonatal rat brain. CA, catecholamine, BZD, benzodiazepine, GABA, γ-aminobutyric acid, ACh, acetylcholine.

MECHANISMS OF ACTION OF BENZODIAZEPINE COMPOUNDS

Mature Brain

Specific, high-affinity binding sites for BZD compounds were identified in mammalian brain in 1977 (Squires and Braestrup, 1977; Mohler and Okada, 1977), and the interaction of the drugs with these binding sites was found to correlate highly with their pharmacologic actions (Braestrup and Squires, 1978). However, even before identification of BZD binding sites, BZD compounds had been shown to potentiate the actions of γ-amino butyric acid (GABA), the primary inhibitory neurotransmitter in mammalian brain (Polc et al., 1974; Suria and Costa, 1975). The BZD recognition site was subsequently shown to be part of an oligomeric complex that also contains the recognition site for GABA and forms a chloride channel as well (Tallman and Gallager, 1985; Haefely, 1990). The BZD binding site associated with the GABA receptor is generally referred to as the central-type BZD binding site, and the GABA receptor linked to the BZD binding site and chloride channel appears to be a low-affinity GABA receptor (the $GABA_A$ receptor). It is now quite evident that different subsets of the $GABA_A$/BZD receptor complex exist in the brain, as defined by binding, protein chemistry, and molecular cloning studies (Hebebrand et al., 1988; Olson et al., 1990). These subsets of receptors are differentially distributed in different brain regions.

GABA-containing synapses and GABA receptors are widely distributed in the CNS (Belin et al., 1980; Young and Chu, 1990), and GABA is the major transmitter in the mammalian CNS that mediates rapid-timescale inhibition. It does so via functional activation of chloride channels. The action of GABA on chloride channels is facilitated by BZD compounds (Schwartz, 1988). This GABA-mediated inhibition is also modulated by a variety of other drugs, such as the barbiturates, steroid anesthetics, adrenal and gonadal steroid metabolites, ethanol, and various convulsants (Schwartz, 1988). There are multiple drug-receptor sites on the complex through which the various compounds influence the function of the complex. However, all of the various binding sites on the complex are allosterically related. For example, while BZD binding to its recognition site facilitates the action of GABA on the chloride channel, the presence of GABA at its recognition site facilitates the binding of BZDs to their site.

Benzodiazepines also bind to sites in peripheral organs, and this binding has a different pharmacologic profile from central-type BZD binding sites. These peripheral-type BZD receptors have also been identified on both glial (Shoemaker et al., 1982; Owen et al., 1983) and neural elements (Anholt et al., 1984) in the CNS. Binding associated with peripheral-type sites has been localized to the mitochondrial outer membrane in the adrenal gland (Anholt et al., 1986) and observed to be associated subcellularly with mitochondrial fractions in the brain (Doble et al., 1987; Marangos et al., 1982). Such a location suggests a role for these BZD receptors in the regulation of cellular metabolism. This mitochondrial BZD binding site does not appear to be linked to GABA receptors. Recent evidence has indicated, however, that the drugs used to classify the peripheral-type site may also bind to a neuronal membrane-located BZD binding site that is associated with GABA receptors (Gee et al., 1987). Thus, there are multiple sites, and variations of sites, in the mature brain to which BZD compounds can bind. These drugs, therefore, are capable of influencing myriad CNS functions.

Development of Benzodiazepine-Receptor Systems

Clearly then, any perturbation of such a widely distributed receptor system during critical stages of development could be expected to have considerable impact on the developing CNS. Specific, high-affinity recognition sites for BZDs have been identified in rat brain by the third week of gestation (Braestrup and Nielsen, 1978), and autoradiographic analysis indicates that they appear earliest in the brain stem and hypothalamus but are present in all areas of the rat CNS by birth (Schlumpf et al., 1983). Postnatally, the density of BZD binding sites increases rapidly, reaching adult densities between the second and fourth week in the rat (Braestrup and Nielsen, 1978; Candy and Martin, 1979). The affinity of these sites for BZD ligands is similar across all ages studied. These BZD binding sites have been identified in human fetal tissue by 12 weeks conceptual age (Aaltonen et al., 1983; Brookshank et al., 1982), indicating an early appearance of these sites in the human brain.

Analysis of the development of BZD-binding subtypes in rat brain indicates that the appearance of receptor heterogeneity is regionally specific (Chisholm et al., 1983). Thus, BZD binding sites showing a high affinity for the triazolopyridazine Cl 218-872 (the so-called Type I sites) are not present in either the cerebral cortex or cerebellum at birth in the rat but appear postnatally. However, in the hippocampus, adult proportions of both high- and low-affinity sites are present at birth and are maintained throughout postnatal development. Protein chemistry studies have also indicated age-dependent heterogeneity of the BZD/GABA receptor complex (Vitorica et al., 1990; Eichinger and Sieghart, 1986).

Available evidence supports the hypothesis that occupancy of BZD receptors in the rat brain during early stages of development can influence the action of GABA on receptive cells. The outgrowth of GABA fibers during development in the rat (Lauder et al., 1986) correlates well with the pattern of development of BZD recognition sites (Schlumpf et al., 1983). Functional interrelationships between GABA and BZD binding sites appear to be present from very early developmental stages. Thus, diazepam (DZ) facilitates GABA-mediated chloride uptake in synaptoneurosomal preparations of rat cerebral cortex from at least gestational day 19 (Kellogg and Pleger, 1989), and GABA can modulate BZD binding in rat brain tissue as early as 16 days of gestation (Mallorga et al., 1980). Evidence that GABA-mediated inhibition of cells is present during fetal development has come from studies carried out on cultured fetal mouse cerebral cortex (Crain and Bornstein, 1974). Electrophysiologic measurements indicated the presence of tonic inhibition, and since GABA antagonists reversed the inhibition, GABA receptors probably mediated the inhibition. Recent work has demonstrated that exposure of fetal rat neuronal cultures to DZ decreases the uptake of 2-deoxyglucose by the cultured cells, an indication that neural activity was decreased by the exposure (Daval et al., 1988). This observation is consistent with the observations that DZ can facilitate GABA-mediated chloride uptake and that GABA-mediated inhibition is present in fetal cerebral cortex.

Therefore, the rat brain during late fetal stages would appear capable of responding to any DZ that might reach it. Benzodiazepines are lipid-soluble, low-molecular-weight compounds that cross the placenta very rapidly in humans, reaching concentrations in cord blood that are twofold to threefold higher than in maternal blood following maternal administration (Mandelli et al., 1975). That DZ reaches the fetal brain during in utero exposure of rats is indicated by the presence of the drug in the brain at birth following 8 days of exposure (Simmons et al., 1983). Its presence at fetal central-type BZD receptors in the brain is indicated by the observation that the affinity of BZD binding sites for a radiolabeled BZD ligand is markedly decreased in fetal as well as maternal brain 1 hr following administration of DZ to the dam on the twentieth day of gestation, indicating competition from DZ for the binding site (Kellogg, 1988). Since rat brain development at birth is comparable to human brain development around 20 weeks of gestation (Dobbing, 1974), the developing human brain is also likely to be very susceptible to BZD compounds from the second trimester (i.e., after the appearance of specific BZD binding sites).

Peripheral-type BZD binding sites have been identified in rat brain by gestational day 16 (Schlumpf et al., 1989). Furthermore, prenatal exposure to drugs considered to bind to such sites induces profound effects on specific aspects of cerebral metabolism measured in young adult and aging prenatally exposed animals, suggesting the presence of functional peripheral-type BZD binding sites during the fetal exposure period (Miranda et al., 1989; Miranda et al., 1990a,b).

It is difficult, however, to prove that the functional consequences of in utero *exposure* to psychoactive drugs are definitively related to their in utero *action* on specific receptors in the fetal brain. The possibility that the drugs exert their primary effect in the maternal compartment, and that responses mediated there are then secondarily transmitted to the fetal compartment must be considered. Certainly for a drug such as cocaine, which markedly influences placental circulation, the consequences of in utero exposure may reflect the impact of hypoxia on brain development as well as a direct effect of cocaine on the fetal brain. The effects of prenatal exposure to BZDs, however, appear to result from a direct action of the drugs on the fetal brain. The evidence, presented below, will illustrate the similarity of the consequences measured in the young adult of both in utero exposure and acute exposure to BZDs. Furthermore, the observation of differential effects after exposure to other compounds that interact at the BZD binding site but that affect the receptor differently (i.e., drugs that are anxiolytic or neutral) underscores the likelihood of a direct action of these compounds on the fetal brain. And finally, studies showing similar consequences of late gestational versus early postnatal exposure (via direct injection to the rat pup) further support a direct action of the drug during in utero exposure.

BEHAVIORAL CONSEQUENCES OF PRENATAL EXPOSURE TO BENZODIAZEPINES

To evaluate the consequences of prenatal exposure to psychoactive drugs, we hypothesized that the consequences should include those effects seen in adults during an acute exposure to the compound in question. That the selection of tests could be crucial in identifying the effects of developmental exposure to drugs that interact with very select neural mechanisms was evident from reports showing, for example, that prenatal exposure to DZ at doses as high as 200 mg/kg/day had little effect on the organism when evaluations were made using a general test battery (Butcher and Vorhees, 1979).

In adult experimental animals, BZDs influence behavioral responses to three types of environmental stimuli: conflict, novelty, and nonreward (Paul et al., 1986). Such stimuli have been proposed to activate neural systems inhibiting ongoing behavior, and this inhibition results in enhanced arousal and attention (Gray, 1982). Benzodiazepines release behavior suppressed by these stimuli. Clinically, BZDs are used in the management of anxiety. The responses in both humans and animals that are affected by these drugs in all likelihood serve the same purpose. Anxiety in humans can be considered the cognitive and emotional component of a behavioral alarm/arousal system that serves to prepare the organism for danger (Breier and Paul, 1990). Whereas anxiety is a normal emotion, anxiety can also become pathologic. The concept that normal adaptive responses may lead to disease states was put forth in Selye's theories of stress responses (Selye, 1976). He identified specific dysfunctions that result from faulty adaptive responses to challenge, and called these diseases of adaptation. Darwin proposed in 1872 that an animal's expressive behavior, as well as its anatomical structure, had an evolutionary history and that emotional behavior, in particular, served a functional role and affected an organism's chances of survival [see Darwin (reprinting), 1955]. Therefore, considering the effects of BZDs in humans and animals, BZD compounds can be viewed as influencing behaviorally adaptive responses. Recent work has emphasized the importance of the BZD/GABA receptor complex in adaptive behaviors and in normal and pathologic anxiety (Breier and Paul, 1990).

As demonstrated below, prenatal exposure to DZ has indeed been shown to alter several behaviors that are influenced by acute DZ exposure. These behaviors include those that reflect behavioral arousal/attention, that are responses to novel stimuli, and that can be considered adaptive behaviors. Importantly, as will be demonstrated, the precise effects of the prenatal exposure depend on the age of the organism at testing.

Effects on Arousal/Attention

Studies demonstrating an effect of prenatal DZ on arousal/attention are summarized in Table 1. Postnatal development of locomotor activity in rats (tested in isolation) follows a pattern characterized by a sudden increase in activity around postnatal days 14 to 15, followed by an abrupt decline in activity on days 17 to 18 (Campbell et al., 1969). This characteristic developmental pattern has been hypothesized to result from asynchronous rates of maturation of arousal and arousal-inhibitory structures. In early studies, we observed that prenatal exposure to DZ (2.5, 5, or 10 mg/kg) over gestational days 13 to 20 attenuated this developmental phase of locomotor hyperactivity, in a manner related to prenatal exposure dose (Kellogg et al., 1980). There was

TABLE 1   Prenatal Diazepam Exposure: Indication of Effect on Arousal/Attention Systems

Study	Prenatal exposure days and dose (mg/kg)	Measure	Postnatal age at observation	Effect
Kellogg et al. (1980)	13–20 2.5, 5, 10	Developmental locomotor hyperactivity	15–17 days	Attenuated
		Locomotor activity	18–20 days	Normal
		Acoustic startle response (ASR)	12–20 days	Decreased
		Noise-facilitated ASR	12–20 days	No facilitation
Kellogg (1988)	13–20 1.0, 2.5, 10	ASR	70–90 days	Normal to enhanced
		Noise-facilitated ASR	70–90 days	Normal
Schlumpf et al. (1989)	14–20 2.5	Olfactory-guided behaviors	6–10 days	Attenuated
Livezey et al. (1985)	14–20 (rats) 5 to 7.5	Sleep EEG	4 months	Enhanced light slow-wave sleep EEG
Livezey et al. (1986)	20–53 (cats) .32 to.6 to.15	Post-reinforcement EEG	1 year	Absent

no effect of the exposure on locomotor activity measured at ages after the normal phase of hyperactivity. Similar results have been reported from other laboratories (Shore et al., 1983). It is unlikely that the effects of prenatal DZ reflected a direct effect of persisting drug. Whereas DZ can be detected in the brain up to day 10 postnatal age following prenatal exposure over gestational days 13 to 20 to DZ at 2.5 mg/kg/day, the levels are very low (3.2 pmol/100 mg; parent compound and metabolites), and no drug is detectable on postnatal day 20 (Simmons et al., 1983). Thus, prenatal exposure to DZ may have interfered with the development of specific arousal or arousal-inhibitory mechanisms. Interestingly, postnatal exposure to DZ (10 mg/kg) over days 3 to 18, a period that included the phase of normal developmental hyperactivity, produced hyperactivity at days 20 and 35 (Frieder et al., 1984). Taken together, these results suggest that neural systems underlying developmental hyperactivity, arousal, and arousal-inhibitory systems may include the BZD/GABA receptor complex.

Because the effects of prenatal exposure to DZ on locomotor activity were evident only during the phase of locomotor hyperactivity, other tests were selected in order to detect an effect of the exposure on behavioral arousal at other ages. The acoustic startle reflex (ASR), elicited in mammalian species by an intense but brief acoustic stimulus, is potentiated when the startle stimulus is presented against a steady background white noise (Ison and Hammond, 1971). This effect of background noise has been presumed to result from noise-produced activation of arousal mechanisms. The neural pathway mediating the ASR is reasonably well characterized (Davis et al., 1982), and several of the nuclei involved in the reflex and in its modification are densely populated with BZD receptors (Kuhar and Young, 1979).

Acute exposure of adult rats to DZ attenuates the effect of noise on the ASR (Kellogg et al., 1991b). Prenatal exposure to DZ delayed the development of noise facilitation of the ASR (Kellogg et al., 1980). In control rats, noise potentiation of the ASR was evident at 12 days postnatal age and became pronounced after 16 days. By 20 days of age, the magnitude of noise potentiation was similar to that observed in adults. No noise potentiation of the ASR was evident over this period in rats exposed in utero to DZ. Noise potentiation of the ASR was, however, evident in young adult prenatally exposed rats (Sullivan and Kellogg 1985; Kellogg, 1988). Thus, in a second test, prenatal exposure to DZ appeared to interfere with the development of arousal-related mechanisms.

Indication of an influence of prenatal exposure to DZ on arousal/attention mechanisms in adult animals has, however, been observed by others. For example, high-amplitude post-reinforcement electroencephalogram (EEG) synchronization that is triggered by a rewarded lever press is sensitive to the tone of brainstem arousal systems. One-year-old cats exposed in utero to DZ showed interference with post-reinforcement synchronization of the EEG measured over the parieto-occipital cortex (Livezey et al., 1986). Evidence of altered EEG synchronization during sleep was observed in 4-month-old rats exposed in utero to DZ (5.0–7.5 mg/kg/day) during late gestation (Livezey et al., 1985). These animals spent more of their slow-wave sleep time in lighter slow-wave sleep, a measure of impaired synchronization.

That the effects of prenatal exposure to DZ on locomotor activity and on noise potentiation of the ASR were most evident during early developmental periods, whereas the effects on EEG synchronization were evident in young adult animals, may indicate that the exposure affected different neural systems that underlie behavioral arousal. Developmental locomotor hyperactivity is normally transient; thus, the attenuation of activity during this period following prenatal exposure to DZ might reflect an earlier maturation of arousal-inhibitory versus arousal systems. This earlier maturation of inhibitory systems could also explain the absence of noise potentiation of the ASR, observed over a similar period. The interference with EEG synchronization in the adult may reflect interference with cerebral cortical arousal systems, systems that may become active as an organism reaches young adulthood. Post-reinforcement synchronization of the EEG is related to an animal's learning ability (Marczynski et al., 1981); thus, prenatal exposure to DZ may influence cognitive aspects of arousal/alarm reactions. The possibility that this aspect of the alarm reaction emerges with maturity will be expanded in the presentation of the effect of prenatal DZ exposure on environment-related social interaction.

Effects on Behavioral Responses to Novel Stimuli

As discussed, BZDs attenuate behavioral responses to novel stimuli. That novel environments may elicit behaviorally adaptive responses is demonstrated by the observation, for example, that the latency of an orienting response to an auditory stimulus varied with an animal's familiarity with the environment (Richardson et al., 1988). Whereas the behavioral responses discussed (developmental locomotor hyperactivity and noise-facilitated ASR) could be considered responses to novel stimuli, the importance of novelty to the responses cannot be determined, since responses to other stimuli were not evaluated.

We examined the effect of prenatal exposure to DZ on behavioral responses to novelty, selecting environment-related social interaction as our test (Kellogg et al., 1991a). In this test, we measured the behavioral interaction between two rats (strangers to each other) placed in a neutral environment that was either familiar or unfamiliar to both rats. The amount of time that adult male rats spend in social interaction (SI) is a function of their familiarity with the environment, such that SI is decreased in an unfamiliar (novel) environment (File, 1988; Primus and Kellogg, 1990a). This decrease in SI in the unfamiliar environment is considered to represent an adaptive response to an anxiogenic situation (File, 1980). A total SI score is obtained by adding the time in a 7.5-min session that the rats spend engaging in the following behaviors: sniffing, following, grooming each other, pushing, jumping over each other, wrestling.

As can be seen in Fig. 2, adult (60-day-old) male rats exposed in utero to vehicle (40% propylene glycol, 10% ethyl alcohol) exhibited more SI in the familiar than the unfamiliar environment, a response typically seen in naive adult rats (Kellogg et al., 1991a). In contrast, there was no difference in SI in the two environments in adult male rats exposed in utero to DZ (1.0 or 2.5 mg/kg/day). Interestingly, rats prenatally exposed to DZ spent significantly more time in SI in the unfamiliar environment and significantly less time in SI in the familiar environment than did vehicle-exposed rats in the respective environments. The effects of prenatal DZ exposure have been described by some as reflecting an increase in emotionality or anxiety (Marczynski and Urbancic, 1988). The results of the SI test suggest, however, that such interpretations might not be correct. The increase in SI observed in the unfamiliar environment in DZ-exposed rats (Kellogg et al., 1991a) is similar to the response observed in control rats after acute exposure to DZ (Primus and Kellogg, 1990b). In this respect then, the adult rats exposed in utero to DZ could be considered to be in an anxiolytic state. However, the in utero exposure, but not adult acute exposure, also led to *decreased* SI in the familiar environment, a change considered to reflect enhanced anxiety in naive adult rats (File, 1988). It does not seem logical to propose, however, that the early exposure led to both an anxiolytic and an anxiogenic state. Furthermore, as observed and illustrated in Fig. 3, acute treatment with DZ (1.0 mg/kg) does not change SI behavior in either environment in adult male rats exposed in utero to DZ at 2.5 mg/kg. In contrast, acute DZ treatment increased SI in the unfamiliar environment in rats exposed in utero to vehicle. Perhaps the results of the SI test reflect an effect of the prenatal exposure on cognitive processing of environmental information, consistent with the observations on post-reinforcement EEG synchronization previously discussed.

An impact of the unfamiliar environment on SI normally appears with the onset of puberty in male rats (Primus and Kellogg, 1989). Thus, SI is decreased in the

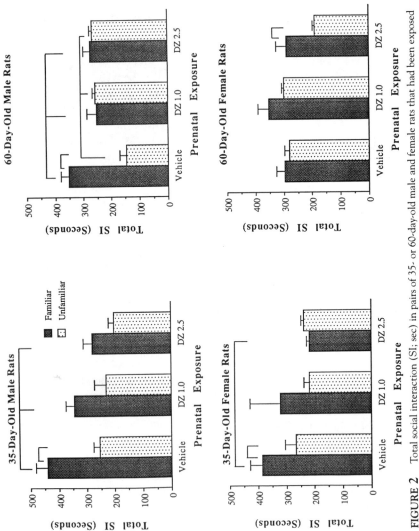

FIGURE 2 Total social interaction (SI; sec) in pairs of 35- or 60-day-old male and female rats that had been exposed in utero (over gestational days 14 to 20) to DZ (1.0 or 2.5 mg/kg/day to the dam) or vehicle. Data are presented as mean ±SEM. Brackets indicate significance between designated groups ($p < 0.05$). 60-day-old data from Kellogg et al. (1991a).

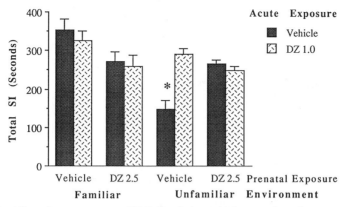

FIGURE 3   Effect of acute exposure to DZ (1.0 mg/kg) or vehicle on total social interaction (SI; sec) in pairs of 60-day-old male rats tested in a familiar or unfamiliar environment as a function of prenatal exposure to DZ (2.5 mg/kg) or vehicle over gestational days 14 to 20. * Significant difference between acute exposures ($p < 0.05$).

unfamiliar environment relative to the familiar environment at 35 days but not at 28 days, and there is no difference in the amount of SI in the familiar environment at the two ages. We recently measured SI at 35 days in male rats exposed prenatally to DZ or vehicle (Primus and Kellogg, 1990, unpublished observations). As indicated in Fig. 2, SI was markedly decreased in the unfamiliar environment relative to SI in the familiar environment in vehicle-exposed male rats at 35 days ($p < 0.002$). Social interaction was also significantly decreased in the unfamiliar environment in male rats prenatally exposed to DZ at 1.0 mg/kg ($p < 0.03$), but there was no significant difference in SI between the familiar and unfamiliar environment in rats exposed in utero to DZ at 2.5 mg/kg. Further evaluation indicated that the apparent dose-related influence of pre-natal DZ exposure on the magnitude of the difference in SI between environments was the result primarily of a dose-related decrease in SI in the familiar environment in rats exposed to DZ relative to those exposed to vehicle. In the familiar environment, SI in rats exposed to DZ at 2.5 mg/kg differed significantly from SI in the same environment in vehicle-exposed rats ($p < 0.002$), and the difference between SI in the familiar environment between vehicle-exposed controls and rats exposed to DZ at 1.0 mg/kg approached significance ($p < 0.07$). In contrast, there was no significant difference at this age in SI in the unfamiliar environment between any prenatal-exposure groups. Thus, an effect of prenatal exposure to DZ on SI in the familiar environment appears at an earlier age than does the effect on SI in the unfamiliar environment (i.e., at 35 days versus the young adult). Different neural systems, perhaps influenced by pubertal events, may underlie SI in the different environments. To evaluate the possibility that pubertal development may be a factor in determining the consequences of prenatal DZ exposure, we examined the role of specific pubertal events in the normal developmental changes that take place in SI in the familiar and unfamiliar environments, respectively.

Intact gonadal function was shown to be necessary for the pubertal emergence in male rats of decreased SI in the unfamiliar environment (Primus and Kellogg, 1990a) as well as for the pubertal emergence of a modifying effect of DZ on this response (Primus and Kellogg, 1990b). The marked effect of prenatal exposure to DZ on SI in the unfamiliar environment at a young adult age (60 days) but not at 35 days of age (early puberty) suggests that the early exposure affected neural mechanisms sensitive to the organizing influence of the gonadal androgens. In addition to the emergence of decreased SI in a novel (unfamiliar) environment, SI in the familiar environment also decreases from 35 to 60 days of age (Primus and Kellogg, 1989). Furthermore, the behavioral composition of SI in male rats changes from 35 to 60 days (Primus and Kellogg, 1989); at 35 days, rats spend a greater portion of SI in physical interactions (grooming, pushing, wrestling), whereas at 60 days, SI is primarily defined by "curiosity" interactions (sniffing and following). That gonadal androgens may contribute to the decrease in SI and the change in behavioral composition (physical versus curiosity) that takes place over pubertal development is supported by the observation that an adult amount and composition of SI in the familiar environment is seen in 35-day-old male rats made precocious by exposure to testosterone propionate over days 14 to 30 (Primus and Kellogg, 1990b). Therefore, SI in both the familiar and unfamiliar environments changes over pubertal development in male rats, and gonadal androgens appear to influence these changes. It is possible, however, that different neural systems may be involved in the changes taking place in response to the respective environments.

That there may be an interaction between the consequences of prenatal DZ exposure and an organizing influence of gonadal hormones on SI is further suggested by examining the effect of the exposure on SI in female rats. As was observed in male rats, SI behavior also changes in the female rat during puberty. As illustrated in Fig. 2, young adult (60-day-old) female rats (vehicle exposed) do not demonstrate environment-specific SI, whereas they do at 35 days of age. Interestingly, the amount of SI in the familiar environment decreases over this period in female rats, whereas the magnitude of SI in the unfamiliar (novel) environment does not change. This specific change in the SI response in the familiar environment accounts for the lack of a differential impact of environment on SI in adult female rats. Thus the change in response in female rats from early to late adolescence implicates an influence from ovarian hormones on the organization of this behavioral response. The effect of prenatal DZ exposure on SI in the adult rat was observed to be sexually dimorphic (Kellogg et al., 1991a). Whereas the exposure abolished environment-specific SI in adult male rats, exposure to DZ at 2.5 mg/kg induced environment-related SI in female rats tested at 60 days. Female rats exposed in utero to this same dose of DZ showed no environment-specific SI at 35 days, and the lack of difference in SI between the two environments appeared to be related to an effect of the exposure on SI in the familiar environment. Interestingly, at 35 days of age, SI in male and female rats (as a function of prenatal exposure) is similar. The sexually dimorphic aspects of the influence of prenatal DZ exposure on SI in the adult may reflect sexually dimorphic influences of specific gonadal hormones during pubertal development. The effect of prenatal DZ exposure on SI in females also suggests the possibility of a masculinizing influence of the early exposure on specific neural systems, in that SI in adult female rats prenatally exposed to DZ at 2.5 mg/kg now resembles that of normal male rats.

Perinatal DZ exposure has also been shown by others to have a different effect on nonreproductive sexually dimorphic behaviors in male rats versus female rats (Guillamon et al., 1991). In addition, prenatal exposure to oxazepam enhanced maternal aggression toward a male intruder, a behavior specific to adult females (Laviola et al., 1991). Furthermore, acute BZD treatment enhanced maternal aggression in lactating dams exposed in utero to either vehicle or oxazepam. Again, the effect of the prenatal exposure to BZDs resembled the effect of an acute exposure.

Thus, specific behavioral responses to the challenge of novelty (such as the presence of a stranger or an unfamiliar environment) appear to emerge during adolescent development, and prenatal exposure to DZ alters the emergence of these responses. The analysis, therefore, of normal development of environment-related SI and the effects of prenatal exposure to DZ on this development has supported the suggestion that cognitive aspects of the alarm reaction emerge, and are perhaps organized, during pubertal development and that in utero exposure to DZ interferes with mechanisms in this maturation.

Effects on Complex Behaviors

Prenatal exposure to DZ might indeed influence cognitive processing of environmental information. For example, we have observed that prenatal DZ exposure alters temporal acuity in the auditory system (Kellogg et al., 1983b). Studies in humans have shown that individual differences in auditory temporal acuity (auditory fusion) are related to speech perception in adults (Trinder, 1979) and to language and learning disabilities in children (McCrosky and Kidder, 1980). Temporal acuity is defined in rats by the effectiveness of brief (3–24 msec) silent periods (gaps) in white noise on inhibition of the ASR (Ison, 1982).

Developmentally, gaps become effective in inhibiting the ASR in normal, untreated rats between 25 and 28 days of age (Fig. 4). Sensitivity to the silent periods is stable from 28 through 56 days. However, at 70 days there is an increase in the sensitivity to the inhibitory effects of gaps. The most pronounced effects of prenatal exposure to DZ were evident in rats at 70 days. Other than a developmental delay (from 28 to 35 days) in the onset of gap detection in animals exposed in utero to a high dose of DZ (10 mg/kg), no effect of the early exposure was observed until 70 days. At 70 days, prenatally exposed rats showed both altered sensitivity to gap detection (Table 2) and altered inhibition to gaps of long (greater than 10 ms) duration. Hence, the effects of prenatal DZ exposure were expressed as the animals reached the age when a normal change in sensitivity to gap detection takes place (a change in the threshold for auditory temporal resolution). Again, it appears that changes in the processing of environmental information occur as organisms reach young adulthood and that early developmental exposure to DZ interferes with mechanisms involved in these maturational changes.

Work from other laboratories has demonstrated that early developmental exposure

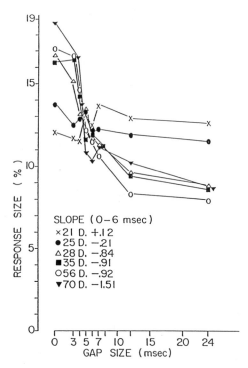

FIGURE 4   Development of auditory temporal acuity in rats. The response at each gap size was averaged across 8 trials and then normalized by dividing each response by the total sum of the responses. The response at any gap duration, therefore, represents a percentage of the total response. Increasing inhibition of the acoustic startle response by increasingly larger gaps in background noise will result in an increased proportion of the total response being present in the control (0 gap) trial. The slope of the response curve measured over gaps up to 6 msec in duration provides an index of sensitivity to the interrupted noise. As illustrated, marked changes in sensitivity to gap detection (temporal acuity) occur between 25 and 28 days of age and between 56 and 70 days. It is at 70 days that alterations in gap detection become most evident following prenatal exposure to DZ. From Kellogg et al. (1983b).

to DZ interferes with learning and retention of a choice discrimination task in 2- to 3-month-old rats. Exposure to DZ (5 and 10 mg/kg) over gestational days 4 through 19 induced deficits in performance on a complex simultaneous-choice discrimination maze without interfering with simple motor learning or with learning a simple successive discrimination task (Gai and Grimm, 1982). Prolonged gestational exposure to DZ at a lower dose (2.5 mg/kg) did not lead to any interference in maze performance. However, exposure to DZ (10 mg/kg) limited to gestation days 14 to 20 did induce impairment in complex maze performance, as did direct postnatal exposure of rat pups to DZ (10 mg/kg) over days 2 through 17 (Frieder et al., 1984). The similarity between the effect of late gestational and postnatal exposure to DZ on complex maze performance supports the hypothesis that in utero exposure to DZ induces effects by direct action on the fetal brain.

TABLE 2   Effect of Prenatal Diazepam
Exposure on Auditory Temporal Acuity
in 70-Day-Old Male Rats[a]

Exposure[b]	Slope (0–6 msec)[c]
Uninjected	−1.51
Vehicle	−1.68
DZ 2.5 mg/kg	−2.44
DZ 10.0 mg/kg	−0.96

[a]From Kellogg et al. (1983b).
[b]Over gestational days 13 to 20.
[c]The slope of gap-induced inhibition of the ASR measured over 0 to 6 msec provides an index of sensitivity to the interrupted noise. Relative to the control groups, animals exposed to DZ at 2.5 mg/kg/day exhibited an enhanced sensitivity, whereas animals exposed to DZ at 10 mg/kg/day demonstrated a decreased sensitivity.

Summary of Behavioral Consequences of Prenatal Exposure to Benzodiazepines

Behavioral studies have demonstrated pronounced effects in rats following exposure to BZDs during late gestation (and early postnatal life). Clearly, the consequences of early BZD exposure include effects also induced by acute BZD exposure during adulthood, suggesting that during fetal development, BZD compounds interact with the same systems that these drugs interact with in the mature organism. Many of the behaviors reported here and affected by the early exposure reflect an influence of the early exposure on behavioral alarm/arousal systems. These behaviors subserve adaptive functions for the organism. Of particular note is the observation that many of the consequences of early BZD exposure are delayed in their expression until young adulthood. The behaviors that are altered in juvenile rats during their second to third week of life seem related to an effect of the early exposure on the rate of maturation of arousal-inhibitory versus arousal systems. Thus, the effects of early DZ exposure on developmental locomotor hyperactivity, on development of the ASR and noise facilitation of the ASR, as well as on olfactory-guided behaviors (Schlumpf et al., 1989), may be related to differential rates of development of arousal-regulatory systems in the exposed animal. However, over the period of 4 to 8 weeks of age few consequences of late gestational exposure to DZ have been reported. The delayed appearance of many of the effects of prenatal BZD exposure until young adulthood suggests that neural systems underlying these behavioral responses undergo a change during pubertal development, a change that involves systems influenced by the prenatal exposure. Thus, the BZD/GABA receptor complex appears to be a major site through which the developmental organization of adaptive behaviors can be influenced.

NEURAL AND HORMONAL CONSEQUENCES OF PRENATAL
EXPOSURE TO BENZODIAZEPINES

Again, the likelihood of observing a biologic effect of in utero exposure to psycho-active drugs is enhanced by investigating the effects on systems known to be affected by acute exposure to the drug in question. An abundance of literature has demonstrated that BZD compounds influence neural and physiologic responses to environmental challenges (stressors). In this section, the effects of prenatal exposure to BZDs on selected neural and hormonal systems will be evaluated.

Effects on Hormonal and Brain Catecholamine Responses to Stressors

As presented earlier, BZD compounds attenuate alarm responses. A major physiologic component of an organism's alarm responses, as initially described by Selye (1976), is activation of the hypothalamic–pituitary–adrenal (HPA) axis, which leads to elevation of plasma corticosteroid (CS) levels. This activation of the HPA axis, along with activation of the sympathetic nervous system, aids the organism in meeting the demands of challenge and serves to minimize deviations from a homeostatic state. The neural regulation of the HPA axis is complex (Reisine et al., 1986), but one neural input mediating a major influence on the release of corticotropin-releasing factor (CRF) from the hypothalamus is the norepinephrine (NE) projection to the hypothalamus from cell bodies in the brain stem (Plotsky et al., 1989). Recent evidence suggests that the NE projection to the paraventricular nucleus of the hypothalamus facilitates hypothalamic CRF secretion (Plotsky et al., 1989; Leibowitz et al., 1989).

Benzodiazepine administration to naive rats attenuates the CS response to stressors (Lahti and Barsuhn, 1975; LeFur et al., 1979; DeBoer et al., 1990), whereas exposure to anxiogenic BZD receptor ligands enhances stressor-induced increases in plasma CS (Pellow and File, 1985). Furthermore, BZDs attenuate stressor-induced increases in NE turnover in the hypothalamus (Corrodi et al., 1971). Evidence from our laboratory suggests that BZD/GABA receptors may be located presynaptically on NE terminals in the hypothalamus, thereby providing a mechanism mediating the influence of BZDs on the function of hypothalamic NE projections (Harary and Kellogg, 1989). Selective destruction of NE terminals in the paraventricular nucleus of the hypothalamus prevents stressor-induced increases in plasma CS (Morton et al., 1990); therefore, the attenuating influence of BZDs on the CS response to stressors could result from an action of the drugs on the NE projection to the hypothalamus. Considering these relationships, we predicted that prenatal exposure to DZ would affect the hypothalamic NE-plasma CS responses to stressors.

Initial studies indicated a marked and selective effect of prenatal exposure to DZ on the NE projection to the hypothalamus, sparing NE projections to the cerebral cortex and hippocampus (Simmons et al., 1984a; Kellogg and Retell, 1986). Norepinephrine levels, turnover rate, and depolarized release in the hypothalamus were reduced in a dose-related manner by prenatal exposure to DZ. All of the effects of the early exposure observed on the NE projection to the hypothalamus were prevented by coexposure in utero to the central-type BZD receptor antagonist, Flumazenil, which had no effect when administered alone. Hence, the long-term effects of the exposure on this neural pathway did appear to be mediated by in utero interaction of DZ with central-type BZD receptors. Interestingly, as was observed for many of the behavioral studies, the effect of prenatal DZ exposure on the hypothalamic NE projection did not become evident until after 35 days postnatal age, i.e., during pubertal development (Fig. 5). Other laboratories have also reported a delayed expression of the effect of prenatal DZ exposure on the hypothalamic NE projection in that, compared to con-

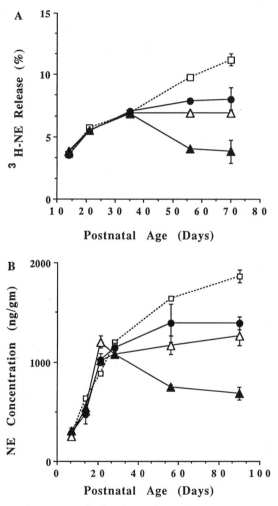

FIGURE 5   Developmental time-course for the effects of prenatal exposure (over gestational days 13 to 20) to DZ (1.0 (●), 2.5 (△), or 10.0 (▲) mg/kg to the dam) or vehicle (no drug; □) on hypothalamic NE projections. (A) Changes in depolarized release of [^{3}H]norepinephrine (NE), induced by incubation in 25 mM potassium, from isolated rat hypothalamus. From Kellogg and Retell (1986). (B) Changes in NE concentration in the hypothalamus. From Simmons et al. (1984a).

trols, β-adrenergic receptors were significantly decreased in the hypothalamus at 90 days but not at 60 days following exposure to DZ at 1 mg/kg over gestational days 7 to 20 (Rothe and Langer, 1988).

Therefore, since function within the hypothalamic NE projection system was so affected in adult rats that had been exposed in utero to DZ, it seemed probable that stress responses related to function within this system would likewise be influenced. Indeed, stressor-induced changes in both hypothalamic NE neurons and plasma CS levels were altered in DZ-exposed adult rats (Simmons et al., 1984b). Restraint stressor-induced increases in plasma CS levels (over high basal levels achieved by housing the animals overnight in a novel area) were attenuated by prenatal exposure to DZ, an effect that was prevented by coexposure to the BZD antagonist, Flumazenil. Furthermore, the stressor-induced increase in hypothalamic NE utilization was prevented by the prenatal exposure. In fact, the NE utilization rate in prenatally exposed rats appeared to decrease in response to the stressor. The direction of the changes in the hypothalamic NE response to the stressor observed following prenatal DZ exposure was consistent with the effect on plasma CS levels. Thus, if activity in hypothalamic NE projections facilitates release of CRF and thereby activates the HPA axis, then the prevention of stressor-induced increases in hypothalamic NE turnover correlates with an attenuated CS response to the stressor.

In addition to the effects of prenatal exposure to DZ on the HPA axis, prenatal exposure to DZ (0.2 or 2.0 mg/kg) also has been reported to alter functioning of the hypothalamic–pituitary–thyroid system (Fujii et al., 1983). Serum thyroxine levels were decreased in both male and female rats at 3 and 6 weeks of age. In addition, at 10 weeks of age, prenatally exposed female, but not male, rats demonstrated an enhanced thyrotropin response when administered thyrotropin-releasing hormone (TRH). Basal thyroxine levels were unchanged in either male or female rats at this age by the early exposure. Thus, as was observed with the effect of prenatal DZ exposure on plasma CS, it was the hormonal response to stimulation in adult female rats that was altered by the early exposure, and not basal functioning of the system. The hypothalamic–pituitary–thyroid system has been implicated in the mediation of stress responses. For example, thyroid hormone modulation of brain glucocorticoid receptors during development has been postulated as a mechanism for mediating the effects of early handling on the development of the adrenocortical stress response (Meaney et al., 1987). Early handling also alters the density of BZD receptors (Bodnoff et al., 1987); thus changes in function of the hypothalamic–pituitary–thyroid axis by early DZ exposure might be predicted. However, it is not clear why the effect of prenatal DZ exposure on the response to TRH in adult rats was sexually dimorphic.

Perinatal exposure to DZ (via implantation into the pregnant dam on gestational day 8 of DZ-containing silastic capsules that yield an average release of 5 mg/kg/24 hr) also interferes with stressor-induced changes in select dopamine (DA) projection systems (Deutch et al., 1989). Specifically, the exposure reduced the magnitude of the foot-shock-induced increase in DA turnover in the prefrontal cortex. Basal DA turnover was unaffected in this region. On the other hand, the investigators observed altered basal DA turnover, but not stressor-induced increases in DA turnover, in the nucleus accumbens. Both prefrontal cortex and nucleus accumbens receive their DA projection from the ventral tegmental area. Again, as is the effect of prenatal exposure to DZ on NE projection systems, the effect of prenatal exposure on DA-defined neural systems appears to be regionally selective. Expression of the effects of prenatal DZ exposure on specific DA neural populations also appears to be delayed. The effects of prenatal DZ exposure observed at 60 days of age could not be separated from the effects observed in implantation controls, but were clearly different at 90 days (Gruen et al., 1990a).

Effects on the GABA/Benzodiazepine Receptor Complex

Recent evidence has demonstrated that function at the GABA/BZD receptor complex changes in response to challenging environmental conditions (Havoundjian et al., 1986a). Furthermore, we have demonstrated that challenge-induced changes in function at the complex are influenced by experience (Primus and Kellogg, 1991a). The GABA/BZD complex, therefore, appears to be an important participant in an organism's response to challenge. As such, one would expect an influence from pre-natal BZD exposure on functioning of this receptor complex, since the exposure alters so many responses to environmental challenge.

Many studies have examined the effect of prenatal DZ exposure on binding characteristics of the BZD recognition site on the complex. However, the effects observed on these measures have not been consistent. Most of the studies report no effect of prenatal DZ exposure (at various doses, gestational periods, and routes of exposure) on BZD binding in early postnatal life (Braestrup et al., 1979; Massotti et al., 1980; Kellogg et al., 1983a). One study, however, reported an increase in BZD binding over the first 3 weeks postnatal age following in utero exposure to DZ (at 20 mg/kg; Shibuya et al., 1986). We have observed a 20% decrease in maximal binding density (B_{max}) and a 25% increase in the equilibrium dissociation constant (K_D) in the forebrain of 3-month-old rats exposed in utero to DZ (Kellogg et al., 1983a). Taken together, these two characteristics suggest that prenatal DZ exposure might induce a subsensitivity at the BZD recognition site that becomes apparent in the adult organism. Livezey et al. (1985) reported a decrease in B_{max} of BZD binding limited to the thalamus of 1-year-old prenatally exposed rats; however, this change in B_{max} was accompanied by a decrease in K_D of equal magnitude. Since a decrease in K_D reflects an increase in affinity of the binding site for the ligand, the change in affinity could counteract any impact of a decrease in receptor density. Consistent with a minimal effect of prenatal DZ exposure on BZD binding characteristics in the adult, early DZ exposure also did not change BZD binding site heterogeneity as measured by CL 218–872 displacement (Kellogg, 1988).

The influence of prenatal DZ exposure on binding characteristics of the GABA recognition site on the complex has not been extensively evaluated. Binding to the high-affinity GABA site, using [³H]-muscimol (a GABA receptor agonist), was not altered by prenatal DZ exposure (Kellogg et al., 1983a; Massotti et al., 1980; Rothe et al., 1988). However, the GABA receptor that is linked to a BZD binding site is probably the low-affinity GABA receptor. Whereas an earlier report measuring low-affinity GABA receptors using a variation on [³H]-muscimol binding reported no effect of prenatal DZ exposure (Rothe et al., 1988), another report suggests that binding to this receptor may be altered in 90-day-old male rats following prenatal DZ exposure via silastic tubing capsules implanted into the pregnant dam (Gruen et al., 1990b). [³H]-Bicuculline (a GABA receptor antagonist) binding, which is thought to measure binding to the low-affinity GABA receptor, was significantly reduced in the cingulate cortex by the early exposure, whereas binding was not altered in the other regions examined (cerebellum, prefrontal cortex, and hypothalamus). On the other hand, GABA displacement of [³H]-bicuculline binding was not altered in the cingulate cortex by the exposure, whereas it was decreased in the hypothalamus. It is not clear that either of these observations implicates a decrease in GABA receptor density in the affected regions; a decrease in [³H]-bicuculline specific binding or in GABA displacement could reflect changes in affinity for the compounds at the recognition site. The data obtained may indicate, though, that specific aspects of binding to the GABA recognition site are differentially influenced in different brain regions by the early exposure.

We recently examined bicuculline inhibition of GABA-mediated chloride (Cl^-) uptake into synaptoneurosomes from the cerebral cortex (Fig. 6). In prenatally exposed male rats, low concentrations of bicuculline produced a greater inhibition of Cl^- uptake stimulated by GABA at a 50% effective concentration (EC_{50}) (10 μM) (Bitran et al., 1991). The prenatal exposure did not affect bicuculline inhibition of Cl^- uptake stimulated by a higher concentration of GABA (25 μM). These data indicate that the precise effect of prenatal DZ exposure on bicuculline action at the GABA receptor may depend on the status of functioning of the complex at the time of analysis. The effect of prenatal DZ exposure on the action of bicuculline at the GABA receptor corresponded to an enhanced sensitivity to bicuculline-induced seizures in the prenatally exposed rats. In addition, the early exposure to DZ also led to an increased sensitivity to seizures induced by the administration of drugs that interact at other sites on the receptor complex (DMCM, methyl-6,7-dimethoxy-4-ethyl-β-carboline-3-carboxylate an anxiogenic BZD receptor ligand, and picrotoxin, a drug that acts at the Cl^- channel).

The somewhat inconclusive data on the effect of prenatal DZ exposure on [³H]-bicuculline binding and GABA displacement thereof (Gruen et al., 1990b) may reflect the fact that it is not the GABA binding site per se that is affected by the exposure. Rather, the available data suggest that prenatal exposure to DZ when the complex is evolving may result in long-lasting changes in functional interactions at the complex. The GABA/BZD receptor is a complicated structure, and it seems quite likely that interferences in function of this receptor would be expressed in a manner other than as a simple change in the density of recognition sites. In fact, the impact of environmental challenges on this receptor complex most reliably involves changes in function of the Cl^- channel, as indicated by changes in Cl^- enhancement of BZD and t-butylbicyclophosphorothionate (TBPS) binding and changes in GABA-gated Cl^- flux (Havoundjian et al., 1986a; Schwartz et al., 1987).

A

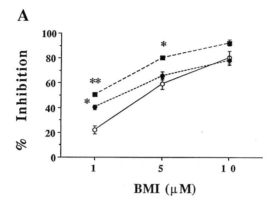

BMI (μM)

B

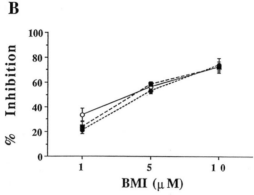

BMI (μM)

FIGURE 6   Percentage inhibition by bicuculline methiodide (BMI) of GABA-stimulated $^{36}Cl^-$ influx in synaptoneurosomes from the cerebral cortex of adult control males (○) and adult males prenatally exposed (over gestational days 14 to 20) to DZ at 1.0 mg/kg (●) or 2.5 mg/kg (■). The efficacy of BMI was evaluated against (A) 10 μM GABA or (B) 25 μM GABA. Significance: $^{*}p < 0.05$ and $^{**}p < 0.01$ relative to controls (prenatally uninjected and vehicle injected combined). Basal GABA-stimulated Cl⁻ uptake did not differ among the groups, and the average values were (in nmol/mg protein) 14.13 ± 1.73 at 10 μM and 23.57 ± 2.1 at 25 μM. From Bitran et al. (1991).

Changes in these indices of function at the GABA/BZD receptor complex have been reported in animals following early developmental exposure to BZDs. We have recently observed a dose-related change in the basal sensitivity of Cl⁻ uptake to GABA in adult male rats exposed in utero to DZ (Kellogg et al., 1991a). Furthermore, the early exposure interfered with the ability of DZ added in vitro to the synaptoneurosomal preparation to facilitate GABA-stimulated Cl⁻ uptake (Fig. 7). No effect of the prenatal exposure on GABA-stimulated Cl⁻ uptake was apparent in female rats. Miller et al. (1989) reported a decrease in TBPS binding (an index of function of the Cl⁻ channel; Havoundjian et al., 1986b) and a marked decrease in muscimol-stimulated Cl⁻ uptake in 4-week-old chick cerebral cortex following *in ovo* exposure to lorazepam (2 mg/kg) over embryonic days 8 to 18. The more profound

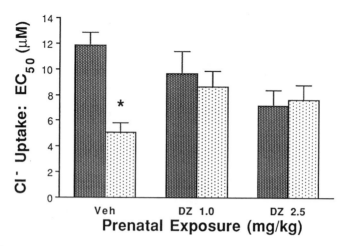

Prenatal Exposure (mg/kg)

FIGURE 7 GABA-stimulated $^{36}Cl^-$ uptake into synaptoneurosomal preparations from cerebral cortex of 60- to 90-day-old male rats exposed in utero (over gestational days 14 to 20) to DZ (1.0 or 2.5 mg/kg to the dam) or vehicle. Data presented as the EC_{50} (μM) for GABA stimulation. Uptake was evaluated in the presence of GABA only (dark bars) or in the presence of GABA plus DZ (10 μM) (light bars) added in vitro. *Significant difference between GABA only and GABA plus DZ, $p < 0.001$. From Kellogg et al. (1991a).

effect of the BZD exposure observed in chicks compared to young adult rats may relate to differences in drug clearance between the two species, which could influence the bioaccumulation of the drugs during the exposure period. These authors recently reported a similar effect of prenatal lorazepam exposure on muscimol-stimulated Cl^- uptake in 42-day-old mice (Chesley et al., 1991). Several studies have demonstrated, therefore, that functional interactions among the various components of the receptor complex may be vulnerable to BZD exposure during early development. Based on this information, one could predict that the early exposure would alter the responsiveness of the complex to environmental challenges. Indeed, we have recently completed evaluation of challenge-induced enhancement of Cl^--facilitated BZD binding and of GABA-stimulated Cl^- uptake in adult male rats exposed in utero to vehicle or DZ (2.5 mg/kg). In vehicle-exposed control rats, a challenge of either 15 min of restraint or 10 min of forced swimming increased the efficacy and potency of Cl^- facilitation of BZD binding and increased maximal GABA-stimulated Cl^- uptake (Kellogg et al., in preparation). However, there was no change from basal function in either of these two measures in response to the challenges in adult male rats exposed in utero to DZ. Clearly then, the in utero exposure interfered with the ability of the receptor complex to respond to environmental challenge.

Mechanisms that could account for this failure of the complex to respond to challenge in prenatally exposed animals have not yet been delineated. We are currently evaluating the effect of prenatal DZ exposure on subunit composition of the GABA/BZD receptor complex. A change in composition could underlie the altered responsiveness to challenge. We have also evaluated the responsiveness of the complex as a function of pubertal development, as this information could elucidate mechanisms underlying the effects of prenatal DZ exposure. Neither Cl^--facilitated BZD binding nor GABA-mediated Cl^- uptake were influenced in the cerebral cortex by the challenge of a novel environment in drug-naive juvenile, 28-day-old male rats, whereas challenge-induced changes in receptor function were clearly apparent in 60-day-old rats (Primus and Kellogg, 1991b). Thus, responsiveness of this receptor complex to environmental challenge appears to emerge with pubertal development. Juvenile castration led to altered response of the complex to challenge in adult rats, suggesting that gonadal function may influence the development of receptor responsiveness. We have recently observed that the GABA/BZD receptor in the cerebral cortex of male rats (exposed in utero either to DZ or vehicle) does not respond to a forced-swim challenge at an early pubertal age (35 days; Kellogg et al., in preparation). Interestingly, this lack of response of the GABA/BZD complex to challenge at 35 days of age mirrors the inability of DZ to modify environment-related SI at the same age (Primus and Kellogg, 1990b). It may be that the GABA/BZD receptor complex in the cerebral cortex does not become a participant in an organism's response to challenge until late pubertal development. Thus, the lack of responsiveness of the complex in prenatally exposed adult rats would appear to reflect a failure in the emergence during puberty of changes in the complex that normally lead to responsiveness to stressors in the adult.

Effects on Cellular Metabolism

Acute administration of BZDs to naive adult animals alters cellular metabolism. For example, BZD compounds increase brain glucose uptake and transiently increase glycolysis (Gey, 1973; Young et al., 1969). Whereas such effects have been considered "nonspecific" effects of BZD exposure, the association of a BZD receptor population with mitochondrial membranes (the peripheral-type BZD binding site) implicates that population in the modulation of mitochondrial function, including intracellular metabolism. Thus, BZD-induced changes in indices of cellular metabolism quite likely result from specific drug-receptor interactions. Prenatal exposure to DZ, which interacts with both central- and peripheral-type BZD binding sites, may then lead to alterations in neural metabolism.

As an initial test of this hypothesis, we measured levels of thiobarbituric acid (TBA)-reactive material in brain regions of young adult rats following prenatal exposure to DZ and other BZD receptor ligands. These TBA-reactive materials reflect cellular metabolic activity that leads to an increase in formation of a variety of free radical species. A recent study has shown that free radicals influence function of the GABA/BZD complex in adult rats (Schwartz et al., 1988). This influence raises the possibility of an interaction between free radical metabolism and function of this receptor complex.

Prenatal exposure to DZ at 1.0 mg/kg/day led to markedly elevated levels of TBA-reactive material in all brain regions of 3- to 4-month-old rats, whereas exposure to a higher dose of DZ (2.5 mg/kg/day) led to an increase only in the hippocampus (Fig. 8,

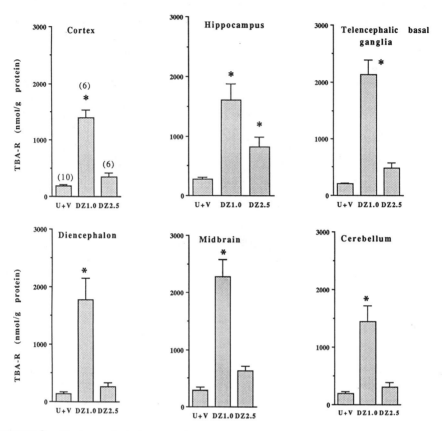

FIGURE 8 Changes in thiobarbituric acid-reactive (TBA-R) material in six brain regions of 3-month-old rats (male and female combined) following in utero exposure (over gestational days 14 to 20) to DZ (1.0 or 2.5 mg/kg to dam). The control group consisted of animals born to both uninjected (U) or vehicle-injected (V) dams. Numbers in parentheses indicate number of animals in each group. *Significant difference ($p < 0.05$) compared to control (U + V) group. From Miranda et al. (1989).

Miranda et al., 1989). Additionally, prenatal exposure to either a central-type BZD receptor agonist (clonazepam) or antagonist (Flumazenil) also induced a marked increase in TBA-reactive material in all brain regions, as did exposure to a peripheral-type BZD receptor ligand (PK 11195). Concurrent exposure to DZ at 2.5 mg/kg attenuated the effects of exposure to the central-type antagonist but did not alter the effect of exposure to the peripheral-type ligand. Therefore, prenatal exposure to drugs specific to different populations of BZD binding sites resulted in significant increases in this index of cellular metabolism measured in the brains of young adult rats.

It is of particular interest that the effect of prenatal DZ exposure on brain levels of TBA-reactive material did not become apparent until after 8 weeks postnatal age (Fig.

9). Such a delayed effect of the early exposure on metabolism could reflect interference from the exposure with specific metabolic processes that regulate free-radical formation, processes that may undergo a change with the establishment of maturity.

As indicated, prenatal exposure to DZ at 1.0 mg/kg/day over gestational days 14 to 20 had a more profound effect on TBA-like material in adult brain than exposure in utero to DZ at 2.5 mg/kg/day. In addition, exposure to either a central-type agonist or antagonist resulted in a similar consequence. Furthermore, exposure to a peripheral-type ligand led to the most marked increase in TBA-reactive material in adult brain. We have suggested (Miranda et al., 1989) that these effects could result from the possible displacement, by the exogenous compounds, of an endogenous substance from central- to peripheral-type BZD sites during the prenatal exposure period. The consequences of the early exposures may also point to a coordinated development of the two types of binding sites. This coordinated development may be critical for the organization of cellular metabolism.

This latter possibility was further emphasized by results from a study that evaluated the impact of prenatal exposure to BZD compounds on cellular energy metabolism in young adult rats using ^{31}P nuclear magnetic resonance spectroscopy (Miranda et al., 1990a). Exposure to DZ (1.0 and 2.5 mg/kg/day) led to a decrease in intracellular pH in the brains of adult rats. This decrease appeared to be related primarily to action of the drugs in utero at the central-type receptor, since the effect of DZ was attenuated by coexposure to the central-type antagonist Flumazenil, and exposure to a peripheral-type ligand exerted only a minimal effect on pH. The effect of prenatal exposure to DZ on energy metabolism, however, appeared to involve action at both types of receptor sites. For example, exposure to DZ only at 1.0 mg/kg and not at 2.5 mg/kg led to a significant decrease in phosphocreatine utilization, an index of high-energy phosphate utilization. Exposure to the central-type antagonist (Flumazenil) alone dramatically reduced phosphocreatine utilization, whereas in utero exposure to a peripheral-type ligand was without effect. However, concurrent exposure to the ineffective dose of DZ (2.5 mg/kg) and the peripheral-type ligand (PK 11195) led to an effect similar to that observed after exposure to the low dose of DZ (i.e., 1.0 mg/kg). These results suggest the possibility that there is an interaction during development between the central- and peripheral-type BZD binding sites that is important to the proper organization of cellular energy metabolism.

As discussed earlier, young adult male rats exposed in utero to DZ show an enhanced susceptibility to seizures induced by drugs that interfere with function of the GABA/BZD receptor complex (Bitran et al., 1991). This enhanced susceptibility may relate in part to an altered state of energy metabolism. Acute administration of GABA receptor antagonists, which leads to seizures, increases the utilization of brain high-energy phosphates and decreases cellular pH (Petroff et al., 1986; Young et al., 1985). The changes in energy metabolism that result from in utero exposure to DZ may sensitize the rats to the action of specific chemoconvulsants.

Prenatal exposure to DZ also interferes with the changes in metabolism that accompany the aging process (Miranda et al., 1990b). During aging in control rats (3 to 26 months of age), intracellular pH decreased, levels of TBA-reactive material

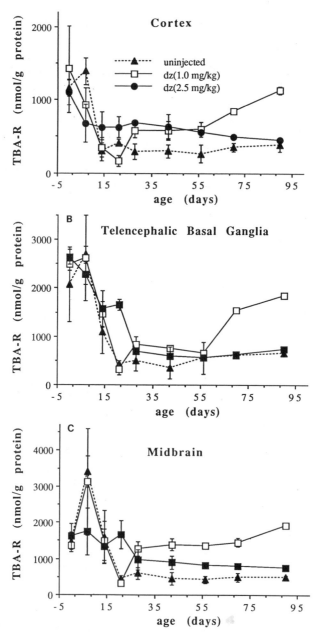

FIGURE 9   Levels of thiobarbituric acid-reactive (TBA-R) material in three brain regions, measured from birth (day 0) to 90 days of age, in uninjected control rats or in rats exposed in utero (over gestational days 14 to 20) to DZ (1.0 or 2.5 mg/kg to the dam). TBA-reactive material varied significantly as a function of age from birth to 28 days ($p < 0.001$) and as a function of age by drug interaction over days 42 to 90 ($p < 0.002$). From Miranda et al. (1989).

increased, and phosphocreatine utilization decreased. Prenatal exposure to DZ induced dose-related and aging-related alterations in levels of TBA-reactive material (Fig. 10) and in intracellular pH. The data illustrated in Fig. 10 indicate that the consequences of prenatal exposure to DZ are similar at 3 and at 24 months of age. Furthermore, the levels of TBA-reactive material measured in the exposed rats at 24 months mirror region-specific levels of this material measured in control rats at 24 months. Such data could be interpreted to suggest that prenatal exposure leads to early aging. However, analysis of TBA-reactive material at 18 months of age demonstrated no difference between DZ-exposed and control rats at this age. A similar aging profile was observed for the effect of prenatal DZ exposure on intracellular pH. The measurement of TBA-reactive material from birth to aging has indicated, therefore, that the processes of development and aging can interact with the effects of prenatal DZ exposure on cellular metabolism in a manner that supports the existence of biologic transition phases during the life span of an organism.

Thus, the prenatal chemical environment can influence cellular metabolism into aging. These studies on the effects of prenatal DZ exposure on cellular metabolism in the brain demonstrate that the impact of early drug exposure can be expressed at many different levels of function. An interference in cellular energy metabolism could be a contributing factor to exposure-induced interference in responses to stressors. Stress responses are made to meet the demands of the stressor and to maintain normal functioning. If cells cannot metabolically meet the increased demands, disordered responses could result.

Summary of Neural and Hormonal Consequences of Prenatal Benzodiazepine Exposure

The influence of prenatal exposure to DZ on various responses to environmental stressors strongly suggests that the early DZ exposure interferes with organization of an organism's stress responses. Not only are behavioral responses to environmental challenge altered, but classic hormonal and underlying neural responses are likewise altered by early developmental exposure to DZ. In addition, the exposure results in altered cellular metabolism, which could affect neural responsiveness to environmental challenges. As proposed by Selye (1976), the brain participates in an organism's general adaptive responses through regulation of the HPA axis. However, we have proposed that the brain may also have built-in systems that are sensitive to stressors that may potentially alter its function (Kellogg and Amaral, 1978). If normalcy (or homeostasis) for the brain is a level of excitability and metabolism that is optimal for processing of environmental input, then the brain should respond appropriately to minimize the effect of stressors on these normal functions. Logical reasoning suggests that all of an organism's responses to challenge are somehow integrated. Since the function of adaptive or regulatory systems is to maintain normalcy, the functioning of adaptive systems is largely inconspicuous as long as they are working properly. The apparent disruption of an organism's responses to challenges following in utero exposure to BZD compounds suggests that one or more adaptive systems was influenced

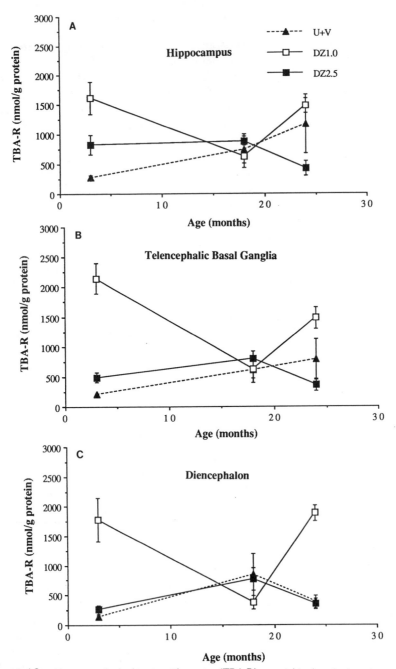

FIGURE 10   Changes in thiobarbituric acid-reactive (TBA-R) material in three brain regions, measured from 3 to 24 months of age following in utero exposure (over gestational days 14 to 20) to DZ (1.0 or 2.5 mg/kg to the dam). The control group consisted of animals born to both uninjected (U) or vehicle-injected (V) dams. From Miranda et al. (1990b).

by the exposure. It is not clear at this time whether the effect of prenatal DZ exposure on, for example, both the hypothalamic NE projection and responsiveness of the GABA/BZD complex in the cerebral cortex result from an early influence on both systems or whether the effect on either system may have been secondary to another effect. Recent evidence from our laboratory has shown that lesions of NE terminals in the paraventricular nucleus of the hypothalamus interfere with the stressor-induced response of the GABA/BZD complex in the cerebral cortex of naive adult rats (Kellogg et al., 1991c). Extrapolating from this observation to the effects of prenatal exposure to BZDs leads to the suggestion that the interference in stress responses that resulted from prenatal DZ exposure could conceivably result from an influence on this one system. Whether this is an accurate suggestion remains to be tested. It is also more than likely that because of the widespread distribution of the GABA/BZD receptor complex in the brain, drugs that act at this complex during development are capable of modifying many independent neural systems. Whether the GABA/BZD receptor complex itself is at the center of the integration of organized responses to stressors is not clear; however, the data certainly support this complex as a critical target for disrupting the normal development of these responses.

CLINICAL IMPLICATIONS

Studies conducted over the past 10 years that have examined the effects of prenatal exposure of experimental animals to DZ and other BZDs have demonstrated lasting and pronounced consequences of the exposure. Furthermore, the effects that have been reported following prenatal exposure to BZDs define a profile that is consistent with interference by the early exposure in integrated stress responses. Considering the extent of the effects that have been reported, it is surprising that the clinical literature is so sparse where defined effects of early exposure to BZD compounds are concerned.

The earliest clinical reports of any effects of early exposure to DZ indicated that the use of DZ during pregnancy and labor induced hypotonia, hypothermia, and respiratory depression in the newborn (Cree et al., 1973). This effect of DZ was termed the *floppy infant syndrome* (Gillberg, 1977; Speight, 1977) and is probably related to a direct action of drug persisting in the neonate. Other early reports described malformations (such as cleft palate) resulting from prenatal DZ exposure (Safra and Oakley, 1975; Saxen, 1975); however, these findings have not been widely replicated. In experimental animals, cleft palate has been reported only after prenatal exposure to doses of DZ that exceeded 100 mg/kg (Tocco et al., 1987). More recently, a prospective study on more than 10,000 births in Sweden defined several dysmorphic signs (telecanthus, epicanthus folds, and high arched palate) in infants whose mothers had taken BZDs throughout pregnancy (Laegreid et al., 1989, 1990). The occurrence rate was 2.3 per 1000 births. Such stigmata of embryofetopathy indicate an interaction of DZ with developmental processes taking place during the embryonic stage. The mechanisms whereby psychoactive drugs can influence embryonic development are not understood. In rats, specific BZD receptors appear in the brain and periphery at the beginning of the fetal period (see Development of Benzodiazepine-Receptor Systems). Exposure of experimental animals to DZ that is limited to the fetal period of development is not associated with any malformations and with only minimal effects on growth. In addition to the dysmorphic signs reported in the studies from Sweden, the authors also linked DZ exposure throughout pregnancy with a syndrome characterized by hypotonia combined with hyperexcitability, delayed motor and mental development, and later perceptual disorders and learning disabilities. It is tempting to compare some of these effects observed in children with the consequences of prenatal DZ exposure described in animals. However, the limited number of clinical studies that have defined such effects, the small numbers of children reported showing the disorders, and the wide range of dosages taken by the mothers makes any such comparisons unreliable.

The experimental literature, however, has provided important information that should guide future clinical investigations and increase the probability of linking later consequences with the early exposure. The experimental literature has indicated that the consequences of prenatal exposure to psychoactive drugs will most likely be very specific to the particular class of drugs, and that such drugs will most likely influence those systems that are affected in the adult organism following acute exposure to a given drug. If we consider, therefore, the primary features of the consequences of in utero exposure of experimental animals to BZDs, several implications for the clinical consequences of early exposure to these compounds emerge:

1. The major effects of prenatal exposure to DZ have a late onset of appearance, often not observed until late adolescence and young adulthood. Interestingly, the clinical literature indicates that many psychiatric disorders have a similar time-course of development. A study, for example, reported that the peak 5-year period for first episode appearance of schizophreniform disorders was 16–20 years of age and the peak 5-year period for first episode bipolar disorder and major depression was 21–25 (Iacono and Beiser, 1989). The similarity to the time-course for appearance of behavioral disorders observed in rats following DZ exposure in utero suggests the hypothesis that perinatal insults at the molecular level may very well be a contributing factor in psychiatric disorders. In support of this hypothesis, an influence of prenatal alcohol exposure in the development of schizophrenia has been suggested (Lohr and Bracha, 1989). However, factors accounting for the delayed onset of schizophrenia, for example, have remained elusive. In order to develop a possible hypothesis, a second feature of prenatal DZ exposure in animals must be considered.

2. Prenatal exposure to DZ interferes with adaptive responses to environmental challenge. Thus, exposed animals appear to process environmental information inappropriately. There is a widespread consensus that many forms of illness result from interactions of predisposing vulnerabilities of individuals and stressors (Elliott and Eisdorfer, 1982). Stress has also been considered to play a role in development of behavioral disorders, including schizophrenic disorders (Asarnow et al., 1989; Breier et al., 1991). Prenatal DZ exposure may very well be considered to be a predisposing factor that influences an individual's response to stressors. Several studies reported in this chapter demonstrate that as animals go through puberty and reach young

adulthood, changes take place in their response to environmental stimuli. The timely appearance during puberty of specific behavioral responses to challenge may reflect an organization of behavior that benefits the survival and reproductive success of an organism. Thus, as rats go through puberty, there is an increase in the sensitivity to gap detection (improved auditory fusion), and there is a change in environment-specific social behaviors observed between two strangers.

Furthermore, there is growing evidence that gonadal factors may influence the emergence of these nonreproductive behaviors during adolescence. Prenatal exposure of rats to DZ alters many of these behaviors as measured in the adult. These observations suggest that the mechanisms or systems altered by the in utero exposure to DZ undergo a change during adolescence, and that the early exposure to DZ impairs the ability of the organism to make the appropriate changes. We have reported a change in the responsiveness of the BZD/GABA receptor complex in the cerebral cortex to challenge as the animal matures from late juvenile ages to adulthood. This complex may be a major participant in an organism's response to environmental stressors. There is also evidence suggesting that this complex in the cerebral cortex may not become a participant until late adolescence, young adulthood. Thus, such a model as presented here could account for the delayed onset of psychiatric symptoms in young adulthood. The symptoms do not appear earlier because younger organisms may use different neural systems in coping responses or because specific systems may undergo a reorganization during puberty. Early developmental insults could interfere with either of these changes. Undoubtedly, as observed in animals following prenatal DZ exposure, earlier evidence of the impairment could be detected with appropriate testing.

3. A final feature to consider is that there may be some sexual dimorphism to the consequences of early DZ exposure. Whereas altered neural and behavioral responses to stressors have been identified in both male and female rats exposed in utero to DZ, nonreproductive behaviors that are normally sexually dimorphic in adult animals seem to be clearly affected by the early DZ exposure. Thus prenatal DZ exposure markedly altered SI in adult male rats (Kellogg et al., 1991a), and postnatal DZ exposure had a pronounced effect on continuous reinforcement responding (CRF) in adult male but not female rats (Guillamon et al., 1991). Both SI and CRF are normally sexually dimorphic in rats, with male rats showing environment-specific SI and having much higher response rates than females. Prenatal exposure to alcohol has also been shown to affect sexually dimorphic behaviors in the adult (McGivern et al., 1984). Even on a test that was not found to be sexually dimorphic, behavior on the elevated plus maze, male rats exposed in utero to DZ were dramatically affected, whereas no effect of the exposure was apparent in females (Kellogg et al., 1991a). Clinical observations have also indicated that the ratio of males:females for first contact for treatment of schizo-phrenic symptoms is 3:1 (Iacono and Beiser, 1989). Clearly though, consequences of prenatal exposure have been identified in female rats, particularly when behaviors typical for females, such as maternal aggression, were evaluated. Likewise, there are clinical disorders, such as depression, which tend to be more prevalent in females than in males. Different neural systems in male versus female rats may underlie responses to environmental stressors, and the neural systems utilized by male and female rats may be differentially susceptible to the action of DZ. Our initial studies do indicate that SI,

for example, undergoes different changes during puberty in the male versus the female rat, suggesting that pubertal organization of adult behaviors may be sexually dimorphic. These observations raise the possibility that the action of DZ in utero may have interacted with some sexually dimorphic perinatal organizational influence, such as perinatal sex hormones (Toran-Allerand, 1986).

The experimental observations suggest, therefore, the need to reevaluate clinical approaches to identification of long-term effects following perinatal exposure to BZDs. Clearly, tests in the neonate or the very young child would not seem to provide the clearest picture of the consequences. The difficulties, however, of linking effects that emerge 20 years after exposure with the initial insult are indeed numerous. One major difficulty in conducting a clinical study is the problem of identifying a population exposed to BZDs during specific in utero periods. As indicated by Laegreid et al. (1990), even perusal of obstetric records may be of little help in identifying women who consumed BZDs during pregnancy. Prescription of BZDs was noted in only one of the 10 records of BZD-positive mothers (determined by evaluation of maternal blood samples taken during the first trimester). It is critical, however, to obtain definitive information on the possibility that prenatal exposure to psychoactive drugs, in particular drugs such as the BZDs and alcohol, which can influence function at the BZD/GABA receptor complex, may be a contributing factor to the development of psychiatric illness. The cost of such illness to society is extreme, both in monetary terms and in terms of wasted human potential.

CONCLUSIONS

Experimental studies that have examined the effects of prenatal exposure to BZD compounds have demonstrated marked interference in a variety of behavioral and neural functions as a consequence of the exposure. In particular, early exposure to these drugs appears to interfere with organized responses to environmental stressors in the adult. It may be argued that effects visible in adulthood and old age result from developmental exposure to a drug, in the complete absence of persisting drug, only if the drug altered or reorganized some aspect (or aspects) of brain function. As presented in the introduction, the final organization of brain and behavior results from the interaction during development of three interacting influences: the genome, the internal chemical environment, and the external environment of the organism. Considering that BZD compounds exert a marked influence on function of GABA at the BZD/GABA receptor complex, even during late gestational ages in the rat, the influence of prenatal exposure to these drugs, therefore, implicates a developmental role for GABA, or the BZD/GABA receptor complex, in the organization of integrated stress responses. Such a hypothesis would lead to the prediction that exposure over a similar period to a variety of drugs could result in similar consequences, since in addition to BZDs, barbiturates, alcohol, volatile anesthetics, synthetic and natural steroids have all been shown to influence function of the BZD/GABA receptor complex (Schwartz, 1988). Some of the consequences of prenatal exposure to DZ have been observed following prenatal exposure to ethanol and to barbiturates (Detering et al., 1981; Middaugh et al., 1981). However, results from our laboratory have indicated that prenatal exposure to phenobarbital does not lead to the same consequences as seen after prenatal DZ exposure (Harary, 1985; Sullivan and Kellogg, 1984, unpublished observations). It must be kept in mind that the subunit composition of the BZD/GABA$_A$ receptor complex may vary in different regions, and that differences in composition may dictate differences in pharmacologic sensitivity (Olsen et al., 1990). The BZD compounds may also interact with binding sites not linked to the GABA receptor (peripheral-type sites), and some of the consequences of early BZD exposure may reflect action of the drugs at these sites. Thus, the developing brain may be especially vulnerable to BZDs because of the multiple receptor sites for these drugs, sites that reside on neural systems critical to the expression of organized stress responses.

ACKNOWLEDGMENTS

The author gratefully acknowledges the time and efforts of the following individuals who have contributed so much to the work reported here: Daniel Bitran, Jane A. Chisholm, Ronnie Guillet, Norma Harary, Jon Inglefield, James R. Ison, Rajesh Miranda, Richard K. Miller, Gloria Pleger, Renee Primus, Todd M. Retell, Monica Rodriquez-Zafra, Steven M. Shamah, Roy D. Simmons, Alan T. Sullivan, Merritt Taylor, Donna M. Tervo, and Joseph P. Wagner. Research reported in this paper was supported by PHS Grants MH 31850 and DA 07080 and by Research Scientist Development Award MH 00651. Diazepam, Clonazepam, and Flumazenil (Ro 15-1788) were generously supplied by Peter Sorter, Hoffmann-LaRoche, Nutley, New Jersey. PK 11195 was a gift from G. LeFur, Pharmuka Laboratories, Gennevillers, France.

REFERENCES

Aaltonen, L., Erkkola, R., and Kanto, J. (1983). Benzodiazepine receptors in the human fetus. *Biol. Neonate* **44**, 54–57.

Anholt, R. R. H., Murphy, K. M. M., Mack, G. E., and Snyder, S. H. (1984). Peripheral-type benzodiazepine receptors in the central nervous system: Localization to olfactory nerves. *J. Neuroscience* **4**, 593–603.

Anholt, R. R. H., Petersen, P. L., DeSouza, E. B., and Snyder, S. H. (1986). The peripheral benzodiazepine receptor: Localization to the mitochondria outer membrane. *J. Biol. Chem.* **261**, 576–583.

Asarnow, R. F., Asarnow, J. R., and Strandburg, R. (1989). Schizophrenia: A developmental perspective. *In* "The Emergence of a Discipline: Rochester Symposium on Developmental Psychopathology" (D. Cicchetti, ed.), pp. 189–219. Lawrence Erlbaum, Hillsdale, New Jersey.

Belin, M. F., Gamrani, H., Agnera, M., Calas, A., and Pujol, J. F. (1980). Selective uptake of [^{3}H]gamma-aminobutyrate by rat supra- and subependymal nerve fibers: Histologic and high-resolution radioautographic studies. *Neuroscience* **5**, 241–254.

Bitran, D., Primus, R. J., and Kellogg, C. K. (1991). Gestational exposure to diazepam increases sensitivity to convulsants that act at the GABA/benzodiazepine receptor complex. *Eur. J. Pharmacol.* **196**, 223–231.

Bodnoff, S. R., Suranyi-Cadotte, B., Quirion, R., and Meaney, M. J. (1987). Postnatal handling reduces novelty-induced fear and increases [^{3}H]flunitrazepam binding in rat brain. *Eur. J. Pharmacol.* **144**, 105–107.

Braestrup, C., and Nielsen, M. (1978). Ontogenetic development of benzodiazepine receptors in the brain. *Brain Res.* **147**, 170–173.

Braestrup, C., and Squires, R. (1978). Pharmacological characterization of benzodiazepine receptors in the brain. *Eur. J. Pharmacol.* **48**, 263–270.

Braestrup, C., Nielsen, M., and Squires, R. F. (1979). No changes in rat benzodiazepine receptors after withdrawal from continuous treatment with lorazepam and diazepam. *Life Sci.* **24**, 347–350.

Breier, A., and Paul, S. M. (1990). The GABA$_A$/benzodiazepine receptor: Implications for the molecular basis of anxiety. *J. Psychiatr. Res.* **24**, (Suppl. 2), 91–104.

Breier, A., Wolkowitz, O. M., and Pickar, D. (1991). Stress and schizophrenia. In "Advances in Neuropsychiatry and Psychopharmacology, Vol. 1: Schizophrenia Research" (C. A. Tamminga and S. C. Schulz, eds.), pp. 141–152. Raven Press, New York.

Brooksbank, B. W. L., Atkinson, D. J., and Balazs, R. (1982). Biochemical development of the human brain. III. Benzodiazepine receptors, free gamma-aminobutyrate (GABA) and other amino acids. *J. Neurosci. Res.* **8**, 581–594.

Butcher, R. E., and Vorhees, C. V. (1979). A preliminary test battery for the investigation of the behavioral teratology of selected psychotropic drugs. *Neurobehav. Toxicol.* (Suppl. 1), 207–212.

Campbell, B. A., Lytle, L. D., and Fibiger, H. C. (1969). Ontogeny of adrenergic arousal and cholinergic inhibitory mechanisms in the rat. *Science* **166**, 635–637.

Candy, J. M., and Martin, I. L. (1979). The postnatal development of the benzodiazepine receptor in the cerebral cortex and cerebellum of the rat. *J. Neurochem.* **32**, 655–658.

Chesley, S., Lumpkin, M., Schatzke, A., Galpern, W. R., Greenblatt, D. J., Shader, R. I., and Miller, L. G. (1991). Prenatal exposure to benzodiazepine—I. Prenatal exposure to lorazepam in mice alters open-field activity and GABA$_A$ receptor function. *Neuropharmacology* **30**, 53–58.

Chisholm, J., Kellogg, C., and Lippa, A. (1983). Development of benzodiazepine-binding subtypes in three regions of rat brain. *Brain Res.* **267**, 388–391.

Cole, J. O., and Charello, R. J. (1990). The benzodiazepines as drugs of abuse. *J. Psychiatr. Res.* **24**, 135–144.

Corrodi, H., Fuxe, K., Lidbrink, P., and Olsen, L. (1971). Minor tranquilizers, stress, and central catecholamine neurons. *Brain Res.* **29**, 1–16.

Crain, S. M., and Bornstein, M. B. (1974). Early onset in inhibitory functions during synaptogenesis in fetal mouse brain cultures. *Brain Res.* **68**, 351–357.

Cree, J. E., Meyer, J., and Hailey, D. M. (1973). Diazepam in labour: Its metabolism and effect on the clinical condition and thermogenesis of the newborn. *Br. Med. J.* **4**, 251–255.

Darwin C. (1955). "The Expression of the Emotions in Man and Animals." Philosophical Library, New York Press, Chicago.

Daval, J-L., De Vasconcelos, A. P., and Lartaud, I. (1988). Morphological and neurochemical effects of diazepam and phenobarbital on selective culture of neurons from fetal rat brain. *J. Neurochem.* **50**, 665–672.

Davis, M., Gendelman, D. L., Tischler, M. D., and Gendelman, P. M. (1982). A primary acoustic startle circuit: Lesion and stimulation studies. *J. Neurosci.* **2**, 791–805.

DeBoer, S. F., Van Der Gugten, J., and Slangen, J. L. (1990). Brain benzodiazepine receptor-mediated effects on plasma catecholamine and corticosterone concentrations in rats. *Brain Res. Bull.* **24**, 843–847.

Detering, N., Collins, R. M., Hawkins, R. L., Ozand, P. T., and Kavahasan, A. (1981). Comparative effects of ethanol and malnutrition on the development of catecholamine neurons: A long-lasting effect in the hypothalamus. *J. Neurochem.* **36**, 2094–2096.

Deutch, A. Y., Gruen, R. J., and Roth, R. H. (1989). The effects of perinatal diazepam exposure on stress-induced activation of the mesotelencephalic dopamine system. *Neuropsychopharmacology* **2**, 105–114.

DeWitt, H., Johanson, C. E. and Uhlenhuth, E. H. (1984). The dependence potential of benzodiazepines. *Curr. Med. Res. Opin.* **8**, 48–59.

Dobbing, J. (1974). Prenatal nutrition and neurologic development. In "Early Malnutrition and Mental Development" (J. Crovido, L. Hambraeus, and B. Vahlquist, eds.), pp. 96–110. Almquist and Wiksel, Stockholm.

Doble, A., Malgouris, C., Daniel, M., Daniel, N., Imbault, F., Basbaum, A., Uzan, A., Gueremy, C., and LeFur, G. (1987). Labeling of peripheral-type benzodiazepine-binding sites in human brain with [³H] PK 11195: Anatomical and subcellular distribution. *Brain Res. Bull.* **18,** 49–61.

Eichinger, A., and Sieghart, W. (1986). Postnatal development of proteins associated with different benzodiazepine receptors. *J. Neurochem.* **46,** 173–180.

Elliott, G. R., and Eisdorfer, C. (1982). "Stress and Human Health: Analysis and Implications of Research." Springer, New York.

File, S. E. (1980). The use of social interaction as a method for detecting anxiolytic activity of chlordiazepoxide-like drugs. *J. Neurosci. Methods* **2,** 219–238.

File, S. E. (1988). How good is social interaction as a test of anxiety? *Anim. Models Psych. Disord.* **1,** 151–166.

Frieder, B., Epstein, S., and Grimm, V. E. (1984). The effects of exposure to diazepam during various stages of gestation or during lactation on the development and behavior of rat pups. *Psychopharmacology* **83,** 51–55.

Fujii, T., Yamamoto, N., and Fuchino, K. (1983). Functional alterations in the hypothalamic–pituitary–thyroid axis in rats exposed prenatally to diazepam. *Toxicol. Lett.* **16,** 131–137.

Gai, N., and Grimm, V. E. (1982). The effect of prenatal exposure to diazepam on aspects of postnatal development and behavior in rats. *Psychopharmacology* **78,** 225–229.

Gee, K. W., Brinton, R. E., and McEven, B. S. (1987). Regional distribution of a RO5-4864 binding site that is functionally coupled to the γ-aminobutyric acid/benzodiazepine complex in rat brain. *J. Pharmacol. Exp. Ther.* **244,** 379–383.

Gey, K. (1973). Effect of benzodiazepines on carbohydrate metabolism in rat brain. In "The Benzodiazepines" (S. Grattini, E. Mussini, and L. O. Randall, eds.), pp. 243–256. Raven Press, New York.

Gillberg, C. H. (1977). "Floppy infant syndrome" and maternal diazepam. *Lancet* **2,** 244.

Gray, J. A. (1982). "The Neuropsychology of Anxiety: An Enquiry into the Functions of the Septo-Hippocampal System." Oxford University Press, New York.

Griffiths, R. R., and Sannerud, C. A. (1987). Abuse of and dependence on benzodiazepines and other anxiolytic/sedative drugs. In "Psychopharmacology: The Third Generation of Progress" (H. Meltzer, ed.), pp. 1535–1541. Raven Press, New York.

Gruen, R. J., Deutch, A. Y., and Roth R. H. (1990a). Perinatal diazepam exposure: Alterations in exploratory behavior and mesolimbic dopamine turnover. *Pharmacol. Biochem. Behav.* **36,** 169–175.

Gruen, R. J., Elsworth, J. D., and Roth, R. H. (1990b). Regionally specific alterations in the low-affinity GABA_A receptor following perinatal exposure to diazepam. *Brain Res.* **514,** 151–154.

Guillamon, A., Cales, J. M., Rodriguez-Zafra, M., Perez-Laso, C., Caminero, A., Izquierdo, M. A. P., and Segovia, S. (1991). Effects of perinatal diazepam administration on two sexually dimorphic nonreproductive behaviors. *Brain Res. Bull.* **25,** 913–916.

Haefely, W. E. (1990). The GABA_A-benzodiazepine receptor: Biology and pharmacology. In "Handbook of Anxiety, Vol. 3: The Neurobiology of Anxiety" (G. D. Burrows, M. Roth, and R. Noyes, eds.), pp. 165–188. Elsevier Science Publishers, Amsterdam.

Harary, N. (1985). "The Relationship of Benzodiazepine-Binding Sites to the Norepinephrine Fiber System in the Developing and Adult Rat." Doctoral dissertation, University of Rochester, New York.

Harary, N., and Kellogg, C. K. (1989). The relationship of benzodiazepine-binding sites to the norepinephrine projection in the hypothalamus of the adult rat. *Brain Res.* **492,** 293–299.

Havoundjian, H., Paul, S. M., and Skolnick, P. (1986a). Acute, stress-induced changes in the benzodiazepine/gamma-aminobutyric acid receptor complex are confined to the chloride ionophore. *J. Pharmacol. Exp. Ther.* **237,** 787–703.

Havoundjian, H., Paul, S. M., and Skolnick, P. (1986b). The permeability of gamma-aminobutyric acid gated chloride channels is described by the binding of a cage convulsant, [³⁵S]t-butylbicyclophosphorothionate, *Proc. Natl. Acad. Sci. U.S.A.* **83,** 9241–9244.

Hebebrand, J., Friedl, W. and Propping, P. (1988). The concept of isoreceptors: Application to the nicotenic acetylcholine receptor and the gamma-aminobutyric acid_A/benzodiazepine receptor complex. *J. Neural Trans.* **71,** 1–9.

Iacono, W. G., and Beiser, M. (1989). Age of onset, temporal stability, and eighteen-month course of first-episode psychosis. In "The Emergence of a Discipline: Rochester Symposium on Psychopathology" (D. Cicchetti, ed.), pp. 221–260. Lawrence Erlbaum, Hillsdale, New Jersey.

Ison, J. R. (1982). Temporal acuity in auditory function in the rat: Reflex inhibition by grief gaps in noise. J. Comp. Physiol. Psychol. 96, 945–954.

Ison, J. R., and Hammond, G. (1971). Modification of the startle reflex in the rat by changes in the auditory and visual environments. J. Comp. Physiol. Psych. 75, 435–452.

Kellogg, C. K. (1988). Benzodiazepines: Influence on the developing brain. Prog. Brain Res. 73, 207–228.

Kellogg, C. K., and Amaral, D. G. (1978). Neurotransmitter regulation of stress responses: Relationship to seizure induction. In "Cholinergic–Monoaminergic Interactions in the Brain" (L. L. Butcher, ed.), pp. 291–304. Academic Press, New York.

Kellogg, C. K., and Pleger, G. L. (1989). GABA-stimulated chloride uptake and enhancement by diazepam in synaptoneurosomes from rat brain during prenatal and postnatal development. Dev. Brain Res. 49, 87–95.

Kellogg, C. K., and Retell, T. M. (1986). Release of [^{3}H]norepinephrine: Alteration by early developmental exposure to diazepam. Brain Res. 366, 137–144.

Kellogg, C., Tervo, D., Ison, J., Parisi, T., and Miller, R. K. (1980). Prenatal exposure to diazepam alters behavioral development in rats. Science 207, 205–207.

Kellogg, C. K., Chisholm, J., Simmons, R. D., Ison, J. R., and Miller, R. K. (1983a). Neural and behavioral consequences of prenatal exposure to diazepam. Monogr. Neural Sci. 9, 119–129.

Kellogg, C., Ison, J. R., and Miller, R. K. (1983b). Prenatal diazepam exposure: Effects on auditory temporal resolution in rats. Psychopharmacology 79, 332–337.

Kellogg, C. K., Primus, R. J., and Bitran, D. (1991a). Sexually dimorphic influence of prenatal exposure to diazepam on behavioral responses to environmental challenge and on γ-aminobutyric acid (GABA)-stimulated uptake in the brain. J. Pharmacol. Exp. Ther. 256, 259–265.

Kellogg, C. K., Sullivan, A. T., Bitran, D., and Ison, J. R. (1991b). Modulation of noise-potentiated acoustic startle via the benzodiazepine/GABA receptor complex. Behav. Neurosci. In press.

Kellogg, C. K., Taylor, M. T., and Inglefield, J. (1991c). Stressor-induced changes in GABA receptor function: Role of hypothalamic norepinephrine projections. Soc. Neurosci. Abstr. 17, 80.

Kuhar, M. J., and Young, W. S. (1979). Radiohistochemical localizations of benzodiazepine receptors in rat brain. J. Pharmacol. Exp. Ther. 212, 337–346.

Laegreid, L., Olegard, R., Walstrom, J., and Conradi, N. (1989). Teratogenic effects of benzodiazepine use during pregnancy. J. Pediatr. 114, 126–131.

Laegreid, L., Olegard, R., Conradi, N., Hagberg, G., Wahlstrom, J., and Abrahamsson, L. (1990). Congenital malformations and maternal consumption of benzodiazepines: A case-control study. Dev. Med. Child Neurol. 32, 432–441.

Langley, J. N. (1906). On nerve-endings and on special excitable substances in cells. Proc. R. Soc. Lond. 78 (Series B), 170–194.

Lahti, R. A., and Barsuhn, C. (1975). The effect of various doses of minor tranquilizers on plasma corticosteroids in stressed rats. Res. Commun. Chem. Pathol. Pharmacol. 11, 595–603.

Lauder, J. M., Han, V. K. M., Henderson, P., Verdoorn, T., and Towle, A. C. (1986). Prenatal ontogeny of the GABAergic system in the rat brain: An immunocytochemical study. Neuroscience 19, 465–493.

Laviola, G., De Acetis, L., Bignami, G., and Alleva, E. (1991). Prenatal oxazepam enhances mouse maternal aggression in the offspring, without modifying acute chlordiazepoxide effects. Neurotoxicol. Teratol. 13, 75–81.

LeFur, G., Guillout, F., Mitrani, N., Mizoule, J., and Uzan, A. (1979). Relationships between plasma corticosteroids and benzodiazepines in stress. J. Pharmacol. Exp. Ther. 211, 305–308.

Leibowitz, S. F., Diaz, S., and Tempel. D. (1989). Norepinephrine in the paraventricular nucleus stimulates corticosterone release. Brain Res. 496, 219–227.

Livezey, G. T., Radulovacki, M., Isaac, L., and Marczynski, T. J. (1985). Prenatal exposure to diazepam results in enduring reactions in brain receptors and deep slow wave sleep. Brain Res. 334, 361–365.

Livezey, G. T., Marczynski, T. J., and Isaac, L. (1986). Enduring effects of prenatal diazepam on the behavior, EEG, and brain receptors of the adult cat progeny. Neurotoxicology 7, 319–333.

Lohr, J. B., and Bracha, H. S. (1989). Can schizophrenia be related to prenatal exposure to alcohol? Some speculations. *Schizophren. Bull.* **15**, 595–603.

Mallorga, P., Hamburg, M., Tallman, J. F., and Gallager, D. W. (1980). Ontogenetic changes in GABA modulation of brain benzodiazepine binding. *Neuropharmacology* **19**, 405–408.

Mandelli, M., Morselli, P. L., Nordio, S., Pardi, G., Principi, N., Sereni, F., and Tognoni, G. (1975). Placental transfer of diazepam and its disposition in the newborn. *Clin. Pharmacol. Ther.* **17**, 564–572.

Marangos, P. J., Patel, J., Boulenger, J., and Clark-Rosenberg, R. (1982). Characterization of peripheral-type benzodiazepine binding sites in brain using [³H]Ro 5-4864. *Mol. Pharmacol.* **22**, 16–32.

Marczynski, T. J., and Urbancic, M. (1988). Animal models of chronic anxiety and "fearlessness." *Brain Res. Bull.* **21**, 438–490.

Marczynski, T. J., Harris, C. M., and Livezey, G. T. (1981). The magnitude of postreinforcement EEG synchronization (PRS) in cats reflects learning ability. *Brain Res.* **204**, 214–219.

Massotti, M., Alleva, F. R., Balazs, T., and Guidotti, A. (1980). GABA and benzodiazepine receptors in the offspring of dams receiving diazepam: Ontogenetic studies. *Neuropharmacology* **19**, 951–956.

McCroskey, R. L., and Kidder, H. C. (1980). Auditory fusion among learning disabled, reading disabled, and normal children. *J. Learn. Disabil.* **13**, 18–25.

McGivern, R. F., Clancy, A. N., Hill, M. A., and Noble, E. P. (1984). Prenatal alcohol exposure alters adult expression of sexually dimorphic behavior in the rat. *Science* **224**, 896–898.

Meaney, M. J., Aitken, D. H., and Sapolsky, R. M. (1987). Thyroid hormones influence the development of hippocampal glucocorticoid receptors in the rat: A mechanism for the effects of postnatal handling on the development of the adrenocortical stress response. *Neuroendocrinology* **45**, 278–283.

Mellinger, G. D., and Balter, M. B. (1983). Psychotherapeutic drugs: A current assessment of prevalence and patterns of use. *In* "Society and Medication: Conflicting Signals for Prescribers and Patients" (J. P. Morgan and D. V. Kagan, eds.), pp. 137–154. D. C. Heath, Lexington, Massachusetts.

Middaugh, L. D., Thomas, T. N., Simpson, L. W., and Zemp, J. W. (1981). Effects of prenatal maternal injections of phenobarbital on brain neurotransmitters and behavior of young C57 mice. *Neurobehav. Toxicol. Teratol.* **3**, 271–275.

Miller, L. G., Roy, R. B., Weill, C. L., and Lopez, F. (1989). Persistent alterations in GABA$_A$ receptor binding and function after prenatal lorazepam administration in the chick. *Brain Res. Bull.* **23**, 171–174.

Miranda, R., Wagner, J. P., and Kellogg, C. K. (1989). Early developmental exposure to benzodiazepine ligands alters brain levels of thiobarbituric acid-reactive products in young adult rats. *Neurochem. Res.* **14**, 1119–1127.

Miranda, R., Ceckler, T., Guillet, R., and Kellogg, C. K. (1990a). Early developmental exposure to benzodiazepine ligands alters brain ³¹P NMR spectra in young adults rats. *Brain Res.* **506**, 85–92.

Miranda, R., Ceckler, T., Guillet, R., and Kellogg, C. K. (1990b). Aging-related changes in brain metabolism are altered by early developmental exposure to diazepam. *Neurobiol. Aging* **11**, 117–122.

Mohler, H., and Okada, T. (1977). Benzodiazepine receptor: Demonstration in the central nervous system. *Science* **198**, 849–851.

Morton, K. D. R., Van de Kar, L. D., Brownfield, M. S., Lorens, S. A., Napier, T. C., and Urban, J. H. (1990). Stress-induced renin and corticosterone secretion is mediated by catecholaminergic nerve terminals in the hypothalamic paraventricular nucleus. *Neuroendocrinology* **51**, 320–327.

Nutt, D. J. (1990). Pharmacological mechanisms of benzodiazepine withdrawal. *J. Psychiatr. Res.* **24**, 105–110.

Olsen, R. W., Bureau, M., Khrestchatisky, M., MacLennan, A. J., Chiang, M.-Y., Tobin, A. J., Xu, W., Jackson, M., Sternini, C., and Brecha, N. (1990). Isolation of pharmacologically distinct GABA-benzodiazepine receptors by protein chemistry and molecular cloning. *In* "GABA and Benzodiazepine Receptor Subtypes" (G. Biggio and E. Costa, eds.), pp. 35–49. Raven Press, New York.

Oppenheim, R. W., and Haverkamp, L. (1986). Early development of behavior and the nervous system. *In* "Handbook of Behavioral Neurobiology" (E. M. Blass, ed.), pp. 1–33. Plenum Press, New York.

Owen, F., Poulter, M., Waddington, J. L., Mashal, R. D., and Crow, T. J. (1983). [³H]Ro 5-4863 benzodiazepine binding in kainate lesioned striatum and in temporal cortex of brains from patients with senile dementia of the Alzheimer type. *Brain Res.*, **278**, 373–375.

Paul, S. M., Crawley, J. N., and Skolnick, P. (1986). The neurobiology of anxiety: The role of the GABA/benzodiazepine receptor complex. In "American Handbook of Psychiatry, Vol. 8, Biological Psychiatry" (P. A. Berger and H. K. H. Brodie eds.), pp. 581–596. Basic Books, New York.

Pellow, S., and File, S. E. (1985). The effects of putative anxiogenic compounds (FG 7142, CGS 8216 and Ro 15-1788) on the rat corticosterone response. Physiol. Behav. **35**, 587–590.

Petroff, O. A. C., Prichard, J. W., Ogino, T., Avison, M., Alger, J. R., and Shulman, R. G. (1986). Combined ^{1}H and ^{31}P nuclear magnetic resonance spectroscopic studies of bicuculline-induced seizures in vivo. Ann. Neurol. **21**, 185–193.

Plotsky, P. M., Cunningham, E. T., and Widmaier, E. P. (1989). Catecholaminergic modulation of corticotropin-releasing factor and adrenocorticotropin secretion. Endocr. Rev. **10**, 437–458.

Polc, P., Mohler, H., and Haefley, W. (1974). The effect of diazepam on spinal cord activities: Possible sites and mechanisms of action. Naunyn-Schmiedeberg's Arch. Pharmacol. **284**, 319–337.

Primus, R. J., and Kellogg, C. K. (1989). Pubertal-related changes influence the development of environment-related social interaction in the male rat. Dev. Psychobiol. **22**, 633–643.

Primus, R. J., and Kellogg, C. K. (1990a). Gonadal hormones during puberty organize environment-related social interaction in the male rat. Horm. Behav. **24**, 311–323.

Primus, R. J., and Kellogg, C. K. (1990b). Developmental influence of gonadal function on the anxiolytic effect of diazepam on environment-related social interaction in the male rat. Behav. Pharmacol. **1**, 437–446.

Primus, R. J., and Kellogg, C. K. (1991a). Experience influences environmental modulation of function at the benzodiazepine (BZD)/GABA receptor chloride channel complex. Brain Res. **545**, 257–264.

Primus, R. J., and Kellogg, C. K. (1991b). Gonadal status and pubertal age influence the responsiveness of the benzodiazepine/GABA receptor complex to environmental challenge in male rats. Brain Res. In press.

Randall, L., and Kappell, B. (1973). Pharmacologic activity of some benzodiazepines and their metabolites. In "The Benzodiazepines" (S. Garattini, E. Mussini, and L. O. Randall, eds.), pp. 27–51. Raven Press, New York.

Reisine, T., Affolter, H-L., Rougon, G., and Barbet, J. (1986). New insights into the molecular mechanisms of stress. Trends Neurosci. **9**, 574–579.

Richardson, R., Siegel, M. A., and Campbell, B. A. (1988). Unfamiliar environments impair information processing as measured by behavioral and cardiac orienting responses to auditory stimuli in preweanling and adult rats. Dev. Psychobiol. **21**, 491–503.

Rothe, T., and Langer, M. (1988). Prenatal diazepam exposure affects β-adrenergic receptors in brain regions of adult rat offspring. J. Neurochem. **51**, 1361–1366.

Rothe, T., Middleton-Price, H., and Bigl, V. (1988). The ontogeny of GABA receptors and glutamic acid decarboxylase in regions of the rat brain. Neuropharmacology **27**, 661–667.

Safra, M. J., and Oakley, G. P. (1975). Association between cleft lip with or without cleft palate and prenatal exposure to diazepam. Lancet **2**, 478–480.

Saxen, I. (1975). Associations between oral clefts and drugs taken during pregnancy. Int. J. Epidemiol. **4**, 37–44.

Schlumpf, M., Richards, J. G., Lichtensteiger, W., and Mohler, H. (1983). An autoradiographical study of the prenatal development of benzodiazepine-binding sites in rat brain. J. Neurosci. **3**, 1478–1487.

Schlumpf, M., Ramseier, J., Abriel, J., Youmbi, M., Baumann, J. B., and Lichtensteiger, W. (1989). Diazepam effects on the fetus. Neurotoxicology **10**, 501–516.

Schwartz, R. (1988). The GABA$_A$ receptor-gated ion channel: Biochemical and pharmacological studies of structure and function. Biochem. Pharmacol. **37**, 3369–3375.

Schwartz, R. D., Wess, M. J., Labarca, R., Skolnick, P., and Paul, S. M. (1987). Acute stress enhances the activity of the GABA receptor-gated chloride ion channel in brain. Brain Res. **411**, 151–155.

Schwartz, R. D., Skolnick, P., and Paul, S. M. (1988). Regulation of γ-aminobutyric acid/barbiturate receptor-gated chloride ion flux in brain vesicles by phospholipase A$_2$: Possible role of oxygen radicals. J. Neurochem. **50**, 565–571.

Selye, H. (1976). "Stress in Health and Disease." Butterworth, Boston, Massachusetts.

Shibuya, T., Watanabe, Y., Hill, H. F., and Salafsky, B. (1986). Developmental alterations in maturing rats caused by chronic prenatal and postnatal diazepam treatments. Japan J. Pharmacol. **40**, 21–29.

Shoemaker, H., Morelli, M., Deshmukh, P., and Yamamura, H. I. (1982). [³H]Ro 5-4864 benzodiazepine binding in the kainate-lesioned rat striatum and Huntington's diseased basal ganglia. *Brain Res.* **248**, 396–401.

Shore, C. O., Vorhees, C. V., Bornschein, R. L., and Stemmer, D. (1983). Behavioral consequences of prenatal diazepam exposure in rats. *Neurobehav. Toxicol. Teratol.* **5**, 565–570.

Simmons, R. D., Kellogg, C. K., and Miller, R. K. (1983). Prenatal diazepam: Distribution and metabolism in perinatal rats. *Teratology* **28**, 181–188.

Simmons, R. D., Kellogg, C. K., and Miller, R. K. (1984a). Prenatal diazepam exposure in rats: Long-lasting, receptor-mediated effects on hypothalamic norepinephrine-containing neurons. *Brain Res.* **293**, 73–83.

Simmons, R. D., Miller, R. K., and Kellogg, C. K. (1984b). Prenatal exposure to diazepam alters central and peripheral responses to stress in adult rat offspring. *Brain Res.* **307**, 39–46.

Speight, A. N. P. (1977). Floppy-infant syndrome and maternal diazepam and/or nitrazepam. *Lancet* **2**, 878.

Squires, R., and Braestrup, C. (1977). Benzodiazepine receptors in brain. *Nature* **266**, 732–734.

Sullivan, A. T., and Kellogg, C. K. (1985). Dose-related effects of diazepam on noise-induced potentiation of the acoustic startle reflex: Disclosure of several underlying mechanisms. *Soc. Neurosci. Abstr.* **11**, 1289, No. 377.4.

Suria, A., and Costa, E. (1975). Action of diazepam, dibutyryl cGMP, and GABA on presynaptic nerve terminals in bullfrog sympathetic ganglia. *Brain Res.* **87**, 102–106.

Tallman, J. F., and Gallager, D. W. (1985). The GABAergic system: A locus of benzodiazepine actions. *Annu. Rev. Neurosci.* **8**, 21–44.

Tocco, D. R., Renskers, K., and Zimmerman, E. F. (1987). Diazepam-induced cleft palate in the mouse and lack of correlation with the H-2 locus. *Teratology* **35**, 439–445.

Toran-Allerand, C. D. (1986). Sexual differentiation of the brain. *In* "Developmental Neuropsychobiology" (W. T. Greenough and J. M. Juraska, eds.), pp. 175–212. Academic Press, New York.

Trinder, E. (1979). Auditory fusion: A critical test with implications in differential diagnosis. *Br. J. Audiol.* **13**, 143–147.

Vitorica, J., Park, D., Chin, G., and de Blas, A. L. (1990). Characterization with antibodies of the γ-aminobutyric acid$_A$/benzodiazepine receptor complex during development of the rat brain. *J. Neurochem.* **54**, 187–194.

Woods, J. H., Katz, J. L., and Winger, G. (1987). Abuse liability of benzodiazepines. *Pharmacol. Rev.* **39**, 251–413.

Young, A. B., and Chu, D. (1990). Distribution of GABA$_A$ and GABA$_B$ receptors in mammalian brain. *Drug Dev. Res.* **21**, 161–167.

Young, R. L., Albano, R. F., Charnecki, A. M., and Demcsak, G. (1969). Effect of diazepam on regional levels of glucose and malate in the central nervous system. *Fed. Proc.* **28**, 444.

Young, R. S. K., Osbakken, M. D., Briggs, R. W., Yagel, S. K., Rice, D. W., and Goldberg, S. (1985). ³¹P NMR study of cerebral metabolism during prolonged seizures in the neonatal dog. *Ann. Neurol.* **18**, 14–20.

— 13 —

Drug Effects on Sexual Differentiation of the Brain: Role of Stress and Hormones in Drug Actions

ह॒

Annabell C. Segarra and Bruce S. McEwen

Laboratory of Neuroendocrinology
Rockefeller University
New York, New York

INTRODUCTION

Sexual differentiation is a developmental processes by which genetic sex and gonadal hormones determine the sex of the reproductive tract and various secondary sex characteristics, and influence the development of brain structure and function, including a variety of behaviors. In mammals, testosterone is the primary signal, and the function of the genetic sex of the animal is to determine the sex of the gonad. When the genetic sex is male, testes develop and testosterone is secreted. Giving testosterone to genetic females during the same developmental period when testosterone is secreted in males, masculinizes and defeminizes the genetic female, indicating that the phenotypic sex is largely independent of the genetic sex and is a function of the presence or absence of testosterone secretion.

Because testosterone is secreted by the testes in response to signals from the hypothalamus via the pituitary gland, there are a number of ways in which environmental factors can influence the process of sexual differentiation. Drugs, such as nicotine, alcohol, and cannabinoids, alter the process of sexual differentiation and may do so via at least five interrelated routes: (1) by altering the secretion of testosterone; (2) by interfering with actions of testosterone; (3) by altering developmental influences of neurotransmitter systems, which are synergistically involved along with testosterone in the process of sexual differentiation; (4) by altering the secretion of other hormones, such as adrenocorticotropic hormone (ACTH) and glucocorticoids, which interfere with sexual differentiation; and (5) by altering the behavior and/or physiology of the mother. In fact, psychological or physical stress to the mother or newborn infant may modify the process of sexual differentiation.

323

This chapter will consider the mechanisms by which the brain develops in a male or female direction. It will also show how this process can be affected by maternal exposure to stressors and drugs during pregnancy. The first part describes the process of sexual differentiation and the role sex steroids play in the process; it is followed by a discussion of the sexually dimorphic body, and neuroanatomical and behavioral traits that are most frequently used as indicators of "maleness" and "femaleness."

The second part of the chapter describes how stress and drugs of abuse administered during gestation affect the process of sexual differentiation of the brain. It begins by considering why stress can affect sexual differentiation during pregnancy and how the effects of drugs and stress are transferred to the fetus. In the course of this discussion, we will consider why the use of animals in this area of biomedical research is so important, but at the same time introduces methodological problems of extrapolation to humans. This is followed by a summary of the effects of stress, ethanol, nicotine, and cannabinoids on sexual differentiation of the brain. Each drug will be analyzed separately in terms of its effects on behaviors, neuroanatomical traits, and hormonal milieu. The correlation between these drugs and changes in the neuronal circuitry and/or brain neurochemicals will also be presented. A central theme is that these drugs act as stressors, and some of their effects in animal experiments may be mediated by stress hormones. The chapter concludes with a brief summary of similarities and differences among prenatal stress and drug manipulations.

BASIC FEATURES OF SEXUAL DIFFERENTIATION

Role of Genetic Factors, Particularly the Y Chromosome

As we have noted, sex determination involves two entities: the chromosomes and the sex steroids. In most vertebrates, the presence of a Y chromosome in the embryo will shift development in the male direction, but the absence of a Y chromosome will result in a female. Recent studies indicate that a gene present in the Y chromosome appears to be responsible for maleness. Evidence indicates that this gene may be involved in initiating a cascade of events that ultimately results in the development of testes and the production of the testosterone in mice (for review see Cooke, 1990).

Although the presence of a Y chromosome will determine the external genitalia, it is the sex steroid milieu that will determine whether the organism exhibits behavioral and neuroanatomical traits that pertain to male or female. Depending on the timing of this hormonal secretion, different degrees of masculinization and defeminization may ensue. Thus individuals may develop with male external genitalia but exhibit female behavioral and even morphological traits. In some cases, these may be attributed to hormonal imbalances during the fetal or neonatal developmental stage. Thus, gender determination involves not only development of the reproductive tract and secondary sexual characteristics, but also development of the neuronal circuitry and neurochemistry responsible for sexually dimorphic behaviors and traits.

Primacy of the Sex Steroids, Particularly Testosterone, for Brain Sex Differences

Sex steroids are the agents responsible for this sexual imprinting of the genital tract as well as the brain. In the absence of sex steroids, the female developmental pathway is favored. Experiments in the late 1940s showed that all gonadless rabbit fetuses, regardless of their genetic sex, acquire a female genital apparatus (for review see Jost, 1985). This implies that the testes prevent the expression of the inherent program for femaleness of the body and produce maleness, since the ovaries of fetal and neonatal females are generally nonsecretory. The major androgen secreted by the fetal testis and responsible for masculinization of the genital tract is testosterone.

Testosterone enters target cells and crosses the blood–brain barrier of fetuses. In brain cells it may act in its original form or, more frequently, it is first converted by intracellular enzymatic systems into either dihydrotestosterone (DHT-5-α-androstane-17-OH-3-one) by a 5α-reductase system; or into estradiol by an aromatase system. The active steroid then binds to specific receptor systems in order to be effective. In human subjects or mice, where the receptor protein is defective, neither testosterone nor DHT exerts any masculinizing activity. As a result, the whole body acquires feminine features despite the testicular production of testosterone (Bardin et al., 1973) a condition known as testicular feminization.

Sexual differentiation of the brain in male rats occurs mainly during the last trimester of gestation [gestation day (GD) 14–21] and extends into the first 2 weeks of postnatal life. It encompasses two main processes: *masculinization* and *defeminization.* Masculinization is the enhancement of "male behaviors," such as male sexual behavior and aggressive behavior. Defeminization is the suppression of traits such as feminine sexual behavior and reproductive hormonal cyclicity. The process of masculinization occurs mainly prenatally, whereas defeminization occurs neonatally. Both processes are androgen dependent (see Fig. 1).

Organizational and Activational Influences of Sex Hormones on the Brain and Behaviors

The effects of sex steroids on sexual development can be divided in two types: *organizational* and *activational.* The organizational effects are those that occur early in development, i.e., prenatally, neonatally, and prepubertally. Levels of sex steroids during this period, by laying out the correct circuitry and neurochemistry, will determine the gender pathway to follow. Activational effects occur mainly during and after puberty. Sex steroids levels during this period will invoke the instructions set up during the organizational phase.

Phoenix et al. (1959) showed that sex steroids exert their effect on reproduction by organizing neuronal pathways responsible for certain behaviors. Plasma testosterone and estradiol are detectable early during gestation in the fetal rat (see Table 1). The estradiol is of maternal origin, since fetal ovaries are mainly nonsecretory, where-

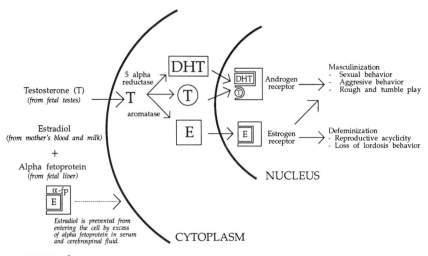

FIGURE 1   Role of sex steroids during sexual differentiation of the male brain in rats.

as androgens are secreted by the fetal testis (see Fig. 1). Androgens masculinize the brain by a mechanism that involves in situ aromatization of testosterone to estrogen in specific neural centers. The fact that serum testosterone levels are more elevated in neonatal males than in females (Corbier et al., 1978; George and Ojeda, 1982; Ward and Weisz, 1984) and the fact that serum androstenedione levels are considerably lower than testosterone in both males and females suggests that testosterone is the physiologically important substrate for aromatization in the brain at this time. Maternal estradiol plays no role in the process of sexual differentiation of the brain owing to circulating levels of alpha-fetoprotein. Alpha-fetoprotein is a ubiquitous protein produced by the fetal liver. It is present in rat and mouse serum (Nuñez et al., 1976) and in brain tissue, where it acts as an estrogen-binding protein. It affords female fetuses protection against the defeminizing effect of estrogens present in the fetal, neonatal, and maternal circulations, including those that may be reaching the pups through the mother's milk.

TABLE 1   Sex Differences in Sex Steroid Levels in the Rat[a]

Steroid	Fetal	Neonatal	Adult	Elevated levels[b]
Estradiol	f = m	f = m	f > m	PN1–2; 9–23 (m and f)
Progesterone	f = m	f = m	f > m	PN20 (m and f)
Testosterone	m > f	m > f	m > f	GD18; PN1 (m only)

[a]Levels indicated as being equal or greater in males (m) or females (f).
[b]Elevated levels indicate days postnatal (PN) or during gestation (GD) when levels are elevated in m and/or f. From Corbier et al. (1978); Döhler and Wuttke (1975); and McEwen et al. (1977).

Brain Sexual Differentiation Involves Two Pathways of Testosterone Action

Testosterone exerts its effects on the developing brain by interacting with two receptor systems, androgen and estrogen receptors (Fig. 1). Aromatization appears to mediate defeminization in the rat, whereas a combination of aromatization and 5α-reduction are involved in masculinization. Moreover, the hypothalamus, preoptic area, and amygdala of the newborn rat contain high levels of aromatizing enzyme activity (Naftolin et al., 1975). Estrogens and aromatizable androgens are effective agonists in this process when given to newborn female or castrated male rats systemically and intracerebrally (Nadler, 1973; Hayashi and Gorski, 1975; Döcke and Dörner, 1975). Competitive inhibitors of aromatization such as ATD (1,4,6 androstatriene-3,17-dione and androst-4-ene-3,6,17-trione) block defeminization in males and in testosterone-treated females (Clemens and Gladue, 1978; Davis et al., 1978; McEwen et al., 1979; Vreeburg et al., 1977). Antiestrogens such as MER-25 (McDonald and Doughty, 1973) and CI628 (McEwen et al., 1977) attenuate the defeminization of females by testosterone and block the defeminization of the female rat by estrogen (McEwen et al., 1979). In addition, agents that interfere with aromatization or estradiol action attenuate or block the effects of testosterone. For example, barbiturates have been reported to attenuate some of the masculinizing effects of testosterone by competing directly with testosterone for the hydroxylation reactions of the aromatization enzyme complex (Arai and Gorski, 1968; Clemens, 1974).

Importance of Mother–Pup Interactions

In addition to hormonal influences, maternal–pup interactions are important in the development of malelike and femalelike behavior. The best-studied example is maternal discrimination. Male and female offspring of some primate species and of rats are treated differentially by their mothers (Mitchell and Brandt, 1970; Moore and Morelli, 1979). It has been proposed that such differential treatment serves to exaggerate gene-based sex differences in behavior (Beach, 1976). Dams spend more time licking the anogenital region of their male pups than that of their females pups (Moore and Morelli, 1979). This discrimination of pup sex has important implications for the development of a number of sex differences in behavior later in life. Clearly, for the purpose of this chapter, drug or stress influences on maternal behavior may affect sexual differentiation.

Diversity of Patterns of Sexual Differentiation among Vertebrate Species

Among vertebrates, there are a variety of mechanisms that contribute to sexual differentiation of the brain. Critical or sensitive periods may encompass different developmental periods during perinatal development. The roles of sex steroids also vary according to the species under study. In birds, for example, the ovary has the major role in differentiation. In the absence of the hormone, the male pattern of organization predominates in the syrinx, genital tubercle, and with respect to copulatory behavior; exposure to ovarian hormones during a developmentally critical period feminizes these traits. In songbirds, androgen promotes singing behavior (Adkins-Regan, 1987; Nottebohm, 1980). In primates, androgens also play a major role in the process of sexual differentiation of the brain. Treatment of fetal female rhesus monkeys with androgens causes severe masculinization of both morphology and various behaviors, including copulation and a wide array of social patterns (for review see McEwen, 1983). In rats testosterone functions as a prohormone; its aromatization product, estradiol, actively mediates the masculinization and defeminization of behavioral potentials during perinatal development; in primates the aromatization step is not required (for review see Ward and Ward, 1985).

In addition, variation exists in terms of the defeminization and masculinization. For example, in species like the rat, mouse, and guinea pig, masculinization and defeminization are events that are separable from each other in time and with respect to the hormones and neural substrates involved. In other species, such as the rhesus monkey and ferret, defeminization has not been demonstrated (Goy and McEwen, 1980). Another difference is the pathway and the receptor system involved in the process. If the process depends on aromatization and 5α-reduction of testosterone, as is the case in the rat (Fig. 1), androgen receptors (i.e., whose natural ligands are testosterone and DHT) are involved in masculinization, and estrogen receptors (estradiol) are effective in masculinizing and defeminizing the neural substrate. However, the strategy employed varies according to the species under study. In hamsters, estrogens are involved in masculinization (Paup et al., 1972), whereas in guinea pigs and rhesus monkeys, 5-α-DHT is the main effective masculinizing agent (Goldfoot and ver der Werff ten Bosch, 1975; Goy, 1978). Several investigators have suggested that the hormone responsible for organization of the neural substrate as male or female will also be the one responsible for the activation of male or female behavior during puberty and/or adulthood. For example, if aromatizable androgens masculinize, then they also activate masculine sexual behavior in adulthood. This appears to be the case with defeminization. In species like the rat, where defeminization does occur, it appears to involve the aromatization of testosterone to estradiol during the perinatal critical period, and it is estradiol that in adult females normally activates the feminine neural pathway (McEwen, 1981).

After the critical periods for masculinization and defeminization, ovarian secretions may have a developmental influence. Estrogen may have a positive role in organizing the neural substrate for ovulation, even though high amounts of estradiol defeminize and even though estradiol is thought to mediate the normal defeminizing effects of testosterone (see McEwen et al., 1982 for review).

Summary of the Endocrinological Determinants of Sex Differences

Sex steroids are determining factors in the process of sexual differentiation of the brain. Largely independent of the genetic sex, it is the presence or absence of androgens during perinatal development that will determine whether the fetus acquires male behavioral and certain neuroanatomical traits. Testosterone produced by the fetal testis will cross the blood–brain barrier and enter the brain cells. In the brain, there are regions or areas, such as the hypothalamus, preoptic area, and limbic brain that contain enzymes that convert testosterone to estradiol and 5α-DHT.

Androgen and estrogen receptors are present in the rat brain before birth. Androgen receptors start increasing about a week after birth, whereas estrogen receptors increase markedly in concentration following the day of birth. This increase appears to be an important element in the timing of defeminization in the rat. Progestin receptors are also present at birth and increase during the first 2 weeks of life. Their inducibility by estradiol is not an immediate feature of their initial appearance in neural tissue; rather estrogen inducibility emerges during the second postnatal week of life. Both males and females produce the same complement of enzymes and contain the same amounts of androgen and estrogen receptors. It is the occupancy of these receptors that differs. Since only the male secretes testosterone at the critical time in development, both androgen and estrogen receptors will show a higher level of occupancy in the rat fetal brain of the male rat. Thus the brains of normal genetic females will develop in a very different hormonal milieu from those of males (see Table 1).

Sexually Dimorphic Traits Used as Indexes for Demasculinization and Feminization

Because so many traits are affected by the process of sexual differentiation, and, in particular, by the presence or absence of testosterone during development, it is useful to briefly summarize what they are. Most pertain to rats and mice, because these species are most frequently used in experimental studies of sexual differentiation and studies of the effect of stress and various drugs during gestation. However, it must be pointed out that not all sexually dimorphic behaviors are subject to both organizational and activational effects.

Anatomical/Neuroanatomical and Neurochemical Traits

Body Weight   A slight sex difference in body weight is apparent at birth, males weighing more than females; this sex difference is maintained throughout life. Sex steroids contribute to this sex differences in body weight by affecting feeding behavior and body weight gain.

Anogenital Distance   The distance between the anus and the urethra is used as an index to separate males from females at birth, since this distance is about three times as large in newborn males as that in females. The greater distance observed in males is dependent on fetal androgen levels.

Brain Preoptic Area: Synaptic Organization and Volume   As early as 1973 a sex difference in the neuropil of the preoptic area (POA) was reported in rat, the number of nonamygdaloid synapses on dendritic spines in the POA being greater in females than in males (Raisman and Field, 1973). The authors suggest that this difference could be related to the ability of the females to maintain a cyclic pattern of gonadotrophin release and/or behavioral estrus, since the preoptic area is involved in the control of ovulation and mating behavior. Subsequently, a nucleus in the medial POA was found to be sexually dimorphic (Gorski et al., 1978); the nucleus was termed the sexually dimorphic nucleus of the preoptic area (SDN-POA). The volume of this nucleus is larger in male than in female rats (Gorski et al., 1978; 1980). The spatial location of this sexually dimorphic nucleus, within a region of the rat critical for neuroendocrine function, suggests that this nucleus may be related to reproductive function. The dimorphic development and differentiation of the male SDN-POA is dependent on local androgen metabolism in the perinatal rat brain (Döhler et al., 1982; Jacobson et al., 1981). Differences between the sexes become apparent at postnatal day 1.

Brain: Cerebral Laterality   At 6 days of age, male Long–Evans rats have thicker right cerebral cortices in several regions, whereas females do not show any asymmetry (Diamond et al., 1983). Males show a thicker right hippocampus (Diamond et al., 1982) and exhibit a significant right-greater-than-left pattern in several cortical areas that are not present in females. There is evidence that suggests that the normal development of cerebral laterality is dependent on circulating levels of testosterone in utero (Geschwind and Galaburda, 1987).

Brain: Corpus Callosum   A sex difference in the rat corpus callosum has been reported, males exhibiting a larger area than females (Berrebi et al., 1988). This sex difference appears to depend, in part, on neonatal testosterone (Fitch et al., 1990) and ovarian hormones (Fitch et al., 1991).

Brain: Arcuate Nucleus   Approximately twice as many spine synapses occur in the arcuate nucleus of the female rat than in that of the male rat. Neonatal castration of male rats that produced "feminized males" caused an increase in number of spine synapses to almost the same level as those of the female; conversely, a decrease in the incidence of spine synapses to the male level was found in the female rats treated with TP on postnatal day 5 (Matsumoto and Arai, 1980).

Brain: Ventromedial Nucleus   The ventromedial nucleus (VMN) of the hypothalamus is also sexually dimorphic; the male VMN has a larger volume (Matsumoto and Arai, 1983) and a greater number of shaft and spine synapses in the ventrolateral area than the VMN in females (Matsumoto and Arai, 1986). In terms of neuronal projections to the midbrain, electrophysiological studies have revealed three subsets of neurons in females and two subsets in males; this extra group of neurons in females consists of faster-conducting axons (Sakuma and Pfaff, 1981). These sex differences appear during perinatal development and can be abolished and/or reversed by neonatal steroid manipulations (Matsumoto and Arai, 1986).

A sex difference in dendritic spine density in VMN neurons is observed in castrated and in estrogen-treated rats, with females exhibiting higher dendritic spine density and a greater response to estrogen priming than that of males. These studies suggest that the sex difference observed in dendritic and soma spine density and in the response to estrogen treatment is owing to an organizational effect of sex steroids (Segarra and McEwen, 1991).

TABLE 2   Sex Differences in Brain Neurochemicals in the Rat[a]

Substance	Difference	Brain area	Reference
5-HIAA/5-HT	f > m	HYP, POA whole brain	Carlsson et al. (1985) Carlsson and Carlssn (1988)
5-HT	f > m	SDN-POA	Simerly et al. (1984) Simerly et al. (1985)
GABA	m > f	mPOA, VMN	Frankfurt et al. (1985)
Glutamate	f > m	mPOA	Frankfurt et al. (1985)
CCK	m > f	mPOA, VMN, LPA	Frankfurt et al. (1985)
PreproENK mRNA	f > m	VMN	Romano et al. (1990) Segarra et al. (1991b)

[a]HYP, hypothalamus; mPOA, medial preoptic area; SDN-POA, sexually dimorphic nucleus of the POA; VMN, ventromedial nuclei; LPA, lateral preoptic area; PreproENK mRNA, preproenkephalin mRNA; GABA, gamma-aminobutyric acid; CCK, cholecystokinin; 5-HIAA, 5-hydroxyindoleacetic acid; 5-HT, serotonin.

Brain: Neurochemical Sex Differences   See Table 2.

Anterior Pituitary Peptides and Hormones

The anterior pituitary of adult male rats contains a higher concentration of substance P (SP) (Yoshikawa and Hong, 1983) and met-enkephalin (Hong et al., 1982) immunoreactivity than that of females. In both studies, the sex-related difference is attributable largely to gonadal hormones; estrogen suppresses, whereas androgen increases the levels of these substances. However, unlike met-enkephalin, the anterior level of SP is not regulated by sex hormones in adulthood. Neonatal castration of males caused a decrease in SP in adult males, which was restored by neonatal testosterone replacement. Administration of testosterone propionate (TP) to females on postnatal days 2, 4, and 6 caused a significant elevation of the SP after maturation (Yoshikawa and Hong, 1983). Results of this study must be interpreted cautiously since the authors do not take into account the stage of the estrous cycle of the females.

Behavior

A description of those sexually dimorphic behaviors used most commonly to assess maleness and femaleness following prenatal manipulations with drugs, hormones, or prenatal stress is described in this section. The reader is referred to several excellent reviews by Beatty, 1979, and by Pfaff and Zigmond, 1971, for a more extensive coverage of this topic.

Sexual Behavior   The repertoire of copulatory behaviors in rats is sexually dimorphic.

Female Sexual Behavior   Female sexual behavior can be subdivided into three major constituents: attractivity, proceptivity, and receptivity (Beach, 1976). Attractivity alludes to those attributes that assist the male in identifying a sexually receptive female, such as odors or pheromones. Proceptivity as defined by Beach (1976) is "appetitive activities shown by females in response to stimuli received from males," (p. 114), which in rats includes ear wiggling and hop darting. Receptivity or lordosis behavior in rats is characterized by concave arching of the back and raising of the head and of rumps in response to mounting from a male. The number of times the female exhibits lordosis in response to being mounted by a male and the degree to which she arches her back serve as indexes of sexual receptivity (for review see Beach, 1976).

Male Sexual Behavior   In male rats precopulatory activities involve mainly sniffing and licking of the anogenital region of the female, and pursuit of the female. This is followed by a series of mounts and intromissions (mounts with penile insertion), culminating in ejaculation. The male then enters the postejaculatory interval, a period characterized by sexual inactivity and genital autogrooming, which lasts for 2 to 5 min. These repertoires or series of behaviors are reinitiated, and this pattern is continued until the male reaches sexual exhaustion. When analyzing male sexual behavior, the following are used as indexes of male copulatory behavior: the frequency of mounts and intromissions; and the latency between the first mount and intromission of a series and the ejaculatory and postejaculatory latency. It is recommended that the testing session end (1) after 15 min if the male fails to mount or intromit, (2) after 30 min if the male fails to ejaculate, or (3) after the male completes two ejaculatory series and initiates a third (for reviews see Sachs and Barfield, 1976; Sachs and Meisel, 1988).

Open-Field Behavior   An open-field testing arena consists of a floor surface of approximately 70 mm^2 divided into grids, surrounded by 20-cm walls, which is placed in a soundproof room with illumination provided by a 10-W bulb suspended above the center of the field. The number of boxes entered, the frequency of grooming and rearing, the number of fecal pellets and urinations bouts are recorded for the first 3 min and the subsequent 3 min. A sex difference in open-field behavior has been reported for rats 60 days or older (Beatty, 1979; Gray and Lalljee, 1974, Pfaff and Zigmond, 1971; Quadagno et al., 1972). Females tend to ambulate, groom, and rear more than males, whereas they urinate and defecate less than males.

Self-Grooming   Male and female rats groom approximately to the same extent, but a sex difference has been observed in the components of genital grooming; males groom their genitals significantly more than do females (Moore and Rogers, 1984). This sex difference appears to be related to differences in testosterone availability during the peripubertal period (Moore, 1986a).

Activity   A sex difference in running-wheel activity has been described in rats. This sex difference is dependent on sex steroid levels, since castration reduces activity in both sexes and abolishes the sex difference. The level of running in castrates is positively related to estrogen levels in both sexes (Roy and Wade, 1975).

Taste Preference   Female rats exhibit a higher preference for sweet nutritive (glucose) and nonnutritive (saccharin) solutions than do males, when offered a choice between these solutions and tap water (Valenstein et al., 1967). This sex difference in saccharin preference seems to be determined by a synergistic interaction between estrogen and progesterone (Wade and Zucker, 1969; Zucker, 1969). Ovarian hormones are important in the establishment of saccharin preference; once developed, the preference persists after ovariectomy. Female rats also exhibit a greater preference for salt solutions than do males (Krecek et al., 1972; Krecek, 1973). Several studies suggest that testicular androgens exert an organizational role in the establishment of the adult pattern of salt preference and that neonatal ovarian secretions may also play a role, since neonatal ovariectomy reduces salt preference (Krecek, 1973; 1976; Krecek et al., 1972).

Aggression In laboratory or domestic rats, a sex difference in aggressive behavior has been reported, with males usually fighting more than females (for review see Beatty, 1979). Gonadal hormones exert an organizational and activational effect on aggressive behavior, even though some measures of agonistic behavior, such as initiation of attacks, are more susceptible to hormonal control than others.

Play Sex differences in social play has been described in numerous species of animals including humans (for review, see Braggio et al., 1978). Prepubertal male rats spend about twice as much time as females engaged in rough-and-tumble play. Neonatal treatment with TP increased the amount of rough play in both sexes.

Maternal Behavior Maternal care patterns in rodents can be divided into pup-directed activities such as transport, retrieval, licking, and nursing of pups and non-pup-directed activities such as nest-building and maternal aggression (for review see Numan, 1988). Although maternal behavior is exhibited mainly by females with litters or by those who have recently given birth, pup-induced maternal behavior can be induced in both male and female virgin rats by repeated exposure to young pups.

Reactivity to Shock Female rats are more responsive to electric shock than are males, as measured by their lower threshold response (Beatty and Beatty, 1970; Pare, 1969) and shorter escape latencies (Beatty and Beatty, 1970; Davis et al., 1976). There is disagreement regarding the activational role of gonadal hormones in both sexes: both androgens and estrogens may play a role.

Feeding Behavior In many mammals, including humans, the male eats more than the female. Estrogen acts to control feeding and body weight by reducing intake after the body weight exceeds a certain "set point." In the cycling female rat, feeding and food-motivated behavior vary with the estrous cycle; eating and weight are lowest when estrogen levels are highest (Tartellin and Gorski, 1973). Androgens, on the other hand, promote feeding behavior and weight gain. A single injection of TP or estradiol benzoate (EB) shortly after birth increases body weight of females in an irreversible manner (Bell and Zucker, 1971).

Active Avoidance In these tests the animal is placed in a testing arena, and must learn to "avoid" an electric shock. The testing arena can be of three types:

1. One-way—The testing box consists of two compartments, one that is always safe, and the other that is potentially dangerous. In this test and in the shuttle box, a cue alerts the animal that shock is impending.
2. Two-way ("shuttle box")—Both sides of the compartment are potentially dangerous. The animal must run to the opposite side to avoid the shock.
3. Free operant (Sidman avoidance)—There is no cue to signal the impending shock; instead, it is programmed to occur briefly every few seconds. Usually a lever press postpones the next shock by an interval longer than the intershock interval.

In the rat, females generally perform better than males during acquisition of each of these tasks (Phillips and Deol, 1977; Barrett and Ray, 1970; Levine and Broadhurst, 1963). Female performance varies with the stage of the estrous cycle: the best performance is exhibited during proestrus, and the worst, during diestrus. The sex difference observed in avoidance acquisition seems to be determined by androgen levels during the prenatal period (Beatty and Beatty, 1970).

Passive Avoidance   Female rats exhibit inferior performance in tests of passive avoidance compared to male rats (Beatty et al., 1973; Denti and Epstein, 1972), but the role that sex steroids play in the establishment of this sex difference is unclear.

Maze Learning   Most studies in maze learning in rats have found that males outperform females. The tasks that are most sensitive to sex differences are complex mazes such as the Lashley III maze and Hebb–Williams maze. However, as Beatty (1979) points out, females may make more "errors" than males because their greater level of activity and exploratory behavior translate into errors, rather than because they are less able to process spatial information. Testosterone levels during early postnatal development are important in establishing these sex differences since neonatal TP treatment thoroughly masculinizes female performance, whereas neonatal castration feminizes male performance (Dawson et al., 1975; Joseph et al., 1978; Stewart et al., 1975).

Depression   Duration of immobility after the forced swimming test (Hilakivi and Hilakivi, 1987) and the Porsolt test can be used as indexes for an animal model of depression. Results obtained in such studies showed that the duration of immobility is higher in males than in females, suggesting a sex difference in this animal model of depression (Hilakivi and Hilakivi, 1987).

GENERAL CONSIDERATIONS IN THE ACTIONS OF DRUGS ON THE DEVELOPING BRAIN

Why Do Drugs of Abuse Have Such Detrimental Effects When Administered during Gestation?

Prenatal exposure to drugs is devastating to the fetus because its tissues and organs, including its nervous system, are still forming and are particularly susceptible to environmental influences. Since brain cells are born and differentiated during fetal and neonatal life, drugs administered throughout the developmental period may alter cell number, the formation of neural circuits, as well as affect the processes of differentiation and expression of developing transmitter systems. In addition to having direct effects on nerve cells, drugs may also stimulate or inhibit the production of hormones that have important developmental effects, including sex hormones, thyroid hormones, and the stress-related hormones. The only way in which such effects of drugs and hormones can be identified and studied mechanistically is through the use of research animals such as rats and mice. In addition to all of the problems in studying the human fetus, it is well recognized that assessment of prenatal drug transfer in the human organism is problematic. First, the degree and pattern of maternal addiction can lead to fluctuating drug levels within and between mother and fetus. Second, many addicted women who agree to participate in prenatal drug studies and/or rehabilitation programs frequently use more than one drug at the same time. They are also malnourished and may not know or truthfully indicate the actual dose taken. These factors complicate even further an accurate assessment of the effects of the drug.

Animal Models in Studies of Prenatal Stress and Drugs

When using animals as models to study the effect of prenatal stress and drugs on the process of sexual differentiation of the brain, certain criteria must be established. First of all, the studies must be consistent in terms of dosages, methods of administration, and the treatment should be administered during the same period in gestation. Initial studies in this area encompassed GD (gestation day) 3–21, whereas subsequent studies limited treatment to GD 14–21, the period characterized by masculinization of the brain. This is the time frame currently used by most investigators in this field. Differences in mode of administration of the drug will also affect the results obtained. When the treatment involves injections of the dam, the investigator must take into consideration the effect the stress of injections might have on the dam and consequently on its progeny. In many cases the use of an osmotic minipump (Alzet, Palo Alto, California) can overcome this problem. Although subcutaneous insertion of the pump may be stressful, the stress is limited to one session. However, many peptides, such as ACTH, cannot be administered by pump since they are rapidly degraded. Thus many peptides must be injected repeatedly, and in many cases the dam must be injected twice a day to increase the time of exposure to the peptide. As a result, the stress of administration of peptides and other drugs to animals raises questions about the role of stress in the effects obtained.

Litter and maternal effects may also affect the validity of the results obtained. These effects can be reduced by increasing the number of litters and by cross-fostering. Once the pups are born, the animals must be assessed at consistent developmental stages for possible prenatal effects. The investigator must be aware that the age of puberty varies with the strain of rats (for review see Ojeda and Urbanski, 1988). Vaginal opening in female rats is an indicator of the advent of puberty, since evidence demonstrates that it coincides with the day of the first ovulation. The age of vaginal opening may vary from day 31 to day 40, according to the strain studied and to individual variations. However, as there is no external morphological manifestation that can be taken as an indicator of puberty in males, determination of their age of puberty acquisition is more difficult. Testis descent occurs at approximately 3–4 weeks of age, when rats are in the juvenile stage, and this cannot be used as an indication of puberty. To determine puberty acquisition in male rats, one must rely on plasma testosterone levels or on the ability of a male rat to display the complete repertoire of male copulatory behaviors. Peak serum testosterone levels are attained between 50 and 60 days of age, which coincides with the time that males exhibit intromissions and ejaculations (for review, see Sachs and Meisel, 1988). Male sexual behavior should be studied after the male reaches puberty, not before, as is the case in several studies. Males, like females, also exhibit considerable variation in the timing of puberty acquisition among strains and among individuals.

Transplacental Transfer of Drugs

Transfer of drugs from the mother to the fetus is accomplished via the placenta, where maternal and fetal blood systems meet. The placenta provides the embryo with nourishment and respiratory gases, eliminates its wastes, and is active in the secretion of many steroid and peptide hormones. The placenta acts as a barrier, similar to the blood–brain barrier. Its permeability is determined by the number of layers of tissue interposed between fetal and maternal blood and by the relative relationship of fetal and maternal blood flow (for review see Petraglia et al., 1990). Transfer of materials through the placenta is achieved by several mechanisms, depending on the nature of the substance being transported. The placenta may act as a passive filter where substances such as gases diffuse from the uterine arterial blood to the chorioallantoic membrane. Nutrients such as glucose and lactic acid traverse the placenta via facilitated diffusion, whereas passage of amino acids, antibodies, many vitamins, and ions is carrier mediated.

Placental transfer of the vast majority of drugs is accomplished by passive diffusion. As extraneous compounds to the organism, drugs seldomly are subjected to specialized transport processes; however, antimetabolite drugs may be actively transported. Water-soluble drugs may cross the lipid membranes by diffusion if they are nonionized and if their molecular weight is less than 100. On the other hand, lipid-soluble drugs up to a molecular weight of 600 to 1000 readily diffuse. Fully ionized drugs such as quaternary ammonium compounds certainly appear to diffuse, albeit slowly, across the placenta, although they cannot cross the blood–brain barrier.

Additionally, drugs can induce changes in the placenta, and the placenta can alter the drugs. As this is a metabolically active organ, xenobiotic metabolism may be induced by drugs such as those present in cigarette smoke. This implies that many drugs and other foreign substances may be metabolically altered during transplacental transfer. On the other hand, some foreign substances, such as ethanol, may impair placental transfer. For example, a decrease in fetal leucine uptake, protein synthesis, and glucose transfer has been observed in rats treated prenatally with ethanol.

EFFECTS OF STRESS AND DRUGS
ON BRAIN SEXUAL DIFFERENTIATION

Assessment of prenatal drug abuse on sexual differentiation of the brain is a complicated matter as drugs of abuse may act as stressors, and stress alters the process of sexual differentiation of the brain. Additionally, drugs can act directly on the brain substrate and neuroendocrine axis, altering the neuronal circuitry and specific neurochemicals.

Many studies of prenatal drug abuse analyze the short-term effects of drug intake, measuring neonatal parameters and looking for congenital defects. However, it is of utmost importance to extend studies into adulthood. In many studies the offspring do not exhibit any morphological irregularities. This is the case for many treatments that alter sexual differentiation of the brain: i.e., both at birth and during early development, males treated with many of the drugs of abuse to be discussed are usually indistinguishable from control males; yet, after they reach puberty, although they are morphologically similar to nontreated males, they present behavioral anomalies. In some cases, neuroendocrinological anomalies are also evident.

A common denominator of all these prenatal drug and stress treatments is that fetal testosterone levels are altered. In rats, a peak in testosterone levels at GD18 is necessary for a male rat to exhibit characteristic behavioral patterns that distinguish it from female rats. Prenatal treatment with ethanol, cannabinoid, nicotine, opioids, and stress, which demasculinize the male progeny, also reduce this fetal testosterone peak.

In this section the effect of all these agents and how they interact with each other during the process of sexual differentiation of the brain will be considered. The focus will be mainly on males, since they are more susceptible to the effects of prenatal drug treatments. Drug effects on sexual differentiation of females will be discussed as well, in those cases in which sufficient data have been gathered.

Stress

Stress refers to a variety of environmental challenges that threaten homeostasis and the health of an organism. Common stressors include physical trauma, heat, cold, exercise, pathogens, and a variety of psychological states, including fear and anxiety. Stressors triggers a stress response from the organism, which involves output of epinephrine and secretion of various pituitary hormones, including activation of the hypothalamic–pituitary–adrenal (HPA) axis. The catecholamines, opiate neuropeptides, and glucocorticoids secreted by the adrenal gland enable the organism to deal with the stressors by mobilizing energy depots, increasing cardiovascular tone, and decreasing sensitivity to pain.

Prenatal Stress

The pioneering work of Ward in the early 1970s (Ward, 1972, 1977) indicated that prenatal stress has a detrimental effect on sexual differentiation of the rat brain.

This study opened a whole field of research, which has provided valuable insights into brain plasticity. It is also testimony to the sensitivity and fragility of the brain during early development.

Defining Stressors and Their Effects on Behavior　The most common forms used in prenatal stress studies are immobilization or restraint stress, extreme temperature, social crowding, light, isolation, malnutrition, audition, and conditioned emotional responses. The stress regimen used by Ward (1972, 1977) and Ward and Ward (1985) consisted of placing the pregnant rats in plexiglass restrainers for 45 min under a floodlight that delivers 200 foot-candles of light. This stress regimen was administered three times daily from GD14 to 21.

Ward (1972) found a decrease in the masculine sexual behavior of adult male rats that had been stressed prenatally. When tested with vigorous, sexually active males, these males also exhibited an increase in female sexual behavior. Other investigators using stressors such as social crowding (Dahlöf et al. 1977), malnutrition (Rhees and Fleming, 1981), and conditioned emotional response (Masterpasqua et al. 1976) have obtained similar results.

The demasculinization and inhibition of defeminization observed in male rats stressed in utero is known as the *prenatal stress syndrome*. This syndrome is characterized by a reduction in male sexual behavior. The decrease in sexual behavior is manifest as a reduction in the number of males that mount and/or intromit when tested for copulatory behavior. Few prenatally stressed males exhibit the complete repertoire of male sexual behavior; but those that do, perform as well as control males.

With "therapy," some of these males exhibit unconventional copulatory behavior. Therapy consists of housing males with females (Dunlap et al., 1978) or of injections of TP over a prolonged period during adulthood (Ward, 1977). After therapy, some males continued to show no copulatory behavior: others exhibited only female copulatory behavior. Another group exhibited male sexual behavior; about 40% were bisexual and exhibited male or female sexual behavior depending on the sex of the stimulus animal (Ward, 1977; Ward and Ward, 1985).

Prenatal stress affects other sexually dimorphic behaviors such as sex differences in behavioral lateralization. The number of males that exhibited behavioral lateralization was decreased by prenatal stress, whereas the number of females exhibiting behavioral lateralization was increased. This has been interpreted as demasculinization of male and defeminization of female offspring (Alonso et al., 1991). However, not all sexually dimorphic behaviors were modified. Prenatally stressed adult male and female rats were tested for active avoidance. As pointed out previously, females normally perform better than males. Both control and prenatally stressed female groups outperformed the males, and there was no significant difference in the performance of prenatally stressed males versus that of control males (Meisel et al., 1979).

Anatomical and Neuroanatomical Consequences of Prenatal Stress Syndrome　Sexually dimorphic neuroanatomical traits are also affected by in utero stress. For example, in prenatally stressed males, the volume of the sexually dimorphic nucleus of the SDN-POA was reduced, indicating that there had been blockade of masculinization or of defeminization of the SDN-POA (Anderson et al., 1986).

At birth, body weight, anogenital distance, and testis weight of prenatally stressed males are typically reduced (Dahlöf et al., 1978; Ward and Ward, 1985), but this effect is transient. Once prenatally stressed males reach adulthood, they are indistinguishable from control males in plasma androgen levels and size of androgen-dependent structures such as the penis, testis, or epididymis (for review see Ward and Ward, 1985). These males are also spermatogenically active, as revealed by histological examination of the testis (Ward, 1984).

Mechanisms Involved in Effects of Prenatal Stress   During fetal development, a peak in plasma testosterone is observed at GD18 and 19 in male rats but is absent in females. Prenatally stressed males exhibit a shift in this testosterone peak and in the peak of 3α-hydroxysteroid dehydrogenase (3α-OHSD) activity, to GD16 and 17 (Ward and Weisz, 1980; Weisz et al., 1982; Orth et al., 1982). These experiments suggest that testosterone levels must peak at GD18 in order to effectively masculinize the brain. If testosterone peaks earlier (GD17), the brain substrate is embryologically less developed and is less capable of responding to the inductive "cue" of testosterone.

There is evidence that suggests that in rats, the effect of prenatal stress on fetal testosterone levels and, ultimately, on sexual differentiation of the brain, may be partially mediated by the opiate system. First, similar to prenatal stress, prenatal exposure to opiates was found to decrease male sexual behavior (Ward et al., 1983) and fetal testosterone levels (Singh et al., 1980). Second, stress precipitates the release of beta-endorphin (Guillemin et al., 1977). Third, if pregnant rats are injected with the opiate receptor blocker naltrexone before each stress session, there is no shift in 3α-OHSD activity that occurs in prenatally stressed males (Ward et al., 1983).

Some Prenatal Stress Effects on Sexual Development Are Found in Females   In prenatally stressed females, no effect on female sexual behavior has been reported (Ward and Ward, 1985). However, the reproductive system of these rats is physiologically impaired (Herrenkohl and Whitney, 1976; Herrenkohl, 1979). Females exposed to stress in utero experience fewer conceptions, more spontaneous abortions, and have fewer viable young than nonstressed females (Herrenkohl, 1979).

In prenatally stressed females, where no effect on sexual behavior was found, there is an increase in the steady-state concentration of dopamine in the arcuate nucleus (Moyer et al., 1977, 1978). Since the arcuate nucleus in involved in the regulation of gonadotropin secretion, these alterations may be partially responsible for the decreased numbers of progeny.

Postnatal Stress

Some investigators argue that the effects of prenatal stress on sexual differentiation of the brain may be owing to maternal factors operating neonatally. However, experiments in which prenatally stressed males are cross-fostered to normal mothers at birth demonstrated that the syndrome still persists (Herrenkohl and Whitney, 1976).

Although the process of sexual differentiation is known to continue throughout the first 2 weeks of life in the rat, neonatal stress does not have any major effect on this process. Rats exposed to stressful handling procedures for the first 10 days after birth showed normal adult sexual patterns (for review see Ward and Ward, 1985). That neonatal stress has no effect on sexual differentiation of the brain is further supported by the fact that combining prenatal and postnatal stress produces no effect different from that of the prenatal treatment alone (Ward and Ward, 1985).

It is possible that neonatal stress has no effect on sexual differentiation of the male rat brain because of the hyporesponsiveness of the HPA axis during the first 14 days after birth (Sapolsky and Meaney, 1986). The first 2 weeks of life in rats are characterized by a lack of or diminished response to stressors. This has been termed the *stress nonresponsive period* or *stress hyporesponsive period*. During this stage, the stimulus necessary to induce an adrenocortical response to stress is unusually high. It is possible that the development of adult concentrations of glucocorticoid receptors in the hippocampus results in the adult sensitivity of the HPA axis to negative-feedback inhibition, and therefore in the adult capacity to promptly end adrenal secretion at the end of stress. Thus, low corticotropin-releasing hormone (CRH) and ACTH levels and the pituitary's reduced sensitivity to CRF are all likely to contribute to the attenuated ACTH secretion during the stress hyporesponsive period (Sapolsky and Meaney, 1986).

Summary

The eventual sexual behavior a rat displays as an adult appears to be crucially dependent on the time in ontogeny when developing neural pathways are exposed to changing hormonal levels, drugs, or environmental influences such as stress. Prenatal stress will differentially affect the reproductive behavior and physiology depending on the sex of the animal. Males will become less "malelike" and more "femalelike" in terms of behaviors and sexually dimorphic traits. Although there is evidence that the brain is a target of prenatal stress, it is interesting to note that although many prenatally stressed males do not perform sexually, their reproductive system presents no morphological or physiological anomalies. Prenatally stressed females, on the other hand, exhibit reduced fertility and fecundity, but no effect has been observed in their copulatory behavior.

Adrenocorticotropic Hormone

Adrenocorticotropic hormone (ACTH) is a 39-amino acid peptide derived from the prohormone proopiomelanocortin (POMC) secreted by the adenohypophysis. Activation of the HPA axis during the stress response involves secretion of ACTH by the anterior pituitary gland. Since prenatal stress decreases male sexual behavior, and ACTH is one of the main hormones involved in the stress response, then prenatal ACTH administration would be expected to decrease male sexual behavior. All studies in this area, with the exception of one (Chapman and Stern, 1977), report a decrease in male sexual behavior following prenatal ACTH administration in rats (Rhees and Fleming, 1981; Stylianopolou, 1983; Segarra et al., 1991b) and mice (Harvey and Chevins, 1984).

Prenatal Adenocorticotropic Hormone Administration

Effects on Sexual Behavior Perinatal administration of $ACTH_{1-24}$ alters the sexual behavior of rodents (Segarra et al., 1991b; Rhees and Fleming, 1981; Stylianopolou, 1983). The effect varies with the dosage administered (see Table 3). At a low dosage alterations in mount and intromission latencies and in mount frequency were observed (Segarra et al., 1991b). At a higher dosage, prenatal ACTH treatment decreases the number of males that ejaculate (see Table 3). These results indicate that prenatal ACTH treatment demasculinizes the male progeny.

Prenatal treatment with ACTH also feminizes the male progeny in other ways. When rats are offered a choice between a saline solution or tap water, females normally exhibit a greater preference for the saline solution than do males, as noted above. Prenatal ACTH administration increased the amount of saline solution intake in male rats to levels comparable to those of females, abolishing the sex difference (Segarra, 1988). In addition, males prenatally treated with ACTH display a higher lordosis quotient and lordosis ratio when tested for female sexual behavior (Rhees and Fleming, 1981).

TABLE 3 Effects of ACTH Administration on Sexual Differentiation

Species	ACTH dose[a]	Effect[a]	Reference
Prenatal			
Rat	8 IU (s.c.)	None	Chapman and Stern (1977)
Mice	1 IU (s.c.)	None	Harvey and Chevins (1984)
	8 IU (s.c.)	↑ ML, ↑ IL	Harvey and Chevins (1984)
Rat	20 IU (i.m.)	↑ MTS, ↑ INT, ↓ males EJAC	Rhees and Fleming (1981)
Rat	20 μg/kg/day (i.p.)	↑ IL	Segarra et al. (1991b)
	500 μg/kg/day (i.p.)	↓ Males exhibiting sex behavior	Segarra et al. (1991b)
Rat	8 IU (s.c.)	↓ Males exhibiting sex behavior	Stylianopoulou (1983)
Prenatal and postnatal			
Rat	20 μg/kg/day prenatal (i.p.) 10 μg/kg/day postnatal (s.c.)	None	Segarra et al. (1991b)
Postnatal			
Rat	20 μg/kg/day (s.c.)	None	Segarra et al. (1991b)
	250 μg/kg/day (s.c.)	↓ ML	

[a]INT, number of intromissions before ejaculation; MTS, number of mounts before ejaculation; EJAC, ejaculations; ML, mount latency; IL, intromission latency; i.p., intraperitoneal; s.c., subcutaneous; i.m., intramuscular; ↑, increase; and ↓, decrease.

Effects on Sex Ratio and Anatomical Traits   Sex ratio and anogenital distance in male rats is unaffected by prenatal ACTH treatment. As far as birth weight is concerned, there are investigators who report a decrease (Rhees and Fleming, 1981) whereas others find no effect (Stylianopolou, 1983; Segarra et al., 1991b).

The few studies that have looked at the effect of prenatal ACTH treatment on sexual differentiation of the brain have focused mainly on the behavior of the progeny. At present, there is no information as to the effect of ACTH on sexually dimorphic neuroanatomical structures such as the SDN. Data on fetal neurotransmitter and hormonal levels of in utero ACTH-treated rats are also lacking. There are, however, some data on adult males treated prenatally with ACTH: testosterone, prolactin, and ACTH plasma levels are not affected; however, a trend toward decreased testis weight and lower spermatogenic activity is observed in these males prenatally-treated with ACTH (Segarra et al., 1991b).

Possible Mechanisms Involved in Effects of Prenatal ACTH Administration   Adult plasma testosterone levels of males prenatally treated with ACTH are normal, indicating that the decrease in sexual behavior is not the result of low androgen levels during adulthood. However, it is possible that fetal testosterone levels are transiently altered by prenatal ACTH administration. There are several interactive neuronal systems involved in the regulation of testosterone secretion at the level of the POA during ontogeny. One of these is the serotonergic (5-HT) transmitter system. Synaptic contacts between serotonergic terminals and immunoreactive luteinizing hormone-releasing hormone (LHRH) elements have been described in the medial preoptic area (Kiss and Halasz, 1985). Although ACTH is known to decrease gonadotropin secretion in vivo, studies in vitro indicate that it is unable to directly modulate gonadotropin secretion at either the brain or pituitary level (Mann et al. 1985). However, ACTH and several of its fragments stimulate the neuronal maturation of serotonergic neurons in vitro (Azmitia et al., 1987). Thus, it is possible that ACTH may indirectly alter fetal testosterone levels during the critical period for sexual differentiation of the brain via the serotonergic system. Evidence for this stems from the increase observed in 5-HT levels (Segarra et al. 1991b) and in 5-HT1B receptors (Segarra et al., 1990) in the POA of males prenatally treated with ACTH. Further experiments, measuring effects of prenatal ACTH treatment on circulating androgen levels and on 5-HT levels in the POA of fetuses, could resolve these questions.

Effects in Females   Prenatal ACTH treatment alters the process of sexual differentiation in females also. Females prenatally treated with ACTH exhibit a delay in vaginal opening indicating a delay in puberty acquisition (Segarra, 1988). Prenatal ACTH treatment, in contrast to prenatal stress, appears to affect sexual behavior in females. Preliminary studies seem to indicate a decline in the copulatory behavior of females prenatally treated with ACTH (Alves and Strand, 1990). Results on the effect of prenatal ACTH on female anogenital distance at birth are controversial. Stylianopolou (1983) reports an increase in anogenital distance in females prenatally treated with ACTH, whereas Segarra et al. (1991b) find a decrease. This discrepancy of results could be attributed to differences in the rat strain and/or dosage used.

Postnatal ACTH Administration

Postnatal administration of ACTH does not alter the number of males that exhibit sexual behavior; all males treated postnatally with ACTH completed two ejaculatory series and initiated a third (see Table 2). In fact, these males showed a decrease in mount latency (Segarra et al., 1991b). Prenatal and postnatal ACTH administration at a low dosage had no effect on the sexual behavior of male rats (see Table 2).

Interestingly, a decrease in the number of males that ejaculated was observed in males prenatally treated with saline when compared to those postnatally treated with saline or ACTH, suggesting that the stress of injecting the dam twice a day is sufficient to induce alterations in the copulatory behavior of male rats (Segarra et al., 1991b).

Summary

Perinatal ACTH administration induces permanent changes in the hypothalamic–pituitary–gonadal axis. The prenatal response to ACTH treatment is demasculinization of the male progeny, as evidenced by a decrease in male copulatory activity. Prenatal ACTH treatment decreases the number of males that exhibit sexual behavior. This effect appears to be mediated at least in part by the serotonergic system, in which an increase in 5-HT levels in the medial POA of adult males was observed. Receptor density in the VMN and POA are currently being assessed, and preliminary data seem to indicate that 5-HT_{1B} receptors in the POA might be involved in mediating the effects of ACTH on sexual differentiation of the brain (Segarra et al., 1990).

In contrast, postnatal treatment with ACTH did not affect the number of males that completed two ejaculatory series and initiated a third (Segarra et al., 1991b). These results are similar to Ward's findings (1972), in which decreased sexual behavior was observed in prenatally stressed males. In fact, in the Segarra et al. (1991b) study, postnatal ACTH treatment decreased mount and intromission latencies, implying enhanced sexual performance. These data provide further evidence that the window for masculinization is restricted to the late prenatal period.

It must be emphasized that the testes of all of these ACTH-treated animals were active in spermatogenesis and steroidogenesis, suggesting the neural substrate as the target for prenatal ACTH administration rather than the gonads themselves. Whether the effects observed are the result of a direct effect of ACTH or mediated by glucocorticoids remains to be elucidated. Increases in fetal ACTH and glucocorticoids have been reported after direct (Nagel, 1969) or indirect (maternal) stress (Cohen et al., 1983). At present, there is evidence to indicate that glucocorticoids can cross the placental barrier (Zarrow et al., 1970), and one report that suggests that ACTH cannot (Dupouy et al., 1980), but this last alternative cannot be totally excluded.

Alcohol

Alcohol is a potent teratogenic substance, and as such it is the primary cause of mental retardation in the western world. It is estimated that as many as 2% of all babies born alive are afflicted with the fetal alcohol syndrome (FAS) in some degree (Schenker et al., 1990). The FAS is characterized by growth retardation, hormonal imbalances, cranial/facial abnormalities, and other congenital anomalies of the heart, kidney, and musculoskeletal system (for review see Schenker et al., 1990; Miller, 1981, 1985; Kakihana et al., 1986; Mendelson, 1986; Claren and Smith, 1978). Neurobehavioral anomalies such as hyperactivity, attentional deficits, language and coordination difficulties, and deficiencies in cognitive and fine motor skills may also be present (Jones and Smith, 1975). In addition, prenatal alcohol increases fetal resorption (Kronoick, 1976; Chernoff, 1977; Randall et al., 1977, 1990), reduces total DNA content of fetal and neonatal brain (for review see Kakihana et al. 1980) and lowers catecholamine content and dopamine–hydroxylase activity in the adrenals of the offspring (Lau, 1976). Moreover, anatomical studies of humans and rodents prenatally exposed to ethanol show that their brains are smaller, contain ectopic cell clusters, and display irregularities in brain cortical development (Miller, 1981, 1985, 1988). For a more extensive coverage of this subject, see reviews by Conry (1990), Riley (1990), and Mendelson (1986).

Alcohol consumption has a detrimental effect on the reproductive axis of a wide variety of species, including humans. Female primates that consume high doses of alcohol exhibit amenorrhea, atrophy of the uterus, decreased ovarian mass, and a significant depression of luteinizing hormone (LH) levels (Mello et al. 1983). Likewise, alcoholic men are frequently impotent, have overt testicular atrophy, and are often feminized (Van Thiel, 1984). The hypogonadism observed in alcoholic men is a direct consequence of alcohol consumption, whereas liver disease acting concurrently with alcohol and/or acetaldehyde is responsible for the feminization (for review see Van Thiel, 1984).

Prenatal Ethanol Administration

Methodology   Ethanol is administered mainly as an ethanol liquid diet. In this form it provides approximately 35% of the total calories. This mode of administration avoids the stress of injections and resembles the pattern of alcohol consumption in humans. However, it increases dosage variability among animals, since the actual dosage depends on the amount ingested.

Because of the high energy value of ethanol, it displaces other food in the diet and decreases nutrient intake, resulting in undernutrition. Many investigators have argued that some of the deleterious effects of ethanol, particularly on growth and development, may be a result of inadequate nutrition. However, a study by Weinberg (1985) indicates that increasing the dietary protein levels does not attenuate the major adverse effects of alcohol on fetal development. Nevertheless, to ensure scientific accuracy, the majority of the studies employ two controls. The first control consists of a group fed an isocaloric diet. This control enables the investigator to assess whether the effects may be attributed to the restricted caloric intake due to the high caloric value of ethanol or to a direct effect of ethanol. The second control is fed normal rat chow to assess the effects, if any, of an isocaloric diet.

A problem encountered when reviewing the literature on prenatal alcohol effects is the inconsistency in results. In many instances this can be attributed to the variability in dosages and/or differences in the developmental period of exposure to the drug.

Effects on Male Sexual Behavior Result of studies on the effect of prenatal ethanol treatment on male sexual behavior are inconsistent. Hard et al. (1984) reported no difference in male sexual behavior of males prenatally treated with ethanol, but indicated that when these same males were tested for feminine sexual behavior, there was an increase in the number of males that exhibited lordosis. These results would seem to indicate that in males, defeminization is inhibited, but that masculinization proceeds normally.

On the other hand, Udani et al. (1985) and Parker et al. (1984) state that prenatal and postnatal ethanol treatment decreased copulatory activity in males treated from GD12 to PN10, indicating demasculinization of the male progeny. A careful examination of these last two studies indicates that they contain the same data and are by the same authors, so they will be considered as one study and will be referred to as the Parker et al. (1984), naming the first one published. Differences in methodology might contribute to the apparent contradiction of these results with those of other investigators. The dosage used in the Hard et al. (1984) study was half the dosage used by Parker et al. (1984); and treatment was confined to the prenatal period, whereas treatment in the Parker et al. (1984) study extended into the first 2 weeks of postnatal life. Additionally, in the Parker et al. (1984) paper, a decrease in the intromission frequency was the only sexual behavior measurement altered by prenatal ethanol treatment. However, a decrease in intromission frequency is usually taken as an indication of increased penile sensitivity and not of decreased sexual behavior, as suggested by the authors. Thus further studies must be conducted in this area before any valid conclusions can be made.

Other Behaviors Prenatal alcohol treatment abolishes many sex differences in behaviors, as previously noted in the section on "Basic Features of Sexual Differentiation." Males treated prenatally with alcohol showed an enhanced preference for saccharin solutions and needed more trials to learn a Lashley III maze when compared to control males. Additionally, females treated prenatally with ethanol exhibited a decrease in saccharin preference when compared to control females and a decrease in the number of trials to complete a Lashley III maze (McGivern et al., 1984). These results indicate that alcohol abolishes sex differences in saccharin preference and maze learning by blocking masculinization and/or defeminization of the male progeny and by enhancing masculinization of the female progeny.

Social play or play fighting among juvenile rats is also sexually dimorphic, as noted above, males showing higher levels of play than females (Meaney and Stewart, 1981). Males prenatally treated with alcohol exhibited a decrease in the number of "pins" as compared to control males, whereas females displayed an increase in the number of pins as compared to control females (Meyer and Riley, 1986). Thus, in social play, as in saccharin preference and maze learning, alcohol exposure during development reduces sex differences by causing males to behave in a more femalelike manner and females to behave in a more malelike way. Thus alcohol treatment completely reversed the sex difference normally observed in saccharin-preference taste tests, in Lashley III maze learning, and in play fighting among juveniles rats.

Effects on Neuroanatomy Demasculinization of the male progeny by prenatal ethanol treatment is also manifest in sexually dimorphic neuroanatomical traits. Prenatal ethanol treatment decreased total callosal area in male rats and enhanced the normal right-greater-than-left asymmetry in the anterior neocortical volume of 3-day-old rats (Zimmerberg and Reuter, 1990). In addition, the size of the sexually dimorphic nucleus of the POA, which is reduced in prenatally stressed males, is also smaller in males exposed to ethanol in utero and during the first 2 weeks of life (Rudeen, 1986).

Effects on Testosterone Levels Prenatal alcohol exposure, as prenatal stress, abolished both the prenatal (GD18) and postnatal (day 1) testosterone surges, which are required for the brain to differentiate in the male direction (McGivern et al., 1988). This absence of a testosterone peak appears to be related to a marked insensitivity of the rat testis to LH, as suggested by in vitro studies (McGivern et al., 1988). In addition, adult males treated prenatally with ethanol had lower plasma testosterone levels and showed a decrease in the weight of androgen-sensitive structures such as testis and prostate (Parker et al., 1984). The plasma LH titers were also reduced; however, they exhibited a normal LH response to LHRH infusion, indicating that it is not the sensitivity to LHRH that has been lost (Handa et al., 1985). These results are very different from those obtained with prenatal stress, in which no effect on adult levels of testosterone or on androgen-dependent structures at maturity was observed. In addition, ethanol treatment of male rats decreased the number of androgen receptors in the anterior pituitary, hypothalamus, and cortex of adult rats (Chung, 1989).

A large number of studies indicate that alcohol causes alterations in sex steroid metabolism in the liver (Gordon et al., 1976; Southren and Gordon, 1976). One possibility is that ethanol, by altering nicotinamide adenine dinucleotide:reduced nicotinamide adenine dinucleotide (NAD:NADH) ratios in the liver, may increase plasma estradiol levels. Estradiol is a potent teratogenic agent. Numerous studies have indicated that administration of exogenous estrogens during the first trimester of gestation increases the risk for morphologic abnormalities, development of neoplastic disease, impaired cognitive function, and development of affective disorders.

Anogenital Distance One study reports a decrease in birth weight and in anogenital distance of male rats treated prenatally with ethanol (Parker et al., 1984; Udani et al. 1985), whereas others found no effect (Hard et al., 1984; McGivern et al., 1984, 1988). In some cases this discrepancy can be attributed to difference in dosages, as is the case of the Hard et al. (1984) study, in which a 16% ethanol solution was the dose administered, less than half the dose regularly used in other prenatal ethanol studies (i.e., a 35% solution).

Effects on Females Prenatal alcohol administration delayed vaginal opening; increased plasma prolactin levels (which persisted into adulthood); and decreased LH levels the day of vaginal opening in female rats (Esquifino, 1986). In addition, exposure of rat pups to high cyclic blood alcohol concentrations during PD 4–10 decreased spatial navigation abilities in females when tested in the Morris water maze, but had no effect on male performance (Kelly et al., 1988).

Mechanisms of Prenatal Ethanol Action   There are various mechanisms by which ethanol may produce the effects observed in FAS. First, several studies have indicated that ethanol and acetaldehyde (a metabolic product of ethanol) can directly affect fetal growth and development (Brown et al., 1979; O'Shea and Kaufman, 1979). Ethanol is known to affect transplacental transfer of glucose and amino acids (Snyder et al., 1976; Jones et al., 1981; Gordon et al., 1985) and to cause hormonal imbalances early in pregnancy that greatly enhance risk for fetal abnormalities (for review see Randall et al., 1990; Cicero, 1981).

Another action of alcohol is on the development of neurotransmitter systems. Prenatal exposure to ethanol alters the development of several systems, such as sero-tonin, dopamine, and noradrenaline (for review see Druse, 1986). Various studies suggest that ethanol plays a modulatory role in the gamma-aminobutyric acid (GABA)/benzodiazepine receptor complex (Kuriyama et al., 1987). Ethanol, like benzodiazepines, reduces tension, anxiety, and irritability in humans (Cameron, 1975; Williams, 1966). A low dose of ethanol behaves like diazepam in preventing stress-induced changes in dopamine metabolism, which is reversed by the application of a partial inverse benzodiazepine agonist (RO 15-4513) (Fadda et al., 1987).

Alcohol exposure during early development also influences the secretion of hor-mones, particularly the adrenal steroids. Many of the characteristics of the FAS, such as growth and developmental delays, are similar to those found following perinatal corticosteroid evaluations. In general, administration of high doses of alcohol in both animals and humans elevates plasma levels of corticosterone and cortisol, respectively. In addition, prenatal ethanol increases brain corticosterone levels in rats and en-hances the pituitary–adrenal response to certain stressors.

A major site of ethanol action appears to be at the HPA axis (for review see Cicero, 1981). Alcohol stimulates the release of ACTH from the pituitary, which in turn enhances secretion of corticosteroids from the adrenal cortex. Alcohol actions on ACTH release appear to be dose related (i.e., ingestion of progressively higher doses of alcohol are correlated with increasing levels of ACTH and corticosteroids). Prelimi-nary studies indicate that alcohol-induced increments in plasma corticosteroids are the result of increased secretion of corticotropin-releasing hormone (CRH) from the hy-pothalamus (for further discussion see Cicero, 1981), but it is not clear whether alcohol is acting directly on CRH neurons in the hypothalamus or through modula-tion of afferent connections to the hypothalamus, which may modulate CRH secre-tion.

Postnatal Ethanol Administration

In some cases it is difficult to differentiate between prenatal and postnatal ethanol effects. In some studies ethanol treatment of the dam starts 4–5 weeks before mating and is continued throughout the entire gestation period (Esquifino et al., 1986), and even during the first 2 weeks of postnatal life (Parker et al., 1984).

In one such study, rat pups were exposed to high cyclic blood alcohol concentra-tions during PD 4–10. When the pups were tested in the Morris water maze, females showed decreased spatial navigation abilities, and males were unaffected (Kelly et al., 1988).

Summary

Prenatal exposure to ethanol inhibits masculinization and enhances some feminine traits of male rats. However, contrary to what is observed with prenatal stress, the reproductive capability of the female also seems to be impaired by prenatal ethanol.

Nicotine

Nicotine is one of the most widely used drug in human society. The alkaloid [1-methyl-2-(3-pyridyl-pyrrolidine)], commonly referred to as nicotine, is responsible for the dependency of regular smokers on cigarettes. During gestation, smoking has been reported to affect fetal motor behavior as shown by a reduction of fetal movements (Kelly et al., 1984).

Many of the detrimental nicotine effects on fetal development persist into adulthood. An example is the behavioral abnormalities seen in children of mothers who smoked during pregnancy. Symptoms of minimal brain dysfunction such as short attention span, hyperactivity, decrements in reading ability, and difficulties in social adjustments may be present in these children. Nicotine is also known to reduce fertility in humans, as evidenced by a trend toward reduced fertility in humans with increasing number of cigarettes smoked per day (Howe et al., 1985; Baird and Wilcox, 1985).

Prenatal Nicotine Administration

Several methods are employed to assess the effects of prenatal nicotine exposure. Nicotine has been administered by exposing the pregnant dams to cigarette smoke (0.3 mg/kg/day; Bernardi et al., 1981); by dissolving nicotine in the drinking water (40 μg/ml; Peters and Tang, 1982; 165 μg/ml; Jansson et al., 1989); by introducing an osmotic minipump subcutaneously (25 μg/100g/hr; Lichtensteiger and Schlumpf, 1985); or via intraperitoneal injections (0.25 mg/kg twice daily; Segarra and Strand, 1989).

Results of prenatal nicotine studies show that prenatal nicotine administration decreases male sexual behavior in male rats (Segarra and Strand, 1989). Similar to the data with prenatal stress and prenatal ACTH, the decline in sexual behavior is manifest as a decrease in the number of males that mounted or intromitted. The performance of the males prenatally treated with nicotine that exhibited sexual behavior was similar to that of the saline-treated males. The only significant difference found was a decrease in the number of intromissions (Segarra and Strand, 1989).

Prenatally nictoine-treated rats have been tested for other sexually dimorphic behaviors, such as taste preferences and open-field behavior. Males treated prenatally with nicotine exhibited an increase in rearing when tested for open-field behavior (Peters and Tang, 1982). When rats are tested for taste preferences, females exhibit a predilection toward saccharin and saline solutions. Lichtensteiger et al. (1988) reported that male rats treated prenatally with nicotine exhibited an increase in saccharin preference that resembled that of female rats. On the other hand, when males prenatally treated with nicotine were tested for saline preference, a decrease in the amount of saline solution intake was observed (Segarra, 1988). Thus, nicotine reduces masculinization of male progeny with regard to some sexually dimorphic behaviors, such as rearing and saccharin preference, but not with regard to others, such as salt preference.

One study reported a decrease in birth weight in animals treated prenatally with nicotine (Lichtensteiger and Schlumpf, 1985), whereas others found no effects (Segarra and Strand, 1989; Jansson et al., 1989). These differences in results can be attributed in part to the higher dosage used in the first study. However, there is agreement in two studies that anogenital distance, a morphologic sex difference, was reduced in males prenatally treated with nicotine (Lichtensteiger et al., 1988; Segarra and Strand, 1989). Sexually dimorphic neuroanatomical traits have not as yet been studied.

Females appear to be less susceptible to prenatal nicotine administration than do males. When females were tested for sexually dimorphic behaviors, such as saccharin (Lichtensteiger and Schlumpf, 1985) and salt (Segarra, 1988) taste preference, open-field behavior, (Peters and Tang, 1982; Segarra, 1988) and two-way avoidance (Genedani et al., 1983) no trend toward masculinization was observed. The only significant effect observed was an increase in ambulation (Segarra, 1988). Thus sexually dimorphic behaviors are unaffected by prenatal nicotine administration. However, females prenatally treated with nicotine exhibited a delay in vaginal opening and had higher ovary weights than control females (Segarra, 1988).

Mechanisms of Prenatal Nicotine Action

Nicotine administration causes a wide variety of changes in the neuroendocrine system. Increases in ACTH, vasopressin, β-endorphin, prolactin, and LH have been reported (for review see Fuxe et al., 1989). These changes appear to be mediated by the activation of nicotinic cholinergic as well as dopaminergic receptors in the tubero-infundibulum and median eminence. Thus the detrimental effect of nicotine on the reproductive axis may be exerted directly on LHRH and prolactin secretion or indirectly, via the HPA axis and its effects on the reproductive neuroendocrine system.

The release of ACTH and of glucocorticoids caused by nicotine administration can be accomplished by several pathways. Nicotine may act on nicotinic cholinergic receptors within the hypothalamus and POA to stimulate release of CRH (Fuxe et al., 1989). Alternatively, nicotine may prompt the release of catecholamines from the adrenal medulla by activating the adrenergic receptors of ACTH-containing cells. Many smokers who do not show a tolerance to the ACTH-releasing activity of nicotine may be exposed to glucocorticoid hypersecretion. Such a state can cause hormonal changes, specifically on the HPG axis, similar to those following prenatal stress.

Levels of LHRH are also affected by nicotine administration. Nicotine may interact directly or indirectly with cholinergic receptors that regulate LHRH systems originating in the POA. Evidence for this is provided by the effect of nicotine on LH secretion. Nicotine is known to increase dopamine utilization and release in the lateral palisade zone (LPZ), which inhibits the release of LHRH (Fuxe and Hökfelt, 1969).

As with prenatal stress and prenatal ethanol treatment, males treated prenatally with nicotine exhibit a decrease in fetal testosterone levels at GD18 (Lichtensteiger and Schlumpf, 1985). However, contrary to what was observed with other prenatal manipulations, the nicotine-induced decrease in plasma testosterone levels appears to persist into adulthood (Segarra and Strand, 1989). Prolactin plasma levels were also increased in adult males by prenatal nicotine exposure (Segarra and Strand, 1989).

Prenatally, nicotine may alter sexual differentiation of the brain by interacting with nicotinic or muscarinic receptors. The hypothalamus and POA are the brain areas with the highest density of nicotinic receptors (Clarke et al., 1988; Harfstrand et al., 1988). In vitro receptor autoradiography has revealed specific binding sites for [³H]nicotine in rat lower brainstem and spinal cord from GD12; these spread rostrally between GD14 and 15 (Schlumpf and Lichtensteiger, 1987).

Summary

Nicotine affects many hormonal systems; however, lasting alterations following prenatal treatment have been documented only for functions linked with the reproductive system. Prenatal nicotine administration decreases male sexual behavior in male rats. Several other sexually dimorphic behaviors, such as saccharin preference and two-way avoidance behavior, are also altered by prenatal nicotine administration. Sexual differentiation of the female brain seems to be unaffected by prenatal nicotine administration.

Nicotine effects on sexual differentiation of the brain appear to be mediated by a reduction in testosterone levels during GD18. These results are similar to those observed with prenatal stress and prenatal ethanol treatment, in which demasculinization and feminization of the adult male rat is preceded by a decrease in testosterone levels in utero. In contrast to results obtained with other prenatal manipulations, nicotine treatment before birth decreased testosterone levels during adulthood.

The effects of prenatal nicotine administration on sexual differentiation of the brain appear to involve several neuroendocrine systems, involving gonadotropin, prolactin, and ACTH secretion. These effects may occur in part by nicotine interacting with nicotinic cholinergic receptors. In addition, nicotine-induced increases in CRH and ACTH secretion leads to glucocorticoid hypersecretion, which alters many reproductive processes. Moreover, nicotine increases dopamine turnover in the lateral palisade zone (LPZ), causing inhibition of LHRH release, which can result in decreased LH and testosterone levels.

Cannabinoids

Delta-9-tetrahydrocannabinol (Δ^9-THC) is the major psychoactive component of the plant, *Cannabis sativa*, commonly known as marijuana. Other ingredients are Δ^8-THC and cannabinol. Cognitive impairment in humans has been associated with marijuana use. Distractibility, fragmentation of thought, and difficulty in solving problems are thought to be mediated by cannabinoid receptors located in the brain cortex and hippocampus. In addition, *Cannabis* is known for its effects as an analgesic and antinociceptive agent and is used therapeutically to ameliorate nausea and vomiting caused by cancer chemotherapy. Prenatal marijuana exposure induces certain neuroendocrine responses, such as decreased plasma concentration of thyrotropin and decreased gonadotropin and prolactin release in rodents and in primates, which have been reported following prenatal marijuana exposure (Howlett et al., 1990).

In addition, prenatal cannabinoid administration decreases brain weight at birth (Walters and Carr, 1986), and studies indicate that excessive marijuana use can cause aginglike degenerative changes in the brain similar to those caused by stress and glucocorticoids (Landfield, 1981).

Cannabinoid effects on the reproductive capability of animals, including humans, are well documented. In males, marijuana decreased androgen production (Landfield, 1981), sperm count, and induced chromosomal abnormalities during spermatogenesis (Dalterio and Bartke, 1979, 1981). Cannabinoid effects on females include abnormal estrous cycles and increased diencephalic endogenous opioids (Kumar et al., 1990).

Methodology

In the majority of prenatal cannabinoid studies, THC is dissolved in sesame oil and administered orally. Although subcutaneous administration of THC is poorly absorbed, intraperitoneal injection produces higher blood cannabinoid levels at a faster rate than oral or subcutaneous administration. At present there is no scientific rationale as to why investigators choose to administer cannabinoids orally. If the reason is to avoid the stress of injection, one wonders whether intraperitoneal administration of THC, properly injected, would be more stressful than oral administration by intubation. The allegation by Abel (1980) that damage to the fetus and peritonitis can occur when injecting pregnant rats and/or mice is not valid, as previous prenatal studies have shown.

Dosages in prenatal cannabinoid studies vary and so does the time of administration. In several studies, cannabinoid administration was started 2–5 weeks before mating. This makes it difficult to assess which effects are the result of prenatal exposure of the fetuses and which effects may arise from chromosomal damage or behavioral or neuroendocrine anomalies resulting from long-term treatment of the dams with cannabinoids before impregnation.

Cannabinoids, like ethanol, decrease food and water intake. However, unlike controls in prenatal ethanol studies, a pair fed isocaloric control is absent in most prenatal cannabinoids studies. In addition, cannabinoids applied during gestation may be stored in the maternal tissue and released into the milk during lactation, extending the developmental period of exposure to the drug into neonatal life.

Cannabinoid exposure during pregnancy induces changes in the behavior of the dam. Alterations in maternal behavior, with an increase in cannibalism (Abel, 1975), have been reported following prenatal cannabinoid exposure. Cross-fostering drug-free pups to mothers treated with THC during gestation caused these pups to weigh less and rear less in an open field at 28 to 30 days of age (Abel et al., 1979).

Effects of Prenatal Cannabinoid Exposure on Sexual Differentiation of the Brain

There are very few studies of the effect of prenatal cannabinoid exposure on sexual differentiation of the brain. When one tries to compare these few studies with the rest of the literature on drugs and sexual differentiation of the brain, one encounters several methodological differences. First, mice instead of rats have been used as the animal model for most studies of cannabinoids. Second, in the majority of studies of sexual differentiation of the brain, treatment has been confined to the last trimester of gestation [i.e., GD14–21]. However, in studies of prenatal cannabinoids, females were treated with cannabinoids 2–5 weeks before mating. Perinatal treatment was then confined to GD18 and 19 and postnatal days 1–6. Although days 18 and 19 of gestation are critical for this process, confining treatment to these 2 days during gestation makes it difficult to compare this study with other studies and make any generally valid conclusions.

Effects on Sexual Behavior Male mice treated perinatally with Δ^9-THC (50 mg/kg; GD21 through PN 1–6) exhibited a decrease in sexual behavior when compared to controls, as measured by an increase in mount latency and number of mounts (Dalterio and Bartke, 1979).

Effects on Testosterone Levels Both cannibinol and Δ^9-THC administered from GD12 to 19 (50 mg/kg, orally) decreased testosterone levels in 16-day male fetal mice (Dalterio and Bartke, 1981). However, testosterone and LH levels of adult male mice treated prenatally with cannabinoids are not affected. These results are similar to those obtained with prenatal stress in male rats, in which testosterone levels were reduced at GD18 but were normal in the adult animal.

Effects on Sex Ratio and Anatomical Traits A decrease in body weight at birth of rodents treated prenatally with cannabinoids has been reported by some investigators (Walters and Carr, 1986), whereas others found no effect (Pace et al., 1971; Wright et al., 1976; Dalterio and Bartke, 1981). Anogenital distance was also unaffected by prenatal THC administration (Dalterio and Bartke, 1981).

Mechanism of Prenatal Cannabinoid Administration Prenatal cannabinoid administration (crude marihuana extract 20 mg/kg of THC) decreased D_2-dopaminergic receptor levels and tyrosine hydroxylase activity in the striatum of 20- and 40-day-old rats (Walters and Carr, 1986). Dopaminergic activity enhances male sexual behavior. Whether this decrease in D_2-receptor levels can be correlated with decreased male sexual behavior remains to be investigated.

Summary

Prenatal cannabinoid exposure has been found to decrease male sexual behavior in male mice. This demasculinization is similar to effects of other drugs mentioned previously and is accompanied by a decrease in testosterone levels during day 18 of gestation. However, when these animals reach adulthood, their plasma testosterone levels are normal.

At present, very few studies have assessed the effect of prenatal cannabinoid exposure on sexual differentiation of the brain. Thus a discussion of the effect of prenatal cannabinoid exposure on sexually dimorphic neuroanatomical and behavioral traits and on possible mechanisms of action must await further research.

DISCUSSION

This chapter has examined how drugs and stress administered during development can alter the process of sexual differentiation of the brain. Our thesis has been that, in order to understand how drugs affect sexual differentiation, it is necessary to exploit further the information obtained in the developmental neuroendocrine field during the last three decades. This information includes the time of appearance and pattern of secretion of gonadal and adrenal hormone systems during prenatal and early postnatal development; and, indeed, sex and stress steroids and their receptor systems have been among the main characters in this story. We have also emphasized that in order to delineate clearly how sexual differentiation is affected by prenatal drug administration, the drugs must be used at consistent and well-defined times during development; moreover, dosages must be consistent between studies. Unfortunately, many studies have failed to accomplish this. Therefore, in the interests of future research, a major point of this chapter has been to describe the processes involved during sexual differentiation of the brain, to present and criticize the studies that have been done; and to make suggestions for the future study of how drugs interfere with this process.

Similarities and Differences between Prenatal Drug Studies

The organizational effects of sex steroids are exerted mainly during perinatal development. Therefore, one common effect of prenatal manipulations that result in decreased male sexual behavior (such as prenatal stress, prenatal administration of cannabinoids, nicotine, and ethanol) is to decrease fetal testosterone levels. In some cases, as with prenatal nicotine and prenatal cannabinoid exposure, this decrease in testosterone levels persists into adulthood. In other cases, such as with prenatal stress and prenatal ethanol and opioid administration (Ward et al., 1983), the decrease in testosterone levels occurs only during fetal development.

A second common denominator of prenatal drug studies is that their administration to animals acts as a stressor and activates the HPA axis (Table 4). As noted, prenatal stress in males produces a decrease in sexual behavior and in fetal testosterone levels. Before extrapolating results from animal studies to humans, it will be important to assess whether the same drugs are stressors in humans or whether the specific requirements of giving drugs to experimental animals introduces a separate stressful component.

Although prenatal drug administration may be stressful, several studies have demonstrated that certain drug treatments produce effects beyond their actions to activate those mediated by the HPA axis. Ethanol provides a good example. Experiments using in vitro cell culture systems have shown that ethanol causes growth retardation and retards differentiation in a dose-dependent manner (Brown et al., 1979), indicating that ethanol can act directly on developing mammalian tissue.

Many of the differences in results obtained following perinatal manipulations may be attributed to differences in the drug dosage and to the particular window of time during development when the drugs were given. The brain may be embryologically

TABLE 4 Effects of Drugs and Stress Administered during
Gestation on Male Sexual Behavior and Testosterone
Levels in Rats

Prenatal manipulations	Sexual behavior	Fetal T[a]	Adult T
Stress	Decreased	Decreased	No change
ACTH	Decreased	Not known	No change
Cannabinoids	Decreased	Decreased	Decreased
Ethanol	Not clear	Decreased	Decreased
Nicotine	Decreased	Decreased	Decreased

[a]T, testosterone level.

competent to respond to some substances, endogenous or exogenous, at one stage in development, and yet be nonresponsive, or respond completely differently, to these same agents at other developmental stages. This is the case with prenatal stress, as well as prenatal ACTH and nicotine administration. We have discussed in this chapter that the eventual sexual behavior of the adult rat appears to be crucially dependent on the time in ontogeny when developing neural pathways are exposed to changing hormonal levels, drugs, or environmental influences such as stress.

Problems

Besides the issue of stress as a mediator and possible confounding factor in the actions of drugs on sexual differentiation, there are a number of other technical problems in understanding mechanisms involved in their effects. For example, it is not easy to distinguish the direct effects of drugs on the developing fetus or infant from the indirect effects of the drug mediated through the mother. The amount of drugs to which a fetus is exposed will vary depending on maternal and fetal metabolic rates, dosage and mode of drug administration, amount of body fat, and the ionization state of the drug, among others. For example, embryos and fetuses from humans and most animal species are incapable of metabolizing ethanol; thus, they are dependent on maternal metabolism of ethanol for the metabolism and elimination of ethanol from fetal circulation and amniotic fluid (Clarke et al., 1989).

Another problem that arises in these types of studies is that detrimental effects of a drug may be exerted simultaneously or sequentially on several developing systems. Prenatal ethanol and prenatal cannabinoid exposure, for example, are associated with mental retardation of the progeny, reflecting widespread effects (that are not confined to sexual differentiation) on cerebral development. On the other hand, we have also noted that the long-term effects of prenatal stress and prenatal administration of ACTH and on the progeny are limited mainly to sexually dimorphic behaviors and/or traits of the male offspring.

Another methodological issue is whether the decreased sexual behavior produced by prenatal drug administration might be attributed to the decrease in body weight observed at birth. However, as we have discussed, there are several studies in which no significant differences in body weight at birth were found, as is the case with prenatal ACTH and nicotine administration, but a decrease in male copulatory behavior was observed.

Mechanisms

The mechanisms by which prenatal manipulations mediate changes in adult copulatory behavior vary according to the treatment. Although ethanol, cannabinoids, and nicotine decrease fetal testosterone levels, the steps from decreased fetal testosterone levels to decreased male sexual behavior are still not fully elucidated. There are several interactive neuronal systems involved in the regulation of testosterone secretion at the level of the POA during ontogeny, such as the serotonergic and noradrenergic systems. Drug effects on development of synaptic contacts between various neurotransmitter systems and LHRH elements in the POA and arcuate nucleus may indirectly alter fetal testosterone levels during the critical period for sexual differentiation of the brain. Investigators have looked at various neurotransmitter systems that would be good candidates to mediate these changes. In prenatally stressed males, the endogenous opioid system has been implicated in mediating changes in fetal testosterone levels, whereas males prenatally treated with nicotine exhibit alterations in catecholamine metabolites and in the serotonergic system. The decrease in sexual behavior of males treated prenatally with cocaine is associated with the noradrenergic system, and alterations in several neurotransmitter systems (such as the serotonergic, dopaminergic, and noradrenergic) were observed in males treated prenatally with ethanol. However, it is still too early to make any general statement, since it is possible that other neurotransmitter systems that have not been studied may be involved.

Although many of these prenatal treatments have the same end result, namely, blockade of masculinization or defeminization of the male progeny, the mechanisms by which these effects are achieved may be quite different. This is illustrated by prenatal nicotine and prenatal ACTH administration. The evidence indicates that prenatal nicotine effects are mediated mainly by the neuroendocrine system, as indicated by decreased testosterone and prolactin levels during adulthood, whereas the prenatal ACTH "target" is mainly neural, as evidenced by increased serotonergic levels and receptors in the POA and VMN, respectively.

Additionally, many of these drugs can alter the neuronal circuitry during development via other neural targets. For example, prenatal stress is known to reduce the size of the SDN of the POA. Neurogenesis in this SDN area occurs as late as GD18 (Jacobson and Gorski, 1981). This period overlaps with that found by Ward and Weisz (1980), in which there is a sex difference in fetal plasma levels of androgens. Prenatal stress and drugs could directly alter processes such as neurogenesis, neuronal death, synaptogenesis, or sprouting.

Differences in Drug Effects on Males and Females

The drug treatments reviewed in this article are most effective on males; typically, drug treatment reduces the display of reproductive behaviors, while allowing the reproductive physiology of the adult animal to remain functional. Typically, in females, if a drug has an effect, then the reproductive physiology of females is hindered, but their reproductive behavior is unaffected.

Combined Drugs Have More Potent Effects

Combinations of drugs sometimes produce different effects than either drug alone. The combined effect of THC and alcohol produced a greater fetotoxic impact than either drug alone; moreover, the combinations had a greater impact than that resulting from the additive effects of these drugs given at different times (Abel, 1985; Abel and Dintcheff, 1986). However, a comparable effect was not seen for the combination of THC and phenobarbital (Abel et al., 1987). Moreover, when THC and chlordiazepoxide hydrochloride (Librium) were each administered to pregnant mice, neither drug alone affected implantations, but THC significantly increased resorptions. Given together, THC and chlordiazepoxide decreased fetal weight, but there was no synergism between the two drugs for any of the measures examined (Abel and Tan, 1987).

In conclusion, in order to understand how the process of sexual differentiation is altered by drugs, we first need to understand the process of sexual differentiation. Unfortunately we are still ignorant as to many of the mechanisms and participants involved in regulating this important process. A great need exists to increase research in this area since drug addiction continues to escalate in our society. The detrimental effects of drug abuse during gestation result in an emotional, clinical, and economic burden on our society. It is hoped, as more research and social programs geared to this particular problem are developed, the teratological effects of prenatal drug abuse on children will be minimized.

ACKNOWLEDGMENTS

Research by the authors described in this article was supported by a MARC Faculty Fellowship 1F 34 GM 13001 and a Porter Fellowship—American Physiological Society (to ACS), and by NIH Grant NS 07080 (to BMc). We should also like to thank Ganya Alvarado, Marithelma Costa, Catalina Montealegre and Jorge L. González for their help in bibliographic work and other assistance in preparing this manuscript.

REFERENCES

Abel, E. L. (1975). Suppression of maternal behavior in the mouse by Δ⁹-tetrahydrocannabinol. *Fed. Proc.* **34**, 2968.
Abel, E. L. (1980). Prenatal exposure to Cannabis: A Critical Review of Effects on Growth, Development and Behavior. *Behav. Neural Biol.* **29**, 137–156.
Abel, E. L. (1985). Alcohol-enhancement of marijuana-induced fetotoxicity. *Teratology* **31**, 35–40.
Abel, E. L., and Dintcheff, B. A. (1986). Increased marijuana-induced fetotoxicity by a low dose of concomitant alcohol administration. *J. Stud. Alcohol* **47**, 440–443.
Abel, E. L., and Tan, S. E. (1987). Effects of delta-9-tetrahydrocannabinol, chlordiazepoxide (Librium), and their combination on pregnancy and offspring in rats. *Reprod. Toxicol.* **1**(1), 37–40.

Abel, E. L., Day, N., Dintcheff, B. A., and Ernst, C. A. (1979). Inhibition of postnatal maternal performance in rats treated with marihuana extract during pregnancy. *Bull. Psychonomic Soc.* **14,** 353–354.

Abel, E. L., Tan, S. E., and Subramanian, M. (1987). Effects of delta-9-THC, phenobarbital, and their combination on pregnancy and offspring in rats. *Teratology* **36**(2), 193–198.

Ader, R. I., and Conklin, P. M. (1963). Handling of pregnant rats: Effects on emotionality of their offspring. *Science* **142,** 411–412.

Adkins-Regan, E. (1987). Hormones and sexual differentiation. *In* "Hormones and Reproduction in Fishes, Amphibians, and Reptiles." (D. O. Norris and R. E. Jones, eds.), pp. 1–29. Plenum Press, New York.

Alonso, J., Castellano, M. A., and Rodriguez, M. (1991). Behavioral lateralization in rats: Prenatal stress effects on sex differences. *Brain Res.* **539,** 45–50.

Alexis, M. N., Kitraki, E., Spanou, K., Stylianopoulou, F., and Sekeris, C. E. (1990). Ontogeny of the glucocorticoid receptor in the rat brain. *Adv. Exp. Med. Biol.* **265,** 269–276.

Alves, S. E. and Strand, F. L. (1990). Effects of prenatal administration of ACTH and nicotine on the subsequent sexual behavior of female rats. *Abstract. Soc. for Neuroscience*, St. Louis, MO.

Anderson, R. H., Fleming, D. E., Rhees, R. W., and Kinghorn, E. (1986). Relationships between sexual activity, plasma testosterone, and the volume of the sexually dimorphic nucleus of the preoptic area in prenatally stressed and non-stressed rats. *Brain Res.* **370,** 1–10.

Angevine, J. B., and Sidman, R. L. (1961). Autoradiographic study of cell migration during histogenesis of cerebral cortex in the mouse. *Nature* **192,** 766–768.

Arai, Y., and Gorski, R. A. (1968). Critical exposure time for androgenization of the rat hypothalamus determined by anti-estrogen injections. *Proc. Soc. Exp. Biol. Med.* **127,** 590–593.

Arai, Y., and Matsumoto, A. (1978). Synapse formation of the hypothalamus and arcuate nucleus during postnatal development in the female rat and its modification by neonatal estrogen treatment. *Psychoneuroendocrinology* **3,** 31.

Archer, J., and Blackman, D. (1971). Prenatal psychological stress and offspring behavior in rats and mice. *Dev. Psychobiol.* **4,** 193–248.

Arnold, A. P., and Breedlove, S. M. (1985). Organizational and activational effects of sex steroids on brain and behavior: A reanalysis. *Horm. Behav.* **19,** 469–498.

Azmitia, E. C., and de Kloet, E. R. (1987). ACTH neuropeptide stimulation of serotonergic neuronal maturation in tissue culture: Modulation by hippocampal cells. *Prog. Brain Res.* **72,** 311–318.

Baird, D. D., and Wilcox, A. J. (1985). Cigarette smoking associated with delayed conception. *J.A.M.A.* **253,** 2979–2983.

Bardin, C. W., Bullock, L. P., Sherins, R. J., Mowszowicz, I., and Blackburn, W. R. (1973). Androgen metabolism and mechanism of action in male pseudohermaphroditism: A study of testicular feminization. *Rec. Prog. Horm. Res.* **29,** 65–105.

Barrett, R. J., and Ray, O. S. (1970). Behavior in the open field, Lashley III maze, shuttle box, and Sidman avoidance as a function of strain, sex, and age. *Dev. Psychol.* **3,** 73–77.

Barron, S., and Riley, E. P. (1985). Pup-induced maternal behavior in adult and juvenile rats exposed to alcohol prenatally. *Alcohol Clin. Exp. Res.* **9,** 360–365.

Beach, F. A. (1976). Sexual attractivity, proceptivity, and receptivity in female mammals. *Horm. Behav.* **7,** 105–138.

Beach, F. A., and Jordan, L. (1956). Sexual exhaustion and recovery in the male rat. *Q. J. Exp. Psychol.* **8,** 121–133.

Beatty, W. W. (1979). Gonadal hormones and sex differences in nonreproductive behaviors in rodents: Organizational and activational influences. *Horm. Behav.* **12,** 112–163.

Beatty, W. W., and Beatty, P. A. (1970). Hormonal determinants of sex differences in avoidance behavior and reactivity to electric shock in the rat. *J. Comp. Physiol. Psychol.* **73,** 446–455.

Beatty, W. W., Gregoire, K. C., and Parmiter, L. L. (1973). Sex differences in retention of passive avoidance behavior in rats. *Bull. Psychon. Soc.* **11,** 71–72.

Bell, D. D., and Zucker, I. (1971). Sex differences in body weight and eating: Organization and activation by gonadal hormones in the rat. *Physiol. Behav.* **7,** 27–34.

Bernardi, M., Genedani, S., and Bertolini, A. (1981). Sexual behavior in the offspring of rats exposed to cigarette smoke or treated with nicotine during pregnancy. *Rivista di Farmacologia e Terapia, XII* **3,** 197–203.

Berrebi, A. S., Fitch, R. H., Ralphe, D. L., Denenberg, J. O., Friedrich, V. L., and Denenberg, V. H. (1988). Corpus callosum: Region-specific effects of sex, early experience, and age. *Brain Res.* **438,** 216–224.

Bloom, E., Matulich, D. T., Lan, N. C., Higgins, S. J., Simons, S., and Baxter, J. (1980). Nuclear binding of glucocorticoid receptors. Relations between cytosol binding, activation, and the biological response. *J. Steroid Biochem.* **12,** 175.

Braggio, J. T., Nadler, R. D., Lance, J., and Miseyko, D. (1978). Sex differences in apes and children. In "Recent Advances in Primatology" (D. J. Chivers and J. Herbert, eds.), Vol. 1 (Behavior), pp. 529–531. Academic Press, New York.

Brown, N. A., Goulding, E. H., and Fabro, S. (1979). Ethanol embryotoxicity: Direct effects on mammalian embryos *in vivo. Science* **206,** 573–575.

Butte, J. C., Kakihana, R., Farham, M. L., and Noble, E. P. (1973). The relationship between brain and plasma corticosterone stress response in developing rats. *Endocrinology* **92,** 1775.

Cameron, D. (1975). The psychopharmacology of social drinking. *J. Alcoholism* **9,** 50–55.

Carlsson, M., and Carlsson, A. (1988). A regional study of sex differences in rat brain serotonin. *Prog. Neuro-Psychopharmacol. Biol. Psychiatr.* **12,** 53–61.

Carlsson, M., Svensson, K., Eriksson, E., and Carlsson, A. (1985). Rat brain serotonin: Biochemical and functional evidence for a sex difference. *J. Neural Transm.* **63,** 297–313.

Chapman, R. H., and Stern, J. M. (1977). Maternal stress and pituitary–adrenal manipulations during pregnancy in rats: Effects on morphology and sexual behavior of male offspring. *J. Comp. Physiol. Psychol.* **92**(6), 1074–1083.

Chen, J. J., and Smith, E. R. (1979). Effects of perinatal alcohol on sexual differentiation and open-field behavior in rats. *Horm. Behav.* **13,** 219–231.

Chernoff, G. F. (1977). The fetal alcohol syndrome in mice: An animal model. *Teratology* **15,** 223–230.

Chung, K. W. (1989). Effect of ethanol receptors in the anterior pituitary, hypothalamus and brain cortex in rats. *Life Sci.* **44,** 273–280.

Cicero, T. J. (1981). Neuroendocrinological effects of alcohol. *Annu. Rev. Med.* **32,** 123–142.

Clarke, D. W., Smith, G. N., Patrick, J., Richardson, B., and Brien, J. F. (1989). Activity of alcohol dehydrogenase and aldehyde dehydrogenase in maternal liver, fetal liver, and placenta of the near-term pregnant ewe. *Dev. Pharmacol. Ther.* **12,** 35–41.

Clarke, P. B. S., Schwartz, R. D., Paul, S. M., Pert, C. B., Pert, A. (1988). Nicotinic binding in rat brain: Autoradiographic comparison of [³H]acetyl-choline, [³H]nicotine, and [¹²⁵I]α-bungarotoxin. *J. Neurosci.* **5,** 1307–1315.

Clarren, S. K., and Smith, D. W. (1978). The fetal alcohol syndrome. *N. Engl. J. Med.* **298,** 1063–1067.

Clemens, L. G. (1974). Neurohormonal control of male sexual behavior. In "Advances in Behavioral Biology, Vol. 11 Reproductive Behavior," (W. Montagna and W. A. Sadler, eds.), pp. 23–53. Plenum Press, New York.

Clemens, L. G., and Gladue, B. A. (1978). Feminine sexual behavior in rats enhanced by prenatal inhibition of androgen aromatization. *Horm. Behav.* **11,** 190.

Cohen, A., Chatelain, A., and Dupony, J. P. (1983). Late pregnancy maternal and fetal time-course of plasma ACTH and corticosterone after continuous ether inhalation by pregnant rats. *Biol. Neonate* **43,** 220–228.

Conry, J. (1990). Neuropsychological deficits in fetal alcohol syndrome and fetal alcohol effects. *Alcohol Clin. Exp. Res.* **14**(5), 650–655.

Cooke, H. (1990). The continuing search for the mammalian sex-determining gene. *Trends in Genetics* **6**(9), 273–275.

Corbier, P., Kerdelhue, B., Picon, R., and Roffi, J. (1978). Changes in testicular weight and serum gonadotropin and testosterone levels before, during, and after birth in the perinatal rat. *Endocrinology* **103**(6), 1985–1991.

Csaba, G. (1986). Receptor ontogeny and hormonal imprinting. *Experientia* **42,** 750.

Dahlgreen, I., Matuszczyk, J., and Hard, E. (1991a). Sexual orientation in male rats prenatally exposed to ethanol. *Neurotoxicol. Teratol.* **13,** 267–269.

Dahlgreen, I. L., Eriksson, C. J. P., Gustafsson, B., Harthon, C., Hård, E., and Larsson, K. (1991b). Effects of chronic and acute ethanol treatment during prenatal and early postnatal ages on testosterone levels and sexual behaviors in rats. *Pharmacol. Biochem. Behav.* **33**, 867–873.

Dahlöf, L. G., Hård, E., and Larsson, K. (1977). Influence of maternal stress on offspring sexual behavior. *Anim. Behav.* **25**, 958–963.

Dalterio, S., and Bartke, A. (1979). Perinatal exposure to cannabinoids alters male reproductive function in mice. *Science* **205**, 1420–1422.

Dalterio, S., and Bartke, A. (1981). Fetal testosterone in mice: Effect of gestational age and cannabinoid exposure. *J. Endocrinol.* **91**, 509–514.

Davis, H., Porter, J. W., Burton, J., and Levine, S. (1976). Sex and strains differences in lever press shock escape behavior. *Physiol. Psychol.* **4**, 351–356.

Davis, P. G., Chaptal, C. V., and McEwen, B. S. (1978). Independence of the differentiation of masculine and feminine sexual behavior in rats. *Horm. Behav.* **12**, 12.

Dawson, J. L. M., Cheung, Y. M., and Lau, R. T. S. (1975). Developmental effects of neonatal sex hormones on spatial and activity skills in the white rat. *Biol. Psychol.* **3**, 213–229.

Denti, A., and Epstein, A. (1972). Sex differences in the acquisition of two kinds of avoidance behavior in rats. *Physiol. Behav.* **8**, 611–615.

Diamond, M. C., Murphy, G. M., Akiyama, K., and Johnson, R. E. (1982). Morphological hippocampal asymmetry in male and female rats. *Exp. Neurol.* **76**, 553–565.

Diamond, M. C., Johnson, R. E., Young, D., and Singh, S. S. (1983). Age-related morphological differences in the rat cerebral cortex and hippocampus: Male-female; right-left. *Exp. Neurol.* **81**, 1–13.

Diaz, J. D., and Samson, H. (1980). Impaired brain growth in neonatal rats exposed to ethanol. *Science* **208**, 751–753.

Döcke, F., and Dörner, G. (1975). Anovulation in adult female rats after neonatal intracerebral implantation of oestrogen. *Endokrinologie* **65**, 375–377.

Döhler, K. D., and Wuttke, W. (1975). Changes with age in levels of serum gonadotropins, prolactin, and gonadal steroids in prepubertal male and female rats. *Endocrinology* **97**(4), 898–907.

Döhler, K. D., Coquelin, A., Davis, F., Hines, M., Shryne, J. E., and Gorski, R. A. (1982). Differentiation of the sexually dimorphic nucleus in the preoptic area of the brain is determined by the perinatal hormone environment. *Neurosci. Lett.* **33**, 295–298.

Druse, M. J. (1986). Effects of perinatal alcohol exposure on neurotransmitters, membranes, and proteins. In "Alcohol and Brain Development" (J. West, ed.), pp. 343–372. Oxford University Press, New York.

Dunlap, J. L., Zadina, J. E., and Gougis, G. (1978). Prenatal stress interacts with prepubertal social isolation to reduce male copulatory behavior. *Physiol. Behav.* **21**, 873–875.

Dupouy, J. P., Chatelain, A., and Allaume, P. (1980). Absence of transplacental passage of ACTH in the rat. Direct experimental proof. *Biol. Neonate* **37**, 96–102.

Esquifino, A. I., Sanchis, R., and Guerra, C. (1986). Effect of prenatal alcohol exposure on sexual maturation of female rat offspring. *Neuroendocrinology* **44**, 483–487.

Fadda, F., Mosca, E., Niffoi, T., Colombo, G., and Gessa, G. L. (1987). Ethanol prevents stress-induced increase in cortical DOPAC: reversal by RO 15-4513. *Physiol. Behav.* **40**, 383–385.

Fitch, R. H., Berrebi, A. S., Cowell, P. E., Schrott, L. M., and Denenberg, V. H. (1990). Corpus callosum: Neonatal hormones and sexual dimorphism in the rat. *Brain Res.* **515**, 111–116.

Fitch, R. H., Cowell, P. E., Schrott, L. M., and Denenberg, V. H. (1991). Corpus callosum: Ovarian hormones and feminization. *Brain Res.* **542**, 313–317.

Frankfurt, M., Fuchs, E., and Wuttke, W. (1984). Sex differences in gamma-aminobutyric acid and glutamate concentrations in discrete rat brain nuclei. *Neurosci. Lett.* **50**, 245–250.

Frankfurt, M., Siegel, R. A., Sim, I., and Wuttke, W. (1985). Cholecystokinin and substance P concentrations in discrete areas of the rat brain: Sex differences. *Brain Res.* **358**, 53–58.

Fride, E., Dan, Y., Gavish, M., and Weinstock, M. (1985). Prenatal stress impairs maternal behavior in a conflict situation and reduces hippocampal benzodiazepine receptors. *Life Sci.* **36**, 2103–2109.

Fuxe, K., and Hökfelt, T. (1969). Catecholamines in the hypothalamus and the pituitary gland. In "Frontiers in Neuroendocrinology" (W. F. Ganong and L. Martini, eds.), pp. 47–96. Oxford University Press, Oxford.

Fuxe, K., Andersson, K., Eneroth, P., Harfstrand, A., and Agnati, L. F. (1989). Neuroendocrine actions of nicotine and of exposure to cigarette smoke: Medical implications. *Psychoneuroendocrinology* **14**(1, 2), 19–41.

Genedani, S., Bernardi, M., and Bertolini, A. (1983). Sex-linked differences in avoidance learning in the offspring of rats treated with nicotine during pregnancy. *Psychopharmacology* **80**, 93–95.

George, F. W., and Ojeda, S. R. (1982). Changes in aromatase activity in the rat brain during embryonic, neonatal, and infantile development. *Endocrinology* **111**(2), 522–529.

Gerlach, J. L., and McEwen, B. S. (1972). Rat brain binds adrenal steroid hormone: Radioautography of hippocampus with corticosterone. *Science* **175**, 1133.

Geschwind, N., and Galaburda, A. M. (1987). "Cerebral Lateralization: Biological Mechanisms, Associations and Pathology." MIT Press, Cambridge, Massachusetts.

Goldfoot, D. A., and van der Werff ten Bosch, J. J. (1977). Mounting behavior of female guinea pigs after prenatal and adult administration of the propionate of testosterone, dihydrotestosterone, and androstanedione. *Horm. Behav.* **6**, 139.

Gordon, G. G., Southren, A. L., Altman, K., Rubin, E., and Liebr, C. S. (1976). The effect of alcohol (ethanol) administration on sex hormone metabolism in normal men. *N. Engl. J. Med.* **295**, 793–797.

Gordon, B. H. J., Streeter, M. L., Rosso, P., and Winick, M. (1985). Prenatal alcohol exposure: Abnormalities in placental growth and fetal amino acid uptake in the rat. *Biol. Neonate* **47**, 113–119.

Gorski, R. A. (1988). Sexual differentiation of the brain: Mechanisms and implications for neuroscience. *In* "From Message to Mind" (S. S. Easter, Jr., K. F. Barald, and B. M. Carlson, eds.), pp. 256–271. Sinauer Associates, Sunderland, Massachusetts.

Gorski, R. A., Gordon, J. H., Shryne, J. E., and Southam, A. M. (1978). Evidence for a morphological sex difference within the medial preoptic area of the rat brain. *Brain Res.* **148**, 333–346.

Gorski, R. A., Harlan, R. E., Jacobson, C. D., Shryne, J. E., and Southam, A. M. (1980). Evidence for the existence of a sexually dimorphic nucleus in the preoptic area of the rat. *J. Comp. Neurol.* **193**, 529–539.

Goy, R. W. (1978). Development of play and mounting behavior in female rhesus monkeys virilized prenatally with esters of testosterone and dihydrotestosterone. *In* "Recent Advances in Primatology" (D. J. Chivers and J. Herbert, eds.), Vol. 1 (Behavior), p. 449. Academic Press, New York.

Goy, R. W., and McEwen, B. S. (1980). "Sexual Differentiation of the Brain." MIT Press, Cambridge, Massachusetts.

Gray, J. A., and Lalljee, B. (1974). Sex differences in emotional behaviour in the rat: Correlation between open-field defecation and active avoidance. *Anim. Behav.* **22**, 856–861.

Guillemin, R., Vargo, T., Rossier, J., Minick, S., Ling, N., Rivier, C., Vale, W., and Bloom, F. (1977). β-endorphin and adrenocorticotropin are secreted concomitantly by the pituitary gland. *Science* **197**, 1367–1369.

Handa, R. J., McGivern, R. F., Noble, E. P., and Gorski, R. A. (1985). Exposure to alcohol *in utero* alters the adult patterns of luteinizing hormone secretion in male and female rats. *Life Sci.* **37**, 1683–1690.

Hannigan, J. H., and Riley, E. P. (1989). Prenatal ethanol alters gait in rats. *Alcohol* **5**, 451–454.

Hard, E., Dahlgren, I. L., Engel, J., Larsson, K., Liljequist, S., Linde, A., and Musi, B. (1984). Development of sexual behavior in prenatally ethanol exposed rats. *Drug Alcohol Depend.* **14**, 51–61.

Härfstrand, A., Adem, A., Fuxe, K., Agnati, L. F., Andersson, K., and Nordberg, A. (1988). Topographical distribution of nicotinic receptors in the rat tel- and diencephalon quantitative receptor autoradiography using [^{3}H]acetylcholine, [^{125}I]α-bungarotoxin and [^{3}H]nicotine. *Acta Physiol. Scand.* **132**, 1–14.

Hart, B. L., and Leedy, M. G. (1985). Neurological bases of male sexual behavior: A comparative analysis. *In* "Handbook of Behavioral Neurobiology: Vol. 7 Reproduction" (N. Adler, D. Pfaff, and R. W. Goy, eds.), pp. 373–422. Plenum Press, New York.

Harvey, P. W., and Chevins, P. F. D. (1984). Crowding or ACTH treatment of pregnant mice affects adult copulatory behavior of male offspring. *Horm. Behav.* **18**, 101–110.

Hayashi, S., and Gorski, R. A. (1974). Critical exposure time for androgenization by intracranial crystals of testosterone propionate in neonatal female rats. *Endocrinology* **94**, 1161–1167.

Herrenkohl, L. (1979). Prenatal stress reduces fertility and fecundity in female offspring. *Science* **206**, 1097–1099.

Herrenkohl, L. R., and Whitney, J. B. (1976). Effects of prepartal stress on postpartal nursing behavior, litter development, and adult sexual behavior. *Physiol. Behav.* **17**, 1019–1021.

Hilakivi, L. A., and Hilakivi, I. (1987). Increased adult behavior "despair" in rats neonatally exposed to desipramine or zimeldine: An animal model of depression? *Pharmacol. Biochem. Behav.* **28**, 367–369.

Hong, J., Yoshikawa, K., and Lamartiniere, C. (1982). Sex-related difference in the rat pituitary [Met⁵]-enkephalin level altered by gonadectomy. *Brain Res.* **251**, 380–383.

Howe, G., Westhoff, C., Vessey, M., and Yeates, D. (1985). Effects of age, cigarette smoking, and other factors on fertility: Findings in a large prospective study. *Br. Med. J.* **290**, 1697–1700.

Howlett, A. C., Bidaut-Russell, M., Devane, W. A., Melvin, L. S., Johnson, M. R., and Herkenham, M. (1990). The cannabinoid receptor: Biochemical, anatomical and behavioral characterization. *Trends in Neuroscience* **13**(10), 420–423.

Jacobson, C. D., and Gorski, R. A. (1981). Neurogenesis of the sexually dimorphic nucleus of the preoptic area of the rat. *J. Comp. Neurol.* **196**, 512–529.

Jacobson, C. D., Csernus, V. J., Shryne, J. E., and Gorski, R. A. (1981). The influence of gonadectomy, androgen exposure, or a gonadal graft in the neonatal rat on the volume of the sexually dimorphic nucleus of the preoptic area. *J. Neurosci.* **1**, 1142–1147.

Jakubovic, A., Hattori, T., and McGeer, P. L. (1973). Radioactivity in suckled rats after giving [¹⁴C]THC to the mother. *Eur. J. Pharmacol.* **22**, 221–223.

Jansson, A., Andersson, K., Bjelke, B., and Eneroth, P. (1989). Effects of combined pre- and postnatal treatment with nicotine on hypothalamic catecholamine nerve terminal systems and neuroendocrine function in the 4-week-old and adult male and female diestrus rat. *J. Neuroendocrinol.* **1**(6), 455–464.

Jones, K. I., and Smith, D. W. (1975). The fetal alcohol syndrome. *Teratology* **12**, 1–10.

Jones, P. J. H., Leichter, J., and Lee, M. (1981). Placental blood flow in rats fed alcohol before and during gestation. *Life Sci.* **29**, 1153–1159.

Joseph, R., Hess, S., and Birecree, E. (1978). Effect of hormone manipulations and explorations on sex differences in maze learning. *Behav. Biol.* **24**, 364–377.

Jost, A. (1985). Sexual organogenesis. *In* "Handbook of Behavioral Neurobiology" (N. Adler, D. Pfaff, and R. W. Goy, eds.), Vol. 7 (Reproduction), Plenum Press, New York. pp. 3–19.

Kakihana, R., Butte, J. C., and Moore, J. A. (1980). Endocrine effects of maternal alcoholization: Plasma and brain testosterone, dihydrotestosterone, estradiol, and corticosterone. *Alcoholism: Clin. Exp. Res.* **4**(1), 57–61.

Kaplan, M. S., and Hinds, J. W. (1977). Neurogenesis in the adult rat: Electron microscopic analysis of light radioautographs. *Science* **197**, 1092–1094.

Kelly, J., Matthews, K. A., and O'Connor, M. (1984). Smoking in pregnancy: Effect on mother and fetus. *Br. J. Obstet. Gynecol.* **91**, 111–117.

Kelly, S. J., Goodlett, C. R., Hulsether, S. A., and West, J. R. (1988). Impaired spatial navigation in adult females but not adult male rats exposed to alcohol during the brain growth spurt. *Behav. Brain Res.* **27**, 247–257.

Kiss, J., and Halasz, B. (1985). Demonstration of serotonergic axons terminating on luteinizing hormone-releasing hormone neurons in the preoptic area of the rat using a combination of immunocytochemistry and high-resolution autoradiography. *Neuroscience* **14**, 69–78.

Krecek, J. (1973). Sex differences in salt taste: The effect of testosterone. *Brain Res.* **10**, 683–688.

Krecek, J. (1976). The pineal gland and the development of the salt intake pattern in male rats. *Dev. Psychobiol.* **8**, 181–188.

Krecek, J., Novakova, V., and Stibral, K. (1972). Sex differences in the taste preference for a salt solution in the rat. *Physiol. Behav.* **8**, 183–188.

Kronick, J. B. (1976). Teratogenic effects of ethyl alcohol administered to pregnant mice. *Am. J. Obstet. Gynecol.* **124**, 676–680.

Kumar, A. M., Haney, M., Becker, T., Thompson, M. L., Kream, R. M., and Miczek, K. (1990). Effect of early exposure to Δ^9-tetrahydrocannabinoid on the levels of opioid peptides, gonadotropin-releasing hormone and substance P in the adult male rat brain. *Brain Res.* **525**, 78–83.

Kuriyama, K., Ohkuma, S., Taguchi, J., and Hashimoto, T. (1987). Alcohol, acetaldehyde and salsolinol-induced alterations in functions of cerebral GABA/benzodiazepine receptor complex. *Physiol. Behav.* **40**, 393–399.

Landfield, P. W., Baskin, R. K., and Pitler, T. A. (1981). Brain aging correlates: Retardation by hormonal–pharmacological treatment. *Science* **214**, 581–584.

Lau, C., Thadani, P. V., Schanberg, S. M., and Slotkin, T. A. (1976). Effects of maternal ethanol ingestion on development of adrenal catecholamines and dopamine-β-hydroxylase in the offspring. *Neuropharmacology* **15**, 505–507.

Lephart, E. D., and Ojeda, S. R. (1990). Hypothalamic aromatase activity in male and female rats during juvenile peripubertal development. *Neuroendocrinology* **51**, 385–393.

Levine, S., and Broadhurst, P. L. (1963). Genetic and ontogenetic determinants of adult behavior in the rat. *J. Comp. Physiol. Psychol.* **56**, 423–428.

Lichtensteiger, W., and Schlumpf, M. (1985). Prenatal nicotine affects fetal testosterone and sexual dimorphism of saccharin preference. *Pharmacol. Biochem. Behav.* **23**, 439–444.

Lichtensteiger, W., Ribary, U., Schlumpf, M., Odermatt, B., and Widmer, H. R. (1988). Prenatal adverse effects of nicotine on the developing brain. *In* "Progress in Brain Research" (G. J. Boer, M. G. P. Feenstra, M. Mirmiran, D. F. Swaab, and F. Van Haaren, eds.), Vol. 73, pp. 137–157. New York, Elsevier Science Publishers B.V.

Mann, D., Evans, D., Edoimioya, F., Kamel, F., and Buterstein, G. M. (1985). A detailed examination of the *in vivo* and *in vitro* effects of ACTH on gonadotropin secretion in the adult rat. *Neuroendocrinology* **40**, 297–301.

Martin, C. E., Cake, M. H., Hartman, P. E., and Cook, I. F. (1977). Relationship between fetal corticosteroids, maternal progesterone, and parturition in the rat. *Acta Endocrinol.* **84**, 167–176.

Masterpasqua, F., Chapman, R. H., and Lore, R. K. (1976). The effects of prenatal psychological stress on the sexual behavior and reactivity of male rats. *Dev. Psychobiol.* **9**(5), 403–411.

Matsumoto, A., and Arai, Y. (1980). Sexual dimorphism in "wiring pattern" in the hypothalamic arcuate nucleus and its modification by neonatal hormonal environment. *Brain Res.* **190**, 238–242.

Matsumoto, A., and Arai, Y. (1983). Sex differences in volume of the ventromedial nucleus of the hypothalamus in the rat. *Endocrinology* (Japan) **30**, 277–280.

Matsumoto, A., and Arai, Y. (1986). Male–female difference in synaptic organization of the ventromedial nucleus of the hypothalamus in the rat. *Neuroendocrinology* **42**, 232–236.

McDonald, P. G., and Doughty, C. (1973). Inhibition of androgen-sterilization in the female rat by administration of an antiestrogen. *J. Endocrinol.* **55**, 455.

McEwen, B. S. (1981). Neural gonadal steroid actions. *Science* **211**, 1303–1311.

McEwen, B. S. (1983). Gonadal steroid influences on brain development and sexual differentiation. *In* "Reproductive Physiology IV; International Review of Physiology" Vol. 27. (R. O. Greep, ed.) Ch. 3, pp. 99–145. University Park Press, Baltimore.

McEwen, B. S., Lieberburg, I., Chaptal, C., and Krey, L. (1977). Aromatization: Important for sexual differentiation of the neonatal rat brain. *Horm. Behav.* **9**, 249–263.

McEwen, B. S., Lieberburg, I., Chaptal, C., Davis, P. G., Krey, L., MacLusky, N. J., and Roy, E. J. (1979). Attenuating the defeminization of the neonatal rat brain: Mechanisms of action of cyproterone acetate, 1,4,6-androstatriene-3,17-dione, and a synthetic progestin, R5020. *Horm. Behav.* **13**, 268.

McEwen, B. S., Biegon, A., Davis, P. G., Krey, L. C., Luine, V. N., McGinnis, M. Y., Paden, C. M., and Parsons, B. (1982). Steroid hormones: Humoral signals which alter brain cell properties and functions. *Rec. Prog. Horm. Res.* **38**, 41.

McEwen, B. S., de Kloet, E. R., and Rostene, W. (1986). Adrenal steroid receptor and actions in the nervous system. *Physiol. Rev.* **66**, 1166.

McGivern, R. F., Clancy, A. N., Hill, M. A., and Noble, E. P. (1984). Prenatal alcohol exposure alters adult expression of sexually dimorphic behavior in the rat. *Science* **224**, 896–898.

McGivern, R. F., Holcomb, C., and Poland, R. E. (1987). Effects of prenatal testosterone propionate treatment on saccharin preference of adult rats exposed *in utero*. *Physiol. Behav.* **39**(2), 241–246.

McGivern, R. F., Raum, W. J., Salido, E., and Redei, E. (1988). Lack of prenatal testosterone surge in fetal rats exposed to alcohol: Alterations in testicular morphology and physiology. Alcohol Clin. Exp. Res. 12(2), 243–247.

Meaney, M. J., and Stewart J. (1981). Neonatal androgens influence the social play of prepubescent rats. Horm. Behav. 15, 197–213.

Meaney, M. J., Stewart, J., Poulin, P., and McEwen, B. S. (1983). Sexual differentiation of social play in rat pups is mediated by the neonatal androgen-receptor system. Neuroendocrinology 87, 85–90.

Meaney, M. J., Sapolsky, R. M., Aitken, D. H., and McEwen, B. S. (1985a). [3H]Dexamethasone binding in the limbic brain of the fetal rat. Dev. Brain Res. 23, 297–300.

Meaney, M. J., Sapolsky, R. M., and McEwen, B. S. (1985b). The development of the glucocorticoid receptor system in the rat limbic brain. I. Ontogeny and autoregulation. Dev. Brain Res. 18, 159–164.

Meisel, R. L., and Ward, I. L. (1981). Fetal female rats are masculinized by male littermates located caudally in the uterus. Science 212, 239–242.

Meisel, R. L., Dohanich, G. P., and Ward, I. L. (1979). Effects of prenatal stress on avoidance acquisition, open-field performance and lordotic behavior in male rat. Physiol. Behav. 22, 527–530.

Mello, N. K., Bree, M. P., Mendelson, J. H., Ellingbore, J., King, N. W., and Sehgal, P. (1983). Alcohol self-administration disrupts reproductive function in female macaque monkeys. Science 221, 677–679.

Mendelson, J. H. (1986). Alcohol effects on reproductive function in women. Psychiatr. Lett. IV(7), 35–38.

Meyer, L. S., and Riley, E. P. (1986). Social play in juvenile rats prenatally exposed to alcohol. Teratology 34, 1–7.

Meyer, L. S., Kotch, L. E., and Riley, E. P. (1990). Alterations in gait following ethanol exposure during the brain growth spurt in rats. Alcoholism Clin. Exp. Res. 14(1), 23–27.

Miller, M. W. (1981). Effects of alcohol on the generation and migration of cerebral cortical neurons. Science 233, 1308–1310.

Miller, M. W. (1985). Cogeneration of retrogradely labeled corticocortical projection and GABA-immunoreactive local circuit neurons in cerebral cortex. Dev. Brain Res. 23, 187–192.

Miller, M. W. (1988). Effect of prenatal exposure to ethanol on the development of cerebral cortex: I. Neuronal generation. Alcohol Clin. Exp. Res. 12(3), 440–449.

Mitchell, G., and Brandt, E. M. (1970). Behavioral differences related to experiences of the mother and sex of infants in the rhesus monkey. Dev. Psychol. 3, 149–158.

Moore, C. L. (1986a). A hormonal basis for sex differences in the self-grooming of rats. Horm. Behav. 20(2), 155–165.

Moore, C. L. (1986b). Sex differences in self-grooming of rats: Effects of gonadal hormones and context. Physiol. Behav. 36, 451–455.

Moore, C. L., and Brandt, E. M. (1979). Mother rats interact differently with male and female offspring. J. Comp. Physiol. Psychol. 96, 123–129.

Moore, C. L., and Morelli, G. A. (1979). Mother rats interact differently with male and female offspring. J. Comp. Physiol. Psychol. 93, 677–684.

Moore, C. L., and Rogers, S. A. (1984). Contribution of self-grooming to onset of puberty in male rats. Dev. Psychobiol. 17, 243–253.

Moyer, J. A., Herrenkohl, L. R., and Jacobowitz, D. M. (1977). Effect of stress during pregnancy on catecholamines in discrete brain regions. Brain Res. 144, 173–178.

Moyer, J. A., Herrenkohl, L. R., and Jacobowitz, D. M. (1978). Stress during pregnancy: Effect on catecholamines in discrete brain regions of offspring as adults. Brain Res. 144, 173–178.

Mukherjee, A. B., and Hodgen, G. D. (1982). Maternal ethanol exposure induces transient impairment of umbilical circulation and fetal hypoxia in monkeys. Science 218, 700–702.

Mullins, R. F., Jr., and Levine, S. (1968). Hormonal determinants during infancy of adult sexual behavior in the female rats. Physiol. Behav. 3, 333–338.

Nadler, R. D. (1973). Further evidence on the intrahypothalamic locus for androgenization of female rats. Neuroendocrinology 12, 110–119.

Naftolin, F., Ryan, K. J., Davies, I. J., Reddy, V. V., Flores, F., Petro, Z., and Kuhn, M. (1975). The formation of estrogens by central neuroendocrine tissues. Rec. Prog. Horm. Res. 31, 295–315.

Nagel, J. (1969). Réponse du tissu hématopoïétique du foie foetal aprées agression directe portée sur le foetus. *J. Physiol.* Paris **61** (Suppl. 2), 361.

Nottebohm, F. (1980). Brain pathways for vocal learning in birds: A review of the first 10 years. In "Progress in Psychobiology and Physiological Psychology" (J. M. Sprague and A. N. Epstein, eds.), Vol. 9, pp. 85–124. Academic Press, New York.

Numan, M. (1988). Maternal behavior. In "The Physiology of Reproduction" (E. Knobil and J. D. Neill, eds.), Vol. 2, pp. 1569–1646. Raven Press, New York.

Nuñez, E., Benassayag, C., Savu, L., Valette, G., and Jayle, M. F. (1976). Serum binding of some steroid hormones during development in different animal species. Discussion of the biological significance of this binding. *Ann. Biol. Anim. Biochem. Biophys.* **16**, 491.

Ojeda, S. R., and Urbanski, H. F. (1988). Puberty in the rat. In "The Physiology of Reproduction" (E. Knobil and J. D. Neill, eds.), Vol. 2, pp. 1699–1737. Raven Press, New York.

Orth, J. M., Weisz, J., Ward, O. B., and Ward, I. L. (1983). Environmental stress alters the developmental pattern of Δ^5-3β-hydroxysteroid dehydrogenase activity in Leydig cells of fetal rats: A quantitative cytochemical study. *Biol. Reprod.* **28**, 625–631.

O'Shea, K. S., and Kaufman, M. H. (1979). The teratogenic effect of acetaldehyde: Implications for the study of fetal alcohol syndrome. *Teratology* **27**, 231–238.

Pace, H. B., Davis, W. M., and Borgen, L. A. (1971). Teratogenesis and marijuana. *Annals of the New York Academy of Sciences,* **191**, 123–132.

Pare, W. P. (1969). Age, sex, and strain differences in the aversive threshold to grid shock in the rat. *J. Comp. Physiol. Psychol.* **69**, 214–218.

Parker, S., Udani, M., Gavaler, J. S., and Van Thiel, D. H. (1984). Adverse effects of ethanol upon the adult sexual behavior of male rats exposed in utero. *Neurobehav. Toxicol. Teratol.* **6**, 289–293.

Paup, D. C., Coniglio, L. P., and Clemens, L. G. (1972). Masculinization of the female golden hamster by neonatal treatment with androgen or estrogen. *Horm. Behav.* **3**, 123.

Pechinot, D., and Cohen, A. (1983). The determination of maternal and fetal rat plasma corticosterone concentration in late pregnancy by competitive protein-binding assay. *J. Steroid Biochem.* **18**, 601.

Perry, B. D., Pesavento, D. J., Kussie, P. H., U'Prichgard, D. C., and Schnoll, S. H. (1984). Prenatal exposure to drugs of abuse in humans: Effects on placenta neurotransmitter receptors. *Neurobehav. Toxicol. Teratol.* **6**, 295–301.

Peters, D. A. V., and Tang, S. (1982). Sex-dependent biological changes following prenatal nicotine exposure in the rat. *Pharmacol. Biochem. Behav.* **17**, 1077–1082.

Petraglia, F., Volpe, A., Genazzani, A. R., Rivier, C., Sawchenko, P. E., and Vale, W. (1990). Neuroendocrinology of the human placenta. *Front. Neuroendocrinol.* **11**(1), 6–37.

Pfaff, D. W., and Zigmond, R. E. (1971). Neonatal androgen effects on sexual and nonsexual behavior of adult rats tested under various hormones regimens. *Neuroendocrinology* **7**, 129–145.

Phillips, A. G., and Deol, G. (1977). Neonatal androgen levels and avoidance learning in prepubescent and adult male rats. *Horm. Behav.* **8**, 22–29.

Phoenix, C. H., Goy, R. W., Gerall, A. A., and Young, W. C. (1959). Organizing action of prenatally administered testosterone propionate on the tissues mediating mating behavior in the female guinea pig. *Endocrinology* **65**, 369–382.

Quadagno, D. M., Shryne, J., Anderson, C., and Gorski, R. A. (1972). Influence of gonadal hormones on social, sexual, emergence, and open-field behaviour in the rat (*Rattus norvegicus*). *Anim. Behav.* **20**, 732–740.

Raisman, G., and Field, P. M. (1973). Sexual dimorphism in the neuropil of the preoptic area of the rat and its dependency on neonatal androgen. *Brain Res.* **54**, 1–29.

Randall, C. L., Taylor, W. J., and Walker, D. W. (1977). Ethanol-induced malformation in mice. *Alcoholism* **1**, 219–223.

Randall, C. L., Ekblad, U., and Anton, R. F. (1990). Perspectives on the pathophysiology of fetal alcohol syndrome. *Alcohol Clin. Exp. Res.* **14**(6), 807–812.

Rhees, R. W., and Fleming, D. E. (1981). Effects of malnutrition, maternal stress, or ACTH injections during pregnancy on sexual behavior of male offspring. *Physiol. Behav.* **27**, 879–882.

Riley, E. P. (1990). The long-term behavioral effects of prenatal alcohol exposure in rats. *Alcohol Clin. Exp. Res.* **14**(5), 670–673.

Romano, G., .Mobbs, C., Lauber, A., Howell, R., and Pfaff, D. (1990). Differential regulation of proenkephalin gene expression by estrogen in the ventromedial hypothalamus of male and female rats: Implications for the molecular basis of a sexually differentiated behavior. *Brain Res.* **536,** 63–68.

Roy, E. J., and Wade, G. N. (1975). Role of estrogens in androgen-induced spontaneous activity in male rats. *J. Comp. Physiol. Psychol.* **89,** 573–579.

Rudeen, P. K. (1986). Reduction of the volume of the sexually dimorphic nucleus of the preoptic area by *in utero* ethanol exposure in male rats. *Neurosci. Lett.* **72,** 363–368.

Sachs, B. D., and Barfield, B. J. (1976). Functional analysis of masculine copulatory behavior in the rat. In "Advances in the Study of Behavior" (J. S. Rosenblatt, R. A. Hinde, E. Shaw, and C. G. Beer, eds.), pp. 91–154.

Sachs, B. D., and Meisel, R. L. (1988). The physiology of male sexual behavior. In "The Physiology of Reproduction" (E. Knobil and J. D. Neill eds.), Vol. 2, pp. 1393–1485. Raven Press, New York.

Sakly, M., and Koch, B. (1981). Ontogenesis of glucocorticoid receptors in anterior pituitary gland: Transient dissociation among cytoplasmic receptor density, nuclear uptake and regulation by corticotropic activity. *Endocrinology* **108,** 591.

Sakuma, Y., and Pfaff, D. W. (1981). Electrophysiologic determination of projections from ventromedial hypothalamus to midbrain central gray: Differences between female and male rats. *Brain Res.* **225,** 184–188.

Sapolsky, R. M., and Meaney, M. J. (1986). Maturation of the adrenocortical stress response: Neuroendocrine control mechanisms and the stress hyporesponsive period. *Brain Res. Rev.* **11,** 65–76.

Sarrieau, A., Vial, M., McEwen, B. S., Broer, Y., Dussaillant, M., Philibert, D. Moguilewsky, M., and Rostene, W. (1986). Corticosterone receptors in rat hippocampal sections: Effect of adrenalectomy and corticosterone replacement. *J. Steroid Biochem.* **24,** 721.

Schenker, S., Becker, H. C., Randall, C. L., Phillips, D. K., Baskin, G. S., and Henderson, G. I. (1990). Fetal alcohol syndrome: Current status of pathogenesis. *Alcohol Clin. Exp. Res.* **14**(5), 635–636.

Schlumpf, M., and Lichtensteiger, W. (1987). Benzodiazepine and muscarinic cholinergic binding sites in striatum and brain stem of the human fetus. *Int. J. Dev. Neurosci.* **5**(4), 283–287.

Segarra, A. C. (1988). Sexual differentiation in rats is altered by perinatal nicotine or ACTH administration. Ph.D. dissertation, New York University, New York, N.Y.

Segarra, A. C., and McEwen, B. S. (1991). Estrogen increases spine density in ventromedial hypothalamic neurons of peripubertal rats. *Neuroendocrinology* **54,** 365–372.

Segarra, A. C., and Strand, F. L. (1989). Perinatal administration of nicotine alters subsequent sexual behavior and testosterone levels of male rats. *Brain Res.* **480,** 151–159.

Segarra, A. C., Mendelson, S. D., Strand, F. L., and McEwen, B. S. (1990). "Decreased Sexual Behavior Induced by Prenatal ACTH is Correlated with Increased 5-HT$_{1B}$ Receptors in the MPN of Male Rats." Presented at the XXXI Meeting—Int. Soc. Psychoneuroendocrinol. (ISPNE) Aug. 1990.

Segarra, A., Angulo, J., and McEwen, B. S. (1991a). Estrogen increases proenkephalin mRNA expression in the VMN of juvenile male and female rats. *Abstr. Soc. Neurosci.* **17,** 528.1.

Segarra, A. C., Luine, V. N., and Strand, F. L. (1991b). Sexual behavior of male rats is differentially affected by timing of perinatal ACTH administration. *Physiol. Behav.* **50,** 689–697.

Silverman, A. J., Krey, L. C., and Zimmerman, E. A. (1979). A comparative study of the luteinizing hormone-releasing hormone (LHRH) neuronal networks in mammals. *Biol. Reprod.* **20,** 98–110.

Simerly, R. B., Swanson, L. W., and Gorski, R. A. (1984). Demonstration of a sexual dimorphism in the distribution of serotonin-immunoreactive fibers in the medial preoptic nucleus of the rat. *J. Comp. Neurol.* **225,** 151–166.

Simerly, R. B., Swanson, L. W., and Gorski, R. A. (1985). Reversal of the sexually dimorphic distribution of serotonin-immunoreactive fibers in the medial preoptic nucleus by treatment with perinatal androgen. *Brain Res.* **340,** 91–98.

Singh, H. H., Purohit, V., and Ahluwalia, B. S. (1980). Effect of methadone treatment during pregnancy on the fetal testes and hypothalamus in rats. *Biol. Reprod.* **22,** 480–485.

Smotherman, W. P., and Robinson, S. R. (1987). Psychobiology of fetal experience in the rat. In "Perinatal Development: A Psychobiological Perspective" (N. A. Krasnegor, E. M. Blass, M. A. Hofer, and W. P. Smotherman, eds.), pp. 39–60. Academic Press, New York.

Snyder, A. K., Singh, S. P., and Pullen, G. L. (1986). Ethanol-induced intrauterine growth retardation: Correlation with placental glucose transfer. *Alcohol Clin. Exp. Res.* **2**, 155–163.

Sorette, M. P., Maggio, C. A., Starpoli, A., Boissevain, A., Greenwood, M. R. C. (1980). Maternal ethanol affects rat organ development despite adequate nutrition. *Neurobehav. Toxicol.* **2**, 181–188.

Southren, A. L., and Gordon, G. G. (1976). Effects of alcohol and alcoholic cirrhosis on sex hormone metabolism. *Fertil. Steril.* **27**, 202–208.

Spencer, R. L., and McEwen, B. S. (1990). Adaptation of the hypothalamic–pituitary–adrenal axis to chronic ethanol stress. *Neuroendocrinology* **52**, 481–489.

Stewart, J., Skvarenina, A., and Pottier, J. (1975). Effects of neonatal androgens on open-field and maze learning in the prepubescent and adult rat. *Physiol. Behav.* **14**, 291–295.

Strand, F. L., Rose, K. J., King, J. A., Segarra, A. C., and Zuccarelli, L. A. (1989). ACTH modulation of nerve development and regeneration. *Prog. Neurobiol.* **33**, 45–85.

Streissguth, A. P., Herman, C. S., and Smith, D. W. (1978). Intelligence, behavior, and dysmorphogenesis in the fetal alcohol syndrome. *J. Pediatr.* **92**, 363–367.

Stumpf, W. E., and Sar, M. (1976). The differential distribution of estrogen, progestin, androgen, and glucocorticoid, in the rat brain. *J. Steroid Biochem.* **7**, 1170.

Stylianopoulou, F. (1983). Effect of maternal adrenocorticotropin injections on the differentiation of sexual behavior of the offspring. *Horm. Behav.* **17**, 324–331.

Tartellin, M. F., and Gorski, R. A. (1973). The effects of ovarian steroids on food and water intake and body weight in the female rat. *Acta Endocrinol.* **79**, 551–568.

Taylor, A. N., Branch, B. J., Liu, S. H., and Kokka, N. (1982). Long-term effects of fetal ethanol exposure on pituitary–adrenal response to stress. *Pharmacol. Biochem. Behav.* **16**, 585–589.

Taylor, A. N., Branch, B. J., Liu, S. H., Wiechman, A. F., Hill, M. A., and Kokka, N. (1981). Fetal exposure to alcohol enhances pituitary–adrenal and temperature responses to ethanol in adult rats. *Alcoholism* **5**, 237–246.

Taylor, A. N., Branch, B. J., Kokka, N., and Poland, R. E. (1983). Neonatal and long-term neuroendocrine effects of fetal alcohol exposure. *Mono. Neural. Sci.* **9**, 140–152.

Thompson, W. R. (1957). Influence of prenatal maternal anxiety on emotionality in young rats. *Science* **125**, 698–699.

Thompson, W. R., and Quinby, S. (1964). Prenatal maternal anxiety and offspring behavior: Parental activity and level of anxiety. *J. Genet. Psychol.* **105**, 359–371.

Toran-Alerand, C. D. (1976). Sex steroids and the development of the newborn mouse hypothalamus and preoptic area *in vitro*: Implications for sexual differentiation. *Brain Res.* **106**, 407–412.

Udani, M., Parker, S., Gavaler, J., and Van Thiel, D. H. (1985). Effects of *in utero* exposure to alcohol upon male rats. *Alcohol Clin. Exp. Res.* **9**(4), 355–359.

Valenstein, E. S., Cox, V. C., and Kakolewski, J. W. (1969). Sex differences in taste preferences for glucose and saccharin solutions. *Science* **156**, 942–943.

Van Dorp, A. W. V., and Deane, H. W. (1950). A morphological and cytological study of the postnatal development of the rat's adrenal cortex. *Anat. Rec.* **107**, 265–281.

Van Thiel, D. H. (1984). Ethyl alcohol and gonadal function. *Hospit. Pract.* **Nov.**, 152–154.

Vito, C. C., and Fox, T. O. (1982). Androgen and estrogen receptors in embryonic and neonatal rat brain. *Dev. Brain Res.* **2**, 97–110.

Vreeburg, J. T. M., van der Vaart, P. D. M., and van der Schoot, P. (1977). Prevention of central defeminization but not masculinization in male rats by inhibition neonatally of oestrogen biosynthesis. *J. Endocrinol.* **74**, 375.

Wade, G. N., and Zucker, I. (1969). Taste preferences of female rats: Modification by neonatal hormones, food deprivation, and prior experience. *Physiol. Behav.* **4**, 935–943.

Walker, C. D., Perrin, M., Vale, W., and Rivier, C. (1986). Ontogeny of the stress response in the rat: Role of the pituitary and the hypothalamus. *Endocrinology* **118**(4), 1445–1451.

Walters, D. E., and Carr, L. A. (1986). Changes in brain catecholamine mechanisms following perinatal exposure to marihuana. *Pharmacol. Biochem. Behav.* **25**, 763–768.

Ward, I. L. (1972). Prenatal stress feminizes and demasculinizes the behavior of males. *Science* **175**, 82–84.

Ward, I. L. (1977). Exogenous androgen activates female behavior in noncopulating, prenatally stressed male rats. *J. Comp. Physiol. Psychol.* **91**(3), 465–471.

Ward, I. L. (1984). The prenatal stress syndrome: Current status. *Psychoneuroendocrinology* 9(1), 3–11.

Ward, I. L. (1985). Prenatal stress and prepuberal social rearing conditions interact to determine sexual behavior in male rats. *Behav. Neuroscience.* 99(2), 301–309.

Ward, I. L., and Ward, O. B. (1985). Sexual behavior differentiation: Effects of prenatal manipulation in rats. *In* "Handbook of Behavioral Neurobiology" (N. Adler, D. Pfaff, and R. W. Goy, eds.), pp. 77–98. Plenum Press, New York.

Ward, I. L., and Weisz, J. (1980). Maternal stress alters plasma testosterone in fetal males. *Science* 207, 328–329.

Ward, I. L., and Weisz, J. (1984). Differential effects of maternal stress on circulating levels of corticosterone, progesterone, and testosterone in male and female fetuses and their mothers. *Endocrinol.* 114(5), 1635–1644.

Ward, O. B., Orth, J. M., and Weisz, J. (1983). A possible role of opiates in modifying sexual differentiation. *In* "Drugs and Hormones in Brain Development" (M. Schlumpf and W. Lichensteiger eds.), pp. 306–316. Karger, Basel, Switzerland.

Weinberg, J. (1985). Effects of ethanol and maternal nutritional status on fetal development. *Alcohol Clin. Exp. Res.* 9(1), 49–55.

Weisz, J., and Ward, I. L. (1980). Plasma testosterone and progesterone titers of pregnant rats, their male and female fetuses, and neonatal offspring. *Endocrinology* 106, 306–316.

Weisz, J., Brown, B. L., and Ward, I. L. (1982). Maternal stress decreases steroid aromatase activity in brains of male and female rat fetuses. *Neuroendocrinology* 35, 374–379.

West, J. R., Hodges, C. A., and Black, A. C. (1981). Prenatal exposure to ethanol alters the organization of hippocampal mossy fibers in rats. *Science* 211, 957–959.

Whalen, R. E., and Edwards, D. A. (1987). Hormonal determinants of the development of masculine and feminine behavior in male and female rats. *Anat. Rec.* 157, 173–180.

Wiener, S. G., Shoemaker, W. J., Koda, L. J., and Bloom, F. E. (1981). Interaction of ethanol and nutrition during gestation: Influence on maternal and offspring development in the rat. *J. Pharmacol. Exp. Ther.* 216, 572–579.

Williams, A. F. (1966). Social drinking, anxiety, and depression. *J. Pers. Soc. Psychol.* 3, 689–693.

Wright, P. L., Smith, S. H., Keplinger, M. L., Calandra, L. C., and Braude, M. C. (1976). Reproductive and teratologic studies with Δ^9-tetrahydrocannabinol and crude marihuana extract. *Toxicology and Applied Pharmacol.* 38, 223–235.

Yoshikawa, K., and Hong, J. S. (1983). Sex-related difference in substance P level in rat anterior pituitary: A model of neonatal imprinting by testosterone. *Brain Res.* 273, 362–365.

Zarrow, M. X., Philpott, J. E., and Denenberg, V. H. (1970). Passage of [^{14}C]4-corticosterone from the rat mother to the foetus and neonate. *Nature* 226, 1058–1059.

Zimmerberg, B., and Mickus, L. A. (1990). Sex differences in corpus callosum: Influence of prenatal alcohol exposure and maternal undernutrition. *Brain Res.* 537, 115–122.

Zimmerberg, B., and Reuter, J. M. (1989). Sexually dimorphic behavioral and brain asymmetries in neonatal rats: Effects of prenatal alcohol exposure. *Dev. Brain Res.* 46, 281–290.

Zucker, I. (1969). Hormonal determinants of sex differences in saccharin preference, food intake, and body weight. *Physiol. Behav.* 4, 595–602.

— Index —

369